D1246867

Mostly nonmetals
p orbitals
(except H, He)

				VIIA	0
IIIA	IVA	VA	VIA	1 **H** 1.008	2 **He** 4.003
5 **B** 10.81	6 **C** 12.01	7 **N** 14.01	8 **O** 16.00	9 **F** 18.99	10 **Ne** 20.18
13 **Al** 26.98	14 **Si** 28.09	15 **P** 30.97	16 **S** 32.06	17 **Cl** 35.45	18 **Ar** 39.95

IB	IIB						
28 **Ni** 58.71	29 **Cu** 63.55	30 **Zn** 65.37	31 **Ga** 69.72	32 **Ge** 72.59	33 **As** 74.92	34 **Se** 78.96	35 **Br** 79.90	36 **Kr** 83.80
46 **Pd** 106.4	47 **Ag** 107.9	48 **Cd** 112.4	49 **In** 114.8	50 **Sn** 118.7	51 **Sb** 121.8	52 **Te** 127.6	53 **I** 126.9	54 **Xe** 131.3
78 **Pt** 195.1	79 **Au** 197.0	80 **Hg** 200.6	81 **Tl** 204.4	82 **Pb** 207.2	83 **Bi** 209.0	84 **Po** (210)	85 **At** (210)	86 **Rn** (222)

METALS   NONMETALS

Inner transition metals
f orbitals

62 **Sm** 150.4	63 **Eu** 152.0	64 **Gd** 157.3	65 **Tb** 158.9	66 **Dy** 162.5	67 **Ho** 164.9	68 **Er** 167.3	69 **Tm** 168.9	70 **Yb** 173.0	71 **Lu** 175.0
94 **Pu** (242)	95 **Am** (243)	96 **Cm** (247)	97 **Bk** (247)	98 **Cf** (249)	99 **Es** (254)	100 **Fm** (253)	101 **Md** (256)	102 **No** (256)	103 **Lr** (257)

DATE DUE

12 DEC 1979   FEB

DEC 12 1979
JUL 5 1984
JUL 9 1984
AUG 29 1985
DEC 26 1985
UNRETURNED
APR 15 1990
APR 23 1990
RETURNED
NOV 04 1990
RETURNED
NOV 13 1991
NOV 26 1991
AUG 27 1993
OCT 07 1992
APR 29 1994
APR 18 1994
RETURNED
OCT 20 1995
JAN 06 1997

DEMCO NO. 3B-298

Science Library

**JOINT UNIVERSITY
LIBRARIES**

NASHVILLE   TENNESSEE

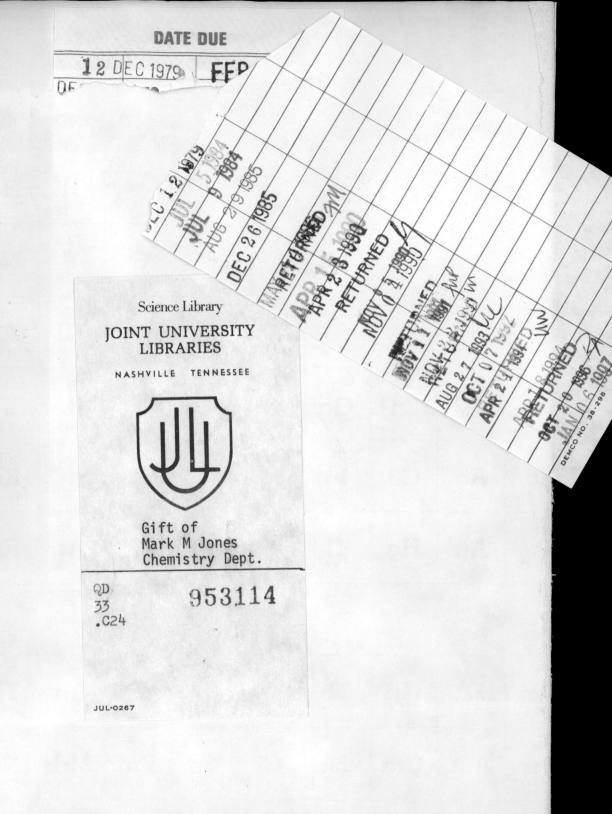

Gift of
Mark M Jones
Chemistry Dept.

QD
33
.C24   953114

JUL-0267

CHEMISTRY:
THE UNENDING FRONTIER

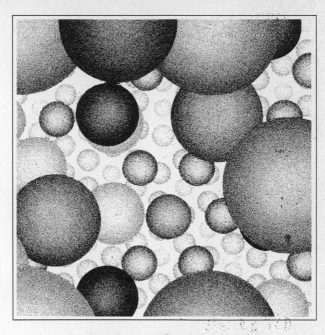

J. Arthur Campbell

Harvey Mudd College
Claremont, California

Goodyear Publishing Company, Inc., Santa Monica, California

Library of Congress Cataloging in Publication Data

Campbell, James Arthur, 1916–
    Chemistry, the unending frontier.

    Includes index.
    I. Chemistry.   I. Title.
QD33. C24   540   77–25214
ISBN 0–87620–133–8

Copyright © 1978 by Goodyear Publishing Company, Inc.
Santa Monica, California 90401

All rights reserved. No part of this book may be
reproduced in any form or by any means without
permission in writing from the publisher.

Current printing (last digit)
10 9 8 7 6 5 4 3 2 1

Library of Congress Catalog Number: 77–25214

Y–1338–6

Designer and Art Director: John Odam
Art Direction and Design Consultant: Tom Gould
Editorial Supervisor: Jackie Estrada
Cover Design by Tom Gould

Illustrators:
    Barry Age
    Mat Antler
    Doug Armstrong
    Rick Geary
    Alice Harman
    Andy Lucas
    Darrel Millsap
    Everett Peck
    Arline Thompson
    Robert Watts
    Delle Willett

Photo Research:
    Linda Rill

Composition: Typothetae Book Composition

JOINT UNIVERSITY LIBRARIES
SCIENCE
LIBRARY
NASHVILLE, TENNESSEE

The only persons free to choose
are those who know the choices.

The persons best prepared to choose
will know the likely outcomes.

PREFACE

Most chemistry students are not going to be chemists. This book is designed for them. My assumptions include the idea that such students wish to understand themselves and the world around them in terms of molecular behavior, a principal point of view in chemistry. My emphasis is on familiar, practical systems with interpretations in terms of clear, direct, and correct scientific ideas and experiments. Environmental and health questions are dealt with throughout the book. Those students who find chemistry attractive enough to wish further study should be able to continue in other courses. But they will need a higher level of mathematical background than is required here. This book requires no calculus and little algebra.

Most of this book follows a much-used sequence. It begins with atomic theory in the early chapters, followed by a discussion of the nucleus (Chapters 3 and 4), the periodic table (Chapter 5), and then proceeds through an introduction to chemical bonding, behavior of gases, rates and mechanisms of reaction, and chemical equilibrium, with separate chapters on general equilibrium and Le Châtelier's principle (Chapter 15), acids and bases (Chapter 16), redox (Chapter 17), solubility (Chapter 18), and thermodynamics—here optional as Chapter 19. The sequence continues with chapters on polymers (Chapter 21) and biochemistry (Chapters 22 and 23).

But, unlike most texts, I have no separate chapter on carbon (organic) chemistry, nor on the chemistry of any other element or group of elements. This material is introduced in contexts that involve knowledge already available to the student—contexts that encourage the development, correlation, and application of chemical ideas. For example, hydrocarbons are used to illustrate the unlimited number of possible chemical compounds (Chapter 8). The structure of graphite is used to introduce the alkenes and unsaturated ring compounds (Chapter 9), including delocalized electrons as contrasted to the localized electrons in diamond and alkenes. Functional groups (ether, ester, amine, amide, and so on) are introduced in terms of interesting compounds, not as a catalog in one place. Nomenclature of both inorganic and organic compounds is discussed as needed throughout the book, but the main ideas are tied together in Appendix B. References to particular ideas, types of compounds, or even individual compounds are readily available through use of the index.

The periodic table is presented early (in both Chapters 1 and 5) as a valuable guide and aid in learning, understanding, and even predicting properties. The chemistry of the individual elements and their families is taken up as needed to provide emphasis and to supply useful, interesting, and long-lasting correlations with what the student already knows and what he or she is apt to meet in the future. For example, iron is discussed in terms of its preparation (as in the blast furnace, Chapter 6) and in terms of rusting (Chapters 14 and 17).

The book is a tapestry of many threads, but some are more continuous than others. Omission of some of the chapters causes little interruption in alternate study patterns. Furthermore, their deletion does not eliminate ideas fundamental or necessary to what follows. At worst, a few minor deletions might be required to bridge the thin places in the total web. The chapters designed for optional deletion are: 4, 11, 14, 21, and 23. If further pruning (for a one-term course, for example) is required, Chapters 6, 9, 12, 16, 17, 18, and 19 could be added to the list.

These chapters are neither dangling threads nor patches whose removal leads to a torn fabric. They are, rather, complements to the other chapters. They make the pattern richer in detail and fill in examples that enhance the overall design. Making these deletions would leave a one-term course providing excellent coverage of and insights into chemistry. And there would be ample additional material for those ever-present students who wish "more."

It is impossible to give individual credit to all who have been instrumental in exploring, selecting, winnowing, and polishing the ideas presented here. I would have to list all my teachers, students, and fellow authors whose questions, suggestions, and ideas have not only penetrated my senses but circulated in my brain. But three institutions merit special

mention: Harvey Mudd College for providing an exceptionally stimulating atmosphere plus a sabbatical to tie things together, the Chinese University of Hong Kong for a perfect sabbatical site plus two fine typists (Grace Poon Suet Lin and Steven Chan Ping Chung, who deciphered my English hieroglyphics as well as they did Chinese ones), and the Rockefeller Foundation for three grandly effective weeks at the Villa Serbelloni. Finally, whereas copy editors can often be a nuisance to authors, I must cite Larry McCombs and Gloria Joyce as an exceptional pair. If you notice any expressions of which Mr. Fowler's *English Usage* would disapprove, please let me know. Please do! But rest assured that the editors have gone through the manuscript with great care, as have my much appreciated reviewers: James P. Birk, Daniel R. Decious, James E. Byrd, Charles H. Carlin, T. Cassen, John H. Harrison, Lawrence P. Elbin, Lavier J. Lokke, and Perry Reeves. Thanks once more to all of them.

May I wish you as much fun in the reading as I had in the writing.

Jarthur Campbell

J. Arthur Campbell
Harvey Mudd College, 1978

CONTENTS

Contents

VISUAL SUMMARY: CHEMICAL BONDS

Section Three
How Do Molecules React? The Kinetics and Mechanisms of Change

Chapter 12
The Sea of Air, and Other Gases 205

Chapter 13
Change, Change, Change: Reaction Rates 225

Chapter 14
Pyres, Pops, and Plagues: Reaction Mechanisms 241

Section Four
How Complete Is a Reaction? Chemical Equilibria

Chapter 15
Constancy Yet Change: Equilibrium States and Steady States 263

Chapter 16
Sour Versus Bitter: Acids and Bases 283

To Mrs. Jimmy, who taught me most of what this book is about

Any chemicals here?

1 DIVE IN!

"Begin at the beginning." One difficulty in following this advice when writing a book is that every reader is different. Each one begins with a different background. You may know that there is a limited number of chemical elements and that these elements combine to form compounds. But some individuals do not know these things. You may know that ozone is made from oxygen, that gasoline is made up mostly of hydrocarbons, that combustion usually produces carbon dioxide, and that incomplete combustion yields carbon monoxide, a deadly poison to humans. You may have noticed the chemical names on bottles of vitamins—vitamin C is ascorbic acid, vitamin B_1 is thiamine, vitamin B_2 is riboflavin—and that an increasing number of foods bear labels identifying the chemicals in them. Then again you may not have noticed. Not every reader has the same background, nor does each one understand equally what the chemical words indicate.

When you look at Figure 1.1, you might see only a carefree diver. But, with a shift in viewpoint, you might describe the diver as wearing nylon shorts and leaping from a wooden board into water. You might even add that the liquid water is continuously evaporating into an atmosphere of nitrogen and oxygen, through which the gaseous water rises and later condenses into white clouds of water droplets. You might have learned that nylon is a synthetic polymer, and that wood is mainly cellulose—a polymeric carbohydrate, $C_n(H_2O)_m$. You might also know that water contains molecules of H_2O and that air is mainly molecules of N_2 and O_2 (nitrogen and oxygen respectively, each containing two atoms per molecule). But somewhere in this sequence your background—and your understanding—may have run out. Nor is this surprising; as you shall find, words and symbols used in chemistry are a very concise means of expressing a large number of experimental observations. They require care and precision in use.

Because of these differences in background, I begin many chapters with questions and/or observations on widely known phenomena. I hope these beginnings will encourage you to search your background for some things you already know, things that will give you a frame of reference for reading and studying the chapter. You need not be able to answer the questions nor to recall all the observations mentioned. Each chapter will further develop the ideas involved.

Certain key ideas in each chapter are set out by colored type, forming a "microchapter" that gives a quick overview of the main ideas.

"Begin at the beginning," the King said, very gravely, "and go on till you come to the end: then stop."—*Lewis Carroll, Through the Looking Glass.*

POINT TO PONDER
Expressions like $C_n(H_2O)_m$ are called *chemical formulas*. C, H, and O are *symbols* used to represent carbon, hydrogen, and oxygen atoms, respectively. This symbolism is discussed in Chapter 2. See if you can understand some of the symbolism even before it is explained. But just ponder, don't fret!

Remember that the Points to Ponder are intended to provide opportunities for you to exercise your mind as creative scientists do. Treat them as puzzles or ideas to stretch your "mental muscles," ideas whose development will continue as you learn more and more about them. After all, solving puzzles and developing ideas can be exciting lifelong activities.

QUESTIONS ARE OFTEN MORE EXCITING THAN ANSWERS.

Figure 1.2
What do scientists do?
(The Bettmann Archive, Inc.)

I strongly encourage you to scan the section headings, the illustrations and tables, and the "microchapter" before starting to read the material intensively. You should next read the problems at the end of the chapter—again, before you read the chapter. The first problem provides a preview of the chapter in the form of a checklist of the new terms and equations presented. Later this checklist and the microchapter can be used for quick review and self-testing. You will probably be able to handle a few of the problems at once; the rest will give a framework for your reading.

So scan the headings, the illustrations and tables, and the microchapter; read the problems and try to think of some possible answers. Then read the text. As you do so, try the exercises in the margin. These exercises cover many of the more obvious objectives of the chapter. You should be able to work out many of them in your head.

The words, symbols, and ideas will sometimes come thick and fast, but I'll try not to drown you. In fact, the idea is to float, flow, and have fun. Dive in!

The White Coat Syndrome

What does Figure 1.2 suggest to you? Most likely the white coat is translated as "scientist," the background as "laboratory." You may envision a person remote from everyday life, engaged in difficult research—research that is unknown by and uninteresting to the general public. Or you may see a "mad scientist," a threat to the safety of humanity.

The man in Figure 1.2 is actually Alexander Fleming, one of the discoverers of penicillin. Born of a poor family, he became the winner of a Nobel Prize. It is true that Fleming, like most scientists, worked in a laboratory remote from public observation, and it is true that his projects interested few outside the laboratory. But occasionally something unexpected happens. Then a keen observer with a lively curiosity may see marvelous surprises unfold.

In Fleming's case, he noticed that one of his bacterial cultures failed to grow where it had become contaminated with a mold. By chance, Fleming's culture was next to a window (open by chance) through which blew a mold from the manure of the horse stables that were (by chance) across the street. Suppose the laboratory had been air-conditioned, hence had no open windows! Or suppose the city zoning laws had forbidden a stable there—the effect might never have been observed.

But chance favors the prepared mind. All the chances just mentioned could have remained just that—chance occurrences. But in this case, the result was noticed, recorded, pondered, further explored, and discussed with others by a person with a well-prepared mind. Even so, penicillin was not used for humans until 13 years after Fleming's initial

Mold

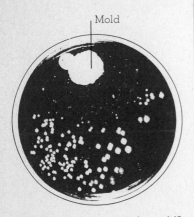

Why no bacteria near the mold?

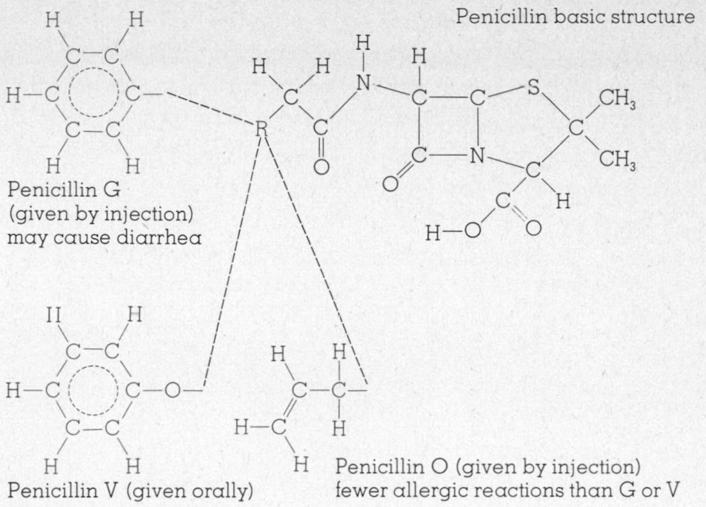

Penicillin basic structure

Penicillin G
(given by injection)
may cause diarrhea

Penicillin V (given orally)

Penicillin O (given by injection)
fewer allergic reactions than G or V

Figure 1.3
Some penicillins. All types of
penicillin have the same basic
formula, but each type has a
unique R group.

discovery in 1928. Many scientists were involved in its development. It then proved so effective as an antibacterial drug that it was quickly added to the list of approved medical treatments. But there were still plenty of problems. My younger daughter, for instance, was placed in serious danger when the penicillin administered to her in 1946 proved to be from a faulty batch. At that time, the chemistry of penicillin was not understood; it could be synthesized only by molds. Its structure was unknown until 1949, when Dorothy Hodgkins determined it.

Once the structure was known, modifications were synthesized, so that there are now many varieties of penicillin. The physician can now match drug to patient to maximize effectiveness and minimize bad side reactions. Much of the drug's detailed chemistry is also understood. We now know that its principal effect is to interfere with the chemical synthesis of bacterial cell walls. Figure 1.3 lists structural formulas for some of the penicillins.

Yet penicillin is still dangerous. Every year several hundred persons who have unusual body chemistries are killed by careless dosage with penicillin. All discoveries by those in "white coats" are both potentially good and potentially bad.

Every chemical—whether called food, medicine, drug, or whatever—is fatal if taken internally by humans in large enough doses. Yet many chemicals are most helpful in moderate amounts, and some are essential if life is to survive. Paracelsus, in the sixteenth century, put it well: "Poisoning is not a matter of chemicals, it is a matter of dose size." This statement has not been contradicted by discoveries made since Paracelsus' time.

POINT TO PONDER
Alexander Fleming once saved another youth from drowning. The youth's family responded by paying for Fleming's education. Some 50 years later, Fleming's penicillin saved the same individual from dying of pneumonia. That man was Winston Churchill, only one of untold millions whose lives Fleming extended. Quite a return on providing for one education!

POINT TO PONDER
There are no good chemicals and there are no bad chemicals. There *are* good and bad uses. And their "goodness" and "badness" may vary from individual to individual.

Atoms, Molecules, and Life

When we looked at Figure 1.1, I gave chemical names for almost everything in the picture except for the diver. But humans also contain chemicals. In 1976, the major chemical elements in an adult could be purchased from a chemical supply company for $4.02.

Table 1.1 lists the chemical elements known to be present in every

ONLY $4.02 ?

Table 1.1   Some Elements Known To Be Present in the Human Body
(number of atoms per 10 million atoms in the body)

Element	Symbol	Number of atoms
Principal Elements[a]		
Hydrogen	H	6,090,000
Oxygen	O	2,560,000
Carbon	C	1,050,000
Nitrogen	N	247,000
Phosphorus	P	18,000
Sulfur	S	6,000
Total principal elements		9,971,000
Major Elements Present as Ions[b]		
Calcium	Ca	22,400
Potassium	K	2,700
Sodium	Na	1,800
Chlorine	Cl	1,200
Magnesium	Mg	600
Total major ions		28,700
Minor Elements		
Fluorine	F	80
Iron	Fe	57
Zinc	Zn	20
Rubidium[c]	Rb	6
Aluminum[c]	Al	3
Bromine[c]	Br	2
Copper	Cu	2
Chromium	Cr	1
Iodine	I	1
Lead	Pb	1
Manganese	Mn	1
Molybdenum	Mo	1
Nickel	Ni	1
Selenium	Se	1
Silicon	Si	1
Tin	Sn	1
Vanadium	V	1
Cobalt	Co	0.1
Total minor elements		180

[a]Note that hydrogen, oxygen, carbon, and nitrogen total to over 99% of the atoms present.
[b]These five major elements (and many of the minor elements) are present as electrically charged atoms: Ca^{2+}, K^+, Na^+, Cl^-, Mg^{2+}. Electrically charged atoms are known as ions.
[c]These elements, and several present at 1 in 10 million, are not known to be essential to human life.

healthy human; other animals have much the same composition. In terms of the main elements present, we could represent the composition of the human body by the chemical formula $C_{105}H_{609}O_{256}N_{25}Ca_2P_2S$. In words, we would say that for each 105 carbon atoms in the body, there are 609 hydrogen atoms, 256 oxygen atoms, 25 nitrogen atoms, 2 calcium atoms, 2 phosphorus atoms, and 1 sulfur atom. As the table shows, other elements are present in smaller amounts. We know that at least one living species needs each of the following elements in small amounts: barium, boron, cadmium, chromium, molybdenum, selenium, strontium, and vanadium. These elements are present in the human body in such tiny amounts that we are not yet sure whether they are present accidentally or are necessary for life.

In all, there are 29 chemical elements currently known to be essential for life in at least one species. This number represents about one-third of the 90 or so elements that exist on the earth's surface.

No one in his right mind would, of course, consider either evaluating or describing humans solely in terms of their elemental composition. Furthermore, no human could survive by eating only the elements listed. In many cases, specific molecules—atoms arranged in well-defined structures—must be provided. The remarkable thing (considering that humans probably contain several hundred thousand different kinds of molecules) is how few kinds of molecules are specifically required in food—probably less than 100. From these few molecules, humans can synthesize the many thousands of other molecules required to maintain life. Some species—molds, for example—do even better. They can survive on half a dozen molecules including carbohydrate (such as breadcrumbs), water, oxygen, and nitrogen. More generally, it is estimated that the more than 1 million species existing could have evolved from an initial living species containing 30 kinds of molecules (20 different amino acids, 5 nucleotides, 1 fatty acid, 2 sugars, glycerol, and choline) plus water and minerals.

As the methods of analytical chemists become more and more sensitive, we will discover additional essential elements and essential molecules, plus more information about the molecules we are made of. Table 1.2 lists some of the simpler molecules currently known to be found in *all* living systems. You probably have never heard of most of these 69 substances. But glance over the list. You will recognize the major classifications (proteins, carbohydrates, fats, DNA, vitamins) and will probably recognize one or two substances in each class. What should impress you most is the small number.

The molecules in Table 1.2 are known as *ubiquitous* (found everywhere) *molecules*. Some of those thought to be *essential* in human food are set in boldface type. To this list can be added all the elements in Table 1.1. Most of these elements occur in food as electrically charged

FROM 30 KINDS OF MOLECULES?

POINT TO PONDER
It is *not* true that if a little bit is good, then more is better. Even the essential chemicals, without which you would die, are lethal in large enough amounts.

EXERCISE 1.1
Which of the ubiquitous molecules listed in Table 1.2 is most
common in humans (that is, has
the highest molecular fraction)?
Does this seem reasonable?

Table 1.2   Some Compounds Found in All Living Systems and Their Percentage by Weight in *E. coli* Bacteria[a,b]

I. Amino Acids (proteins): 59% of Total

Alanine, A	5.5	**Leucine, L**	4.9
Arginine, R	3.1	**Lysine, K**	4.9
Asparagine, N	2.5	**Methionine, M**	2.5
Aspartic acid, D	1.2	**Phenylalanine, F**	1.8
Cysteine, C	1.2	Proline, P	3.1
Glutamic acid, E	2.5	Serine, S	3.7
Glutamine, Q	4.3	**Threonine, T**	3.0
Glycine, G	4.9	**Triptophan, W**	0.6
Histidine, H	0.6	Tyrosine, Y	1.2
Isoleucine, I	3.1	**Valine, V**	3.7

II. Sugars (carbohydrates) and Derivatives: 19% of Total

Glucose	12.5	1,3-Diphosphoglyceric acid
D-Ribose	5.4	3-Phosphoglyceric acid
D-2-Deoxyribose	1.2	2-Phosphoglyceric acid
Glucose-6-phosphate		Phosphopyruvic acid
Fructose-6-phosphate		Pyruvic acid
Fructose-1,6-diphosphate		Lactic acid
Glyceraldehyde-3-phosphate		5-Phosphoribosyl-1-pyrophosphate
Dihydroxyacetone phosphate		Ribose-5-phosphate

III. Lipids (fats) and Precursors: 5% of Total

Glycerol	1.7	α-Glycerol phosphate
Fatty acids	3.4	Ethanolamine
Linoleic and γ-Linolenic acids		

IV. Purines, Pyrimidines, and Derivatives (DNA and RNA): 7% of Total

Uracil	1.4	Adenine	1.7
Cytosine	1.7	Thymine	0.3
Guanine	1.7		

V. Vitamins, Coenzymes, and Precursors

Biotin (oxybiotin)	**Thiamine (vitamin B$_1$)**
Flavin mononucleotide	**Pantothenic acid**
Flavin adenine dinucleotide	**Riboflavin (vitamin B$_2$)**
Diphosphopyridine nucleotide	**Pyridoxine (vitamin B$_6$)**
Triphosphopyridine nucleotide	**Nicotinic acid**
Coenzyme A	**Tetrahydrofolic acid**
Adenosine triphosphate	**Vitamin B$_{12}$**

VI. Miscellaneous

Water	40–90[c]	Phosphoric acid	8.3
Carbon dioxide		Succinic acid	
Ammonia		Fumaric acid	
Glutathione		Acetic acid	
Carbamyl phosphate			

Source: Adapted from H. Morowitz,
Energy Flow in Biology, Academic
Press, New York, 1968.

[a] The numbers refer to how much of the dry material in *E. coli* is made up of the particular compounds; taken as a group, the compounds so listed make up about 97% of the dry material in *E. coli*.

[b] Compounds set in boldface type are essential to humans—they must be present in food, since they cannot be synthesized by the human body.

[c] Percentage of living material by weight.

atoms—that is, as *ions.* Still other substances essential to humans have not been listed in Tables 1.1 or 1.2 because they are not found in *E. coli* bacteria. And as with penicillin, so also with essential foods—either too little or too much leads to illness, even death. The trick is to find the proper dosage.

Don't feel overwhelmed by the unusual names in Tables 1.1 and 1.2. Relax! Float! The point of the tables is to show the small number of essential elements and molecules and of ubiquitous chemicals required by and found in living systems as complicated as humans. The origin and meanings of the names of the chemicals we will explore later. And your ego may be pleased to know that the $4 worth of elements in you would cost more than $6 million if you purchased them in the many molecular forms into which your body has synthesized them. This value ignores many substances that cannot be purchased. You are a skillful synthetic chemist!

The Chemical Viewpoint

Had you first opened this book at Figure 1.3 or Table 1.1 or 1.2, you would probably have recognized the symbols and terms as chemical— perhaps you would even have known that the symbols represent atoms arranged in space somewhat as shown in Figure 1.3. You might even have immediately recognized that the lines between the symbols represent chemical bonds. If not, fine. If so, don't feel offended that we will spend some time developing the details of this symbolism. Determining such structures is an important activity of chemists, and I would like you to come to understand the meanings of the representations.

The discovery and development of penicillin is a rather good example of one of many general scientific approaches (it has been called *strong inference*):

1. Collect data and make observations (especially of surprising occur-rences).
2. Design alternative hypotheses to interpret the results (chemists tend to think in terms of molecular behavior).
3. Design *critical experiments* with alternate possible outcomes that will eliminate one or more hypotheses.
4. Carry out the experiment to get the clearest results possible.
1'. Return to step 1.

The cycle can start anywhere, often by what seems to be chance. No interpretation is ever completely satisfactory, so the cycle is repeated over and over to improve our ability to understand, to predict, to modify, and to control chemical systems.

One of the chemist's roles in all this is to find out what atoms are

EXERCISE 1.2
A man lives on the thirtieth floor of an apartment house. Each morning he goes down in the elevator unassisted, but every evening (though still as healthy and in possession of all his abilities as ever) he needs help in getting to his apartment. Why? (Hint: Analyze the succes-sive requirements for going down and up in an elevator.)

POINT TO PONDER
You have probably been taught
to measure things in English
units: 1 mile = 1760 yards,
1 yard = 3 feet, 1 foot =
12 inches, and so on. But the
conversion factors in English
units are not uniform. A System
of International (SI) units has
now been adopted and almost
all countries are moving to
establish it. (See Appendix A.)
There is only one basic conversion factor, 10. For example,
1 meter = 10 decimeters,
1 decimeter = 10 centimeters,
and 1 centimeter = 10 milli-
meters. This change is called
metrication and should be completed in the United States by
1990. Do you see advantages
in metrication?

present, in what amounts, and how they are combined. These steps have been called *analytical chemistry*. Once a structure is postulated, the chemist attempts to construct a substance of the same structure starting with known structures and using well-understood steps. This activity is *synthetic chemistry*. And in *physical chemistry,* the chemist makes measurements on the properties and structure of a material and studies the correlations between them. Many other titles can be attached to the activities of chemists, and the actual activities usually overlap any set of categories used to describe them.

But, regardless of categories, chemistry and the activities of chemists are most commonly aimed at interpreting observations in terms of behavior at the atomic and molecular levels.

Making Atoms and Molecules Visible

Unfortunately, atoms and molecules are tiny and, for most students, hard to visualize. It takes several million of them arranged side by side to stretch one centimeter (abbreviated cm)—about half an inch. Until quite recently, although much evidence for their existence was known, we had to admit that (with the exception of radioactivity) there were no observations of a single atom or a single molecule. But methods of observation improve and we now have many photographs showing atoms or molecules. The most enlarged (and perhaps least convincing) is Figure 1.4. It is a photograph taken by Albert Crewe of the University

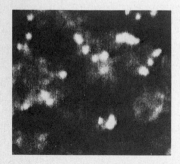

Figure 1.4 (above)
Uranium atoms linked together
by a long-chain molecule.
Magnification 100,000×.
(Courtesy Albert Crewe, Enrico
Fermi Institute, Chicago)

Figure 1.5 (right)
A picture of a platinum needle
tip taken using helium ions.
Magnification 1 million ×.
(Courtesy R. W. Newman,
Florida Institute of Technology)

of Chicago with a microscope that "looks" with electrons rather than visible light. Each of the strung-out bright spots indicates a single atom of uranium (a strong scatterer of electrons). The atoms are held in position by an intervening chain of atoms, mainly carbon, that do not scatter enough electrons to be seen.

More complicated, but perhaps more convincing, are photographs like that in Figure 1.5 showing the much enlarged tip of a fine metallic needle. Each bright spot represents an atom with which helium atoms have collided. The helium atoms (electrically positive after the collision) then travel in straight lines to hit the photographic film and give a picture of the atomic arrangement at the metal tip.

Figure 1.6 is a remarkable electron-microscope picture of hundreds of long, thin molecules of RNA (ribonucleic acid) being synthesized from long, thin molecules of DNA (deoxyribonucleic acid). The DNA pattern is like the quill of a feather, the RNA like the barbs. Both DNA and RNA molecules consist of long strings of some of the chemicals listed in categories II and IV of Table 1.2.

Figure 1.7 is the clearest of the lot. Also taken with an electron microscope, it shows molecules of a tobacco virus packed closely together in a crystal structure. Each more or less spherical object is one molecule of the virus, composed primarily of long chains of the chemicals listed in categories II and IV of Table 1.2.

Figures 1.4, 1.5, 1.6, and 1.7 all result from the use of specialized methods and instruments that give very high magnifications. But all

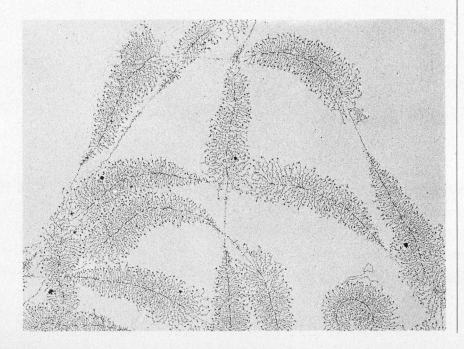

Figure 1.6 (left)
Molecules of RNA (lengthening side chains) being synthesized from molecules of DNA: one step in the mechanism of heredity. Magnification 26,000×. (From O. L. Miller, Jr., *Scientific American*, March 1973, p. 38)

Figure 1.7 (below)
Crystalline tobacco virus. Magnification 160,000×. (Courtesy R. W. G. Wyckoff; from *J. Ultrastruc. Res.*, **2**, 1958, p. 8)

EXERCISE 1.3
How many tobacco virus molecules are nearest neighbors to
(that is, are touching) each one
in the crystal shown in Figure
1.7? (Hint: Remember, you see
only one layer.)

are direct methods where the image due to the atomic or molecular
object is recorded directly onto a photographic film. In many ways the
most satisfying feature of these figures is that each bears out in every
detail the models of atomic and molecular behavior previously reached
by independent and less direct methods.

It is, of course, at once apparent that none of these photos gives full
details of the atomic or molecular structures, but it is most satisfactory
that what is seen matches ideas obtained in other ways. This agreement
strengthens our confidence in the general validity of the more detailed
models, such as Figure 1.3, that are so much a part of chemistry.

Summary

Every object you see, whether "natural" or synthetic, is composed of
atoms and molecules—chemicals. A chemist is a person who interprets
observations in terms of behavior at the atomic and molecular level.
This requires analysis, synthesis, and correlations between measurements and structures. The chemicals themselves are neither good nor
bad, nor is it always possible to predict correctly the outcome of experiments. (Otherwise, why do them?)

Every chemical is dangerous—indeed *all* are poisonous to humans in
large enough doses. But many are highly beneficial, even essential, in
appropriate quantities in appropriate places. One of the functions of the
chemist is to help determine these conditions. And one purpose of this
book is to help you learn the factors that enter into such studies, so that
you can select for yourself the most favorable environment—made up,
of course, of chemicals. Do dive in.

HINTS TO EXERCISES
(Your answers should be more
complete than these hints.)
1.1 Glucose. 1.2 The man is a
midget. 1.3 Six.

Problems

Note: Numerical values (to one significant figure)
and dimensions are given at the ends of most problems requiring numerical answers. For example, Problem 1.10 is followed by [~30,000 extra deaths], which
means that the data yield an increase in deaths of
approximately (indicated by ~) 30,000.

Problems preceded by an asterisk are challenge
problems. They require insight rather than lengthy
calculations. Good luck!

1.1   Define or identify: analytical chemistry, atoms,
      atomic composition, bad chemicals, essential
      element or molecule, good chemicals, ions,
      molecules, physical chemistry, synthetic chem-
      istry, ubiquitous chemicals.

1.2   Does collecting data such as those in Tables 1.1
      and 1.2 involve mostly analytical, physical, or
      synthetic chemistry?

1.3   Tryptophan and histidine are the least com-
      mon of the 20 amino acids in Table 1.2. Does
      this mean they are less essential than the
      others?

*1.4 Describe in your own words the structural
      difference between penicillins G and V (Figure
      1.3). Do not try for scientific accuracy; just
      describe.

1.5   Cite evidence not given in this chapter for the
      existence of molecules.

1.6   Can a product ever be proven safe?

*1.7   How do scientific and nonscientific approaches to problem solving differ?

1.8   In which of the following areas can a "scientific" approach give useful information for deciding governmental and industrial policy? Be specific as to the type of information you have in mind. Name those areas in which scientists should have the final say—that is, veto power. If not scientists, who?

a.   Control of air pollution
b.   Approval of supersonic flights
c.   Equal rights for women
d.   Equality for all human beings
e.   Birth-control methods
f.   Free access to marijuana
g.   Freedom to smoke in public rooms

1.9   One estimate is that air pollution in the United States costs $16 billion per year. What is the cost per person if the population is 220 million? Suggest some of the items probably included in this cost. If this figure is accurate, would you be willing to pay a $200 premium for a new car that has only 20% of the pollution capability of present cars? (Hint: Appendix A discusses solving problems involving very large numbers.) [∼even after 4 years]

*1.10   In 1935, 138,000 Americans died from cancer, compared with 337,000 in 1971. During this period the population doubled. Assuming equal reporting standards (unlikely), how many "extra" cancer deaths occurred in 1971? The President's Council on Environmental Quality in its 1975 report attributed these "extra" deaths to cigarette smoking and to exposure to chemicals, solar and cosmic radiation, and natural asbestos (according to a newspaper story). Would you agree that these four factors should be grouped together? [∼30,000 extra deaths]

1.11   Red meant "stop," green "caution," and white "go" for railroad signal lights of the 1890s. The colored glass lenses got hot from the flame behind them and often broke in rainstorms. Explain why broken red lenses could not be tolerated. This problem led to the development of Pyrex glass, which would not break under sudden temperature changes. Were any other changes made in the signals?

1.12   Raisins can be produced by solar drying of grapes. A treatment with sodium hydroxide (NaOH) followed by bleaching with sulfur dioxide (SO_2) may be used to give a "golden" product. Olives are always treated with sodium hydroxide. Should an advocate of organic gardening eat raisins? Olives?

1.13   Most fish contain 40–80% protein, plus 50–15% lipid (fats), based on dry weights. They have almost no carbohydrate. What is most of the other 5–10% of their dry weight?

1.14   An advertisement states: "X is the only product containing Zippon." Does this statement prove anything about the merits of Zippon?

1.15   Crest toothpaste ads state: "And over 20 clinical studies have proven that Crest reduces cavities an average of 29% better than the same toothpaste without fluoride." Do the results described prove Crest provides protection from "normal" tooth decay?

1.16   How can one person estimate the chemicals in you to be worth $4 and another estimate $6 million, even though both agree on the chemical analysis?

1.17   Many of the complex molecules listed in Table 1.2 "come apart" at temperatures higher than about 105°F. Many of the chemical reactions carried out in living systems can take place only in liquid solutions. How do these observations relate to speculations about the possibility of life on other planets?

Figure 2.1
A heterogeneous system.

2 SIMPLIFY, SIMPLIFY

A fundamental assumption of science is that the complicated objects we find about us are made up of simpler parts. A corollary is that a powerful way to understand the whole is to study the parts and how they interact. You use these ideas almost daily. For example, what do you do if your car won't start (Figure 2.1), if your tennis serve goes sour, if you need to memorize a poem, or if you need to build a doghouse?

We have already seen that a chemist is a person who interprets observations in terms of atomic and molecular behavior. And we have already assumed the existence of atoms and molecules to interpret the figures and discussions in Chapter 1. Let's now consider some simplified methods used in getting from the level of visual observation to that of atoms and molecules.

So the Car Won't Start!

Robert Persig has a great discussion of what to do when the car won't start in his *Zen and the Art of Motorcycle Maintenance*. First one looks at the various systems. Is the electrical system turning over the starter? If so, is the mechanical system turning over the cylinders? If so, is the electrical system firing the spark plugs? If so, is the fuel system supplying any fuel? And so on until an inadequacy is found in one system. (You very much hope it will be in only one system since troubleshooting is much harder when multiple failures occur.)

Now failure in any system can often be interpreted at the molecular level, though few automobile mechanics are trained to go that far in their analysis. Nor need they be, because chemical failures have usually been anticipated rather successfully by the designer and they seldom occur. But let's take a case where a molecular analysis is profitable.

A thorough analysis of the reluctant auto indicates that all systems are go. But the car still won't start. The starter grinds, the fuel line is open, the spark plugs are firing, the exhaust is heavy, but there is no satisfying roar of a motor starting. Neither pouring gas into the air filter nor preventing flooding of the cylinder with gasoline is effective. It's the first really cold day in November, and chances of getting to the football game are decreasing rapidly. Can chemistry come to the rescue? Pour lighter fuel into the air filter you say? The coughs and pleasant

To understand the very large, we must understand the very small.—*Democritus* (470–380 BC).

EXERCISE 2.1
Which of the following are
homogeneous systems: smoke,
distilled water, paper, clothing,
concrete, wood, floor tile, blood?
What do you call the others?

roar that follow change the day. But why? For that, let's proceed with a typical chemical approach to categorizing systems.

Do Let It Phase You

Perhaps the most obvious thing about our surroundings is that they differ from place to place. Your car has many *systems,* each made up of different parts. Even most small objects clearly contain regions with different properties. Such objects are said to be *heterogeneous* (from the Greek *heteros,* "different," and *genos,* "kind").

The different regions may be hard or soft, liquid or gaseous, electrical conductors or insulators, odorous or odorless, varied in color, and so on. Sometimes your unaided senses permit clear distinctions to be drawn. Sometimes specialized equipment is useful. Clearly it is difficult to study and catalog all these variables for a complicated system.

A common technique is to sort out all the similar regions and place them together—at least in thought, if not in actuality. Such a collection (each part having exactly the same properties as any other part) is said to be *homogeneous.* (This time the prefix comes from the Greek *homos,* "same.")

EXERCISE 2.2
You are in your kitchen at home
and have just spilled the contents of a salt shaker into an
open can of coffee beans. Outline two methods for separating
the resulting heterogeneous
system, and discuss their relative
merits. You, of course, wish to
leave no trace of your accident.

Any set of homogeneous regions is called a *phase.* In turn, a heterogeneous system consists of more than one phase. Sorting based on visual and other differences in properties can be used to separate heterogeneous systems into their homogeneous parts. For example, a beaker of ice and water (a heterogeneous system) contains four phases: glass, crystalline water, liquid water, and gas (wet air). (See Figure 2.2). Each phase can easily be separated from the others.

Returning to your balky car, it is clearly heterogeneous. Each of the systems—electrical, mechanical, fuel, and so on—is also heterogeneous. Even a single part like the carburetor is heterogeneous. But it consists of homogeneous smaller parts. These include several types of metal, the incoming liquid fuel, and the gaseous air. It's hard to believe the air is our problem (as long as the filter is unclogged), and the metal parts seem functional because fuel is reaching the engine. So how about the homogeneous gasoline? Back to chemistry again.

Solutions and Pure Substances

You have undoubtedly seen gasoline spilled and you have noted that it evaporates. You may have noticed that, though much of the evaporation occurs quickly, a small fraction remains much longer. Sometimes there is an oily residue quite resistant to evaporation. Liquid water, on the other hand, evaporates completely. This is consistent with the observation that water has a fixed temperature for boiling. Its *boiling*

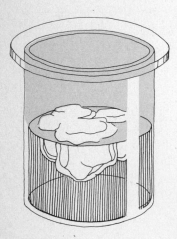

Figure 2.2
Four phases: glass, crystalline water (ice), liquid water, gas (wet air).

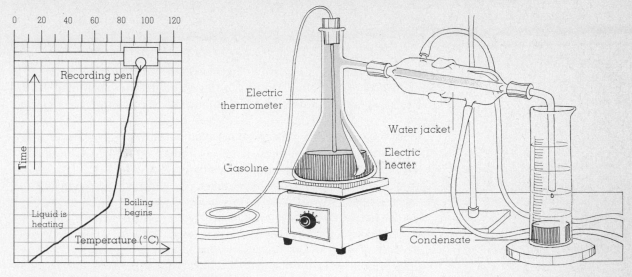

Figure 2.3
Temperature versus time as gasoline is boiled away.

point is 100°C. A thermometer placed in a kettle of boiling water registers 100°C (if the surrounding pressure is about atmospheric) from the start of boiling until the last liquid has evaporated. At lower pressures the boiling temperature is less than the boiling point; at higher pressures, the boiling temperature is greater.

Now, it would be most unwise to fill a kettle with gasoline and try to measure its boiling temperature on a stove. After all, gasoline burns—even explodes—when in contact with air and a flame. But it is possible to heat it in a container that no flame can penetrate (Figure 2.3). The gasoline begins to boil at about 70°C. The boiling temperature then rises as the volume of liquid gets smaller and smaller. Both gasoline and water are homogeneous, yet they differ in boiling behavior. Why so?

A chemist, you recall, tends to interpret observations in terms of molecular behavior. How about an interpretation of boiling and the constancy or lack of constancy of the boiling temperature? Well, boiling involves converting liquid to gas—or, at the molecular level, separating the more tightly packed molecules of the liquid so they become more widely spread out in the gas. This separation requires heating, as you know. In the case of water, boiling of the first molecules occurs at the same temperature as boiling of the later molecules. This is consistent with the idea that water molecules are all alike. Because the molecules of a pure substance are all alike, they boil at a constant temperature. Their rate of boiling is determined solely by how fast heat is supplied to the liquid.

With gasoline you observe that a lower temperature is required to boil the first molecules, and a higher temperature to boil later ones. Thus, the molecules in gasoline must differ from one another—as many other tests confirm. Gasoline must be a *solution*. Because solutions

POINT TO PONDER
Humans have developed many numerical systems of measurement. The rapid rise in international contacts and exchanges is now leading to worldwide standards. (See Appendix A.) You may be most accustomed to the Fahrenheit scale of temperature, on which water boils at 212°F and freezes at 32°F. A much more widely used scale on a worldwide basis (and one about to be adopted in the United States) is the *Celsius* scale. The boiling point of water is 100°C and the freezing point of water is 0°C. We shall use Celsius temperatures in this book. The nature of temperature and temperature scales is discussed in Chapter 12. The conversion equation is

$$°C = (°F - 32) \times 5/9$$

Do you see why? (The fraction 5/9 indicates 5°C for each 9°F, for instance.)

EXERCISE 2.3
Sketch a graph of temperature versus time that would be obtained if the apparatus in Figure 2.3 were used to boil pure water.

POINT TO PONDER
Changes in observed properties reflect changes at the molecular level.

EXERCISE 2.4
The steamship *Titanic* sank after hitting an iceberg. The temperature of the ocean water was −3°C. Was the water pure or a solution? The ice (as is almost always the case with ice) was pure. How so?

contain more than one kind of molecule, the temperature rises as boiling proceeds.

We have so far discussed only boiling. What about freezing? Just as pure water has a fixed boiling point, so also it has a fixed *freezing point,* 0°C (at the standard pressure of 1 atmosphere). And just as the boiling temperature of gasoline varies with the amount that has boiled away, so the freezing point varies with the amount of gasoline that has crystallized.

Thorough study of homogeneous systems shows a very general set of relationships. Some homogeneous systems boil and freeze (or condense and melt) at constant temperatures characteristic of the substances; they are called pure substances. We conclude that a pure substance contains only one kind of molecule.

Other homogeneous systems typically boil and freeze (or condense and melt) over a range of temperatures; such systems are called solutions. A solution contains more than one kind of molecule.

The boiling behavior of gasoline tells us that it is a solution. Even at relatively low temperatures, some of the molecules in liquid gasoline escape from the liquid surface to form a gaseous phase, or vapor. It is this vapor that normally burns and powers the automobile engine. Apparently, the nonstarting engine was so cold that very little vapor was formed, and no combustion could occur in the cylinders.

Lighter fluid is also a solution, but (because of its different molecular composition) it boils or vaporizes at a lower temperature. Even in a cold engine, it forms enough vapor to begin combustion. This warms the cylinders to a temperature sufficient to vaporize gasoline and allow the engine to run normally (Figure 2.4).

Boiling, freezing, condensation, and melting are all *phase changes.* Boiling converts liquids to gases; *condensation* is the reverse. *Freezing* converts liquids to crystals; *melting* is the reverse.

The behavior of the temperature during such phase changes not only serves to identify homogeneous systems as pure substances or as solutions, but also allows the separation of solutions into pure substances.

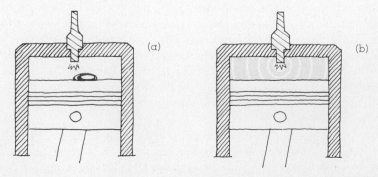

Figure 2.4
(a) Cold gasoline gives too little vapor to burn. (b) Lighter fluid vaporizes and burns even when cold.

Referring to our molecular model again, we see that boiling part of a sample of water to steam and then condensing that steam will give a sample of liquid water with the same properties as the original sample—because the molecules were, and still are, all alike.

But boiling part of a sample of gasoline and then condensing the resulting gas does *not* give a liquid with properties the same as the original sample. The condensate contains a different range of molecules. So, of course, does the remaining original liquid. The phase change has led to a separation of molecules of different types (Figure 2.5).

The earliest known piece of "chemical" equipment is from 3600 BC—a distillation apparatus for separating perfume ingredients. The earliest chemical directions (written by a woman in 1200 BC) tell how to make perfume. Never underestimate the power of women!

There are no completely pure substances. Impurities are always present—from the atmosphere, from unremoved processing materials, even from the container. Modern techniques make it possible to detect impurities in substances previously considered pure. A common measure, which we shall use, is parts per million, ppm. We may use units of weight, molecules, or volume. Thus, an impurity present at 1 ppm in weight means that 1 gram (abbreviated g) of material contains 1 mil-

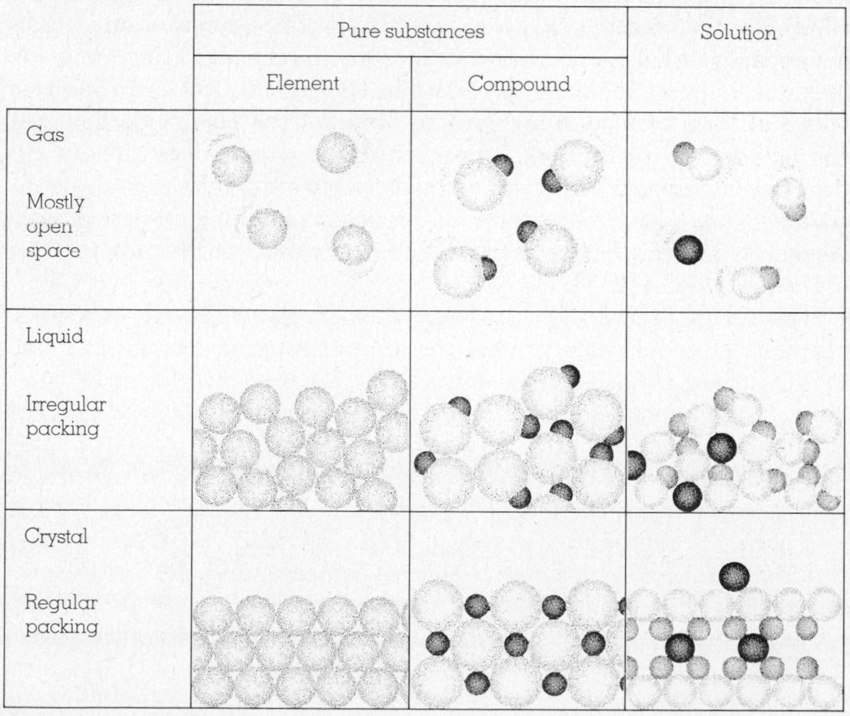

	Pure substances		Solution
	Element	Compound	
Gas Mostly open space			
Liquid Irregular packing			
Crystal Regular packing			

Figure 2.5
Some homogeneous phases (for both pure substances and a solution) in terms of atoms and molecules present.

lionth gram of impurity. One ppm in molecules means that 1 out of every million molecules is an impurity.

Using phase-change techniques, among others, chemists have identified several million "pure" substances. They have learned that many more pure substances exist and that the possible number of pure substances is indefinitely large. They have also observed that the properties of solutions and of heterogeneous substances can often be quite satisfactorily interpreted in terms of the pure substances they contain. Is any further simplification possible? After all, millions of pure substances are a lot to deal with.

We could ask, similarly, now that the car has started, *why* the molecules in lighter fluid boil more readily than those in gasoline, and how gasoline could be "winterized" for easier starting of cars in cold weather. We'll return to such questions later, after gaining more insight into the nature of molecules.

Compounds and Elements
(They Do Not Compound the Difficulties)

You already know from Chapter 1, if not before, that there are only about 90 *elements* found naturally on the earth's surface. (Scientists have synthesized about 15 more, but we shall neglect these most of the time.) Yet the preceding section stated that millions of pure substances have been isolated, more are known to exist, and the potential number is indefinitely large. It should be clear that the more fully we can describe these millions of pure substances in terms of the chemical elements, the simpler our picture will be and the fewer isolated details we will need to remember. The elements known today are given inside the front cover of this book, arranged in the periodic table, which lists similar elements in its columns. The development of this table is a major theme of the first half of this book.

Chemistry, as we know it today, has a history of only about 200 years. Most people date it from the time of Antoine Lavoisier (1743–1794). Among his other contributions, he clarified the nature of combustion, the importance of oxygen, the conservation of matter, and the role of gases. He invented a systematic chemical nomenclature and presented a coherent picture in an innovative 1789 book—its translated title: *Elementary Treatise on Chemistry.*

Figure 2.6 reproduces Lavoisier's table of elements. We no longer consider *light* and *caloric* as chemical elements, and the earthy substances listed at the bottom are now known to be *compounds,* as he probably suspected. But we still use the experimental criterion he used, a criterion suggested by the Englishman Robert Boyle in 1661 in his book, *The Skeptical Chemist.*

Boyle suggested that elements are simple substances that cannot be

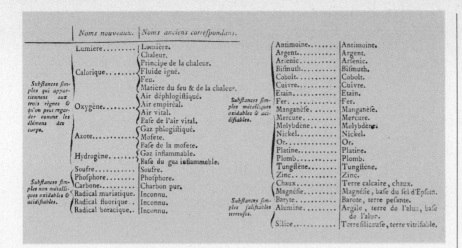

Noms nouveaux.	Noms anciens correspondans.

Figure 2.6
Lavoisier's table of chemical
elements. (From Lavoisier's
Traité Elémentaire de Chimie,
1789)

further decomposed. Any such substance was considered an element until, if ever, a method was found for decomposing it. One or more elements could combine to form a compound, and every compound could be decomposed into two or more elements.

Lavoisier added the vital observation that mass is conserved in such changes. Prior to Lavoisier, laboratory procedure often permitted product gases to escape during heating; or atmospheric gases would react unnoticed with heated solids. In each case the final weight of the solid was different from the initial weight. Lavoisier trapped gases and protected his solids from the atmosphere and so was able to see that the total weight did not change (within his *experimental uncertainty)*.

John Dalton in England had convinced himself by 1805 not only that mass is conserved but also that each pure substance has a constant composition regardless of its origin: The weights of the constituent elements equal the weight of the compound, and the relative weights are always the same. Hence, each compound has a well-defined *constant composition* in terms of the elements making it up.

Interestingly enough it took about 50 more years to convince some of the doubters—for a simple reason. Uncertainties in measurement were so great that the weights seldom did add up exactly. Yet the experimenters hated to admit their techniques and instruments were at fault. Dalton showed greater perception of experimental uncertainties. His generalizations, though based on rather poor data, are still accepted, and they correlate well with modern data.

So Boyle's criterion, presented in 1661, is close to present-day thinking. Let's again summarize Boyle's argument. Any pure substance that can be decomposed into, or formed from, more than one substance is a compound. The pure substances that cannot be further decomposed, nor converted into one another, are chemical elements.

EXERCISE 2.5
Lavosier lists as elements oxygen and hydrogen, gases that you know. He also lists azote. You know this common element also, but by what name? "Phlogisticated air" means air in which something has burned until the air no longer supports combustion.

POINT TO PONDER
Every measurement has some experimental *uncertainty* associated with it. Even if all gross errors and bungling are eliminated, the measurements themselves include an uncertainty because of the imperfection of all measuring instruments. Good experimenters always estimate these uncertainties and list the uncertainty of each measurement. They do this partly by careful calibration of their instruments against known standards, and partly by repeating measurements until the experimental uncertainty in each measured value is rather well known. (See Appendix A.)

And Then Came the Element (Atom) Smashers

Two major discoveries in the last 90 years have forced us to revise Boyle's criterion slightly. First, it has been found that elements can gain and lose electric charges *(electrons)*. Second, it has been found that elements can emit high-energy particles and even convert spontaneously into other elements (they can be *radioactive*). Clearly each of these experimental observations violates Boyle's criterion of the indivisible nature of elements. These discoveries have led to the present model of the *nuclear atom.*

It would be difficult to get into secondary school (or even to read comic books) without running into a model for nuclear atoms. Even television commercials use it: a central, tiny nucleus surrounded by rapidly moving, electrically negative *electrons.* Gain or loss of electrons can give charged atoms, called *ions.* Loss of a high-energy particle from a *nucleus* usually converts it to a nucleus of a different element. You will learn much more about the nuclear atom in Chapters 3 and 4.

So we come to our current definition of an element: A chemical element is a pure substance composed of electrically neutral atoms all containing the same number of negative electrons, hence all also having the same nuclear charge.

In turn, a chemical compound is a pure substance containing more than one kind of element.

The more than 100 known chemical elements combine in various proportions and structures to give the millions of possible compounds.

Figure 2.7
A chemical method of classifying and simplifying observed substances.

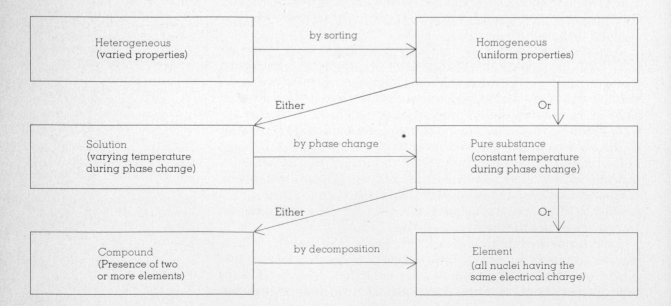

It is a study of this wealth of compounds that provides the information to answer why the lighter fluid helps start the car and how to "winterize" the gasoline. But let's defer that discussion for a few chapters.

Heterogeneous, homogeneous, solution, pure substance, compound, element, nucleus, electrons. What a lot of words, and some of them long and from the Greek! But the purpose of this chapter is to simplify, simplify. Surely finding that the multitude of things we observe are made from only about 100 chemical elements is a simplification in numbers. If we stop our simplification at the level of the chemical elements, you may find it a simplification in ideas as well. Figure 2.7 shows the relationships. Note that all substances fit into one of four mutually exclusive classes: heterogeneous, solution, compound, or element.

Following the outline of Figure 2.7 allows the classification of any system. The outline shows how to separate a system into simpler and simpler parts, and in the end to determine the composition in terms of the chemical elements. Presumably the system could then be reconstructed as often as desired by starting with the elements (or compounds) and reversing the processes. This is, of course, a general outline of the procedure followed in chemical analysis (determining the elementary and compound composition), and chemical synthesis (building up the system from known compounds and elements).

Chemicals Are Conservative (Atoms Are Conserved)

Each change in any system, living or nonliving, is described by the law of the conservation of mass: The total mass of products equals the total mass of reactants.

Chemists do a great deal of measuring of mass, but their symbols are more directly interpreted in terms of atoms. Thus we shall anticipate a future discussion, while again relying on your past experience with the idea of atoms, to point out that atoms are conserved as well as mass. In fact, Dalton's observation of constant composition led to wide acceptance of the idea that not only were atoms conserved, but the mass of each atom remained unchanged during reaction.

You will actually use the idea of *conservation of atoms* more often than the idea of conservation of mass. But the two are clearly tied together by the constant mass of atoms.

Chapter 1 referred to essential foods and ubiquitous compounds in Table 1.2. Note the passage of food through the body in Figure 2.7. Your intake is extremely heterogeneous. The digestive system then converts most of this heterogeneous food into homogeneous solutions or suspended pure substances (often electrically charged species). Most of these nutrients then enter body cells where they are either synthesized

EXERCISE 2.6
Liquefied bottled gas burns with a higher pressure from a full than from a partially emptied tank. Interpret this behavior in terms of the outline in Figure 2.7. (Hint: Is bottled gas a pure substance or a solution?)

POINT TO PONDER
The law of conservation of mass in living systems becomes this: (Nourishment in) = (waste out) + (weight gain by the system). Note the relation to dieting.

EXERCISE 2.7
Glucose is the most abundant of the ubiquitous molecules in humans (not counting H_2O; see Table 1.2). Its formula is $C_6H_{12}O_6$. In what forms do you think most of these atoms leave the human body?

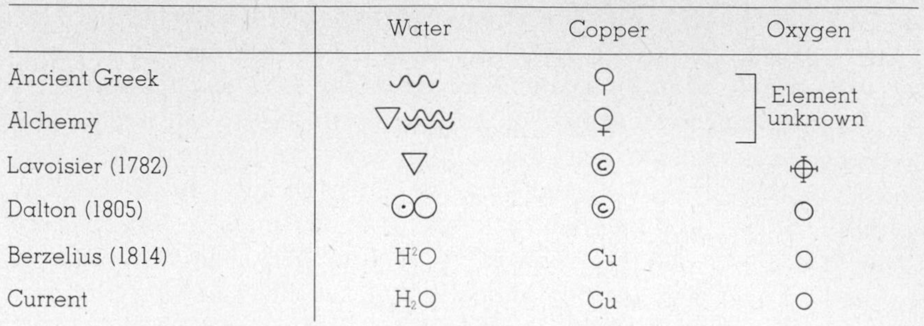

Figure 2.8
Evolution of some chemical
symbols.

	Water	Copper	Oxygen
Ancient Greek	∿	♀	Element unknown
Alchemy	▽∿∿	♀	
Lavoisier (1782)	▽	Ⓒ	⊕
Dalton (1805)	⊙⊙	Ⓒ	○
Berzelius (1814)	H²O	Cu	○
Current	H_2O	Cu	○

into needed substances or are metabolized, with the help of the element oxygen from breathing, to simpler compounds like carbon dioxide (CO_2) or water (H_2O). The waste products and unused food are then eliminated as solutions or as heterogeneous systems. A living system has thousands of ways to maintain itself by decomposing and synthesizing compounds, by causing phase changes, and by sorting. But each is limited by the law of conservation of atoms (and mass).

Chemical Symbols

I have referred several times to H_2O as a chemical symbol for water. Both in Chapter 1 and so far in this chapter I have used Latin letters (you probably call them English) to represent chemicals. The idea of chemical elements, and a unique symbol for each, goes back 2000 years to the Greeks, but our current Latin letter symbols date from J. J. Berzelius in 1814, as Figure 2.8 shows. These letter symbols can carry a great deal of information, but for the moment they will be used as a kind of abbreviation for the element, and especially as a symbolic representation for one atom of an element. Each chemical symbol will represent one nucleus and the electrons surrounding it.

There are, inside the covers of this book, a periodic table of the elements (and a periodic table showing relative atomic sizes) and a list of all the known chemical elements. In the list, the chemical symbol for each element is given (as is an atomic weight, which we shall ignore for the moment). Usually the symbol consists of the first letter of the name of the element and one other letter from the name. Fourteen of the elements are represented by a single letter—B, C, F, H, I, K, N, O, P, S, U, V, W, and Y—but most have two letters. Eleven rather common elements have symbols not derived from their English names; these symbols are based on Latin or other stems—Sb, Cu, Au, Fe, Pb, Hg, K, Ag, Na, Sn, and W. The name, symbol, and original stem for each of these elements are listed in Table 2.1.

Do not sit down and try to memorize the symbols of all the elements.

Table 2.1   Sources of
Chemical Symbols
Not Derived From the English
Name of the Element

Element	Symbol	From
Antimony	Sb	*Stibium*
Copper	Cu	*Cuprum*
Gold	Au	*Aurum*
Iron	Fe	*Ferrum*
Lead	Pb	*Plumbum*
Mercury	Hg	*Hydrargyrum*
Potassium	K	*Kalium*
Silver	Ag	*Argentum*
Sodium	Na	*Natrium*
Tin	Sn	*Stannum*
Tungsten	W	*Wolfram*

Rather, each time a symbol is used, make sure (by referring to the list inside the front cover) that you know the element represented. You will soon be able to recognize the symbols for all the common elements. For a start, why don't you circle the symbols you can identify in the preceding paragraph?

The relative numbers of atoms in a compound are listed as subscripts in the formula of the compound. Thus NaCl describes a compound (sodium chloride, table salt) containing sodium (Na) and chlorine (Cl) atoms in a 1 to 1 atomic ratio. The formula $CaCO_3$ describes a compound (calcium carbonate, the abrasive in most toothpastes) containing calcium, carbon, and oxygen in a 1 to 1 to 3 atomic ratio. (The absence of a subscript indicates only one atom is present.)

Getting more complicated, we come to $MgSO_4 \cdot 7 H_2O$, indicating a compound (hydrated magnesium sulfate, or Epsom salts, a laxative) containing magnesium, sulfur, and oxygen atoms in a 1 to 1 to 4 ratio combined with 7 molecules of water, each water molecule containing hydrogen and oxygen in a 2 to 1 atomic ratio. Similarly $Ca_3(PO_4)_2$ indicates a compound (calcium phosphate, a fertilizer) containing calcium atoms and PO_4 (phosphate) groups in a 3 to 2 ratio, with each phosphate group made up of 1 phosphorus and 4 oxygen atoms.

As a final example, consider penicillin G, represented by the formula $C_{16}H_{18}O_4N_2S$, indicating carbon, hydrogen, oxygen, nitrogen, and sulfur atoms in a 16 to 18 to 4 to 2 to 1 ratio. Compare this condensed formula to the structural formula in the margin (and also in Figure 1.3). It should be clear that the formulas are consistent but that the structural formula contains much more information (and required much more effort to determine). Table 2.2 lists chemical formulas for some common substances. Can you understand them?

But Matter Is Not Enough

The arrows in Figure 2.7 all indicate classifications and changes in material of substances, but that is not all that happens. Nor is matter all that is involved in the changes. Each transition represented by a horizontal arrow involves *energy*. Sorting, phase change, and decomposition all involve energy flows. And just as mass and atoms are conserved, so also is energy: The total energy before reaction equals the total energy afterward.

But there is one very important difference between the manner in which atoms and mass are conserved, and the manner in which energy is conserved. For instance, a plant can absorb sunlight, carbon dioxide, and water to form plant tissue [carbohydrate, $C_n(H_2O)_m$], oxygen, and heat. The tissue (carbohydrate) and the oxygen can then burn or decay to reconstitute the carbon dioxide and water in their original amounts

Penicillin G

(H atoms are indicated by bond lines only.)

EXERCISE 2.8
The formula of all carbohydrates may be written as $C_n(H_2O)_m$, where the subscripts n and m can be replaced by various numbers. Most carbohydrates lose water and turn into a black mass of carbon when heated. (Burned toast is a good example.) Does all this prove that there actually are molecules of water in carbohydrates, similar say to the water in $MgSO_4 \cdot 7 H_2O$?

Table 2.2 Formulas of Some Common Substances

Formula	Substance
$CaCO_3$	Eggshell, pearl
Al	Aluminum metal
H_3BO_3	Boric acid
W	Tungsten metal
$C_{12}H_{22}O_{11}$	Sugar
SiO_2	Quartz
$Fe_2O_3 \cdot n H_2O$	Rust
$KAl(SO_4)_2 \cdot 12 H_2O$	Alum

Figure 2.9
Growth and decay in a leaf.

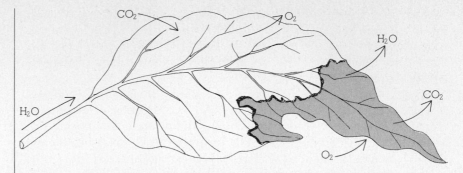

EXERCISE 2.9
A photoflash bulb containing zirconium (Zr) and oxygen (O_2) is weighed before and after being flashed. How do you think the two weights compare?

POINT TO PONDER
Energy is conserved in amount, but its availability for use always decreases with each use.

(Figure 2.9). If the process is carried out in a *closed system* (one sealed to prevent entry or escape of atoms), the mass never changes nor does the number of carbon, hydrogen, or oxygen atoms. Mass and atoms are conserved, and can be returned to a condition identical to the original.

But thousands and thousands of experiments have shown that it is *not* possible to get the energy also back to its original condition. The same amount of energy is present finally as initially, but it was first light and now it is heat. No way is known to reconvert it totally. The process is not reversible as far as energy is concerned. We shall run into this property of energy many times, because energy flows are involved in almost every change.

To generalize, energy can only be used once for any purpose. It cannot be totally recovered and used for that purpose again. It becomes less available with each use. Energy "runs down." Atoms, on the other hand, can in principle always be recovered and used over and over again. We shall also find that electrical charge is conserved. As with atoms, charge can be recycled.

Around and Around and Around We Go— Or, Let's Recycle

The laws of conservation are a great source of confidence to a person interested in recycling. And, as the richest and most available sources of raw materials are exhausted, recycling becomes more and more attractive. Some day you'll decide that old car just isn't worth starting and you may junk it. But it still has most of the materials of a new car. It's just well worn. How about recycling it?

The ultimate suggestion so far for future recycling uses a plasma torch (Figure 2.10). The torch generates a very high temperature and dissociates the junked waste into gaseous charged atoms of the elements. These charged atoms are then sorted out by being passed through a combination of electric and magnetic fields. The separated stream of elements is then cooled, to obtain pure elements. The pure elements may then be used to synthesize desired compounds, solutions, and heterogeneous systems—such as new cars.

Other less dramatic methods of sorting out the elements in waste are already available, and I will discuss some later on. But all suffer, and always will suffer, from two limitations:

1. No recovery system for atoms can be complete.
2. Energy is always required.

Thus recycling depends on two things—a good recovery system for materials, and an available energy source.

Recycling is nothing new. Highly efficient cycles have been in operation for millions of years. You have probably studied the water cycle in nature, perhaps also the oxygen and carbon cycles, maybe even the nitrogen cycle. There are many others, but these are sufficient to illustrate some of the problems and possible solutions.

The atmosphere is an important part of the water, oxygen, carbon, and nitrogen cycles, as Figure 2.11 illustrates. Note that the nitrogen cycle is by far the slowest. The average atmospheric nitrogen atom requires 10^8 years to go around the cycle once. The nitrogen cycle is highly efficient. Almost no nitrogen escapes from the top of the atmosphere. (In fact, the upper atmosphere may gain more nitrogen from the sun than the earth loses into space.) And sunlight has been supplying constant amounts of energy for a long time, so the cycle varies little from year to year.

We saw in Table 1.1 that nitrogen is an essential part of living systems, including humans. In fact, each person needs an average intake of about 50 g of nitrogen per day. There exists on earth about 10^5 square meters (m^2) of agricultural land per person. The nitrogen cycle delivers much less than 50 g per day to this land. Many fields get almost none. It is for this reason that farmers, for thousands of years, have fertilized their fields with nitrogen compounds. In China human waste is called "brown gold." It helps offset the slow cycling rate of nitrogen. Although the water cycle is much more rapid, farmers must often irrigate to compensate for the irregular distribution of water over the earth's surface. Oxygen and carbon requirements are readily met by those two cycles (about 8000 years and 100 years, respectively, for the turnover of atmospheric atoms). Thus the nitrogen cycle is often the limiting one.

The nitrogen limitation continually gets more stringent as the world human population increases—about 2% annually, doubling each 30 to 35 years. For over 50 years the natural fertilizers have been augmented more and more by synthetic ones. Ammonia, urea, and ammonium nitrate are a few synthetic fertilizers widely used today. Without the additional nitrogen supplied by these synthetics, it would be quite impossible to feed the present world population its current diet.

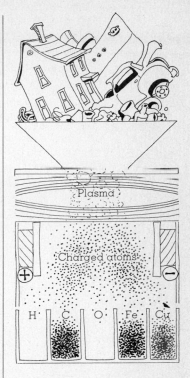

Figure 2.10
The plasma torch at work.

SCIENTIFIC NOTATION
See Appendix A for a discussion of expressions like 2×10^8, the number of people in the United States. You will also find abbreviations such as m and g discussed there.

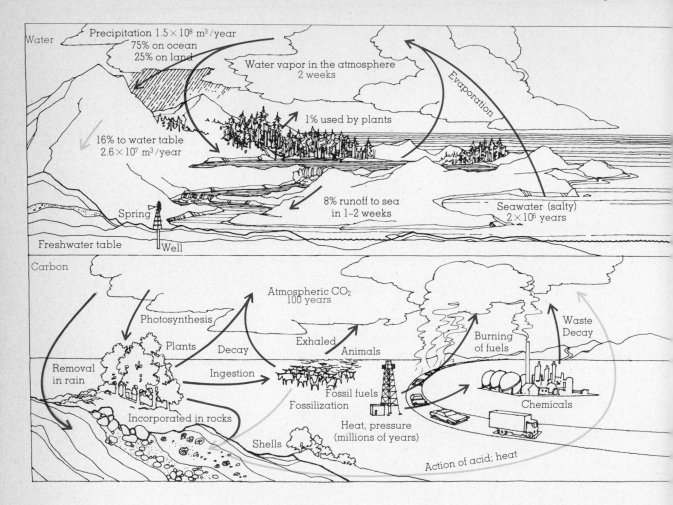

Figure 2.11
Some cycles in nature. Colored
arrows indicate the faster
processes; gray arrows, the
slower ones. Times listed are
average residence times in the
atmosphere.

EXERCISE 2.10
Recycling of paper was much
practiced in the United States
in the early 1930s, yet is less
feasible now. Suggest why.

If the population continues to increase at the present rate, it will
not be long before it becomes impossible to supply even a restricted diet
were all synthetic fertilizers to disappear from the market. "Organic"
gardening is impossible on a nationwide scale unless the population
decreases. And, of course, the energy required for doing the synthesis
gets more expensive annually. Energy increased in price fivefold between
1971 and 1976. For the near future the problem can be handled with
minimum changes in diet, but there is clear need for serious consideration of the problems of even such an efficient cycle as the nitrogen one.

The recycling of such items as aluminum, glass, and cars is less vital
in the long run than is the recycling of nitrogen. Already some communities do a good job on glass and aluminum. These homogeneous substances need only collection, melting, and recasting to be useful again.
Yet the collection costs are such that none of these projects would

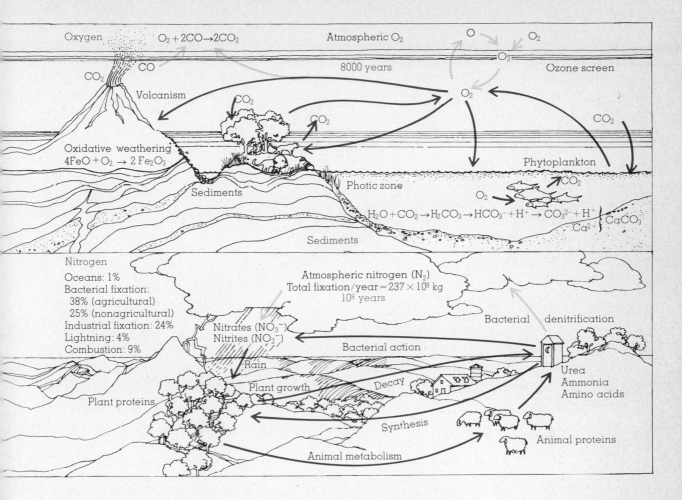

maintain their present operations were it not for the free labor of many
collectors and sorters, usually dedicated groups of citizens. Without
them, both the inefficiency of collection and the cost of transportation
energy would quickly stop the recycling.

The principal problem in car recycling is somewhat different. Cars
are heterogeneous, and until recently no economical sorting process
was available to handle the growing piles of junked cars. But the pile
has finally become large enough that expensive sorters can be kept busy.
The scrap is first shredded in a hammermill. (See Figure 2.12.) The
shreds can be sorted by magnetic fields and blasts of air into iron alloys,
other metals, glass, and fabrics. These are then sufficiently simple to
allow further separation into solutions and pure substances useful in
making new materials.

But there are, as usual, alternatives. For example, initial design

POINT TO PONDER
Science is based on the idea that
identical conditions will lead to
identical events. This is what
scientific laws formulate. Without faith that nature is subject
to law, there can be no science.
Yet there is no proving this
faith.

Figure 2.12
Car recycling. A recycled car
could give a ton of iron, 50
pounds of zinc, 50 pounds of
aluminum, 30 pounds of copper,
20 pounds of lead, 140 pounds
of rubber, and 80 pounds of
glass.

HINTS TO EXERCISES
2.1 Distilled water is homogeneous; others are heterogeneous. 2.2 Sieve; hand pick;
and wash, evaporate. 2.3
Temperature rises until boiling
begins, then remains constant.
2.4 Ocean water (a solution)
can freeze over a range of
temperatures below 0°C to give
pure ice. 2.5 Nitrogen. 2.6 The
more volatile molecules leave
the liquid faster. 2.7 CO_2, H_2O.
2.8 Formulas give atomic ratios;
no. 2.9 Weight is unchanged.
2.10 Labor costs are proportionately higher.

could greatly lengthen the useful life, could make eventual recycling
easier by producing a less heterogeneous system, and could use alternative materials that would make disposal and/or recycling easier. We
could simplify our systems. Or the demand for the product could decrease, thus diminishing the pressure to produce it in the first place.

Summary

Atoms, mass, charge, and energy are conserved in all processes. Because
all systems—heterogeneous, solution, compound, or element—are made
of atoms, we see that any such system can be used, destroyed, and recon-
stituted if we can only recover the atoms. Because they are conserved,
this amounts to keeping track of them and collecting them again.
Energy is also conserved, but it cannot be totally recovered in its original form. Its availability decreases with use.

The classification of matter into heterogeneous, solution, compound,
or element allows us to design schemes to analyze and to synthesize
systems. It also emphasizes the fact that new sources of energy are
always required for any changes to be accomplished.

Problems

Note: Some of these problems contain abbreviations that may be unfamiliar to you—for example, kg for kilogram and kJ for kilojoule. Whenever this occurs look in Appendix A, where numerical calculations and units are discussed.

2.1 Identify or define: boiling point, Celsius temperature, closed system, compounds, condensation, conservation (of atoms, charge, energy, and mass), constant composition, distillation, elements (chemical), experimental uncertainty, formula (chemical), freezing point, heterogeneous, homogeneous, melting point, phase, pure substance, recycling, solution, symbols (chemical), system.

2.2 Seawater is cooled until half of it freezes. The solid proves to be almost pure ice. What has happened to the salt?

2.3 Is a "zero pollution level" possible?

2.4 Classify the following as heterogeneous, solution, compound, or chemical element: water, milk, air, glass, nitrogen, tungsten, iron, rust, flour, toothpaste.

2.5 List methods you would suggest for the following:
 a. obtaining pure water to refill your car battery.
 b. separating peanuts from sugar used by mistake instead of salt;
 c. finding and removing a needle from a swimming pool;
 d. separating 1 kg of table salt and 1 kg of white sand mixed by error;
 e. identifying whether ice from an iceberg is pure H_2O;
 f. deciding whether an eggshell is homogeneous;
 g. producing a sample of water with no air dissolved in it.

*2.6 You are given a colorless liquid known to be either pure alcohol or a 50/50 solution of alcohol and water. How could you find which it is in an ordinary kitchen using only the equipment in the kitchen? (No thermometer is available.)

2.7 Are any of the following recycled in your community: paper, aluminum, bottles, tin cans, plastic, lead storage batteries, used auto oil, copper, cars? If so, by whom?

*2.8 Make a list of additives in bread, hot dogs, catsup, jelly, a canned fruit, and a canned vegetable. Try to identify each by chemical formula and function in the food.

2.9 Is the water you drink chlorinated? Fluoridated? Why or why not?

*2.10 The world consumption of fuel is about 2×10^{17} kJ/year, and one-third of it is used by the 2×10^8 persons in the United States. Estimated unused fossil fuel (about 80% still undiscovered) is $25,000 \times 10^{17}$ kJ. If neither figure changes, how long will fossil fuels last? How long if all the world (4×10^9 people) used energy at the present U.S. rate? Comments? [~ 2000 years if all equal the present U.S. rate]

*2.11 Prove the equation

$$°C = (°F - 32) \times 5/9$$

[Hint: Note that $100/(212 - 32) = 100/180 = 5/9$.]

2.12 The density of water is 1.00 g/cm^3. How much does 1.00 liter (1000 cm^3) of water weigh? What would be the final volume if 200.0 g of water were added to that 1.000 liter? [$\sim 10^3$ cm^3]

2.13 An advertisement for a chewing gum containing aspirin stated that a survey showed it was used by "many doctors." Did a *majority* of doctors in the survey use it?

*2.14 You are boiling potatoes and frying an egg. One of these foods will cook quicker if you increase the burner heat; the other will not. Which is which, and why do they behave differently?

*2.15 In deserts, we often find deposits apparently formed as old lakes dried up. In these deposits we find layers of relatively pure crystals of different chemical compounds. Account for the layering into sorted-out layers.

Figure 3.1
Some uses of nuclear energy.
(Courtesy UPI, American Cancer
Society, Photophile)

3 ATOMS AND THEIR NUCLEI

The discovery of nuclear energy created one of the great controversies of the twentieth century. The introduction of radioactive isotopes into research and medical treatment was hailed as the greatest contribution to the study of living systems since the microscope. The use of nuclear bombs proved to be a decisive way to end hostilities with Japan in 1945, and the existence of these bombs is claimed by many to be the greatest deterrent to major warfare yet found. The prospect of essentially unlimited supplies of nuclear fuel (especially for nuclear fusion) leads many to see nuclear energy as the most promising foreseeable solution to the exhaustion of fossil fuel supplies. Yet the damage caused to living tissues by radiation, the possibilities of nuclear weapons in the hands of irresponsible groups (or even individuals), and the environmental threats by ever greater uses of energy give all thoughtful people pause. Such big problems from such tiny atoms!

Decisions on the use of nuclear energy will almost certainly have long-term effects at least as enduring as the effects of past decisions on the use of other major sources of energy: fire, gunpowder, coal, high explosives, petroleum. And it is highly likely that now, as with each of those past discoveries, the decision in most of the world will be to develop the new sources. Already about one-third of British power comes from nuclear energy. Throughout the world, about 200 nuclear power plants in 19 countries and 6 plutonium recovery plants are operating. One estimate of numbers by 1990 is 800 nuclear power plants and 17 plutonium recovery plants.

Humans, more than any other species, try to control the changes about them. They try to guide these changes in ways presumably beneficial to themselves. This guidance requires energy. Hence the widespread use of all types of available energy, including nuclear energy. Figure 3.2 shows projected energy sources and consumption through the year 2100. We shall discuss the assumptions on which the figure is based later; every projection requires assumptions.

But we have more information than ever before on likely main and side effects, precautions to be taken, and dangers to be minimized. This chapter will explore some of the factors under consideration.

Personal Opinion and Experimental Observations

Because we shall be interested in future projections, I shall outline some of the discoveries that have brought us here. Perhaps historical perspec-

It is the greatest discovery in method which science has made that the apparently trivial, the merely curious, may be clues to an understanding of the deepest principles of nature.
—*G. P. Thomson.*

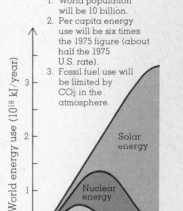

Assumptions:
By the year 2100—

1. World population will be 10 billion.
2. Per capita energy use will be six times the 1975 figure (about half the 1975 U.S. rate).
3. Fossil fuel use will be limited by CO_2 in the atmosphere.

Figure 3.2
A projection of world energy consumption and sources. Note that the use of both fossil and nuclear sources is projected to peak during your lifetime. (From Project Independence Final Report, November 1974, National Science Foundation)

Many winners of Nobel Prizes
in science did the work for
which they were recognized
before the age of 35. Nobel
winners outside science tend to
produce their contributions after
age 40.

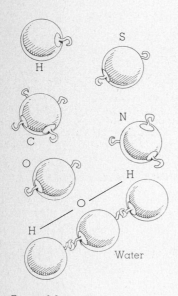

Figure 3.3
Atoms with hooks.

EXERCISE 3.1
Scientists have drawn pictures
like Figure 3.3 to illustrate
atomic ideas. Suggest reasons
they might or might not have
taken the idea of hooks literally.

tive will aid evaluation of future prospects. For example, we shall find
that neither science nor other human activities proceed in smooth or
logical patterns. As with penicillin, so with most discoveries—the noting
and comprehending of chance discoveries are important, often vital.
Experimental observations guide progress as much as, or more than,
prior opinions.

Science does not usually proceed from one triumphant accomplishment to another through carefully planned steps. Like many other
(perhaps all) human endeavors, science explores many paths, uses manifold methods, involves a great variety of persons, and achieves more
errors than successes. At any one moment the paths of hundreds of
researchers on any one problem are almost random; most lead to blind
alleys. Only afterward can we detect and emphasize the connections
between the few results that lead to progress.

Much of the power of science lies in its reliance on experimental evidence rather than personal opinion, no matter how authoritative the
opinion may be. But humans are central to all scientific work. They set
goals. They plan procedures. They evaluate results. They provide the
clinching arguments by comparing interpretations and predictions with
the observed behavior of actual systems. Evidence from the youngest
of researchers may successfully disprove opinions of the most established Nobel Prize winner. Experimental evidence, available to all interested persons, is the heart of scientific advance. Let's look at some of the
evidence for nuclear atoms.

Marbles With Hooks

Dalton's ideas, mentioned in Chapter 2, are consistent with the model
shown in Figure 3.3, in which atoms are hard spheres bearing hooks on
their surfaces. (The Greeks had suggested similar ideas 2000 years before
Dalton, ideas based more on discussion than on experimental evidence.)
The hard spheres seemed consistent with the experimentally observed
conservation of mass and the rigidity and incompressibility of solids.
The hooks seemed consistent with the constant composition of compounds and the difficulty found in separating compounds into chemical
elements.

The Daltonian model includes certain assumptions. All atoms of a
given element (1) have the same mass; (2) are indivisible; (3) cannot be
converted into atoms of another element; (4) combine in fixed proportion with atoms of other elements; (5) show combining proportions that
are simple ratios of whole numbers. All these assumptions are now
known to be incorrect.

Even the popular press assumes that readers know that (1) there are
different kinds of uranium atoms (called *isotopes*); (2) atoms can be
decomposed (by atom "smashers") into simpler particles; and (3) ele-

ments can be converted into one another (for example, uranium into plutonium).

Careful study has also shown that (4) almost never do two different samples of a compound contain *exactly* the same ratio of their combined elements. If you break a crystal of salt, NaCl, in two, what is the chance that each half will have exactly one atom of sodium for every atom of chlorine present? (5) What reader would look at a chemical formula such as that for penicillin G ($C_{16}H_{18}O_4N_2S$) and call the atomic ratios simple? And, of course, much more complicated ratios exist.

Thus all five of Dalton's assumptions are known to be incorrect, yet he is still given great credit for his ideas. Why? Initially the credit rested on his presenting a simpler model than anyone else. His model correlated known data. And it stimulated the performance of many new experiments, all of which (at least for some years) could be interpreted in terms of his model. Furthermore, he marshalled so much experimental evidence (mostly obtained by others) that he was convincing.

The first major attack came from a group of careful analytical chemists, of whom Claude Berthollet is most remembered. Their data convinced them that the composition of a compound depended on how the compound had been prepared. But more careful work showed that they dealt with impure compounds. You can imagine, then, the shock to chemists when, near the end of the nineteenth century, some compounds were discovered that did vary in composition depending on the method of preparation. And some were discovered that differed from a simple atomic ratio in composition. Many such compounds are now known. They are called *Berthollide compounds* in honor of the chemist who believed in but could not prove their existence.

Most Berthollide compounds do not show a simple elementary ratio, refuting both assumption 5 and assumption 4. The compound $Cu_{1.95-1.97}S$ (rather than Cu_2S) is one example. These are often called *defect compounds*. Defect does not refer here to poor match with theory, but to the existence of holes, or defects, in the crystal structures.

However, the great majority of compounds (certainly more than 99%) do have formulas that are constant. Most compounds are made up of integral ratios of atoms within the accuracy of normal chemical synthesis and analysis—say, 1 part in 10,000. So Daltonian theory is quite adequate to describe them, and the chemical formulas you see are very close indeed to correct. But we have had to look for something other than hooks to hold atoms together.

Further Shocks

As Dalton was refining his atomic theory, Benjamin Franklin (using kites to experiment with lightning) and others were studying electricity. They concluded there were two kinds of electrical charge, arbitrarily

POINT TO PONDER
Greatest recognition goes not necessarily to originators, nor to those who are completely "right" on the details, but to those who *convince* their colleagues.

EXERCISE 3.2
It is possible to make zinc oxide that has a formula of ZnO. The Zn to O ratio is found to be 1.0000. Strong heating of this substance gives $Zn_{1.0003}O$. What has probably happened at the atomic level?

A MOST DANGEROUS EXPERIMENT.

POINT TO PONDER
Arrhenius presented his experimental evidence on the existence
of ions in 1884. His supervising
professor refused to accept the
concept of charged atoms
existing in water. Yet, within
4 years, Arrhenius' ideas were
very widely accepted. Such rapid
shifts are not rare in science.
One reason is the relative ease
with which scientists can repeat
experiments and check data in
their own laboratories. Such
checking is much more time-
consuming, sometimes impossible, in the social sciences or
the humanities. One result is
that "schools of thought," based
on the work of one individual or
group, are much less common in
scientific than in other fields.

called *plus* and *minus,* and that oppositely charged objects attract one another, but similarly charged objects repel one another. If both charges are present equally, the system is electrically neutral.

For most of the nineteenth century no understandable connection was found between atoms and electricity. But this was not for lack of trying by hundreds of scientists. Alessandro Volta, for example, discovered that a stack of alternating silver and zinc disks with acid-soaked paper in between each pair could produce electricity. By midcentury Sir Humphry Davy, Michael Faraday, and others had passed an electric current through melted salts and produced chemical elements, including seven not previously known. As a graduate student in the 1880s, Svante Arrhenius tried unsuccessfully to convince his professors that water solutions of salts, acids, and bases contained electrically charged atoms. Later experiments proved he was right.

For the most part, electricity was studied by physicists, atoms by chemists—with little exchange of ideas. But, by 1874, George Stoney had concluded that electricity, like the atom, was made up of particles. He called the unit of charge on these particles the *electron.*

Toward the end of the century it became possible to generate large amounts of electricity and to produce electrical discharges, similar to Franklin's lightning, through gases. Furthermore, photography became generally available. Many physicists took up the study of the beautiful effects seen when electricity is passed through a gas. The modern neon sign was one result of this research—but a minor one compared to the revolution that followed in human understanding of the universe.

Fame and Fortune From Fogged Film

Wilhelm Röntgen was one of those scientists who sealed gases at low pressures in glass tubes, passed electricity through the gases, and photographed the unusual effects. In 1895, while working on his tubes, he was startled to notice a glow in a nearby piece of fluorescent material. The tube itself glowed during use, but Röntgen had surrounded it with a shield, so it was not visible radiation from the tube that caused the fluorescence.

After some puzzling over the matter, Röntgen found that photographic film wrapped in black paper and placed near his tubes became fogged. Furthermore, interposing his hand between the tube and the film gave a shadow image on the developed film of not only the hand but also the bones in it (Figure 3.4). Thus, exposing the film led to varying degrees of fogging.

Further experiments soon showed that his electrical tubes were emitting invisible radiation. Like light, this radiation could fog (expose) photographic film. But unlike light, it passed through both the black

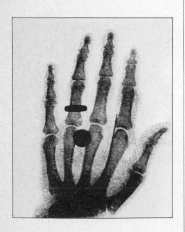

Figure 3.4
This photo of a hand was taken with "Röntgen rays" around the turn of the century.

paper and flesh and even bone. Röntgen called the radiation *x-rays* (from the common use of *x* to indicate an unknown quantity in algebra).

About the same time (1896), Henri Becquerel stored some uranium ores in a cupboard where he also stored some photographic film. When he came to develop the film, he found that it had suffered general exposure, or fogging. This time the radiation was found to be coming from the uranium ore, but it corresponded in many respects to Röntgen's x-rays.

Now the discoveries came thick and fast, and a pattern began to emerge. The gas-discharge researchers had found that negatively charged rays traveled through their tubes in one direction and positive rays in the other. The negative rays were always the same, regardless of the construction of the tube or the nature of the gas in it. But the positive rays (identified by Eugen Goldstein in 1886) varied with the gas.

The uranium researchers found three types of rays from their samples:

1. the x-rays already mentioned;
2. negative rays identical with those found in the gas-discharge tubes;
3. positive rays different from any observed in the gas-discharge tubes.

Initially these three rays were named *gamma* (*γ*) *rays* (electrically neutral, like x-rays), *beta* (*β*) *rays* (negative), and *alpha* (*α*) *rays* (positive). Figure 3.5 shows how the deflection (or lack of it) of the rays when passed between electrically charged plates was used to characterize the charge on the rays.

It did not take much longer to find that alpha, beta, and gamma rays, as well as x-rays, are made up of streams of particles. Both x-rays and gamma rays are made up of high-energy *photons* moving with the speed of light. Beta rays are made of electrically negative particles, which were named *electrons* following Stoney's suggestion. The positive rays in gas discharges are atoms of the gas that have lost one or more electrons. The alpha rays remained a puzzle longer than the rest.

Alpha Particles and Sunlight

By 1905 the Curies (Marie and Pierre) and others had shown that all the very heavy chemical elements (after bismuth in the periodic table inside the front cover) emit one or more *α*, *β*, or *γ* particles. They also showed that the elements change from one to another (undergo *transmutation*) when they emit *α* or *β* particles.

For example, a sample of uranium (whether element, compound, or ore) can emit a series of *α* particles. Loss of one *α* particle converts the uranium atoms to thorium, and loss of a second converts the thorium to radium. A third loss converts radium to radon, a fourth loss converts

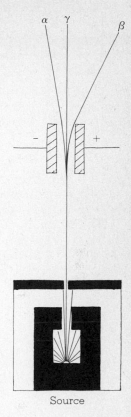

Figure 3.5
Deflections of rays from natural radioactivity by an electric field.

EXERCISE 3.3
What is the function of the piece represented by the slit in the metal plate just above the source in Figure 3.5?

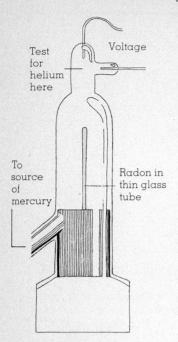

Test for helium here

Voltage

To source of mercury

Radon in thin glass tube

Figure 3.6
Rutherford's apparatus for the identification of α particles as helium nuclei.

EXERCISE 3.4
What is the function of the mercury in the apparatus shown in Figure 3.6?

radon to polonium, and a fifth loss converts polonium to lead. The lead atoms differ from the original uranium atoms by five α particles at least.

It was Ernest Rutherford who in 1907 finally identified α particles. He and J. T. Royds took advantage of the facts that radon is a gas and that its α particles can penetrate thin glass walls although radon itself and the product polonium cannot. They placed gaseous radon—an element Rutherford himself had discovered—in a thin glass tube surrounded by a thick glass tube that could be subjected to an electric discharge, as in Figure 3.6. The α particles were trapped in the outer volume and, when compressed and subjected to an electric discharge, they emitted light. The light corresponded exactly to that observed in sunlight by J. M. Lockyer in 1868 and attributed to an element previously unknown on earth—Lockyer named it *helium*. Rutherford's identification also tallied with the observation of William Ramsay, who had found helium gas trapped in uranium ores. Rutherford assumed correctly that this trapped helium had resulted from the emission of α particles by the uranium throughout the period since the ore had been deposited. (We shall see that the amount of trapped helium provides a valuable measure of the age of such rocks.)

It was clear by this time that atoms are not simple. All atoms can be decomposed into electrons and a residual, positively charged particle. Furthermore, all the heaviest elements spontaneously emit α, β, and/or γ particles and transform into other elements.

The Proof of the Pudding

In 1904, J. J. Thomson (who had done much of the work showing that the streams of negative charge in the discharge tubes were particles and who was Rutherford's teacher) put forward a considerable revision of the Daltonian atom. Thomson postulated that individual electrons exist within atoms as well as in the streams of radiation. He further postulated—consistent with the repulsion between like and the attraction between unlike charged particles—that the electrons are distributed in a more or less uniform way throughout the otherwise electrically positive atom.

Being English, Thomson thought of a similarity between atoms and plum pudding. You might find a similarity to raisin bread more familiar, with the electrons taking the role of the raisins, the dough the role of the positive electricity.

The accumulated experimental evidence of 100 years overwhelmed the details of the Daltonian atom. The discovery (in 1907, by H. N. McCoy and W. H. Ross) that there are at least three different kinds of thorium atom, differing in mass, was a further blow. In 1913 Frederick Soddy coined the word *isotope* to describe such different atoms of a single element. Atoms of a given element need not all be alike; they are not indivisible; one element can change into another; compounds need not have simple atomic ratios. So what about Thomson's pudding model? Many suggestions were made, many experiments designed, much unproductive work undertaken before Rutherford in 1910 suggested to two of his students, Hans Geiger and Ernest Marsden, that they probe atoms by bombarding them with α particles.

The idea was that, just as we can learn about the presence of avalanche-buried skiers by probing snowdrifts with thin poles, or learn about the shapes of objects by bouncing light off them, they could detect the fine structure of atoms by seeing how the atoms scattered a beam of α particles. Alpha particles were known to be quite penetrating (remember, they went through thin glass). They are positively charged, and so would be repelled by positive and attracted by negative charge. They were also known to be heavier than electrons and hydrogen atoms but lighter than other atoms, so they were expected to bounce off rather than drive the other atoms before them. Gold foil was chosen because it can be made very thin and its atoms are heavy. Figure 3.7 shows how simple the experimental setup was. The α source and foil remain stationary, while the fluorescent screen and microscope rotate about the central foil.

The results of this simple experiment changed all thinking about atomic structures. Most of the α particles went through the foil undeflected, indicating that most of the atomic volume is empty space. Some of the α particles were slightly scattered, as had been expected. But some were deflected directly back with almost unchanged energy! It was as though these particles had hit massive objects in the thin gold foil, objects that must have had a large positive charge. Rutherford's comment has been quoted many times: "It was almost as incredible as if you fired a 15-inch shell at a piece of tissue paper and it came back and hit you."

The word *incredible* means "unbelievable," but Rutherford was too good an experimenter to deny the experimental results. The gold foil must contain nearly immovable objects that bounce back some α par-

EXERCISE 3.5
What is the function of the beveled flange in Figure 3.7?

Figure 3.7
Rutherford's method of studying scattering of α particles by thin foil.

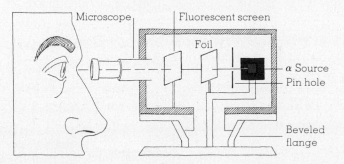

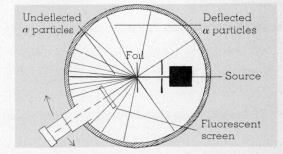

POINT TO PONDER
A further quote from Rutherford: "One day Geiger came to me and said, 'Don't you think that young Marsden, whom I am training in radioactive methods, ought to begin a small research?' Now I had thought that too, so I said, 'Why not let him see if any alpha particles can be scattered through a large angle?' I may tell you in confidence that I did not believe that they would be. Then I remember 2 or 3 days later Geiger coming to me in excitement and saying, 'We have been able to get some of the alpha particles coming backwards. . . .' It was quite the most incredible event that has ever happened to me in my life." After consideration and making calculations, Rutherford "had the idea of an atom with a minute massive center carrying a charge."

ticles. Because most of the α particles are undeflected, these objects in the foil must be very tiny. After some rather straightforward calculations based on the results, Rutherford showed that

1. One of these objects exists in each gold atom.
2. The object contains essentially all the mass of the atom.
3. The object has a positive electrical charge much larger than the positive charge of the α particle.
4. The radius of the object is about 1/100,000 that of the entire atom.

If we could make a scale drawing of an atom as large as a professional football stadium, the central mass would be the size of an ant on the 50-yard line (Figure 3.8)! Each tiny, massive, positively charged center (one to each atom) is called an *atomic nucleus*.

Rutherford published his theory in 1911. Only 7 years after it was suggested, Thomson's pudding fell! The superb student had outdone the excellent master—one of the greatest rewards a teacher can have. Yet Thomson, like Dalton, is recognized as one of the great contributors to scientific experimentation and theory. Each man is admired both for his personal contributions and for the stimulation his ideas provided to others.

In addition Rutherford had, to an exceptional degree, the ability to design experiments that required very modest apparatus but gave data providing deep insight into the nature of atoms. Note how very simple the designs in Figures 3.6 and 3.7 are. He is reported to have been a great supporter of using "baling wire and sealing wax" to build equipment. His original equipment, with the sealing wax still in place, should be viewed by any visitor to Cambridge University.

The Nuclear Atom

Nuclei are now known to be very complicated indeed. More than 100 different particles can be emitted, and the detailed structure is still not understood. But, for our purposes, the model developed by 1932 is quite adequate. This model assumes that three types of particles can be present in an atom: *neutrons* (abbreviated n) and *protons* (p) (individu-

Figure 3.8
The size of an atomic nucleus is to that of an atom as an ant to a football stadium! Small wonder that few of Rutherford's α particles hit anything.

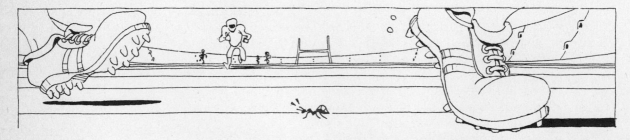

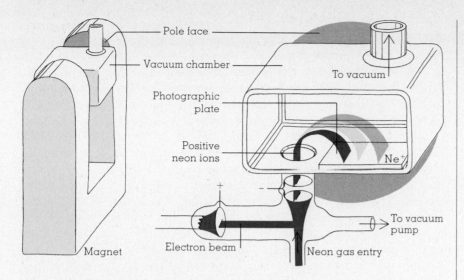

Figure 3.9
Schematic arrangement of a
mass spectrometer.

Pole face

Vacuum chamber

To vacuum

Photographic
plate

Positive
neon ions

Ne⁺

To vacuum
pump

Magnet

Electron beam

Neon gas entry

ally called *nucleons*) in the tiny, massive, positively charged nucleus, and negative *electrons* (*e*) filling out most of the atomic volume. Because the electrons are on the outside of the atoms and are what come into contact when two atoms collide, it should not surprise you to find that most chemistry is interpreted in terms of changes in the electron structures surrounding atoms. I shall discuss these structures at some length a bit later.

But the behavior of the negative electrons around the nucleus is, of course, related to the positively charged nucleus. So let's continue with the nucleus before studying the electrons.

Elements and Isotopes

The very simple setup of Figure 3.5 allows the identification of the sign of the electrical charge on particles. A *mass spectrometer* uses the same principle, but combines both electric and magnetic effects (Figure 3.9). The mass spectrometer allows us to determine not only the sign of the charge but also the ratio of the mass of the particle to its charge.

In the period from 1906 to 1916, Robert Millikan showed that Stoney's assumption of a single unit of charge had been correct. It then became possible to design machines that produced mainly singly charged atoms. These ions were usually obtained by removing a single electron from an electrically neutral gaseous molecule to produce a gaseous ion with a charge of $1+$. When this technique was applied to atoms, it quickly became apparent that atoms of a given element often differ in mass, that isotopes are common.

During the same period it became clear (from scattering experiments

EXERCISE 3.6
The average atomic weight of chlorine was known to be about 35.5 before mass spectrometry showed the existence of chlorine isotopes of mass 35 and mass 37. Which of the two beams, 35 or 37, would be more intense in the mass spectrometer? About what is the ratio of their intensities?

like Rutherford's) that all atoms of a given element have the same positive charge on each nucleus. The number of nuclear unit charges is known as the *atomic number, Z.*

These results led to the present definition of a chemical element: A chemical element is a substance all of whose atoms have nuclei with identical positive electrical charges. Atoms having identical nuclear charges but differing in mass are isotopes of a single element.

Now that atoms were no longer viewed as the simplest building blocks, there was renewed interest in what those simplest particles might be. The electron appeared to be one of them, and even today no one has decomposed a single electron into parts. What about the nucleus? This question is far from settled. But it seems best to use the model first presented in 1932, for it was in that year the neutron was discovered and a simple and powerful nuclear theory was formulated.

The *neutron* was discovered by James Chadwick, who bombarded boron with α particles. Neutrons were one of the products. Neutrons are particles of zero electrical charge. They have just the properties to account for the fact that isotopes have identical nuclear charges but differ in mass.

The lightest, and presumably simplest, nucleus of all is that of the most common hydrogen atoms. These nuclei have a charge of $1+$. Because they are the simplest units of positive electricity, they have been given their own name—*protons.* But there is an isotope of hydrogen named *deuterium,* which is heavier than an ordinary hydrogen atom by one neutron mass—that is, by one neutron. Similar studies of all available elements showed that their isotopes differ from one another in nuclear mass by one or more neutron masses. The nuclear charges are all exact multiples of the charge on a hydrogen nucleus (which charge is identical in size but opposite in sign to that on the electron).

Thus, a logical step was to assume that all nuclei are made up of enough ordinary hydrogen nuclei (called protons) to make up the positive charge. In addition there are enough neutrons (whose masses are almost the same as protons) to give the observed total mass. This idea works remarkably well and is very simple to apply if we assume the masses of both protons and neutrons equal to one (similar to calling the charge on the proton $1+$). Thus, if we call the atomic mass A, the nuclear charge Z, and the number of neutrons N, we get the relationship

$$A = Z + N \tag{3.1}$$

or nuclear mass = number of protons plus number of neutrons.

Table 3.1 illustrates the application of the neutron–proton theory to the compositions of some isotopes. All atoms made from various combinations of three particles—the methods of Chapter 2 have led to simplicity indeed!

Table 3.1 Composition of Some Nuclei in Terms of the Neutron-Proton Theory[a]

Element	A	$=$	Z	$+$	N
Hydrogen	1		1		0
Deuterium	2		1		1
Helium	4		2		2
Oxygen	16		8		8
Fluorine	19		9		10
Chlorine	35		17		18
Chlorine	37		17		20

[a]If the atom is electrically neutral, it contains Z electrons to balance the nuclear charge.

The Basis for Atomic Masses

It was convenient, and certainly simple, in the preceding section to assign a mass of one to protons and neutrons. But you might well have asked "One what?" The answer is one *dalton*. But what is a dalton, other than a unit named after the man who first persuaded the world that atoms exist?

The basic unit for most weighings throughout the world is a chunk of platinum stored near Paris, France, defined as weighing 1 kilogram (kg) exactly (something over 2.2 pounds in English measure). But this is much too big a unit for tiny atoms. However, just as weights throughout the world are compared by ratio to the weight of the chunk of platinum near Paris, so atomic masses have been compared to one another by ratio. One early basis was to assign oxygen atoms a mass of 100. Another system assigned hydrogen (as the smallest) a mass of exactly 1. Later there was a return to oxygen as a standard—this time with its mass defined as exactly 16 (based on the observation that oxygen is about 16 times as heavy as hydrogen and on the desire to keep hydrogen as close to 1 as possible).

The discovery of isotopes and the fact that the isotopic composition of oxygen varies from place to place caused great trouble in exact measurement. So finally worldwide agreement was obtained: The most common isotope of carbon is assigned an atomic mass of exactly 12 daltons, and all other atomic masses are compared to it. This is exactly the same procedure as defining that chunk of platinum as having a mass of 1 kg and comparing other masses to it. The relation of daltons to kilograms will be discussed later, but $1 \text{ kg} = 6.02 \times 10^{26}$ daltons. One dalton is often called 1 *atomic mass unit* (amu).

Mass and Weight

You ordinarily buy things by weight, and you measure your own weight on scales. Yet you know from the astronauts that your weight would become zero in orbit, without your having done any dieting at all! This is because the measurement of weight requires gravity, and the weight of an object varies with the local gravity. Mass is defined in terms of that chunk of platinum near Paris and does not vary with gravity. Technically, what we call *atomic weight* should be called *atomic mass*. It is mass that is measured, quite independent of gravity. But phrases such as *atomic weight* are too well established for easy change.

In a similar way scientists say they are *weighing* something when they compare the object's mass to a standard mass on a chemical balance. The operation really should be called *massing*. Note that scientists, too, can be resistant to using exact language!

POINT TO PONDER
Precision in measurement involves getting the same answer in repeated measurements. *Accuracy* in measurement involves getting the correct answer. Precision depends on the care of the experimenter and the quality of his instruments. Accuracy is impossible without precision, but accuracy also depends on the calibration of the instruments and the reproducibility of the basic standards. The 1-kg weight in France is such a standard. Without such standards, defined with great care to ensure reproducibility, accuracy would be impossible.

EXERCISE 3.7
If $1 \text{ kg} = 6.02 \times 10^{26}$ daltons, how many kilograms in 1 dalton?

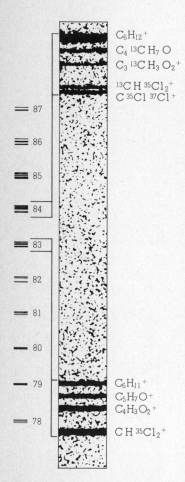

$C_6H_{12}{}^+$
$C_4\,^{13}C\,H_7\,O$
$C_3\,^{13}C\,H_3\,O_2{}^+$

$^{13}C\,H\,^{35}Cl_2{}^+$
$C\,^{35}Cl\,^{37}Cl^+$

— 87

≡ 86

≡ 85

≡ 84

≡ 83

≡ 82

— 81

— 80

— 79

— 78

$C_6H_{11}{}^+$
$C_5H_7O^+$
$C_4H_3O_2{}^+$

$C\,H\,^{35}Cl_2{}^+$

Figure 3.10
Mass spectrum (from masses
78 to 87) for a complicated
molecule that decomposes in
the mass spectrometer into
many kinds of fragments.
Chloroform ($CHCl_3$) was
present as a standard. (From
F. W. McLafferty, *Science*, 151,
1966, 641–649; copyright 1966 by
the American Association for
the Advancement of Science)

Exact Atomic Masses

As with most instruments, so with mass spectrometers: the early measurements had considerable uncertainties in them. Modern mass spectrometers can determine relative masses to 1 part in 100 million (in favorable cases). They can do this not only for atoms but also for molecules. Figure 3.10 shows an example. Table 3.2 summarizes some modern mass measurements of isotopes and fundamental particles. (Note that the mass of the neutron cannot be measured in a mass spectrometer because the charge on a neutron is zero.)

It is clear that all these masses expressed in daltons are very close to whole numbers. These whole numbers are used to identify the isotopes, the number being placed to the left and above the atomic symbol as shown in the table—for example, ^1H or ^{37}Cl. It is also clear that neutrons and protons differ slightly in mass, though each is approximately 1 dalton as assumed in the last section.

We can now determine atomic masses (from the positions of the lines—see Figure 3.10) and relative isotopic abundances (from the relative intensities of the lines) very accurately with mass spectrometers. Most of the average atomic weights given in the list of atomic weights inside the back cover and in the periodic table inside the front cover are obtained from mass spectrometer data. Some are still obtained by actual relative weight measurements on pure chemical compounds. But this is done only if it has not yet proved feasible to obtain gaseous particles containing the elements in question.

You must remember that tabulated atomic weights are average

Table 3.2   Exact Atomic Masses and Mass Defects of Some Particles and Isotopes (in daltons)

	Observed mass $= a$	Sum of masses $Nn + Zp + Ze = b$	Mass defect $= (a - b)$	Mass defect per nucleon $= (a - b)/A$
e^-	0.000548597		0	0
n	1.0086654	1.0086654	0	0
p	1.00727663	1.00727663	0	0
^1H	1.007825	1.007825	0	0
^2H	2.01410	2.016491	0.00239	0.00120
^3H	3.01605	3.025156	0.00911	0.00304
^4He	4.00260	4.0329810	0.03038	0.00760
^{12}C	12^a	12.098943	0.098943	0.00825
^{13}C	13.00335	13.107608	0.10425	0.00802
^{16}O	15.99491	16.130826	0.13592	0.00849
^{35}Cl	34.96885	35.289003	0.32015	0.00915
^{37}Cl	36.96590	37.30633	0.34043	0.00920
^{56}Fe	55.9349	56.4636	0.5287	0.00944
^{238}U	238.0508	239.9861	1.9353	0.00813

^{a}The mass of ^{12}C is defined to be exactly 12.

values and that their constancy depends on all the isotopes of a given element reacting chemically in the same way. If isotopes sort out during chemical reactions, different samples of an element will have different isotopic compositions—hence, different average atomic weights.

The chemistry of isotopes is remarkably similar, but not identical. Isotopes react most differently when the ratio of the isotopic weights differs most from 1. Thus the common isotopes of hydrogen (^{1}H) and deuterium (^{2}H) show larger differences than most. ^{1}H would differ still more from ^{3}H, but ^{3}H (called *tritium*) is very rare indeed and has no measurable influence on most reactions involving hydrogen.

Of greater interest here is the fact that, as shown in Table 3.2, the exact atomic masses do not equal the sum of the masses of the electrons, protons, and neutrons assumed to be present in the atom. The sum of the masses of the electrons, protons, and neutrons in an atom is always greater than the observed mass. A useful number, the *mass defect,* is obtained by taking the difference between the summed particle weights and the observed isotopic weight. The mass defect is, then, the "missing" mass per nucleus. Table 3.2 shows this calculation for several kinds of atoms. Because different nuclei contain different numbers of particles, it is convenient to calculate the *mass defect per nucleon*. This calculation is included in Table 3.2. Figure 3.11 shows the mass defect per nucleon

EXERCISES 3.8
Which would have the more similar chemistry: ^{12}C and ^{13}C, or ^{235}U and ^{238}U?

EXERCISE 3.9
Calculate the mass defect if a neutron is formed from a proton and an electron. Any comment? If not, remember this result when reading the next chapter.

Figure 3.11
Mass defect (binding energy per nucleon) versus atomic number.

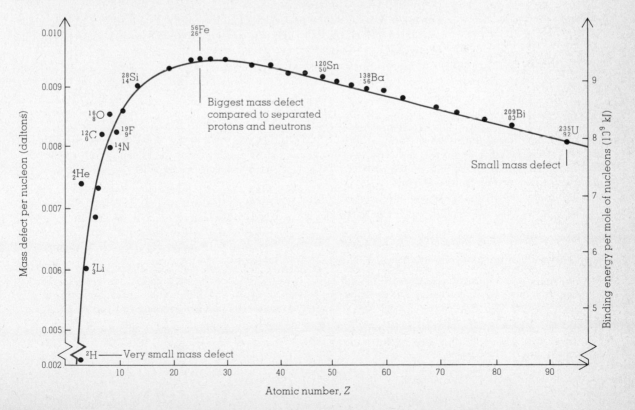

plotted against nuclear charge for many kinds of atoms. It was a curve like this, first plotted in the 1930s, that suggested the nucleus could be a source of tremendous energy release. To understand this we must return once more to the early part of the twentieth century, and to the work of Albert Einstein. This we will do in the next chapter.

Summary

All matter can be thought of as made up of three types of particles: neutrons, protons, and electrons. Every electrically neutral element is made up of atoms, each one of which contains a tiny, massive, positively charged nucleus. The nucleus contains the same number of protons as there are positive charges, Z. Each proton contributes one unit of mass, and there are enough neutrons (number of neutrons = N) also present, each contributing one unit of mass, to make up the total isotopic mass, A. Protons and neutrons, because they are found in the nucleus, are called nucleons. Each nucleus of an electrically neutral atom is surrounded by as many electrons as there are protons in the nucleus, but the electrons have negligible mass compared to that of the neutrons plus the protons. $A = Z + N$.

The observed atomic masses are less than the sum of the masses of the electrons, protons, and neutrons present. This difference is called the mass defect. When mass defect is divided by the number of nucleons and plotted versus Z, a rather regular curve with a single maximum is obtained.

HINTS TO EXERCISES
3.1 No models are true in full details. 3.2 Oxygen gas is driven off. 3.3 To produce a narrow ray. 3.4 To compress the gas. 3.5 To allow the apparatus to rotate. 3.6 Three out of four atoms must be ^{35}Cl. 3.7 One dalton = 1.66×10^{-27} kg. 3.8 Uranium isotopes are more similar. 3.9 Clearly mass is not conserved.

Problems

3.1 Identify or define: alpha (α) ray, atomic number (Z), Berthollide compound, beta (β) ray, dalton (mass unit), Daltonian atomic model, electron, exact atomic mass, gamma (γ) ray, isotope, mass defect, mass spectrometer, neutron, nucleon, nucleus (atomic), photon, proton, x-ray, $A = Z + N$.

3.2 What is a scientific law? How do scientific laws differ, if at all, from governmental laws?

3.3 If a hydrogen atom contains one proton in the nucleus, why don't hydrogen atoms and protons weigh the same? (See Table 3.2.)

3.4 Stable magnesium isotopes have mass numbers (N) of 24, 25, and 26. Which isotope must be most abundant?

*3.5 Choose three of the "lines" at mass 84 in Figure 3.10 and show that their positions are consistent with Table 3.2. [$C^{35}Cl^{37}Cl^+ = 83.95615$]

3.6 Many television sets now have a special glass front, but it is recommended that viewers, particularly children, not be within 2 meters (abbreviated m) of the tube, especially if the set is old. Review Röntgen's experiment to see why.

3.7 Why does the radiologist or dental assistant operate the x-ray tube from the next room, whereas you are directly in front of the tube?

3.8 Calculate the numbers of protons, neutrons, and electrons in each of the following: ^{4}He, ^{18}O, ^{16}O$_2$, ^{23}Na, ^{1}H, ^{80}Br, ^{12}C, ^{3}H, ^{1}H$_2$.

3.9 Is it possible to maintain our present standard of living while cutting down our rate of energy consumption?

*3.10 Describe an experiment to demonstrate that hydrogen is a mixture of two isotopes. Use water as your only source of hydrogen. How could you demonstrate the presence or absence of tritium (^{3}H) in the sample of water?

3.11 Use the "hook model" of Figure 3.3 to predict the formula of one compound each between: H and S, S and C, H and N. There is a compound HCl. Use the hook model to predict the formula of a compound between Cl and C.

*3.12 The most delicate modern balance can detect a weight (mass) change of about 10^{-8} g. (See Appendix A for an explanation of exponents.) How many gold atoms would be present in a sample of that weight? [$\sim 3 \times 10^{13}$ atoms]

3.13 From various evidence, we find that the following molecules exist: H_2O; CO_2, CH_4, NH_3, HCl, and Cl_2O. Draw sphere-and-hook models of these molecules. What is the minimum number of hooks that must be provided for each kind of atom?

*3.14 Suppose that you travel in a time machine back to Ancient Greece. Eager to advance human progress, you decide to explain the atomic theory to Aristotle. What evidence could you cite (or demonstrate experimentally) to convince him that matter is made up of tiny invisible atoms?

*3.15 Scientists are very happy to explain the vast variety of substances on earth as various combinations of fewer than 100 elements. They are even happier to explain the atoms of the different elements as various combinations of only three basic particles (electrons, protons, and neutrons). But they are distressed to keep finding dozens of small particles that can be ejected from nuclei. Why are they distressed? Might a concept of "beauty" have any place in evaluating scientific theories? Is Occam's razor useful here? (Use the index and glossary if words are unfamiliar.)

Figure 4.1
Albert Einstein.
(The Bettmann Archive, Inc.)

4 FROM RELATIVITY TO RADIOACTIVITY

The neutron–proton model is much more than just a simple model for nuclei. It allows interpretation of the source of the energy in nuclear power plants or nuclear explosions. It describes the origins and effects of radioactivity. It provides clues concerning the nature of, and some possible solutions to, the problems of future energy sources. It also helps interpret chemical bonding.

Man always travels along precipices. His truest obligation is to keep his balance.
—*José Ortega y Gasset.*

The One World of Mass and Energy

Albert Einstein was one of the towering scientific geniuses of all time. (See Figure 4.1.) Yet he was so modest as to point out that if he did see farther than some others it was because he stood on the shoulders of the giants who preceded and surrounded him. He made major contributions to at least half a dozen different fields in mathematics and science. Yet he had a terrible time in school. But when he finally got interested, he was superlative. We shall deal here only with his work on mass and energy. If you continue to study science you will run into the work of this remarkable person so often that you'll keep repeating: "Did he do *that* too?"

In 1905, Einstein published his first paper on relativity. It was so abstruse in its symbolism that fewer than 10 people were supposed to be able to understand it. Ten years later he greatly extended his ideas. These ideas were so fundamental in their approach and so widespread in potential application that many scientists were forced to see what was there. Suffice it to say that the theory of relativity as developed by Einstein and others has been a cornucopia of ideas, insights, and guides to experimentation for several generations of scientists. From this cornucopia we here extract only a simple equation:

$$E = mc^2 \tag{4.1}$$

Almost certainly you have seen this equation in books, magazines, and newspapers. You may even remember some of its connotations, particularly that mass can be converted into energy and that this is what happens when a nuclear bomb explodes. It is also what happens when you strike a match, set off dynamite, or fall down and bump your head.

POINT TO PONDER
Just as you express quantities
as a number plus a unit (7 feet,
2 years, or 3 pounds), so do
scientists. Some scientific units
will be new to you. The joule
(abbreviated J), mentioned in
equation 4.2, is probably such a
unit. It may be easier to relate
joules to energies you have
thought about if you learn that
4 joules will raise the temperature of 1 g of H_2O about 1°C.
The exact definition is this:
A joule equals the energy
generated by an electric current
of 1 ampere (I) flowing for
1 second (s) at a voltage of
1 volt (V), $J = I \times s \times V$.
So a joule is a readily measured
quantity of energy.

The Einstein equation states that energy release is always accompanied by a disappearance of mass. Absorption of energy is always
accompanied by a gain in mass. A key fact to remember is that it takes
very little mass change to generate an enormous amount of energy
because the c^2 term (related to the velocity of light) is so very large. If
we express mass (m) in grams and energy (E) in joules, the equation
becomes

$$E \text{ (in joules)} = 9.00 \times 10^{13}\, m \text{ (in grams)} \tag{4.2}$$

This conversion becomes more meaningful if you note that the total
conversion of a single page of paper (mass about 2 g) to energy is equivalent to burning 4 million kg of oil, enough to supply the total energy
requirements of a city of 1 million people for one day. In typical chemical
reactions (say the burning of the paper) less than 1 part per 10 billion
($1/10^{10}$) of mass is converted to energy; this change in mass is much
smaller than can be directly measured by present detection methods.

Einstein was one of the most peaceful men ever to live. Yet, at the
request of other scientists, he wrote a crucial letter (Figure 4.2) to
Franklin Roosevelt pointing out that nuclear bombs were almost
certainly going to be possible—and that the nation obtaining the first
one would almost certainly drive the opponent to defeat in World War
II. From this letter arose the Manhattan Project, a cooperative effort
between Britain and the United States. (A very large number of refugee
scientists from central Europe also played crucial parts.)

Uranium isotopes were separated, plutonium was synthesized, the
bombs were produced, the war did end, and a new source of very large

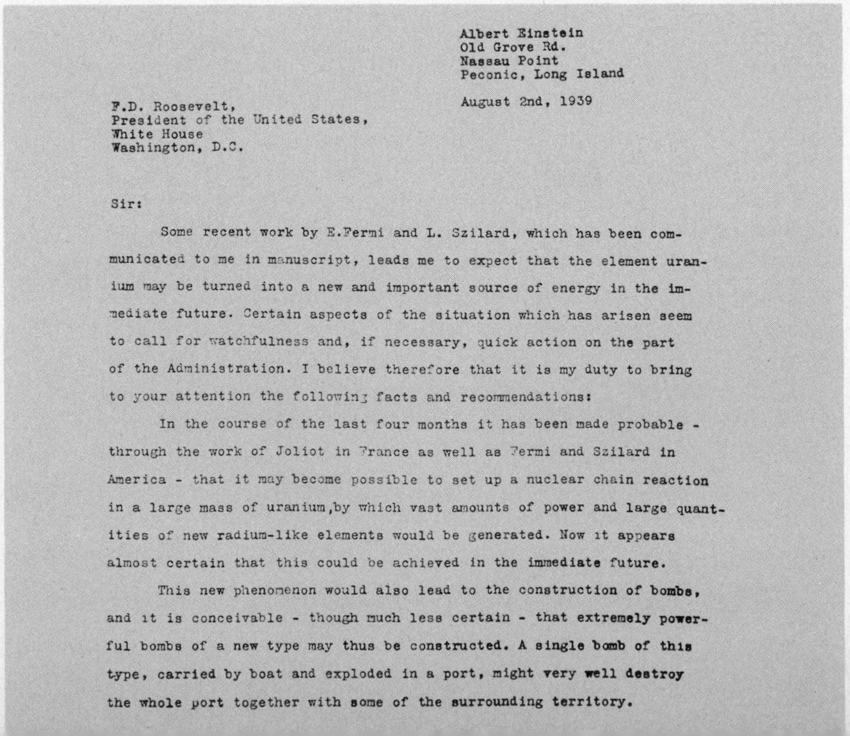

EXERCISE 4.1
An energetic chemical reaction
involving 100 g of reactants
gives off 1000 kJ (kilojoules) of
energy. Calculate the loss in
mass. What fraction of the total
mass is converted to energy?

Albert Einstein
Old Grove Rd.
Nassau Point
Peconic, Long Island

August 2nd, 1939

F.D. Roosevelt,
President of the United States,
White House
Washington, D.C.

Sir:

    Some recent work by E.Fermi and L. Szilard, which has been com-
municated to me in manuscript, leads me to expect that the element uranium may be turned into a new and important source of energy in the immediate future. Certain aspects of the situation which has arisen seem
to call for watchfulness and, if necessary, quick action on the part
of the Administration. I believe therefore that it is my duty to bring
to your attention the following facts and recommendations:

    In the course of the last four months it has been made probable -
through the work of Joliot in France as well as Fermi and Szilard in
America - that it may become possible to set up a nuclear chain reaction
in a large mass of uranium,by which vast amounts of power and large quantities of new radium-like elements would be generated. Now it appears
almost certain that this could be achieved in the immediate future.

    This new phenomenon would also lead to the construction of bombs,
and it is conceivable - though much less certain - that extremely powerful bombs of a new type may thus be constructed. A single bomb of this
type, carried by boat and exploded in a port, might very well destroy
the whole port together with some of the surrounding territory.

Figure 4.2
Part of the Einstein letter. (The
Franklin D. Roosevelt Library)

amounts of energy emerged. The scientific predictions based on graphs such as Figure 4.3 (which you already saw in Chapter 3) were correct. But what were these predictions?

The Effects of Mass Defects

In Chapter 3, we saw that the mass of a nucleus is slightly smaller than the sum of the masses of the separate nucleons (neutrons and protons) that make up the nucleus. This difference in masses is called the *mass defect* of the nucleus. From Einstein's equation, we know that a loss of mass results in a release of energy. Therefore, the size of the mass defect indicates the amount of energy that would be released if separate nucleons combine to form each atomic nucleus. Or, conversely, the mass defect indicates how much energy must be supplied to separate a nucleus into its component nucleons. This energy is called the *binding energy* of the nucleus. According to Einstein's equation, mass defect (m) and binding energy (E) are equivalent concepts—either quantity can be computed from the other by the relationship $E = mc^2$. Note that binding energy per nucleon first increases, then decreases, as Z increases in Figure 4.3.

The mass defect shown in Figure 4.3 has led us to suppose that two forces are important within the atomic nucleus. An attractive force between protons and neutrons is very powerful at extremely short distances, whereas the repulsive electrical force between protons becomes much more significant at slightly greater distances. The nuclear force tends to increase binding energy as Z rises from small to intermediate values; the electrical force tends to decrease binding energy for larger values of Z. In small nuclei with small values of Z, the neutron–proton attraction is more powerful. In large nuclei with large values of Z, the proton–proton repulsion becomes more significant.

High-Z nuclei are relatively unstable because of repulsion between protons. Low-Z nuclei are relatively unstable because there are not enough protons and neutrons to bind one another tightly. The most tightly bound nuclei are those of intermediate Z values.

From this information, scientists predicted the possibility of creating nuclear energy sources. Can you see the reason for this prediction? (Don't feel bad if you don't see it at once. A great many scientists were skeptical of the information we have discussed when it was first published in scientific journals shortly before World War II.)

Fission Doesn't Fizzle

Hundreds of nuclear bombs have now been exploded, and tens of thousands have been made and stored. Six nations—the United States, the Soviet Union, the United Kingdom, France, China, and India—have proved publicly that they know how to make and detonate nuclear

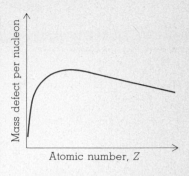

Figure 4.3
Mass defect (binding energy per nucleon) versus atomic number. (You have already seen this graph in Chapter 3.)

POINT TO PONDER
Curves having a single maximum or a single minimum can normally be analyzed in terms of two opposing effects.

EXERCISE 4.2
Look at Exercise 3.9 and comment on the likelihood of the reaction: $n = p + e^-$. (Recall n = neutron, p = proton, and e^- = electron.)

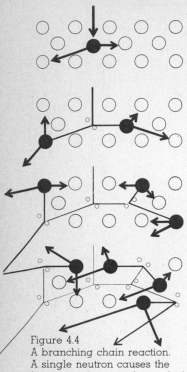

Figure 4.4
A branching chain reaction. A single neutron causes the fissioning of one nucleus, which releases about two neutrons (on the average). Each produces further fission and more neutrons so that the speed of reaction increases rapidly. The energy is liberated in a very short time. Fission products remain.

POINT TO PONDER
The first proof of the possibility of fission from branching neutron-chain reactions was obtained by Enrico Fermi and his colleagues in December 1942. But the effects of fissioning nuclei were first observed in a laboratory in 1934, also by Fermi. In 1934, no one recognized that fission was occurring, for the experiments had been set up to produce elements of high nuclear charge by bombarding uranium ($Z = 92$) with neutrons. Not till 1939 were the radioactive products identified as elements (like barium) of intermediate Z. It is hard to find what you are not looking for.

bombs. Several more countries certainly have the capability, and some may even have bombs in storage. Furthermore, almost all known attempts to explode a bomb have succeeded. Clearly there are no secrets left concerning the fundamental ideas, nor is the path to success a highly restricted one.

In Figure 4.3 you see that the binding energy *per nucleon* is largest for elements with a nuclear charge near that of iron ($Z = 26$). According to Figure 4.3, it should be possible to recombine the neutrons and protons from elements of either high or low nuclear charge to produce elements of intermediate nuclear charge with a net increase in binding, that is loss of mass. This loss of mass would appear as energy. The possibility of such a conversion of an appreciable fraction of mass to a very large amount of energy was the basis of the letter by Einstein. Two possibilities were seen: either decomposition (fission) of high-Z nuclei, or combination (fusion) of low-Z nuclei into nuclei of intermediate Z.

Fission releases energy by breaking up elements of high Z into those of lower Z. Bombs are possible because this breakup process can be triggered by bombarding certain isotopes with neutrons. The usable nuclei (1) undergo fission, (2) release mass as energy, and (3) simultaneously generate several secondary neutrons, which can continue the process in a *branching chain reaction* (Figure 4.4).

The two most difficult problems have been (1) to obtain concentrated samples of the readily fissionable isotopes, such as thorium-233 (^{233}Th), uranium-235 (^{235}U), and plutonium-239 (^{239}Pu), and (2) to assemble several kilograms of the isotope in such a short time that the mass does not blow itself apart before most of the atoms have undergone fission.

Both ^{233}Th and ^{235}U are found in natural thorium and uranium deposits. Materials enriched in the 233 or 235 isotopes can be prepared, as can the pure isotopes. The enriched or purified substances and even the natural materials can be used to generate nuclear power by appropriate modifications in design of the nuclear plant. Plutonium-239 must be synthesized, usually by neutron bombardment of ^{238}U; ^{239}Pu also may be used either as synthesized or in the enriched or pure form.

The second step can be accomplished by placing small pieces of the pure isotope near one another but separated and then surrounding them with chemical explosives. When the explosives are detonated, the small pieces are driven together into one large chunk centered on a source of neutrons. Fission begins and energy and secondary neutrons are released. Since the piece is small and highly concentrated it is likely that the released secondary neutrons will strike fissionable nuclei to continue and amplify the energy release (Figure 4.5). The smallest amount of isotope that will maintain the fission reaction is called its *critical mass*.

The critical mass for a bomb can be as small as a few kilograms. Concentrated larger masses produce correspondingly larger explosions.

At the moment of their release during fission, the neutrons are moving very rapidly—they are called *fast neutrons*. They can be slowed down—for example by collision with carbon or deuterium nuclei, ^{12}C or ^{2}H. Neither of these nuclei absorbs neutrons. They merely bounce them off, absorbing some of the energy and slowing the neutrons in the process. These *slow neutrons* can also cause fission. Because they are moving slowly, it is easier to contain and control them. This can be done in a *nuclear reactor* (Figure 4.6), which is the heart of the nuclear energy plants that generate an increasing amount of the world's energy. In a nuclear plant, the critical mass can be quite large and distributed over a volume of many cubic meters. The slow neutrons wander through this volume and eventually collide with a fissionable nucleus, releasing more neutrons to continue the chain reaction, and produce energy from the loss in mass accompanying fission.

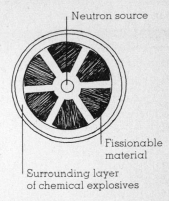

Figure 4.5
A nuclear bomb is detonated by implosion when the chemical explosive goes off.

EXERCISE 4.3
Suppose fission produced only one neutron per nuclear fission. Would nuclear energy be feasible from fission?

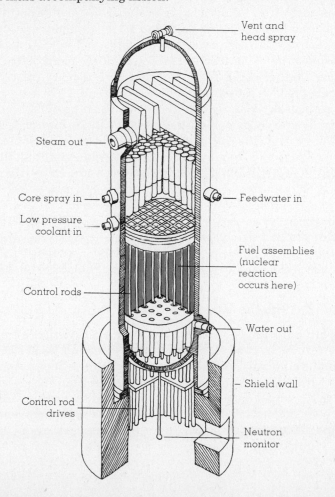

Figure 4.6
Nuclear reactor assembly. The fuel assembly is called a nuclear pile. (Courtesy Bechtel Power Corporation, Los Angeles)

EXERCISE 4.4
After the nuclear bombs
exploded over Hiroshima and
Nagasaki, some scientists
claimed that these sites would
be uninhabitable for 1000 years
because of radioactive contamination. Yet both sites are
crowded cities today. Where
did the radioactive fission
products go?

POINT TO PONDER
Knowledge not only tells us
what we can do, and how; it
also tells us the likely consequences—always a mix of good
and bad potentialities. Thus, the
more knowledge we accumulate,
the more we become aware of
threats. Ignorance *is* bliss. Does
ignorance best fit human
capabilities?

POINT TO PONDER
Many past civilizations have
generated long-term threats to
the survival of their descendants.
The Middle East from the
Mediterranean through Afghani-
stan was heavily forested 5000
years ago. Intensive herding has
created a desert over much of
this area. Both central Africa
and Southeast Asia were cleared
of too much growth hundreds of
years ago, leading to the formation of a stonelike layer of soil
(in a process called *laterization*).
These, and many other environmental changes have been
"given" to us by our ancestors.
The phenomenon is not new.
What is new is our greater (but
not complete) ability to anticipate the long-term effects of our
actions.

Nuclear fission generates three products: heat, neutrons, and product elements called fission products. If the heat is generated very rapidly
(as in a bomb), a shock wave of explosive force is also generated. With
bombs, the greatest effect is the explosive one of blowing things down.
Next in importance is heat radiation, which starts fires and causes
burns. Third is neutron effects on living systems. Least damaging (at
least so far) have been the radiations from the fission products.

Nuclear energy plants so far have an outstanding safety record, far
better per unit of energy produced than either the coal or petroleum
industries. Yet there is great concern among some of the public and
among some knowledgeable scientists concerning the potential dangers.
But it does pay to compare the dangers of different energy sources. For
example, more than three persons per day (on the average) have been
killed in the coal mines of the United States since 1889; a smaller number have been killed in the petroleum industry. There are currently
more than 200,000 former U.S. miners disabled by black lung disease.
Several hundred thousand have had to move home and business as
mines and wells closed down. These deaths and uprootings have been
the result of many accidents (the worst killing about 500 miners) and
of many local closures. Safety standards have risen greatly, but any
newspaper reader knows coal mining is dangerous.

If a nuclear plant were to have the worst possible accident (and
essentially everybody agrees such a plant cannot blow up), the energy
unit would melt, the safety container would rupture, and radioactive
gases and dust could be released. It is conceivable that, if all these things
happened, several hundred thousand people would have to be evacuated
from the surrounding area; several hundred might be killed. The likelihood of such an accident is considerably less than that of a major accident in the other energy-production industries. The biggest differences
probably are that a nuclear accident would be more dramatic and that
it would involve many people not normally considered at risk from
industrial accidents—in other words, not coal miners or oil-well drillers.

Away (?) With the Wastes

There are many (and I am one of them) who believe that a greater cause
for concern is the control, transportation, and long-term storage of the
fissionable materials, the fission products, and the nuclear plants themselves when they have exceeded their useful life. The possibilities of
nuclear blackmail are also widely discussed. Furthermore, some of the
plant components and fission products will be radioactive and lethal to
humans 1000 years after they have been removed from the nuclear
plant. Human experience suggests that future generations should not
be forced to concern themselves with the waste materials of the present

Table 4.1   Some Arguments on the Disposal of Nuclear Wastes

Proposal	Objection
1. Surround them with concrete on the surface of the earth.	1. The concrete will probably deteriorate before the materials are safe.
2. Bury them in abandoned salt mines located in areas free from earthquakes, floods, and other natural forces that might destroy the mine.	2. Transportation to such places is dangerous, and we cannot predict or prevent the occurrence of rare natural disasters.
3. Shoot them into space or into the sun by rocket.	3. What if the rocket disintegrates on takeoff?
4. Encase them in well-designed containers and drop them into the sea.	4. No one knows how to design a container with a sufficiently long life; transportation is a problem; and, big as the ocean is, it would become seriously contaminated if many containers leaked.
5. Drop them into the ocean at spots where the earth's crust is descending, and so carry them toward the center of the earth.	5. No one is sure how this motion occurs, or how to package the waste to ensure that it gets carried down. Could it later be spewed out into the atmosphere by a nearby volcano?

EXERCISE 4.5
Which of the alternatives in Table 4.1 seems safest to you? Suggest another alternative, giving both proposal and objection, if you can.

—so what is needed is a safe long-term storage system. None has yet been agreed on. Table 4.1 lists a few of the proposals and arguments against them.

It does pay to note that every energy source produces wastes. Suppose coal had been discovered recently, rather than hundreds of years ago. With our present environmental knowledge, it is quite likely that attempts would be made to prohibit the burning of coal. One reason is that coal burning produces sulfur dioxide, which is responsible for corrosion (due to acid rain; see Figure 4.7) and health-damaging smog. This has already led to the banning of some coals as fuels.

A second reason is more compelling, perhaps. It applies to all fossil fuels: The carbon dioxide produced has potentially serious long-range effects. The principal effect is increasing the carbon dioxide content of the atmosphere (as coal burning is known to do). This decreases the rate of heat loss to space and leads to an increase in temperature of the atmosphere and of the earth's surface. If unchecked, this could lead to massive melting of glaciers and of arctic and antarctic ice. The resulting rise in ocean level would flood almost all the present large cities, since they tend to be located on low-lying plains. Figures 4.8 and 4.9 show the increasing rate of carbon dioxide production and its rise in atmospheric concentration. It is unlikely that the present concentration will cause serious problems. But the threats of further marked rise are sufficient

Figure 4.7
Effect of acid rain. The statue was erected in 1702. The first photograph was taken about 200 years later in 1908, the second in 1969 after only 61 more years. (Springer-Verlag, New York, 1973; photo by Schmidt-Thomsen, Westfälisches Landesamt für Denkmalpflege, 4400 Münster, Germany)

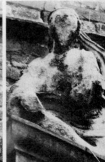

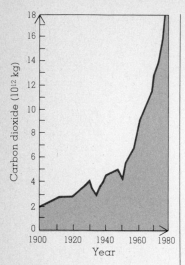

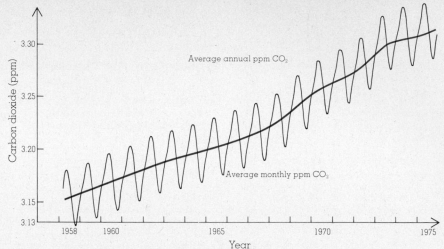

Figure 4.8 (above)
The increasing worldwide rate of carbon dioxide injection into the atmosphere from burning fossil fuels and making cement. Only economic depression or war slows the increase. (Scientific Committee on Problems of the Environment of the International Commission of Scientific Unions)

Figure 4.9 (above right)
Atmospheric carbon dioxide from 1958 to 1975. Peaks occur in spring; valleys occur at the end of the summer growing season. (Data from the Mauna Loa Observatory, Hawaii)

POINT TO PONDER
There are no completely safe human activities, only varying degrees of risk.

EXERCISE 4.6
Figure 4.9 shows that the amount of carbon dioxide in the atmosphere rises in the winter and falls in the summer on top of Mauna Loa in Hawaii where the measurements were made. Why does it rise and fall?

for many to believe there should be little increase in the rate of burning of fossil fuels even if they are available.

To summarize: Every human activity disturbs the environment. In fact, every human activity contains a potential threat to human lives—witness the lethal effect of cars on some drivers, riders, and pedestrians, and of childbirth on some mothers. We must balance presumed gains against presumed losses using the best available information, and then make necessary adjustments as results become apparent and/or better information becomes available.

Is Fusion an Illusion?

The left-hand end of the curve in Figure 4.3 clearly offers even more energy per nucleon than the fission end. Energy release can result when light nuclei join together, or *fuse*. For example, the fusion of deuterium to helium, $^2H + {}^2H = {}^4He$, results in a net loss of mass: $4.0026 - (2 \times 2.01410) = -0.0256$ amu per atom (or grams per mole) of helium.

This is a net loss in mass of about 6 parts per 1000 (0.256/4.00), almost 10 times the conversion of mass in fission processes. Other relative advantages of fusion over fission are

1. the absence of fission products;
2. fewer radioactive by-products;
3. much greater difficulty in assembling, stealing, or detonating the product (which would greatly minimize terrorist use).

Unfortunately, no one yet knows how to make fusion work on a large scale on earth, although there is no question that fusion does occur. Fusion is known to be the main energy source in most stars, including the sun. For example, the sun converts some 4×10^9 kg of mass to energy each *second* by fusion processes. But the stellar processes require enormously large masses (like the sun), with very hot interiors (several million °C), and retention of these conditions for long periods of time (thousands of years). Earth processes must be carried out in small volumes, must be fast yet controllable, and must allow the energy to be converted into transportable forms like electricity.

Optimism among researchers into fusion processes has gone up and down. The fact that each peak creates more optimism than the last suggests that success may come early in the twenty-first century. Current energy sources can bridge this time period without great declines in energy consumption or standards of living in the United States while the developing countries continue to increase their standards of living.

Fusion has one more advantage. We know that there is enough fusionable material (for example, deuterium) to satisfy, for a period of at least 1 billion years, world energy-consumption rates even higher than those of today.

The only other known energy source of comparable importance is sunlight, some of which is already effectively used in photosynthesis. Active research is being expanded rapidly to tap sunlight further. Sunlight is even more advantageous than fusion, for it produces no harmful products—nor any radioactivity at all. It is already available. We just don't use most of it at present, so it radiates back to space without having contributed to our usable energy. The big difficulty with sunlight is that it is so spread out. Almost every human need requires a concentrated energy source. I'll discuss solar energy later, but its study is part of our unending frontier.

Nuclear Reactions

When an atomic nucleus emits one or more particles, it is said to have been radioactive, or to have undergone radioactive decay. Do note the word *decay*. It means that each nucleus is radioactive only once, because the decay converts it to another kind of nucleus. The daughter nucleus may or may not undergo further decay. Let's look at some examples.

I have, so far, mentioned three kinds of particles as coming from nuclei: alpha (α), beta (β), and gamma (γ)—or helium nuclei, negative electrons, and high-energy photons, respectively. We can readily interpret the emission of these three types of particles in terms of our neutron–proton theory. We can make this interpretation very easy by adding one item to our symbolism. Each chemical symbol (H, O, Cl, Na, and so on) represents a unique nuclear charge. But either memory or a table is required to obtain the charge (atomic number) associated with the symbol. We shall often use a small subscript at the lower left to indicate the nuclear charge and a superscript at the upper left to

EXERCISE 4.7
One of the net equations for nuclear fusion in the sun is

$$4\,^1H = \,^4He$$

How would you calculate the fraction of the original mass converted to energy in this reaction? Remember the electrons.

indicate isotopic mass—for example, ^1_1H, $^{16}_8O$, $^{35}_{17}Cl$, $^{23}_{11}Na$, and $^{238}_{92}U$. Emission of an α particle by $^{238}_{92}U$ can be represented in the equation,

$$^{238}_{92}U = {}^4_2\alpha + {}^{234}_{90}? = {}^4_2He + {}^{234}_{90}Th \tag{4.3}$$

where we simply conserve the charges and the masses to identify the daughter nucleus as $^{234}_{90}Th$—that is, the thorium isotope of mass 234 (90 protons + 144 neutrons). Because α, β, and γ particles are known to be helium nuclei, electrons, and photons, we shall usually represent them as ^4_2He, $_{-1}^{0}e$, and $^0_0h\nu$, respectively. Masses are at the upper left, charges at the lower left.

Emission of an electron (β particle) from the nucleus can be represented as in the equation

$$^{234}_{90}Th = {}_{-1}^{0}e + {}^{234}_{91}Pa \tag{4.4}$$

and the daughter isotope is protactinium (Pa) of mass 234.

Most emissions of α or β particles are accompanied by the simultaneous emission of a γ ray, or high-energy photon. The photon ($h\nu$) has neither mass nor charge, so need not appear in the equation representing the conservation of mass and charge. However, more complete equations corresponding to those above would be

$$^{238}_{92}U = {}^4_2He + {}^{234}_{90}Th + {}^0_0h\nu \tag{4.5}$$

$$^{234}_{90}Th = {}_{-1}^{0}e + {}^{234}_{91}Pa + {}^0_0h\nu \tag{4.6}$$

Neutrinos are also emitted but we shall neglect them completely.

Consider now what happens in each nucleus, in terms of the neutrons and protons originally present. Each ^{238}U emits a helium nucleus composed of two neutrons and two protons (charge 2, mass 4). This matches the fact that the product, ^{234}Th, does have two fewer protons and two fewer neutrons than the original ^{238}U.

Now compare ^{234}Th (90 protons, 144 neutrons) with ^{234}Pa (91 protons and 143 neutrons). One neutron has been converted into a proton within the nucleus, and an electron has been expelled.

Many naturally radioactive substances decay through long series of α and β emissions. Figure 4.10 shows two such series. The equations are

$$^{238}_{92}U \rightarrow {}^{206}_{82}Pb + 8\,{}^4_2\alpha + 6\,{}_{-1}^{0}\beta \tag{4.7}$$

$$^{235}_{92}U \rightarrow {}^{207}_{82}Pb + 7\,{}^4_2\alpha + 4\,{}_{-1}^{0}\beta \tag{4.8}$$

Emission of a photon (γ ray) does not affect the nuclear charge or the nuclear mass (except for the conversion of a small amount of mass to energy). Problem 4.3 discusses one other particle commonly emitted by radioactive nuclei.

Perhaps you are wondering why I have not mentioned emission of individual protons and neutrons if those are the particles actually found

EXERCISE 4.8
Cesium-142 ($^{142}_{55}Cs$) decays into barium-142 ($^{142}_{56}Ba$). What particle is emitted? What change occurs in the number of neutrons and protons in the nucleus?

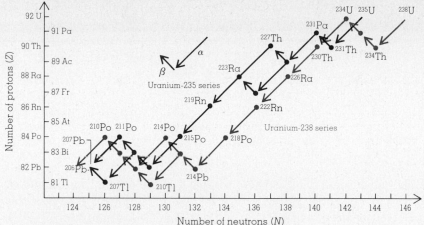

Figure 4.10
Naturally occurring uranium-235 and uranium-238 decay by α series of α and β emissions to different stable isotopes of lead. (Adapted from R. E. Dickerson and I. Geis, *Chemistry, Matter, and the Universe*, Copyright © 1976 by W. A. Benjamin, Menlo Park, Calif.)

in the nucleus. Well, it is true that such emission is rare. We shall settle for the statement that the binding force between individual protons and individual neutrons is so high that their individual escape is difficult.

However, if the nucleus is bombarded from outside by particles such as α particles or neutrons, we find that individual neutrons and protons can be driven out. The original reaction used by Chadwick in his discovery of the neutron was

$$^{10}_{5}B + ^{4}_{2}He = ^{13}_{7}N + ^{1}_{0}n \qquad (4.9)$$

Fission in ^{235}U may be represented by

$$^{235}_{92}U + ^{1}_{0}n = ^{142}_{56}Ba + ^{91}_{36}Kr + 3\,^{1}_{0}n \qquad (4.10)$$

If magnesium is bombarded with neutrons, the following may occur:

$$^{24}_{12}Mg + ^{1}_{0}n = ^{24}_{11}Na + ^{1}_{1}p \qquad (4.11)$$

All these reactions also involve the simultaneous emission of photons (γ particles).

So nuclear reactions are of two types:

1. those in which a radioactive nucleus spontaneously emits a particle;
2. those in which a bombarding particle is required to initiate the nuclear change (almost all nuclei can undergo such changes).

The spontaneous reactions have an interesting property (which you have probably heard of) called a *radioactive half-life*. Each radioactive nucleus behaves quite independently of other nuclei and of almost all outside forces. One result is that the nuclei follow a very simple decay law. If half of the original set of nuclei decompose in 1 hour, the half-life is 1 hour. Then, half of the remainder will decompose in the next hour, half of that remainder in 1 more hour, and so forth. At the end of 4 hours there will be $\frac{1}{2} \times \frac{1}{2} \times \frac{1}{2} \times \frac{1}{2} = \frac{1}{16}$ of the original nuclei.

EXERCISE 4.9
Uranium-238 ($^{238}_{92}U$) decays into stable $^{208}_{82}Pb$ (lead) with an effective half-life of 4.50×10^{9} years. A rock is found containing almost equal quantities of ^{238}U and ^{206}Pb. How old is the rock? What assumptions did you make?

Originally all ^{238}U

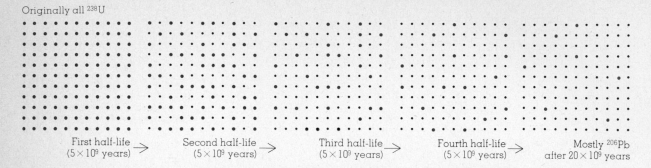

First half-life $\rightarrow$ (5 × 10⁹ years)   Second half-life $\rightarrow$ (5 × 10⁹ years)   Third half-life $\rightarrow$ (5 × 10⁹ years)   Fourth half-life $\rightarrow$ (5 × 10⁹ years)   Mostly ^{206}Pb after 20 × 10⁹ years

Figure 4.11
The conversion of uranium-238 (colored dots) into lead-206 (black dots). The half-life of uranium-238 is 5 × 10⁹ years. Only 1 out of 16 of the original uranium atoms remains after 4 × 5 × 10⁹ = 20 × 10⁹ years. The rest are lead atoms.

If the half-life is 1 minute, there will be $\frac{1}{16}$ of the original nuclei left at the end of 4 minutes. If the half-life is 1000 years, there will be $\frac{1}{16}$ of the nuclei left at the end of 4000 years. Thus, if we find a sample of ^{238}U ore (half-life, 5 × 10⁹ years) that is half decomposed into ^{206}Pb (lead), we know the ore has probably been undisturbed for 5 × 10⁹ years—one half-life (Figure 4.11). Methods are now available for dating objects ranging up to 10¹⁰ years in age, if they contain the appropriate radioactive elements. Table 4.2 lists a few half-lives, many of them used to date archeological objects.

Some Effects of Radioactivity

A principal drawback to the use of nuclear fission is the long half-lives of some of the fission products and of some of the elements produced in the nuclear pile structure by neutron-bombardment reactions. An additional problem is that plutonium, one of the most useful fissionable materials, concentrates in human bone marrow if it is taken into the body. The effect of its α rays on bone marrow is so severe as to make it one of the most dangerous poisons (that is, toxic in one of the smallest doses) of any known substance.

There is still great debate about the effects of radioactive materials on living systems. It is not known, for example, whether there is a threshold dose below which there are no effects or whether the effects of small doses just get too small to measure accurately. Yet it is now generally assumed that there is no "safe *threshold dose.*" This boils down to a discussion of whether there is or is not a *maximum tolerable annual dose*—and, if so, what that dose is. The dose is sometimes called the threshold limit value (*TLV*), and approximates the lowest dose whose effect is detectable. Problem 4.11 gives some data for a discussion of this problem. See also Figure 4.12.

The detailed effects of radiation on living systems are very complicated, but consider what you already know about such systems. The most common molecule in living systems is water—on a weight, volume,

Table 4.2 Some Nuclear Half-lives

Isotope	Particle emitted	Half-life[a]
$^{1}_{0}n$	e^-	7.8×10^2 s
$^{3}_{1}H$	e^-	12.26 yr
$^{14}_{6}C$	e^-	5.6×10^3 yr
$^{40}_{19}K$	e^-	1.28×10^9 yr
$^{90}_{38}Sr$	e^-	28.1 yr
$^{212}_{84}Po$	α	3.04×10^{-7} s
$^{232}_{90}Th$	α	1.41×10^{10} yr
$^{238}_{92}U$	α	4.51×10^9 yr
$^{239}_{94}Pu$	α	2.44×10^4 yr

[a] s = second, yr = year.

or molecular basis. Each atom, including those in the water, consists of a tiny nucleus surrounded by electrons. Thus a highly likely effect would be for high-energy radiation to collide with these electrons, producing residual ions of positive charge and free electrons that might ionize further molecules.

The most common initial ion would presumably be a water molecule minus one electron—that is, H_2O^+. This ion is very unstable and would quickly decompose into H^+ and OH; OH is called *hydroxyl* and is known to be highly reactive. If enough OH molecules are produced, they may combine to give hydrogen peroxide, H_2O_2, also reactive but capable of a long existence. Or OH can react to give highly reactive HO_2, especially if oxygen is present.

Most of the damage is probably caused by these highly reactive molecules rather than by the initial effects of the radiation itself. Thus most protective methods (other than those that minimize the chance of the radiation striking living tissue at all) have attempted to deal with these highly reactive molecules. So far, success has been limited. Minimizing exposure to radiation is the best protection.

Yet you should realize that every living system on earth has been, and is now, continuously exposed to radiation. Some people live in areas with high thorium concentrations; others live at high altitudes where nuclear radiation from the sun and outer space is stronger than at sea level; others work in the upper floors of skyscrapers where the radiation dose is several times higher than on the lower floors. Still others have more x-rays or other radiation treatments. But, in addition to these unusual doses, everyone is exposed to about one neutron per cubic centimeter per second, as well as to high-energy cosmic rays (γ rays). In fact, there is some evidence that these high-energy radiations are an appreciable factor in determining the rate of evolution and that humans could never have evolved had there been no radiation dose at the earth's surface.

Large doses of high-energy (α, β, γ, or neutron) radiation are clearly dangerous. Very large doses are lethal. But there is little evidence that differences in altitude, differences in local natural sources, or present x-ray doses have measurable effects. Whether there is a threshold dose or not, the effects of low doses cannot be sorted out from the effects of the many other toxic and lethal threats to human beings. For example, alcohol intake, cigarette smoking, and automobile collisions cause far more damage and deaths to humans than would any reasonably possible exposure to high-energy radiation in peacetime.

Atom, Atom, Who's Got the Atom?

You read at the beginning of Chapter 3 that radioactivity has been hailed as the greatest contribution to the study of living systems since

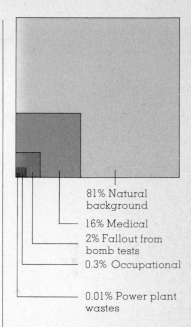

81% Natural background

16% Medical

2% Fallout from bomb tests

0.3% Occupational

0.01% Power plant wastes

Figure 4.12
Sources of radiation exposure to the public. (Data from United Kingdom Atomic Energy Authority Annual Report, London, 1975)

POINT TO PONDER
Vice-President Nelson Rockefeller said in February 1976: "In the 18-year [U.S.] history of commercial nuclear plant operation no accidents have occurred involving public injury. In the same period in the United States alone 848,544 people have been killed by motor vehicles and more than 70 million have been injured by this highly popular invention. Yet, to my knowledge, there is no movement to 'ban the auto.'" Is the analogy apt?

EXERCISE 4.10
Why does the radiation dose received by humans generally increase with altitude?

EXERCISE 4.11
Many more people shorten their lives by smoking cigarettes than could be affected by any probable nuclear plant accident. Does this mean that we are overcautious in our concern about nuclear plants?

Figure 4.13
Tumor outlined by radiations from radioactive phosphorus (^{35}P). (Courtesy U.S. Energy Research and Development Administration)

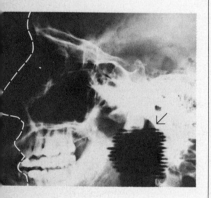

the microscope. Yet the possible applications I have discussed all contain threats to living systems. This dilemma can be partially resolved if you realize that radioactivity allows one to locate single atoms at the moment they are radioactive. One of the most dramatic examples was the discovery of the element mendelevium, element 101; atoms were synthesized and detected *one at a time!*

Usually one works with samples containing hundreds of millions of potentially radioactive atoms. They decay with their fixed half-life. The positions of the decaying atoms can be determined at any time by using photographic film or some other device for detecting radiation. Remember, a potentially radioactive isotope behaves like a stable isotope up to the instant of radioactive decay.

Thus, a photograph can be made by slicing up a tomato that has been fed zinc, some of whose atoms are radioactive. The slice is then placed on a piece of photographic film wrapped in black paper. The radiations penetrate the paper, and (upon developing the film) it becomes clear that the zinc concentrates in the seeds. Magnification may even allow determination of where in the seed the zinc goes.

Figure 4.13 outlines a tumor in a human head. The tumor preferentially picks up radioactive phosphorus injected into the circulating bloodstream, because the growth rate of the tumor is greater than that of the surrounding tissue.

Radioactive isotopes of useful half-life are now known for most of the chemical elements, so that most of the elements can be followed in considerable detail, even in complicated systems. Most of modern medicine would not be where it is today had radioactive isotopes not been available.

Even stable isotopes can be followed with such instruments as mass spectrometers. So *tracers* are available for all the chemical elements.

Summary

Few mathematical statements have had greater influence on humans than Einstein's $E = mc^2$. Both the accomplishment of nuclear fission and the drive to control nuclear fusion are summarized in it. And it has also provided a new context in which to examine and better understand all relationships between mass and energy.

Yet the sizes of mass changes, even in nuclear reactions, are so tiny that we can write nuclear equations by balancing only charge and mass in integral quantities. Such equations allow understanding of nuclear interconversions in terms of the neutron–proton theory—and, when coupled with measurement of radioactive half-lives, give bases for planning the development and dealing with the problems of nuclear energy.

Problems

4.1 Identify or define: branching chain reaction, critical mass, fission product, half-life, joule, neutrons (fast and slow), nuclear (binding energy, decay, equations, fission and fusion, piles), nucleon attractions and repulsions, radioactivity, threshold dose, tolerable dose, tracers, waste (nuclear and other), $E = mc^2$.

4.2 How is it that low-energy neutrons (slow neutrons) can initiate nuclear reactions, but low-energy α and β particles cannot?

4.3 Identify the mass and charge of the particle emitted in the reaction

$$^{22}_{11}\text{Na} = ? + ^{22}_{10}\text{Ne}$$

It is called a *positron* and is formed in many nuclear reactions. Use the neutron–proton theory to account for its mode of formation in the nucleus from neutrons and/or protons.

4.4 In the conversion, $^{238}_{92}\text{U} \rightarrow {}^{206}_{82}\text{Pb}$, how many α and how many β particles are given off? Identify one or two of the unidentified intermediates in Figure 4.10.

*4.5 Curium-238 ($^{238}_{96}\text{Cm}$) decomposes into $^{210}_{82}\text{Pb}$ by emitting a sequence of identical particles. What is the particle? Write nuclear equations for each step and identify the intermediate isotopes.

4.6 Only 3000 atoms of element 104 (half-life of 4.5 seconds) were present in the sample that C. E. Bemis and his colleagues at the Oak Ridge National Laboratory used to establish the existence of the element. About how many of these atoms probably remained at the end of 45 seconds? [~ 3 atoms]

4.7 The age of Scotch whisky, and similar aged drinks, can be checked by determining the tritium content. The half-life of tritium (another name for ^3H) is 12.3 years. What fraction of the tritium level present in your drinking water would you expect in a 50-year-old brandy?

4.8 You are engaged in an anthropological "dig" in Iran and come across some charcoal at a level you estimate to be 23,000 years old. What is your guess as to the level of radioactivity in the ^{14}C (half-life of 5.75×10^3 years, β emitter) compared to modern charcoal? The laboratory reports the relative radioactivity as 50% of modern samples. How do you interpret this? [$\sim \frac{1}{20}$ of modern charcoal]

*4.9 It is much more common to date human archaeological records with radioactive carbon than with uranium. Uranium is used mainly for geological dating. Why? On the other hand carbon is quite useless for the dating of coal, for example. Why?

*4.10 What minimum mass of deuterium must be in a nuclear bomb equivalent to 1 megaton of TNT? One ton of TNT releases about 4×10^9 J (joules). The net hydrogen bomb reaction may be approximated by

$$5\,^2\text{H} = {}^1\text{H} + {}^3\text{He} + {}^4\text{He} + 2\,^1n$$

[~ 20 kg of ^2H]

4.11 The average dose of high-energy radiation in the United States is made up of about 81% natural radiation, 16% medical radiation, and 2% fallout from nuclear bomb testing. Cosmic ray doses (about one-third of the natural radiation) may be twice as great in the top floors of a skyscraper or in a high-altitude city. What is the chance of detecting an increase in disease due to nuclear fallout? Due to medical x-rays? Is there any relation between these facts and the calculation of a maximum tolerable dose of radiation?

4.12 In 1975 in France, 2000 scientists (many employed by the French Atomic Energy Commission) called upon citizens to "refuse to accept nuclear power plants until there is a clear understanding of the risks and consequences of this policy." According to a newspaper account by Martin Brown, they singled out risks in transport and disposal of nuclear material, including obsolete nuclear plants. How would you explain these risks to an adult with no science background? From what you know of the risks would you vote for or against "safe nuclear plants"?

ATOMIC THEORY

Electrons and nucleons...   combine to give atoms.   Each chemical element has its own kind of atoms.

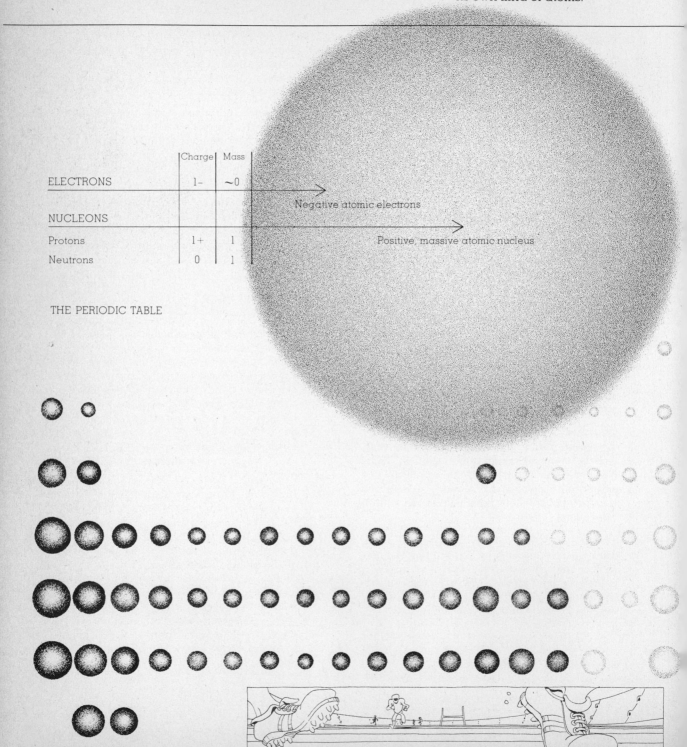

	Charge	Mass
ELECTRONS	1−	~0
NUCLEONS		
Protons	1+	1
Neutrons	0	1

Negative atomic electrons

Positive, massive atomic nucleus

THE PERIODIC TABLE

Each pure substance has its own combination of chemical elements.

Each solution contains two or more pure substances.

Pure substances mingle with solutions to form a heterogeneous universe.

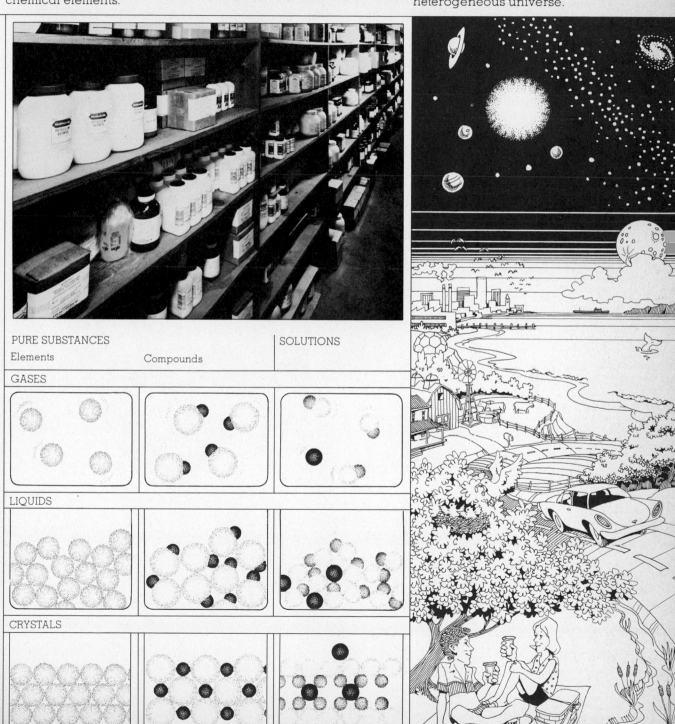

PURE SUBSTANCES

Elements

Compounds

SOLUTIONS

GASES

LIQUIDS

CRYSTALS

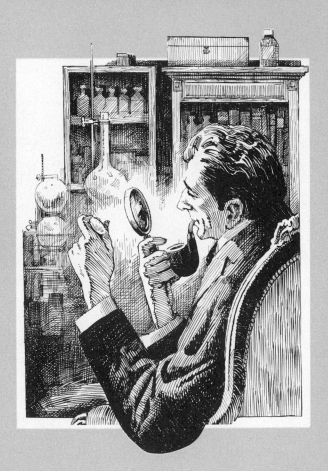

5 ELEMENTS AND THEIR ATOMS

Sherlock Holmes was one of the greatest fictional detectives of all time. His London dwelling, shared with Dr. Watson, was the scene of some of the earliest scientific detective work recorded, often followed by the phrase, "Elementary, my dear Watson." Just as Holmes and other detectives use clues to understand a complicated situation in terms of its elements, so scientists do the same, as outlined in Chapter 2. In Chapter 3 you saw that it is possible to consider all matter as made up of neutrons, protons, and electrons—just three particles. But chemists usually find it preferable to stop one step higher in complexity—that is, at the atomic level—and to discuss changes in terms of the chemical elements. We shall explore some of their reasons in this chapter. You already know a great deal about metals, and we shall build on and extend that knowledge here.

The Periodic Table

In Chapters 3 and 4, I pointed out that every neutral atom must contain a number of electrons equal to its nuclear charge. How are these electrons arranged, and how do they affect observable properties?

You found in Figure 3.11 (reproduced here as Figure 5.1) that plotting mass defect per nucleon against nuclear charge gives a regular curve that measures the tightness of the binding of neutrons and protons in the nucleus. The binding energy increases as more and more neutrons and protons are present to attract one another, but decreases as the repulsion between the like-charged protons increases at larger atomic numbers. It might be reasonable to assume a similar effect with the electrons. Increase in nuclear charge would tend to hold each electron more tightly to the nucleus. But as the number of electrons increases, each negative electron will repel the others, decreasing the binding energy of the electron to the atom. Thus, we might expect a curve much like Figure 5.1 if we plot binding energy of electrons against nuclear charge.

You will remember from the work of Thomson and others that electrons can be driven out of atoms. These electrons (unlike β rays) come from the region outside the nucleus, the nucleus remaining unchanged. The energy required to remove the least tightly bound electrons from a

It isn't what you don't know that hurts you. It's what you do know that isn't so.
—*Will Rogers*.

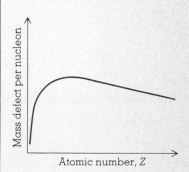

Figure 5.1
Mass defect (binding energy per nucleon) versus atomic number.

67

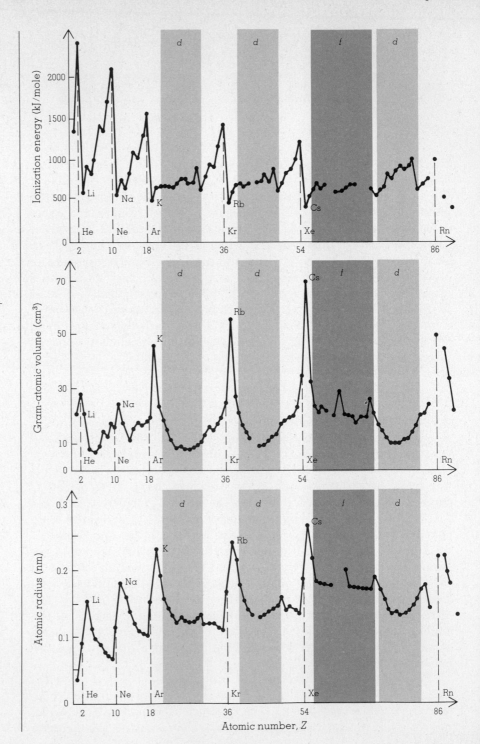

Figure 5.2
First ionization energies of the monatomic gaseous elements.

Figure 5.3
Volume of one atomic weight (in grams) versus atomic number for the elements. Compare with Figures 5.2 and 5.4 and look for correlations.

Figure 5.4
Internuclear distance (atomic radius) in crystals versus atomic number.

gaseous atom is called the *ionization energy* (IE) of the atom. Ionization energies have been measured for all the chemical elements. Values for the monatomic gaseous elements are plotted in Figure 5.2 as a function of nuclear charge. Many maxima and minima are observed in this curve, unlike the single maximum in Figure 5.1. The maxima are at helium (He), neon (Ne), argon (Ar), krypton (Kr), xenon (Xe), and radon (Rn)—with nuclear charges of 2, 10, 18, 36, 54, and 86, respectively. Each maximum is followed by a sharp drop then a fairly regular increase in ionization energy to the next peak. But how can we account for the many peaks and for the other irregularities in the curve? Apparently there are more than two simple effects that control ionization energy. But they repeat as Z increases; note the similarities in the rapidly rising portions of the curve.

Is the curve of Figure 5.2 typical of the variation in properties of elements as nuclear charge increases—or is it an unusual curve? Figure 5.3 shows the volume of one atomic weight (in grams) of each element, plotted against nuclear charge. Figure 5.4 shows a similar plot of internuclear distance of the elements as crystals. Both properties are related to the size of atoms—the closeness with which they pack together—and therefore should be determined by interactions between the electrons.

The shapes of these curves are not identical to each other or to that of Figure 5.2, but the similarities are unmistakable. The curves show a scallop shape, with sharp peaks at nuclear charges very near 2, 10, 18, 36, 54, and 86. Most properties of elements are periodic functions of the nuclear charge.

I would be seriously deluding you if I left you with the impression that the neutron–proton theory of nuclear structure led to the discovery of periodic recurrence of chemical and physical properties of elements. Chemical periodicity was discovered almost 50 years before nuclei were thought of.

Like many other great discoveries, the original periodic table was based on a mistake in interpretation of experimental data. Dalton's atomic theory had stimulated great interest in the determination of relative atomic weights. As data became more available, various scientists looked for correlations between atomic weights and other properties. The early attempts were laughed at by most established scientists. There seemed little more reason to seek a correlation between atomic weights and other properties than to seek one with an alphabetical listing of the elements by name or symbol.

However, by 1865 enough data on atomic weights and other properties became available that the Russian Dmitri Mendeleev, among others, presented many data such as those plotted in Figures 5.2, 5.3, and 5.4—except that he used atomic weight (rather than the then unknown nuclear charge) on the horizontal axis. The periodicity was

EXERCISE 5.1
Will the ionization energy of a second electron from a gaseous atom be greater or less than that of the first electron?

EXERCISE 5.2
Scientists always look at series of related numbers in search of regularities. Do you see any in the 2, 10, 18, 36, 54, 86 set of atomic numbers of the noble gases? There are two quite simple regularities. Can you guess what the next number would be?

POINT TO PONDER
Useful scientific theories do not have to be final and correct in all details. Indeed they are not expected to be. They are judged and used in terms of their ability to describe and correlate known data, and to predict correctly the results of future experiments. A theory is considered especially useful if it provides ideas for new experiments, even though these experiments often lead to modification of the theory.

Table 5.1 Mendeleev's Predictions (Printed in 1871) and the Properties Found for Germanium in 1886

Property	Data predicted	Data found
Element		
Atomic weight	72	72.6
Density	5.5	5.36
Color	Gray	Gray
Preparation	Na + oxide	Na + oxide
Oxide		
Formula	XO_2	GeO_2
Density	4.7	4.703
Soluble in acid	Yes	Yes
Fluoride		
Formula	XF_4	GeF_4
State	Crystal	Crystal

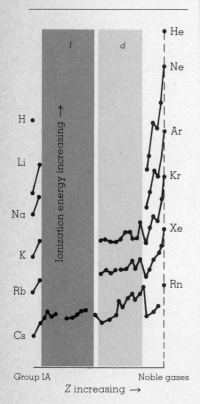

Figure 5.5
Ionization energies in the rows and families of the periodic table.

so marked that others had to take note. Even more convincing to the continuing skeptics was the rapid isolation of several more elements whose properties Mendeleev had predicted based on gaps he discovered in the periodic functions. Table 5.1 shows some of his predictions and the results of the experiments he stimulated. Successful prediction of experimental results is one of the most important tests of a theory, and Mendeleev's periodic table passed this test with flying colors.

But even when predictions prove successful, we cannot be sure that the success is not simply a coincidence. We now know that atomic weight has almost no effect on most properties. (Recall that isotopes of a single element have almost identical properties, even for hydrogen and deuterium where the mass ratio is 1 to 2.)

The coincidence here arises because atomic weights normally increase with an increase in nuclear charge. Each additional proton is (on average) balanced with the addition of one or more neutrons. However, it is nuclear charge, not mass, that affects electron behavior. And it is electron behavior that most directly affects most of the properties we observe. The scientists who had doubted the relevance of atomic weight to properties of elements turned out to be correct. But Mendeleev is remembered both for his successful predictions and for the fact that his theory stimulated the further work that led to a better understanding of atomic properties. His critics were right, but they don't get much credit because their doubts didn't lead anywhere at the time.

The *periodic table* inside the front cover of this book lists the elements in order of atomic numbers (equal to nuclear charge, and to the number of electrons in a neutral atom). The table also lists atomic weights (equal to the average weight of the mixture of isotopes found naturally, with the weight of ^{12}C defined as exactly 12 daltons). You will find the periodic table of great use in correlating observations about properties of the chemical elements.

The Ties That Bind

Figure 5.5 contains the same data on ionization energies as Figure 5.2. But the data are here plotted in the form of the periodic table. Two trends are apparent: (1) The ionization energy usually increases in moving from left to right in a row in the periodic table, and (2) it *always* decreases in descending any of the columns. We may expect, then, that trends in properties will be more regular in the columns than in the rows. This is found to be true, and the members of a given column are called a *family,* or *group,* of elements. Many of the groups have family names— for example: Group IA, the alkalies; Group IIA, the alkaline earths; Group VIA, the chalcogens; Group VIIA, the halogens; Group 0, the noble gases.

The noble gases (Group 0: He, Ne, Ar, Kr, Xe, and Rn) have the highest ionization energies in their respective rows. Each of these elements is gaseous at room temperature and quite unreactive toward other substances. The fact that noble gases have the highest ionization energies is consistent with the fact that each also has the largest nuclear charge of the elements in its row. Yet, within the family, the ionization energy decreases as the nuclear charge increases. The same effect is found in every family. Apparently, increase in the mutual repulsion among the electrons is a larger factor than the increasing nuclear charge in determining changes in ionization energy within a family of elements.

Furthermore, in every case there is a big drop in ionization energy between a noble gas element and the alkali element with the next higher nuclear charge (atomic number). Clearly there is a very large change in the mutual electron repulsion as we move from Group 0 to Group I in the periodic table. You will also note small fluctuations in the curve of Figure 5.5 as the nuclear charge increases along each row.

The full interpretation of these variations is complicated—indeed, it is not fully understood. But the general effects are those I have mentioned: attraction of electrons by the nucleus, and mutual repulsion among the electrons. The observed ionization energy results from the interplay of these two effects. A third effect is brought about by the fact that electrons can group in sets of two *(electron pairs)*. The irregularities in the curve within each row of the table are a result of the formation of such electron pairs. Pair formation increases electron repulsions and leads to a decrease in ionization energy (as from $Z = 7$ to $Z = 8$).

Apparently, certain electron groupings are more stable than others —that is, certain groupings hold the outermost electron more strongly. The results of studies on ionization energy are often presented in the form shown in Figure 5.6. Let's see how it works.

Pigeons Come Home to Roost

Figure 5.6 reminds some people of a pigeon house—that is, a set of roosts for birds. I shall use this analogy. The nucleus corresponds to a source of food. If there is more food (a highly charged nucleus), more birds (electrons) will come to roost. But each roost can hold a maximum of two birds. The birds prefer to be one to a roost (they repel each other), but they prefer even more strongly to be as close as possible to the source of food.

Thus, the first bird to arrive in the pigeon house would perch on the roost labeled 1s. The second bird to arrive would join the first on the 1s perch (its repulsion from the first bird overcome by the chance to be closer to the food). The third and fourth birds would share the 2s roost.

EXERCISE 5.3
Suggest an interpretation for the greater decreases in ionization energy from lithium (Li) to sodium (Na) than from rubidium (Rb) to cesium (Cs). Use your answer to decide which of these pairs of elements has the more similar chemistry.

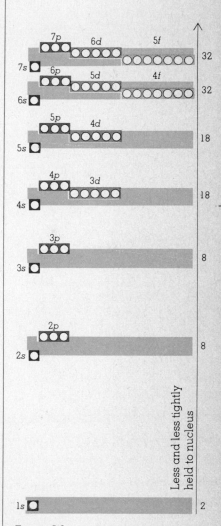

Figure 5.6
Electron energy levels in gaseous atoms.

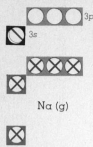

Figure 5.7
Electron structure of gaseous
sodium, $Z = 11$. Magnesium
($Z = 12$) will have two $3s$ electrons, and aluminum ($Z = 13$)
will have a $3p$ electron.

EXERCISE 5.4
How would ionization energy
be represented in Figure 5.6?
Explain why the ionization
energy increases as the nuclear
charge increases.

EXERCISE 5.5
Using Figure 5.6, can you
account for the observed ionization energies (Figure 5.2) of
the elements from lithium, Li
($Z = 3$) through neon, Ne
($Z = 10$)? Does the orbital
diagram account equally well
for the ionization energies of
elements with nuclear charges
from 29 to 36?

The fifth, sixth, and seventh birds would each take one of the $2p$ roosts, and an eighth bird would choose to share one of the $2p$ roosts rather than move to a higher perch. And so it would go.

Figure 5.7 represents the arrangement for 11 birds—or the element sodium (Na) with 11 electrons. Each diagonal stroke represents one occupant; crossed strokes represent a full roost (two occupants). The lower a bird (electron) is on the diagram, the more strongly it is attracted, and the more energy is needed to remove (ionize) it. In the sodium atom, the $3s$ electron is the easiest to ionize. In our analogy, the bird on the $3s$ roost is most easily persuaded to fly to another source of food.

Increasing the nuclear charge from 11 to 12 (the element magnesium, Mg) will increase the attraction between the nucleus and all the electrons. Thus all the "roosts" are found at slightly lower energy levels. Each of the 11 electrons becomes harder to remove. The twelfth electron must share the $3s$ perch. Because this perch is now lower than it was in sodium, we would expect the twelfth electron of magnesium to be somewhat harder to remove than the eleventh electron of sodium. Figure 5.2 shows that this is correct.

If the nuclear charge is increased to 13 (aluminum, Al), all the roosts again move to slightly lower levels. However, the thirteenth electron must go into a $3p$ roost, which is considerably higher than a $3s$ roost. Thus, the thirteenth electron of aluminum is somewhat easier to remove than was the twelfth electron of magnesium. Again, Figure 5.2 confirms that the ionization energy of aluminum is lower than that of magnesium. As the nuclear charge increases to 14 (silicon) and 15 (phosphorus), the additional electrons move into the vacant $3p$ roosts. Each time, because of the increasing nuclear charge, all the roosts move slightly downward and the electrons become slightly harder to remove. Figure 5.2 confirms that the ionization energies of these elements increase slightly.

If the nuclear charge is increased to 16 (sulfur), the sixteenth electron must share a $3p$ roost. Although the increased nuclear charge tends to lower the roosts, the crowding of two electrons into a single roost tends to make the last electron easier to remove. Figure 5.2 shows that the net effect in this case is a decrease in ionization energy. You should verify that the "roost" diagram accounts for the observed ionization energies of chlorine ($Z = 17$) and argon ($Z = 18$).

The scientific name for these electron sites or roosts is *orbitals*. Their labels ($1s$, $2p$, $3s$, and so on) are derived from studies of light spectra made near the end of the nineteenth century. We will use the name and the labels without worrying about their origins.

Of course, orbital diagrams such as Figure 5.6 have been created from data such as that summarized in Figure 5.2. Therefore, it is hardly surprising that we can "explain" the observed ionization energies by

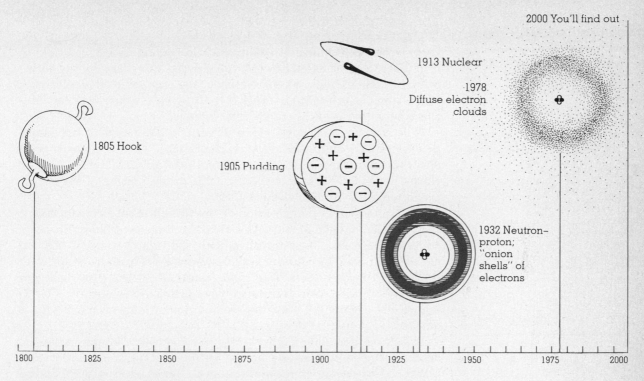

1805 Hook

1905 Pudding

1913 Nuclear

1978 Diffuse electron clouds

2000 You'll find out

1932 Neutron–proton; "onion shells" of electrons

| 1800 | 1825 | 1850 | 1875 | 1900 | 1925 | 1950 | 1975 | 2000 |

Figure 5.8
Cross sections of some atomic models.

using the orbital model. However, the orbital model does provide a useful summary of the data on ionization energy. And, as we shall see later, it proves useful in making predictions about other chemical properties. Increasing nuclear charge tends to hold electrons more strongly; increased electron crowding tends to hold electrons less strongly.

In Figure 5.6, the orbitals are spaced vertically in terms of the relative energies of their electrons with respect to the nucleus. But you, of course, would like to have an idea of where the electrons are in the space around the nucleus. Unfortunately, no one knows. Electrons cannot be accurately located in space. We can only talk about the probability of finding them in various orientations around the nucleus. I will talk about these probable distributions when I discuss compound formation in a later chapter.

For the moment you will not be too far wrong if you assume that the vertical distance in Figure 5.6 is related to the most probable distance from the nucleus. Thus $1s$ electrons are, on the average, closer to the nucleus than $2s$ electrons, $2s$ closer than $2p$, and so forth. An onion model, with shells of unequal thickness, is not too far off for our present use. Figure 5.8 shows some of the atomic models we have discussed. It is more important, both now and later, to remember that electrons are gathered outside the nucleus in groups of equal energies, than it is to worry about their spatial distribution.

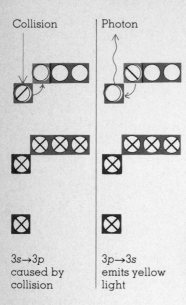

Collision   Photon

$3s\rightarrow 3p$ $3p\rightarrow 3s$
caused by   emits yellow
collision   light

Figure 5.9
Excitation and emission of
yellow light from a sodium
lamp.

This Colorful World

Our pigeon-perch model of electron structure, Figure 5.6, can help to interpret a large fraction of what you observe. For example, the occurrence of color. Almost all visible light is energy emitted as electrons move from one orbital to another. The *difference* in energy of the orbitals is equal to the energy of the light involved.

The sun, our principal source of light, is composed of hot atoms moving rapidly and colliding with great vigor. These collisions drive electrons from their lower (more stable) orbitals into higher ones. As the electrons fall back to the more stable orbitals, some of the energy is emitted in the form of visible light.

The same type of change accounts for the light from an incandescent lamp. Electrons from an outside source of voltage are driven through a tungsten filament. Their collisions with the tungsten atoms heat up the filament, and the continued collisions drive electrons from the lower orbitals of tungsten atoms into higher ones. As electrons fall back into the lower orbitals, some of the energy is emitted as visible light. "Turning off the electricity" stops the bombardment, and no more light is emitted.

Neon, sodium, and mercury lights and fluorescent lamps operate on similar principles. Flashing strobe lights contain xenon. Electrons bombard the gaseous atoms in these lamps (for example, atoms of neon, sodium, or mercury) and drive some of their electrons into higher orbitals. As electrons fall back to the lower orbitals, energy is emitted. It is often in the form of visible light (Figure 5.9).

Gaseous atoms have very well defined energy levels, and they emit only light photons corresponding to the energy differences between their orbitals. For example, the yellow light from sodium lamps is due to electrons falling from the $3p$ to the $3s$ level in gaseous sodium atoms. The lamp is operated so that almost no other light is emitted; thus sodium lamps give a very pure yellow light. Neon undergoes several transitions, but the principal one ($3s$ to $2p$) emits red light. Similarly the principal transition in mercury emits blue light.

Most fluorescent lamps are designed to emit nearly white light, but the gases in their interiors are operating like the others just described. Furthermore, much of the energy is emitted outside the visible spectrum. The inside of the fluorescent tube is coated with a pigment that absorbs these initial radiations, using them to promote its own electrons to higher orbitals. The pigment then emits white light (that is, light of a great many different visible energies) as its electrons more slowly fall back in the many ways possible in the complex orbital structures of pigment.

Some colored objects operate like the fluorescent lamp, absorbing

EXERCISE 5.6
Tungsten light bulbs go dark quickly after the electricity is turned off, whereas fluorescent bulbs glow for an appreciable period. Why the difference?

light of one energy and emitting light of another. But most color is caused by the preferential absorption and reflection of light by molecules of a pigment. The absorbed light, as in the fluorescent tube, raises some of the electrons in the pigment to higher orbitals. But these paths are not readily reversible and the electrons return via paths that convert the energy to heat rather than into visible light. What one sees (the color of the object) is the light that was *not* absorbed, technically called the *complementary color to the absorbed light*.

Perhaps the continuous display of color around you will help you to keep in mind the diagram of Figure 5.6, and to use it and the periodic table in interpreting more and more of your observations. If you are color-blind, it is probably because your eyes lack one of the three pigments most humans have to signal the brain what kind of photon is striking the retina.

Easy Come, Easy Go

We can carry our pigeon-perch analogy one step further. Another source of food may attract pigeons off the outer roosts. This will be easier the farther the roost is from the first source of food. The pigeons might even eat from both food sources. I shall discuss the simultaneous attraction of one electron by more than one nucleus a great deal when I talk about compound formation, but the same effect also occurs in the pure elements. After all, you know that many elements exist as rigid substances at room temperature. It should be clear that the atoms must be tightly bound to one another to give this rigidity. You should see now that this binding together of atoms must involve one or more electrons being attracted by more than one nucleus if the atomic model is to be useful.

The great majority of the elements boil to form gases made up of single atoms. One family (the noble gases) consists of elements all of which are gaseous atoms at room temperatures. The forces between atoms must be minimal in this family. It must be quite difficult for an electron in a noble gas atom to be attracted simultaneously by two nuclei. In fact, the general lack of reactivity in this family, and the high ionization energies, both suggest that every electron in a noble gas atom is held tightly by its own nucleus. Note the orbital patterns of these atoms (Figure 5.10). Helium has an outer set of $1s$ electrons fully exposed to its nucleus. Each of the other noble gases has an outer set of p electrons (six of them) overlying a pair of s electrons. Each electron is strongly attracted to its "own" nucleus.

Further study shows that a majority of unreactive substances are composed of atoms with the same type of orbital occupancy—a set of six p electrons overlying a pair of s electrons. This structure is commonly called a closed set of eight, or is abbreviated sp^3 to indicate that

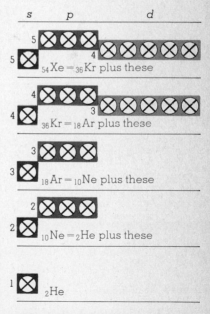

Figure 5.10
Electron structures of the noble gas atoms.

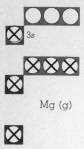

Figure 5.11
Electron structure of gaseous magnesium.

EXERCISE 5.7
How many "reactive" electrons per atom are there for each of the Group II elements? How about Group 0?

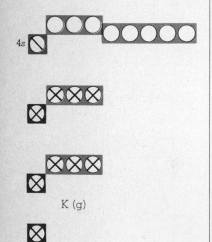

Figure 5.12
Electron structure of gaseous potassium.

one s and three p orbitals are involved, each holding two electrons (eight electrons total). Note that a set of d orbitals is also full in krypton and xenon. The total effect is that all low orbital sets are full in the noble gases, and the electrons are tightly held.

To repeat, sodium has an sp^3 set, as you can see in Figure 5.7, but it has an additional s electron in a higher energy orbital. It is this $3s$ electron that is most exposed, is least tightly bound, and is most responsible for the properties of elemental sodium. All the rest of the electrons in a sodium atom are much more tightly held, so it is the $3s$ electron of sodium that is "easy come, easy go." Consistent with this loosely held electron, sodium is soft (like firm butter), is a good conductor of electricity (electrons are free to move throughout the solid), is a good reflector of light (one characteristic of loosely held electrons is that they reflect light), and reacts vigorously with substances that are able to accept electrons. It has one "reactive" electron per atom.

Magnesium has one more electron than sodium does, so has an outer set of paired s electrons (Figure 5.11). Each is more strongly attached to its nucleus than is the outer s electron in sodium. As a result magnesium is a much harder substance (you have probably seen magnesium ladders), is a poorer conductor of electricity and a poorer reflector of light than sodium, and does not react so vigorously with substances that accept electrons. It has two "reactive" electrons per atom.

Potassium, on the other hand, has a single exposed s electron, as does sodium, and the potassium electron is even less attracted to its nucleus than in the case of sodium (Figure 5.12). Potassium is softer, shinier, a better conductor, and a better giver of electrons than is sodium. Rubidium is still better, and cesium better yet. Francium is a rare element whose properties are not well characterized, but all chemists would bet that francium would excel all its other family members in the types of properties mentioned. Each of the Group I elements has one "reactive" electron per atom. The "reactive" electrons are usually called *valence electrons*. The number of valence electrons in an atom usually equals its group number in the periodic table. Thus noble gases (Group 0) have no valence electrons. The beryllium and zinc families (Groups IIA and IIB) have two valence electrons, four for titanium or tin, seven for manganese or chlorine, and so on. The Group VIII elements, however, usually show only two to four valence electrons.

But there is no need to consider the elements one at a time. Let's return to the periodic table and discuss some of the trends in properties it can summarize so well.

Metals and Nonmetals

One of the most obvious characteristics of objects is the presence or absence of metallic properties. Are the objects shiny, bright, lustrous? Do they have a metallic ring when struck? Are they good conductors of heat and electricity? Can they be hammered into desired shapes? Figure 5.13 shows that the great majority of chemical elements are metallic. The metals cover the left side and lower part of the periodic table. (See also the periodic table inside the front cover of this book.) This positioning is consistent with our previous discussion concerning atomic structures and electron binding. Metals are elements with low ionization energies, so they are found at the left of the rows and at the bottoms of the columns in the tables. Small wonder that chemists would bet that francium would exhibit highly metallic properties!

The other elements, the nonmetals, have higher ionization energies and hold their electrons more tightly. They are listed in the upper right section of the periodic table. Helium could be classed as the "best" non-metal but it and the other noble gases are quite nonreactive. Among the reactive elements, fluorine is clearly the least metallic.

There is a heavy line weaving diagonally through the right portion of the periodic table in Figure 5.13. Elements to its left are usually called metals, those to the right nonmetals. But it should not surprise you to find that the changeover is slow rather than sudden. The borderline elements (such as germanium and antimony) are borderline in proper-ties as well as position. The periodic table inside the front cover shows the borderline in more detail. Borderline elements are sometimes called *metalloids*.

Metals

Almost anyone familiar with the existence of electrons would conclude, as we have, that metals contain many electrons that are loosely held. Some of the other structural features are not so obvious. Although it is true that all metals can be melted to give fluid liquids and boiled to give monatomic gases, they are most commonly known as solids. And if the solids are broken, the tiny glinting surfaces indicate that the solids are crystalline.

I shall not discuss in any detail here how crystal structures are deter-mined, but you should refer to Figures 1.5 and 1.7 for direct visual evi-dence of their existence. If you look very carefully at Figure 1.5, you can see that many of the bright spots (platinum atoms) have six near neigh-bors. This, and much other evidence, indicates that each atom in most of the metallic elements has six near neighbors in a plane surrounding it. Figure 5.14 may remind you of the first time you drew circles with a compass and found that six circles are required just to surround a single circle, all of the same radius. Or you may have made this discovery with coins, where exactly six circular coins will surround a seventh, if all are

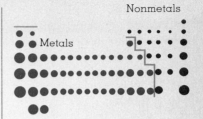

Figure 5.13
Distribution of metals and non-metals in the periodic table.

POINT TO PONDER
Classifications, such as that of the chemical elements into metals and nonmetals, are made by humans for ease in memorizing and correlating information. Classifications are useful, but they should not be assumed to be fixed or unambiguous. Each known property, given enough samples exhibiting it, tends to grade in a continuous fashion from one extreme to the other. The extreme or "typical" form is often rarer than the inter-mediate form. For example, how many typical blondes or brunettes are found in a population? How many typical sunny or stormy days in a year?

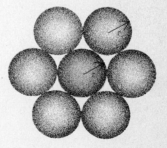

Figure 5.14
Six spheres can pack tightly around a single sphere in one plane if all are the same size (r = radius).

Figure 5.15
Twelve spheres can pack tightly around a single sphere in three dimensions if all are the same size.

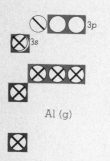

Al (g)

Figure 5.16
Electron structure of gaseous aluminum.

EXERCISE 5.8
Aluminum has three "reactive" electrons. In which group of the periodic table is it? Does this fit any generalization about the relationship of group number and number of reactive (valence) electrons?

Figure 5.17
Glide planes in a crystal allow atomic layers to slide without disruption.

the same size.

In three dimensions, such as we find in crystals, the corresponding number is 12 (Figure 5.15). A maximum of 12 spheres can just surround a thirteenth sphere all of the same size. Most elementary metals have 12 neighboring atoms packed around each atom. So we assume that these metallic atoms act as spheres. Even when the number of neighbors is not 12 (it may be 6 or 8) the atoms appear to act as spheres. We used this concept of spherical atoms in Figure 5.8. Now we see that there is experimental evidence for its validity.

Aluminum is a typical metal with which you are familiar. Each atom has 12 near neighbors to which it is equally attracted or bonded. Yet the electron structure of aluminum (Figure 5.16) has at most 3 loosely held electrons, a pair of $3s$ and a $3p$ electron. If these 3 electrons are to hold 12 atoms together the electrons must be free to move among many positions in the crystal. If they are free to move, the electrical conductivity of the crystal should be high. It is. In fact aluminum is commonly used as an electrical conductor in the high-tension wires you see strung over the countryside.

Though the electrons in crystalline aluminum are free to move, the atoms are not, as you can tell from the rigidity of the crystal. But aluminum can be extruded, even at room temperature, into the many shapes you see—such as beverage cans, rods, wires, window and door frames, and tracks. This relative ease of deformation fits with our deductions of spherical atoms packed in a regular array like that in Figure 5.17. There are almost flat planes (called *glide planes*) between the atomic layers. Motion along glide planes is easy, with little chance of the atoms "getting caught on" one another. Furthermore, the loosely held electrons provide "lubrication" for the deformations.

In fact, atoms of metals act like positively charged spherical particles surrounded by loosely held electrons (Figure 5.18). In electrical conduction, the electrons move past the stationary, positively charged spheres. In deformation, the positively charged spheres move with respect to one another. The second process requires more energy than the first, of course.

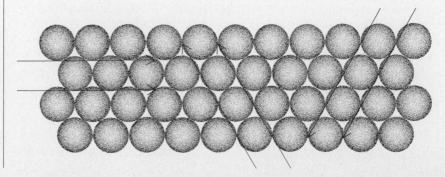

You also know that pieces of metal may be broken by continued bending or stressing at one spot. Car axles sometimes fracture. Wires can be broken by repeated flexing. Airplane wings fail because of metal fatigue. All these effects are due to the movement of the atoms in the metal. But as movement continues it becomes less and less likely that the atoms will return to the ideal packing of 12 near neighbors. Local imperfections arise. Bending tends to cause these imperfections to grow and they become small cracks. Continued bending causes the cracks to grow until finally the piece fractures or "fails"—that is, it breaks in two (Figure 5.19).

Alloys (Mixing Can Maximize)

The resistance of the elementary metals to deformation and breaking can be greatly increased by dissolving other metals in them to give a solution. (Recall Chapter 2.) Two main types of solution, also called *alloys,* can form: *substitutional* and *interstitial* (Figure 5.20). In the first type, the dissolved atoms replace (substitute for) the original atoms; in the second, the dissolved atoms enter the spaces between the original atoms. Alloys often have properties far superior to those of the separate elements.

It should be clear that interstitial alloys normally form only if the dissolving atoms are smaller than the original ones. Carbon dissolved in iron to give "low-carbon" steel is a good example. The small carbon atoms fill in the holes, destroy the smooth glide planes, and make steel much harder and less deformable than pure iron, even though less than 1 carbon atom is present for each 25 iron atoms (1% carbon by weight). Addition of other elements (such as vanadium, molybdenum, silicon, and manganese) can produce steels much harder than any pure metal or any alloy containing only iron and carbon.

One of the difficulties in recycling, especially of cars, is the large number of different alloys used. If only iron–carbon alloys were used (and this is thought possible), recycling would be much easier.

Many alloys are not solutions but are compounds. These inter-

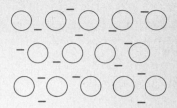

Figure 5.18
Free-electron model of a metal with positive atoms fixed at crystalline positions.

Figure 5.19
Fracture in metals, due to imperfections that became cracks between crystals. (From *Source Book in Failure Analysis*, p. 166, © 1974, American Society for Metals, Metals Park, Ohio)

EXERCISE 5.9
Mercury forms alloy solutions called *amalgams* with many metals. Are amalgams more apt to be interstitial or substitutional alloys?

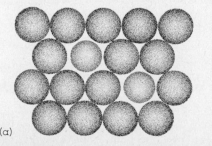

(a)

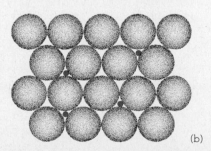

(b)

Figure 5.20
(a) Substitutional alloying;
(b) interstitial alloying.

metallic compounds form if the atoms are sufficiently different in electron structures so that the stable crystal formed is different from that of either of the constituent elements.

Nonmetals

The noble gases, at low temperatures, also form crystals with 12 near neighbors around each atom. But these crystals—unlike those of metals—are transparent, colorless and soft; they are poor conductors of heat and electricity. Consistent with the electron structures already discussed, each atom holds on to its own electrons tightly and they experience little attraction to a second nucleus. The interatomic forces are weak and the electrons are not free to move through the crystal. The "pigeons" (electrons) tend to be localized, not free to wander as in metals.

Crystals of most nonmetals are poor conductors, do not reflect light well, and are not easy to deform. Some, like sulfur and the halogens (Group VIIA in the periodic table) are colored, but the electron transitions are within individual molecules (chlorine, Cl_2, is an example) and do not indicate electrons free to move through the crystal.

Most of the nonmetals have many fewer than 12 near neighbors in their crystals. They do not act like simple spheres packed close together, as do most of the metals and the noble gases. And, in many cases, nonmetals exist as polyatomic molecules in crystal, liquid, and gaseous phases. For example, the elements fluorine, chlorine, bromine, iodine, oxygen, and nitrogen are most commonly found as diatomic (X_2) molecules in all three phases. This means that each atom of these elements can hold tightly one and only one other similar atom. The resulting diatomic molecules then pack together to form a crystal with low melting point and a liquid with only slightly higher boiling point, because the forces between the diatomic molecules are weak. (See Fig-

EXERCISE 5.10
Elementary carbon exists as diamond (a typically nonmetallic substance) and graphite, of which pencil "lead" is an example. Graphite conducts electricity and has a dull luster. What can you conclude about electron structures in graphite and in diamond? Is this consistent with the position of carbon in the periodic table?

Figure 5.21
Possible packing of diatomic molecules (such as iodine, I_2) in (a) a repeating crystalline structure and (b) an irregular liquid structure.

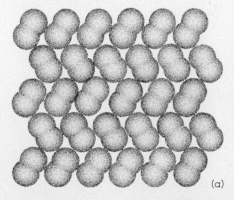
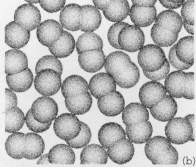

(a)   (b)

ure 5.21.)

I shall discuss the interpretation of both the strong and the weak forces in a later chapter, but it should be clear already that the electrons in atoms of these elements are most commonly attracted strongly only by two nuclei at a time. I shall describe these atoms in terms of overlapping spheres and mutual sharing of *localized electrons*.

The other nonmetals form more complicated arrangements than the diatomic molecules just discussed—for example, S_8 for sulfur, P_4 and As_4 for phosphorus and arsenic, respectively (Figure 5.22). Again the electrons are localized between small numbers of atoms and often bind them together into small molecules.

How Big Are Atoms?

You may have noted in Figures 5.21 and 5.22 that we used spherical segments of different sizes to represent atoms of different elements. You have probably noticed the same variations in molecular models and figures you have seen. How do we find out how big the atoms are?

The sizes of metallic atoms are especially easy to define. We have already seen that there is considerable evidence that the atoms act as spheres. If so, the radius of each spherical atom is just half the distance to the nearest neighbor nucleus (Figure 5.23). This distance can be measured with an accuracy of better than 1%.

Nonmetallic atoms often show at least two internuclear distances, those to the nuclei in the same molecule and those to the nuclei in neighboring molecules (Figure 5.24). I shall define half the distance to the closest nuclei in the same molecule as the *atomic radius* (also called the covalent radius). The distance to neighboring molecules is called the van der Waals radius.

Thus, in most cases, we define the *atomic radius* as half of the shortest distance between nuclei in the crystalline element. In a few cases

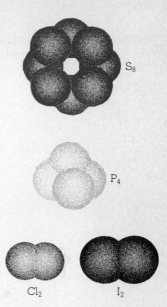

Figure 5.22
Models of molecular sulfur (S_8), phosphorus (P_4), chlorine (Cl_2), and iodine (I_2).

Figure 5.23
The radius, r, of metallic atoms in a crystal is defined as half the shortest internuclear distance.

Figure 5.24 (left)
Covalent (short) and van der Waals (longer) radii of diatomic molecules.

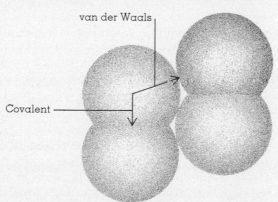

van der Waals

Covalent

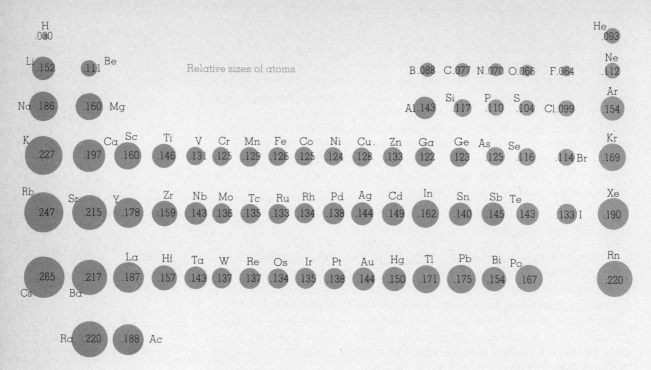

Relative sizes of atoms

Figure 5.25
Atomic size (covalent and metallic radii) as a function of position in the periodic table.

POINT TO PONDER
It is interesting to speculate on the measurement of size. Measuring size requires you to determine where the "edges" are. When atoms are close together, the electron clouds repel one another and a rigid structure forms. Internuclear distances, hence atomic radii, can then be determined. Attempts to determine sizes of individual gaseous atoms by bouncing other particles off them give radii that vary depending on how forceful the collision is. A similar problem occurs if you try to measure the size of a fuzzy tennis ball. You might like to try this, then compare results with someone else.

(hydrogen, oxygen, and nitrogen are good examples) data from compounds are more useful. Why this is so will become clear later. The results of these measurements are listed in Figure 5.25. You will find atomic radii a valuable addition to electron structures in interpreting properties. (Also see the chart inside the back cover of this book.)

Note that there are good correlations between atomic size as measured by internuclear distances and shown in Figure 5.25, and electronic structures as found from ionization energies and correlated with Figure 5.6.

In every column (family of elements) in the periodic table, any change in atomic radius is an increase as nuclear charge increases. Electronic crowding is a somewhat bigger factor than nuclear charge in determining trends in atomic radii in the columns.

As you look from Group IA to Group VIIA, left to right, in the rows of the periodic table, you will see a general decrease in size as nuclear charge increases. The effect of increase in nuclear charge is somewhat more important than increased electron crowding. The long rows show deviations from this simple pattern when the d levels are nearly full and when p electrons begin to appear. Then, for a few elements, there is an increase in atomic radius as nuclear charge increases in the row. This should seem reasonable to you in terms of crowding in the d orbitals, and then starting a new set of p orbitals "outside" the just-filled d set.

The variation in atomic radius gives further insight into the relative looseness with which the elements hold their electrons. The metals tend to have large atomic radii so that the outermost electrons are relatively far from the nucleus and can be attracted to another nucleus rather easily. Metallic atoms can be expected to lose electrons and form positive ions.

The nonmetals have relatively small atomic radii so that their electrons are closer to the nuclei, hence held more tightly. We might expect, and do find, that nonmetals often gain electrons and round out to full spheres the partial spheres shown in Figures 5.22 and 5.24. They can do this by capturing electrons, from metals, for example. In this way positive and negative ions are formed. The processes are illustrated for aluminum and for chlorine in Figure 5.26.

Relative sizes of some of the more common monatomic ions are shown (below and to the right of the elemental atoms) in Figure 5.27, and also inside the back cover of this book. Note that the negative ions are larger and the positive ions are smaller than the corresponding neutral atoms. These relative sizes should "make sense" to you.

Atomic size and ease of loss of electrons increase in moving to the left in the rows of the periodic table, and increase moving down. They remain relatively constant along the upper left to lower right diagonals. This should not surprise you. The line separating the metallic from

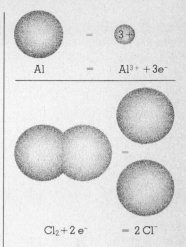

$$Al = Al^{3+} + 3e^-$$

$$Cl_2 + 2e^- = 2Cl^-$$

Figure 5.26
Loss or gain of electrons changes atomic sizes.

Figure 5.27
Atomic and ionic sizes as a function of position in the periodic table.

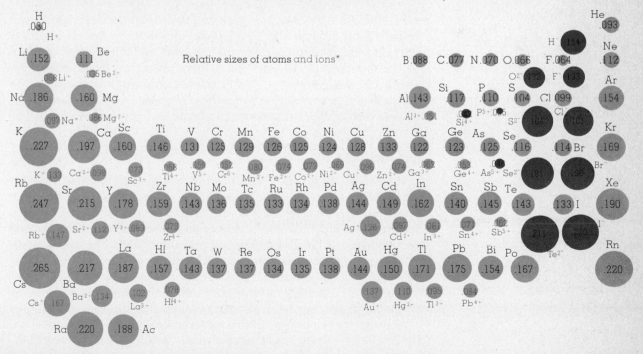

Relative sizes of atoms and ions*

*In nanometers.

Key		
ionic charge		
mp bp (K)		
ionization energy (kJ/mole)		
radius: atom ion (nm)		
rate of reaction with water		

Na (sodium)	Mg (magnesium)	Al (aluminum)
1+	2+	3+
370 1160	920 1390	930 2600
490	730	580
0.186 0.095	0.160 0.065	0.143 0.050
fast	slow	slow
K (potassium)	**Ca (calcium)**	**Sc (scandium)**
1+	2+	3+
340 1030	1120 1760	1700 2750
420	590	640
0.227 0.133	0.197 0.099	0.160 0.081
fast	moderate	slow
Rb (rubidium)	**Sr (strontium)**	**Y (yttrium)**
1+	2+	3+
310 950	1040 1660	1770 3500
410	550	640
0.247 0.148	0.215 0.113	0.178 0.093
very fast	fast	moderate

Figure 5.28
Some trends correlated with the periodic table. Note vertical, horizontal, and diagonal trends. The properties of each element are usually intermediate between those of the pairs of elements surrounding it.

the nonmetallic elements in Figure 5.13 is consistent with this. Figure 5.28 gives some data on elements so you can see the extent of variation in certain properties in the columns and rows and along the diagonals. It is the tendency of one element to have properties intermediate between all its neighboring atoms in the periodic table that makes the table so useful in correlating properties. In most cases, there are eight elements with which comparisons can be made.

The Chemist's Mole

I have talked in this chapter about the properties of elements in terms of their constituent atoms. This is consistent with the chemical approach of interpreting properties at the atomic and molecular level. But it is often impossible to make actual measurements on individual atoms and molecules—much larger samples are normally used. The internationally defined unit of amount of material is the *mole:* A mole is an amount of a substance, of specified chemical formula, containing the same number of formula units (atoms, molecules, ions, electrons, quanta, or other entities) as there are in 12 grams (exactly) of the pure isotope ^{12}C.

According to the best available measurements 1 mole of atoms contains 6.0225×10^{23} atoms. The elemental chemical symbols (H, C, N, Cl, Cu, and so on) commonly stand for 1 mole of atoms as well as for

individual atoms. The number of formula units in a mole is called *Avogadro's number* and is symbolized by N_0. We shall usually work with three significant figures; you will most commonly use 6.02×10^{23} as the value for N_0, Avogadro's number. (See Appendix A for a discussion of significant figures.)

The *atomic weights* listed in the periodic table inside the front cover and in the alphabetical list inside the back cover are not only the average relative weights of the atoms of the elements. They are also the *weight in grams* of a typical sample containing 6.02×10^{23} atoms of the element—that is, of *1 mole of the atoms* of the element.

In the same way *molecular weights,* obtained by adding the weights of the atoms represented in a chemical formula, are not only average relative weights of individual molecules. Molecular weights also give the weight in grams of 1 mole of molecules of the listed formula.

Avogadro's number is a very large number. For example, all the grains of sand on all the beaches of the world total approximately to 1 mole of sand grains—that is, 6×10^{23} grains of sand. Or, if you could measure out 1 mole of water (18 g H_2O), carefully label each molecule, pour the molecules into the ocean, carefully mix the whole ocean system, and then remove a glass of water, you would find about 10 of the marked molecules in the glass.

A mole of atoms of any of the solid or liquid elements would make a small handful of the element, as shown in Figure 5.29. A mole of a gaseous element, at usual room pressure and temperature, occupies about 25 liters, or 1 cubic foot (ft^3). An inflated automobile tire, for example, contains about 1 mole of air, 6.02×10^{23} molecules of gas.

So, although chemists interpret behavior in terms of atomic and molecular properties and most commonly measure amounts of substance in grams, they usually express their measurements finally in moles. The situation is comparable to expressing amounts of eggs and oranges in dozens rather than grams, piles of paper in reams rather

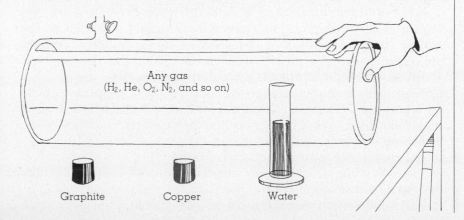

Any gas
(H_2, He, O_2, N_2, and so on)

Graphite   Copper   Water

Figure 5.29
One mole each of several materials in the form found at usual laboratory conditions. The molar volumes of crystalline and liquid elements range from 5 to 30 cm³; the molar volume of gases is 2.5×10^4 cm³.

Figure 5.30
A method of estimating atomic
sizes.

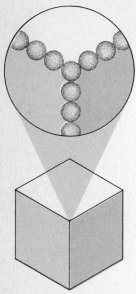

1 mole = 10^{24} atoms of copper
= cube 2 cm on each edge
= $\sqrt[3]{10^{24}}$ = 10^8 atoms along
each edge

than pounds, and money in dollars rather than ounces of gold.

Use of Avogadro's number allows a direct estimate of the size of atoms. For example, 63.54 g of copper (1 mole of Cu atoms) can be formed into a cube 2 cm on a side. The cube contains 6.02×10^{23} atoms, which is approximately 10^{24} atoms (Figure 5.30). The number of atoms along each edge of the cube is the cube root of 10^{24}, or 10^8 atoms, and the diameter of each atom is approximately equal to 2 cm divided by 10^8, or 2×10^{-8} cm. The accepted figure for copper is 2.56×10^{-8} cm, or 0.256 nanometer (nm). (Nanometers and other units of length are discussed in Appendix A; prefixes are given in Table A.2.) The radius of a copper atom, then, is 0.128 nm. Atomic radii vary between 0.030 nm (hydrogen) and 0.265 nm (cesium) as shown in Figure 5.27.

Because ^{12}C atoms are defined as weighing 12 daltons and there are 6.02×10^{23} atoms in 12 grams of ^{12}C, we see that

$$1 \text{ dalton} = \frac{12 \text{ g}}{6.02 \times 10^{23} \text{ atoms } ^{12}C \times 12 \text{ (daltons/atom } ^{12}C)}$$

$$= 1.66 \times 10^{-24} \text{ g} \qquad (5.1)$$

So, 1 g = $1/(1.66 \times 10^{-24})$ = 6.02×10^{23} daltons—that is 1 "mole" of daltons.

Summary

The periodic table correlates in a most useful form the periodic variation of the properties of the elements and their compounds as a function of nuclear charge (atomic number). These variations are a reflection of the periodic recurrence of similar electron structures as the nuclear charge increases. Each increase in nuclear charge results in all the electrons in an atom being more tightly attracted to the nucleus. This causes increased crowding of the electrons, so that finally their mutual repulsions lead to the filling of a new set of orbitals at higher energy levels (lower ionization energy) and at larger average distances from the nucleus. These electrons, in turn, are more tightly attracted as the nuclear charge continues to increase, so the atomic size decreases and the ionization energy increases, until crowding finally forces the filling of still another set of orbitals at yet higher energy levels.

The trends in the families (groups) of elements in the periodic table are consistent with the fact that the atoms get larger as the nuclear charge increases, so that they often hold their electrons less tightly and tend more readily to form positive ions.

For these reasons the most metallic elements are at the lower left of the periodic table (francium, cesium) and the most nonmetallic near the upper right (fluorine).

The noble gases have a full set of orbitals. Except for helium, which

HINTS TO EXERCISES
5.1 Always greater. 5.2 Even, differences repeat twice; 118. 5.3 Larger proportionate increase in number of shielding electrons. 5.4 Above the highest orbital energy. Larger positive nuclear charge attracts electrons more strongly. 5.5 The ionization energy decreases if nuclear charge decreases, if electrons pair in a single orbital, or if a new set of orbitals starts to fill. 5.6 The emission half-lives differ. 5.7 Two; zero. 5.8 Group III. 5.9 Either, depending on the relative sizes. 5.10 Loosely held (metallic) electrons in graphite but not in diamond.

has only two electrons—a filled s orbital—the outermost set is always a set of six p electrons, all tightly held. They are the least reactive group of elements. Many other elements attain the same noble gas electron structure, both as elements and in compounds.

The most common unit in chemistry for amount of material is the mole, 6.02×10^{23} units (usually electrons, atoms, or molecules) of a substance.

Problems

5.1 Identify or define: alkalies, alkaline earths, alloys (interstitial and substitutional), atomic radius (covalent, ionic, metallic, van der Waals), Avogadro's number ($N_0 = 6.02 \times 10^{23}$), electron energy levels, halogens, ionization energy (IE), localized electrons, metals, metalloids, mole, molecular weight, noble gases, nonmetal, orbitals (s, p, d), periodic table, valence electrons.

*5.2 How many moles are there in one atom? [$\sim 2 \times 10^{-24}$ moles/atom]

5.3 Suggest a likely electronic transition that emits light in neon signs. (Transitions never occur between levels of the same letter designation: $s \rightarrow s, p \rightarrow p$, and so on.)

5.4 If you look through gaseous sodium at a white light you see the color blue; gaseous sodium is blue. What electronic transition gives the blue color?

*5.5 Note that data for element 43 (technetium, Tc) are missing in Figures 5.2–5.5. Estimate the missing data and a likely uncertainty in each estimate. Try to find some published data to compare with your predictions.

5.6 Use the ionization energies of Kr and Ar to predict whether Br^- or Cl^- will lose an electron more easily.

5.7 How many impurity atoms are there in a silicon semiconductor chip weighing 2.8 mg (milligrams) and having only 0.001 ppm (parts per million) of impurity in terms of atomic ratios? [$\sim 6 \times 10^{10}$ impurity atoms]

*5.8 Bromine, present in the sea to about 60 ppm by weight, is extracted at a rate of about 10^8 kg/year. Assuming no replenishment and an unchanged rate of removal, how long will it take to reduce the bromine to 50 ppm by weight? The ocean contains about 4×10^{20} kg of water. [$\sim 4 \times 10^7$ years]

5.9 Both ^{90}Sr and ^{137}Cs are fission products. Each produces negative β radiation of similar energy, has a half-life of about 30 years, and is produced in roughly equal amounts. Yet ^{90}Sr is much more of a physiological threat. Why? (Hint: look at the periodic table and Table 1.1, and consider the possible physiological behavior of the elements.)

5.10 An element has two valence electrons. Is it metallic or not? How about one with seven valence electrons? Use the data on atomic size from the chart inside the back cover of this book to predict which of all crystalline elements probably has the highest density—that is, greatest weight (in grams) per unit volume, V, or g/V. Which is probably least dense?

5.11 Would you expect scandium with three valence electrons or sulfur with six valence electrons to be the better conductor of electricity in the crystalline state?

*5.12 How many rows (or layers) of atoms are added per minute to a growing fingernail or hair on average? (Hint: Estimate the rate of growth from your experience in cutting your hair.) [~ 200 layers/minute]

*5.13 How many atoms are there in the ink in the cartridge of the average ball-point pen? (Hint: Estimate the total volume—or weight—of the ink in the cartridge.) [$\sim 10^{20}$ atoms/cartridge]

5.14 You know something about chromium (chrome plate) and tungsten (light filaments). What are you willing to predict about molybdenum? Do the same for cadmium from what you know of zinc, mercury, copper, silver, and gold.

Figure 6.1
Ancient stone, gold, bronze,
and iron pieces.
(Stone implements courtesy
National Audubon Society;
King Tut © Hubertus Kanus,
from Rapho/Photo Researchers;
chariot from Rapho/Photo
Researchers; helmet from
Metropolitan Museum of Art,
Harris Brisbane Dick Fund,
1942)

6 REDUCING TO THE ELEMENTS

Only a small number of the elements are found pure in nature. One reason that gold, silver, and copper have been known and used for so long is that they occur naturally as quite visible pieces of pure metal in many parts of the world. For the same reason, the Stone Age preceded the Iron Age in human development. Yet the Bronze Age preceded the Iron Age, even though pure iron can be found in nature (though very rarely), but bronze never is. How can that be?

You already know a great deal about the chemistry of some of the elements—for example, the role of oxygen in combustion and in human breathing; copper as an electrical conductor; chromium and nickel as corrosion inhibitors; tungsten in incandescent light bulbs; zinc and lead in dry cells and car batteries; neon in signs; sodium and mercury in lamps; graphite and diamond as forms of carbon; gold, silver, and platinum as jewelry metals; hydrogen and oxygen as the elements in water. You have heard of the environmental threats from mercury and lead.

In fact, you almost certainly have knowledge and comprehension of some of the properties of about one-third of the known chemical elements. Examine the list inside the back cover of this book to check this. Then you might consider how little of this information was known by your grandparents, or even by your parents, when they were your age.

Seek and You Shall Find (Detection of the Elements)

Many of the elements can be identified visually if the amount is large enough. Examples are mercury, sulfur, chlorine, iodine, gold, and copper. Even the colorless metallic elements can often be identified with some certainty, as in the case of magnesium, aluminum, chromium, nickel, iron, zinc, lead, and tin. But complete certainty of identification requires a more complete analysis, and the determination of purity requires a still more complete one. The detection of the presence of a substance is called *qualitative analysis,* and the determination of amount or purity is called *quantitative analysis*. Because much, if not most, of our present information about possible environmental threats results from recent great advances in analytical techniques, let's explore some of these techniques.

Man's mind stretched to a new idea never goes back to its original dimensions.
—*Oliver Wendell Holmes.*

EXERCISE 6.1
Gold objects were known before bronze. Why do we not speak of a "Gold Age" between the Stone Age and the Bronze Age?

POINT TO PONDER
Scientific information is increasing, with a doubling period of less than 20 years. This is true not only of information in scientific journals, but of information for the general public as well. The doubling time for misinformation among the public may be even shorter. Each year it becomes more important that valid information on science be widely available.

The simplest conceivable analytical test would involve using a set of chemical reagents, one reagent for each element. Each reagent, when added to an unknown sample, would give a change if and only if its particular characteristic element were present. The extent of the change (such as appearance of a color) would be proportional to the amount of the element present. No such set of reagents is known, nor is such a set likely to be found. The elements occur in many combinations and relatively seldom as pure elements. For example, there are no elementary atoms of sodium in sodium chloride, and no elementary atoms of carbon in carbohydrates like table sugar, even though its formula is $C_{12}H_{22}O_{11}$. And there are several million other compounds containing "carbon" that have no elemental carbon atoms in them.

The chemical name and symbol for an element stand for a nucleus plus some arrangement of electrons around it, not necessarily for only the chemical element itself. When we say that a sample contains copper, we mean that *some* of the nuclei present have a positive charge of 29. In pure copper *all* the nuclei have a charge of 29+. The electron structure can vary and must be determined in each case.

In the same way, the formula H_2O for water says that two-thirds of the nuclei have positive charges of 1 and one-third have positive charges of 8. The whole molecule has zero charge, hence contains 10 electrons.

EXERCISE 6.2
Estimate the number of impurity atoms that must be present in a 0.10 g sample if they can just be detected at a concentration of $1/10^9$. Assume $A = 100$. How does your result compare with the number (4×10^9) of people in the world?

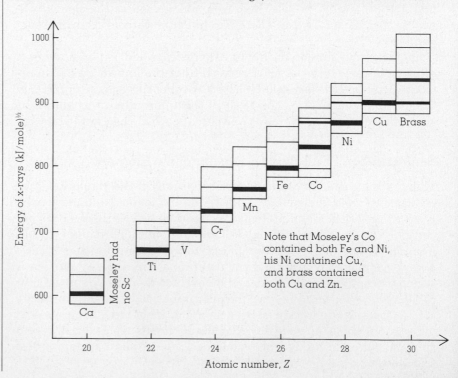

Figure 6.2
Reproduction of Moseley's photographs of x-rays of some elements. Note the regular increase in energy as atomic number increases. Each element emits, in this range, x-rays of two different energies, one intense, one less so.

The formula does not itself give any information on the arrangement of these electrons.

Because chemical analyses really amount to counting kinds of nuclei, you should not be surprised if some of the most sensitive analytical methods rely on properties closely related to nuclear charge. One method, discovered by Henry Moseley in 1911, uses x-rays.

Bombardment of a sample with electrons can drive electrons from the $1s$ level of atoms. When the $2p$ electrons fall into the vacant $1s$ shell, radiation known as K x-rays is emitted. The energy of these x-rays increases in a regular fashion as nuclear charge increases and is hardly affected by the other electrons and nuclei nearby because the $1s$ and $2p$ electrons are so close to their own nucleus. Furthermore, both the energy and amount of the radiation can be detected accurately. This method, called *x-ray analysis,* is good for both qualitative and quantitative analysis. The earliest proof that the periodic table should be based on nuclear charge rather than on atomic weight resulted from the regularities Moseley discovered in K x-rays. (See Figure 6.2.)

A second highly sensitive method of detecting elements is by *neutron bombardment.* Neutrons, being electrically neutral, can enter most nuclei readily. This addition of a neutron usually converts the initial nucleus into a radioactive one. Its characteristic radiation can then be detected and the element identified in tiny amounts regardless of what other elements may be present. The method is called *neutron activation analysis.* It can often detect an element present in less than 1 part per billion, ppb ($1/10^9$). It is neutron activation of the construction materials that makes nuclear fission plants radioactive and difficult to service.

Many methods are known of exciting the outer electrons in order to force the element to emit characteristic radiation—for example, *electron bombardment* of sodium as in a sodium street lamp. The emitted or absorbed radiation is then measured to give both a qualitative and quantitative measure of the element, often to a sensitivity of $1/10^9$. These methods work best if the elements are present as gaseous atoms or ions. The atoms then behave independently of other atoms that may be present and give well-characterized radiation. Colored *flame tests* obtained by injecting the sample into a gas flame are a simple example of this method. (See Figures 6.3 and 6.4.) The test is so sensitive that if you clap your hands near a gas flame it flashes yellow because of the sodium ions from your skin.

There is a story that Robert Bunsen, who invented flame tests, once proved his landlady was using yesterday's meat scraps in today's stew. He secretly sprinkled lithium chloride on the meat scraps and showed that the next day's stew gave a brilliant red flame test characteristic of lithium and not usually found in stews.

Mass spectrometry is another highly sensitive method of analysis

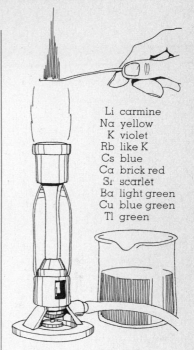

Li carmine
Na yellow
K violet
Rb like K
Cs blue
Ca brick red
Sr scarlet
Ba light green
Cu blue green
Tl green

Figure 6.3
Flame tests for chlorides of some elements.

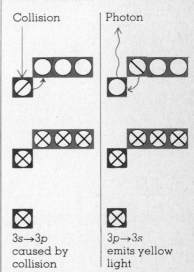

Collision | Photon

$3s \rightarrow 3p$ caused by collision | $3p \rightarrow 3s$ emits yellow light

Figure 6.4
The origin of the flame color of sodium is excitation of the $3s$ electron to $3p$. As the electron falls back to $3s$, yellow light is emitted.

Table 6.1   Composition of the "Clean," Dry Atmosphere (number of molecules per million molecules of air)[a]

Substance	Symbol	Number of molecules
Nitrogen	N_2	780,000
Oxygen	O_2	208,000
Argon	Ar	9,000
Carbon dioxide	CO_2	3,000
Neon	Ne	200
Helium	He	50
Krypton	Kr	10
Hydrogen	H_2	5
Xenon	Xe	0.8

[a]Note that the figures are parts per million, ppm.

(better than $1/10^{10}$ in many cases). It can be used for compounds as well as for elements. (Review pages 41–44.)

There are also *chemical reagents* that can accurately detect quantities of elements as small as $1/10^9$ or smaller. Their use usually requires extensive and careful treatment of the sample prior to analysis, with the final test often being the formation of a colored compound. The depth of the color formed indicates the amount of the element originally present.

And, at the opposite extreme of sensitivity, the miner's pan and nitric acid (gold does not dissolve; fool's gold does) are crude analytical tools. Many techniques are known for finding elements in nature. But let's see what sources one finds, and how the element is then produced from the source.

Occurrence of the Elements

Figure 6.5 shows some natural sources of many of the elements. The problem now becomes how to obtain large quantities of the pure element from each naturally occurring source.

The elements most readily available throughout the world are those occurring in the atmosphere. In spite of local air pollution, the composition of the dry atmosphere is remarkably constant. Table 6.1 gives typical values.

Air is a gaseous solution of electrically neutral molecules. When cooled sufficiently, it forms a liquid solution. The elements are extracted from the atmosphere by liquefying purified air (through compressing and chilling it), and then distilling the liquefied air. (The different ele-

Figure 6.5
Natural sources of the chemical elements.

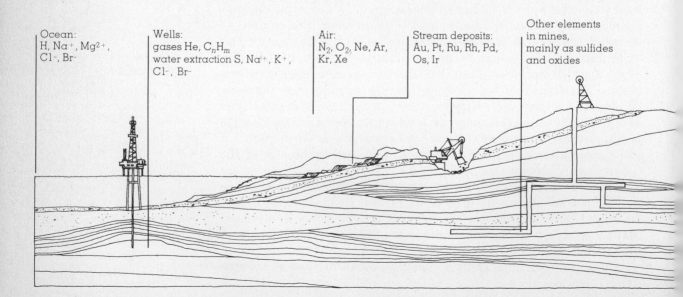

Ocean:
H, Na^+, Mg^{2+}, Cl^-, Br^-

Wells:
gases He, C_nH_m
water extraction S, Na^+, K^+, Cl^-, Br^-

Air:
N_2, O_2, Ne, Ar, Kr, Xe

Stream deposits:
Au, Pt, Ru, Rh, Pd, Os, Ir

Other elements in mines, mainly as sulfides and oxides

Table 6.2   Typical Composition of Seawater (in atoms per billion molecules of H_2O)

Element	Number of atoms	Element	Number of atoms
Chlorine (Cl)[a]	10,000,000	Arsenic (As)	10
Sodium (Na)[a]	8,000,000	Copper (Cu)[b]	10
Magnesium (Mg)[a]	1,000,000	Zinc (Zn)[b]	10
Sulfur (S)	700,000	Manganese (Mn)[b]	5
Calcium (Ca)	300,000	Selenium (Se)	2
Potassium (K)	200,000	Titanium (Ti)[b]	1
Carbon (C)	50,000	Vanadium (V)[b]	1
Bromine (Br)[a]	10,000	Molybdenum (Mo)[b]	1
Boron (B)	7,000	Tin (Sn)	0.4
Strontium (Sr)	5,000	Cesium (Cs)	0.4
Silicon (Si)	5,000	Lead (Pb)[b]	0.4
Fluorine (F)	2,000	Nickel (Ni)[b]	0.4
Nitrogen (N)	1,000	Gallium (Ga)	0.2
Lithium (Li)	1,000	Chromium (Cr)	0.2
Aluminum (Al)	800	Silver (Ag)	0.2
Rubidium (Rb)	100	Cobalt (Co)	0.2
Phosphorus (P)	80	Uranium (U)[b]	0.1
Iodine (I)	10	Cerium (Ce)	0.1
Barium (Ba)	10	Yttrium (Y)	0.1
Iron (Fe)[b]	10	Lanthanum (La)	0.1

[a]Element now mined from the ocean.
[b]Element found concentrated in nodules on the seafloor; may be a mineable resource.

ments boil off at different temperatures.) All the neon for signs, all the argon for filling incandescent light bulbs, and all the krypton for strobe flash lamps are obtained in this way. Increasingly large amounts of pure oxygen are also obtained by distilling liquefied air (over 10 billion kg of oxygen in 1976). About half as much pure nitrogen is obtained by this process. Distillation of liquid air (a solution) to produce pure substances should remind you of the discussion of distillation in Chapter 2.

The next most widespread source of elements is the ocean. Table 6.2 gives a partial analysis of seawater. Humans have produced sea salt, NaCl, for hundreds of years. In fact, even prehistoric man may have learned to block off pools of seawater, let most of the water evaporate, and allow the small residual amount to run back into the sea. He could then scrape up the remaining white crystalline sea salt (which we call sodium chloride) for flavoring, and even for use as money.

You can see there is no visible evidence either in seawater or in table salt for the presence of metallic sodium or elemental chlorine (which has a light greenish-yellow color). Most of the elements in the ocean are not present in their elemental state but as electrically charged atoms (for example, sodium ions and chloride ions). The presence of ions in NaCl is shown by the fact that both aqueous sodium chloride and the liquefied pure salt conduct an electric current. (Ions are charged atoms. When they are free to move, they can carry an electric current much as loosely held electrons do in a metal.) The same ions make up the table

EXERCISE 6.3
Why, when sodium chloride is obtained from seawater, is some residual liquid often allowed to run back into the sea rather than evaporate totally?

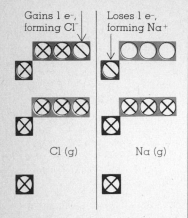

Figure 6.6
Electron structures of gaseous
Na and Cl. Recall that each
diagonal stroke represents one
electron; crossed strokes represent two electrons.

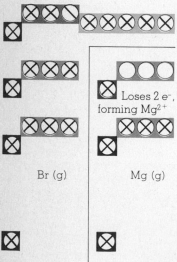

Figure 6.7
Electron structures of gaseous
Mg and Br.

salt crystals, but the crystalline state is too rigid to permit the ions to move, so solid salt is not a conductor of electricity. Examination of Figure 6.6 should suggest to you that the sodium ion is probably Na^+, the chloride is Cl^-.

Most of the other elements, listed in Table 6.2 as present in seawater, are also present as ions. Only magnesium, Mg^{2+}, and bromide, Br^-, are in large enough concentration and valuable enough to make their systematic extraction feasible at the present time. Again the problem arises of converting ions into neutral elements. Note that Figure 6.7 is consistent with the formulas Mg^{2+} and Br^- for these two elements in ionic form. With Mg^{2+} and Br^- there is the additional problem that their concentrations are not large enough to make it feasible to reclaim them by evaporation of seawater as can be done with sodium chloride.

The magnesium is precipitated and filtered out as insoluble magnesium hydroxide, $Mg(OH)_2$. The bromide is converted to gaseous bromine (chlorine is used) and obtained as the liquid element. Magnesium and bromine plants are usually sited on a peninsula swept by strong ocean currents. The water that has been depleted in magnesium and/or bromine can then be dumped into a part of the ocean where it will not immediately reenter the plant.

As Figure 6.5 shows, some elements are found by drilling wells; some are found in alluvial stream deposits (as at mouths of rivers). Ores at the surface can be scooped up by power shovels, as in strip-mining. Other ores are underground and are reached by shafts and tunnels. Or it may be easier to drill wells into which water is pumped, to return to the surface bearing the mineral. Some wells, as with natural gas, may be pressurized enough to force their contents to the surface.

Under the best of circumstances mines tend to be messy, with wells less so. But mining leaves either (1) vast voids underground that may allow the surface to sink, or (2) an altered surface that may remain barren for a long time. Normally there are also large quantities of debris that accumulate and litter the site. All these drawbacks can be overcome. The fundamental question is whether society is willing either to pay the extra price for control and restoration of sites or to do without the materials that are found only in mines. Or mines near population centers can be closed down and those in remote areas (often in developing countries) maintained. Of course, this solution merely transfers the problems to another society and makes the protected society dependent on the unprotected one.

In many cases over the last 20 years, *strip-mining* of coal in the United States has proceeded with minimal long-term alteration of the surface, in contrast to former practice. The additional cost has usually been less than 10 cents per ton of coal and has had little effect on the competitive situation. Better and better techniques are developing. A major future problem in much of the western United States, where very large areas have coal suitable for strip-mining, is the need for water. Mining will require the diversion of water from present agricultural and recreational uses. Political decisions, as much as scientific decisions, will be required in settling on the optimum allocations.

Finding and mining an element does not usually produce a directly useful form of it. Let's consider some of the methods used to convert ores to consumer products. Such processes, of course, require energy and almost always produce wastes in addition to the desired materials. You may be surprised at the amount of material we process that does not end up in a useful form.

From Ore to Element

Most ores contain the elements in ionic form. For example, sulfide and oxide ores contain positively charged atoms of the metals combined with negatively charged sulfide (S^{2-}) and oxide (O^{2-}), respectively. Note in Figure 6.8 that the formulas of these ions are consistent with their position in the periodic table and their expected electron structures. Why not take a few minutes to check that all the ions listed so far (Na^+, Mg^{2+}, Cl^-, Br^-, S^{2-}, and O^{2-}) have the same number of electrons as one or another of the noble gases, whose electron structures are notably stable?

Because the metals occur in nature primarily as their positive ions, we are faced with the general problem of converting positive ions to neutral atoms in order to get the metallic element from its ore. For example, $Na^+ + e^- = Na$; $Zn^{2+} + 2e^- = Zn$. This process of converting an ore into a metal has been known for centuries as a *reduction* to the metal. Today we have generalized the term, and reduction indicates a process of gaining electrons.

Conversely, consistent with the fact that many ores are oxides, the conversion of a metal to its oxide has long been called *oxidation*. Again, we have now generalized the word, and oxidation indicates a process of loss of electrons.

REDUCTION: Gain of electrons by an atom or molecule
OXIDATION: Loss of electrons by an atom or molecule

You already know that the elements on the left of the periodic table are the most metallic (lowest ionization energies, least tightly held electrons). It should not surprise you to find that these highly metallic elements, such as sodium and magnesium, are the most difficult to reduce from their ores. Those elements closer to the right—for example, chromium, iron, and copper (in that order)—become increasingly easier to reduce. Let's look at this sequence in reverse order.

POINT TO PONDER
It is commonly suggested that primitive man was more respectful and careful of his environment than modern man. But the evidence is mixed to say the least. It is likely that the extinction of many large mammals was caused by prehistoric man and that the fall of many early societies was caused by overintensive agriculture. Two points of greater importance may be (1) that our present access to large amounts of energy allows (even encourages) us to do what no early group could, and (2) that modern man has a better idea of the probable results of his actions.

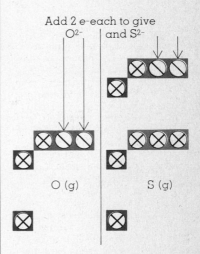

Figure 6.8
Electron structures of gaseous O and S, and of O^{2-} and S^{2-} ions.

EXERCISE 6.4
Use Figures 5.6 (or the periodic table) and 6.8 to predict the formula of calcium ion and the formula of calcium oxide. Repeat for scandium and its sulfide.

POINT TO PONDER
Copper and iron, but not tin, are found as metals in nature; iron is rare. Yet bronze, an alloy of copper and tin, was discovered and used by man almost 1000 years before he used iron. Tin was very likely discovered when primitive humans built their fireplaces with rocks containing the ore cassiterite (SnO_2). Hot carbon reduces SnO_2 to metallic tin, a soft, rather useless metal. Whoever took the step of melting tin and copper together (or reducing a single ore of the two) introduced the Bronze Age and a tremendous change in human life. Bronze is hard and lasting. It makes good tools, adornments—and weapons! Rapid social change began.

EXERCISE 6.5
Write an equation similar to equation 6.1 for the reaction that occurs when HgS is heated in air. Any health problem here?

Copper is commonly found as its sulfide. You know that copper is a rather unreactive element and must therefore hold on to electrons relatively tightly. It does. In fact, it can even get the electrons away from the sulfide ions with which it is combined if the sulfur can get some electrons elsewhere. Heating copper sulfide ore in air produces the desired result, the products being metallic copper and sulfur dioxide. We may write

$$CuS\ (c) + O_2\ (g) = Cu\ (c) + SO_2\ (g) \tag{6.1}$$

The symbols (c) and (g) indicate that the substances are in the *crystalline* and *gaseous* states, respectively. Similarly, (l) indicates *liquid*. Sometimes (s) is used to indicate a *solid,* which may not be crystalline. A substance dissolved in water is indicated by (aq), from *aqueous*. Indicating these states, or *phases,* makes it clear how the separation of products occurs; the sulfur dioxide gas escapes, leaving solid copper.

Now remember that chemical symbols represent not only individual atoms but also moles of substance. Thus, we may read equation 6.1 as "1 mole of copper sulfide reacts with 1 mole of oxygen to give 1 mole of copper and 1 mole of sulfur dioxide." Note that the atoms are conserved; all atoms in the reactants are also present in the products.

Several sulfide ores (for example, HgS and Ag_2S) may be converted to metals by heating in air. But most sulfide ores heated in air are converted to oxides. Sulfur dioxide is also produced and can cause serious environmental changes downwind if it escapes into the atmosphere. I shall later discuss these effects and how to minimize them. Recently, microbes have been used to extract copper from low-grade ores. Some 10% of U.S. copper output is extracted in this way.

One of the most common metallurgical operations is the conversion of oxides to metals. In one method of reducing oxides to metals carbon is both a source of the necessary electrons and a means of effectively removing the oxygen.

Most of the elements, except those in Groups IA, IIA, IIIB, and IVB, can be reduced from their oxides by carbon. Coal is normally the cheapest naturally occurring form of carbon, and the cheapest reducing agent, but it is often converted to coke first. Heating coal in the absence of air converts it to coke and releases a large variety of compounds that can be condensed, purified, and used as raw materials in many industrial processes. I shall mention some of these coke-oven by-products later. These materials almost pay for the cost of the coking process, and most of the coke is then used to reduce iron ore to iron.

There are no naturally occurring reducing agents (sources of electrons) able to reduce the metals of Groups IA, IIA, IIIB, and IVB to their elemental states from their ores. Many of these elements are produced by passing an electric current through a liquefied bath containing ions of the metal. The positive metallic ions gain electrons and are converted to the metal:

$$Na^+ + \quad e^- = Na \qquad (6.2)$$
$$Al^{3+} + 3\,e^- = Al \qquad (6.3)$$

Aluminum is produced in very large amounts in this way, about 10^{10} g per year. It, in turn, may be used as the reducing agent to convert the ores of many other metallic elements into the metals. A typical reaction is

$$Cr_2O_3\,(c) + 2\,Al\,(c) = Al_2O_3\,(c) \mid 2\,Cr\,(c) \qquad (6.4)$$

Although all the ingredients here are indicated as crystalline solids, the actual reaction, once started, generates so much energy (as heat) that both the reacting aluminum and the product metal are liquids. This allows good contact between the reactants and easy separation of the products because they are not soluble in one another.

In actual practice, the reduction of many metals uses special steps suited to the properties of that particular element. The general problem is to provide a source of electrons that can react with the ore to produce the electrically neutral metal, M, in a form easily separated from the other products of reaction: $M^{n+} + n\,e^- = M^0 = M$. I will use the reduction of iron and aluminum to illustrate these problems in more detail. Each of these metals is produced in very large quantities, as a brief examination of your surroundings is bound to indicate. Think how many uses there are for the two metals aluminum and iron.

A Note on Nomenclature

Now that we are getting into discussions of the chemistry of individual elements, an introduction to some chemical names will be useful. You have already seen that, with the exception of helium, the suffixes *-ium* (as in sodium) and *-um* (as in aluminum) indicate metallic elements. The positive ions are given the same elemental names, as in *sodium ion* or *aluminum ion*. However, there may be more than one possible positive ion of a metal—for example, titanium has several, including Ti^{2+} and Ti^{3+}. These can also be written as *titanium (II)* and *titanium (III)*, where the roman numerals indicate the number of electrons lost in forming the ion from the element.

Naming of monatomic negative ions also starts with the name of the element but uses the suffix *-ide* to indicate the monatomic negative ion. I have already mentioned chloride (Cl^-) and bromide (Br^-) ions, as well as oxide and sulfide ions (O^{2-} and S^{2-}, respectively).

The compounds between the two titanium ions and chlorine could be called *titanium(II) chloride* (formula $TiCl_2$) and *titanium(III) chlo-*

EXERCISE 6.6
Rewrite equation 6.4 so it describes the phases during the actual reaction. What would you see if you performed the reaction?

EXERCISE 6.7
Name the compounds CrCl$_2$ and
CrCl$_3$. Also As$_2$S$_3$ and PbO$_2$.

ride (formula TiCl$_3$). These are read as "titanium two chloride" and "titanium three chloride." Similarly the CuS in equation 6.1 is known as *copper(II) sulfide* to distinguish it from copper(I) sulfide, Cu$_2$S.

Red iron ore can be written Fe$_2$O$_3$ and called iron(III) oxide. We deduce the name because three oxide ions, each with two negative charges, give a total negative charge of $6-$. The total charge on the iron ions must be $6+$. Because there are two irons for every three oxides, the charge on each iron must be $3+$ —hence iron(III) oxide. This name serves to distinguish this iron oxide from other iron oxides, such as FeO or Fe$_3$O$_4$. There is only one stable oxide of aluminum, Al$_2$O$_3$, usually called *aluminum oxide.* But you are not expected to know there is only one, so you would probably write *aluminum(III) oxide,* and you would be quite correct.

Aluminum, the "Indestructible" Metal

The only commonly used ore of aluminum is aluminum(III) oxide, Al$_2$O$_3$. The first, and still used, commercial process for reducing aluminum oxide to the metal was discovered by Charles Martin Hall in 1886, 8 months after he graduated from Oberlin College.

Hall, working in the woodshed of the family house, used a frying pan to dissolve the oxide in cryolite (Na$_3$AlF$_6$), a recently discovered aluminum ore. He melted the cryolite and oxide over a blacksmith's forge, and electrolyzed the oxide to aluminum and oxygen using current from electric cells that he made from the jars his mother normally used to can fruit. A very distinguished French chemist, Paul Héroult, made the same discovery 1 week later.

In the subsequent patent suit involving hundreds of millions of dollars, Hall was able to establish his priority of discovery. His college chemistry professor testified that Hall had shown him the first pellets of aluminum 1 week before the old chemistry building was torn down, a date that was easy to verify. Modern dating of priority rights is established by careful maintenance of a research notebook with each page dated and the whole book carefully certified by legal witnesses. (The Hall method of establishing priority is too chancy and too expensive!) Figure 6.9 shows a modern Hall cell.

And what led Hall to look to *electrolysis* of the liquefied ore for the production of aluminum? He did his work in 1886, so he had never heard of metallic ions (proved to exist about 10 years later) or electrons (proved almost 20 years later). But he did know about the work of Michael Faraday. Fifty years earlier, Faraday had prepared sodium and some other hard-to-reduce metals by passing an electric current through their melted salts. (Water must be excluded because all these metals react vigorously with water.) Aluminum could not be prepared in this

POINT TO PONDER
Some people claim that the day of the backyard (or simple) path to discovery is over. Yet in the 1950s Donald Glaser invented the hydrogen bubble chamber (for which he received a Nobel Prize) by watching bubbles form in a mug of beer. In the early 1960s, Frances Crick and James Watson based some of their Nobel Prize winning ideas on molecular models cut out of paper. And in the 1970s the Chinese developed a powerful method of predicting earthquakes, based on using peasants as observers.

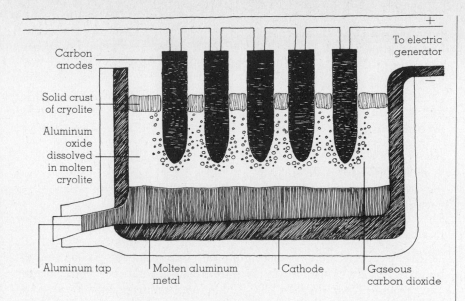

Figure 6.9
A modern Hall cell for producing aluminum metal.

Carbon anodes

Solid crust of cryolite

Aluminum oxide dissolved in molten cryolite

To electric generator

+

−

Aluminum tap

Molten aluminum metal

Cathode

Gaseous carbon dioxide

simple fashion because the common ore, Al_2O_3, will melt only at a temperature of 2050°C—much too high to be practical. However, cryolite
melts at about 950°C, and the Al_2O_3 dissolves in the molten cryolite.

Hall had also learned about the construction of simple electric cells
from zinc metal and a solution of copper sulfate, and he got a sample of
cryolite (among other things he tried) from the college geologist. For
the rest of his equipment, as you have seen, he used what was around
the house. (He also had an education.)

Today we interpret his results by saying that aluminum exists in the
ore as Al^{3+} and so requires three electrons per ion to convert it to the
metal. No chemical occurring in nature will give electrons to aluminum
ions, so we must use an electric generator to supply the electrons. The
process must be carried out in the liquid state, so that the ions are free
to move to the electrode to which the generator supplies electrons. Aluminum oxide itself has too high a melting point for easy melting, so it
must be dissolved in a more easily melted material that is not itself
appreciably decomposed by the electric current. Cryolite meets these
last two criteria. The principal reaction is the decomposition of aluminum oxide into the metal and oxygen:

$$Al_2O_3 \text{ (in liquid } Na_3AlF_6) = 2\,Al\,(\ell) + 1.5\,O_2\,(g) \qquad (6.5)$$

A main cost in reducing aluminum is the electricity. Three moles of
electrons are required per mole of aluminum:

$$Al^{3+} + 3\,e^- = Al\,(\ell) \qquad (6.6)$$

EXERCISE 6.8
Aluminum sinks in molten
$NaAlF_6$, but sodium floats on
molten NaCl. Suggest a reason,
using the chart inside the back
cover of this book.

The cells require a great deal of electric power, with some of it being used to keep the cell contents liquid. A principal reason for conserving and recycling aluminum is to cut down the amount of energy required to process it. The ore itself is not particularly scarce or expensive, nor are the other needed materials.

As with all processes, so with the reduction of aluminum—the simplified chemistry does not tell the whole story. There are three main problems in addition to high energy costs in the aluminum industry:

1. The sludge produced in refining the ore has no known use and so must be disposed of.
2. The electrolysis releases fluoride and carbon monoxide into the atmosphere.
3. Aluminum bars, containers, and foil react very slowly in nature, creating a serious litter problem.

Every process creates waste material, aluminum less so than most because the ore is usually fairly pure and the main by-products are oxygen and carbon dioxide. The sludge (containing mainly Fe_2O_3 and silicates) is usually spread out to dry in the sun in ever-growing piles.

Some carbon monoxide forms when oxygen reacts with the hot carbon electrodes, but there is not much of it. Adequate ventilation in the plant and good venting into the atmosphere keep the concentrations far below toxic levels. The fluoride problem has been more severe, because the bursting bubbles of oxygen create a toxic airborne fluoride dust. Careful control of the fluid temperature and composition and, if necessary, filtering dust from the evolved gases have now minimized this problem.

The problem of the waste metal is more severe. The very property (low corrosion rate) that makes aluminum so "indestructible" creates a litter problem. Waste iron rusts away, wood and paper rot—but not aluminum. This is all the more remarkable when you know that aluminum reacts vigorously with water to produce hydrogen and aluminum oxide. In fact, it is this reaction that stops the corrosion, because aluminum oxide adheres very tightly to the metal surface where it is formed. This thin film of oxide prevents further attack on the metal, because the oxide is one of the most unreactive substances known.

Aluminum oxide is soluble in concentrated sodium hydroxide (as in processing the ore) and in acids. (You may have seen how aluminum pans become shiny if rhubarb, tomatoes, applesauce, or other acid foods are cooked in them.) But the oxide coat quickly forms again, and corrosion stops. It is this oxide coating, incidentally, that can be dyed to give the highly colored aluminum articles and foils with which you are familiar.

The only real solution to disposing of waste aluminum seems to be

EXERCISE 6.9
The oxide coat on aluminum may be dyed in an electrolytic cell containing ions of the dye. These dye ions are discharged at the aluminum electrode and become neutral molecules adhering tightly to the oxide. Should the dye ions be positive or negative? (Hint: Consider the effects of the electrolysis on the surface of the aluminum.)

recycling the waste that can be collected. It may be fortunate that power costs for producing aluminum are so high. Remelting aluminum requires little power—about 5% of that required for reducing the ore. So recycling becomes more and more attractive as power costs become higher and higher. Efficient collection is the main problem.

Living in the Iron Age

Some would argue that we live in a nuclear age, or a computer age. But for contributions to society, both in weight of product and variety of use, the Iron Age it still may be. We make and use about 10^{12} kg of iron every year, which is more than any other processed chemical. Enough iron is produced annually to give every human being his or her weight in iron.

Even if unlimited quantities of all the elements were available, iron would still be the choice for most of the functions it now fills. It is easy to form, it makes alloys of a great range of properties, it has a balance of properties (such as hardness, ductility, malleability, density, and resilience) found in no other substance. It does corrode, but its ores are widespread, their reduction is not difficult, and, as a result, it is the least expensive metal as well as the most widely used.

Iron is the second most abundant element on earth (the earth's core seems to be an iron alloy), consistent with its high nuclear binding energy (Figure 3.11). It is also a fairly common element in the earth's crust, being the fifth most common (after oxygen, silicon, aluminum, and sodium). And there are large, rather pure ore deposits spread widely.

Readily available scrap iron has been collected and recycled for a long time, but only recently has the price of iron risen high enough to make systematic recycling seem feasible (Table 6.3). A principal complication, already mentioned, is the wide range of alloys and metallic mixes (like tin cans) in which scrap iron is found. Simple recycling will require either a simplification in the number of alloys and combinations in which iron is used commercially, or else some new separation techniques.

Should We Blast the Blast Furnace?

It is unquestionable that the manufacture of iron has led to a combination of dangerous and filth-producing industries. No mines are as dirty and potentially dangerous as coal mines, from which comes the material needed to reduce iron ore. No production process puts forth greater clouds of noxious fumes and smoke than that for iron. And no industry has had more safety hazards for its workers than the steel industry. Fortunately, safety standards in much of the world have

WORLD ANNUAL PER CAPITA IRON PRODUCTION.

Table 6.3 Relative Prices of Cold-rolled Steel

Year	Price ($)
1950	100
1960	130
1970	180
1973	200
1975	300

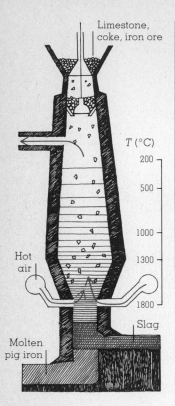

Limestone,
coke, iron ore

T (°C)

200

500

1000

1300

1800

Slag

Hot
air

Molten
pig iron

Figure 6.10
A diagram of a blast furnace
producing metallic iron.

improved so much that the coal–iron–steel combination can achieve a quite respectable record. This is still not true everywhere, as the periodic newspaper stories of mine disasters remind us. But let's look at the fundamental chemistry before discussing the problems and possible solutions further.

Much iron ore is pure enough to be introduced directly into a reduction process. A major problem is that the cheapest reducing agent, coke, is a solid just as iron ore is. Two solids cannot react readily because they cannot come into intimate contact. What is needed is a gaseous reducing agent that can freely approach the solid ore and there form a gaseous product that can leave just as freely.

This problem is solved in the iron blast furnace by the introduction of hot air at the bottom of the tall blast furnace, while a mixture of iron ore, coke, and limestone is introduced at the top (Figure 6.10). In modern blast furnaces, pure oxygen (from liquid air) is sometimes added to raise the oxygen content of the gas and minimize the amount of coke required.

The rising air meets the falling coke, and some of the coke burns. This produces a mixture of carbon dioxide and carbon monoxide gases, plus the heat required to maintain the necessary high temperature. The carbon monoxide gas is a good reducing agent for iron ore, forming liquid iron and carbon dioxide. The carbon dioxide formed in these two steps can react with hot coke to produce more carbon monoxide so the reduction cycle can repeat.

The main overall or net reaction may be represented as

$$3 \, C \, (c) + 2 \, Fe_2O_3 \, (c) = 4 \, Fe \, (\ell) + 3 \, CO_2 \, (g) \tag{6.7}$$

but the mechanism (or individual steps) is better summarized as

$$CO_2 \, (g) + C \, (c) = 2 \, CO \, (g) \tag{6.8}$$
$$3 \, CO \, (g) + Fe_2O_3 \, (c) = 2 \, Fe \, (\ell) + 3 \, CO_2 \, (g) \tag{6.9}$$

Each of the individual "mechanistic" steps involves the reaction of a gas with a solid to produce products, at least one of which is a gas. Thus each step can proceed at a high rate as is required in any successful industrial process.

The limestone (formula $CaCO_3$) is added to form a fluid slag with the impurities in the ore, mainly silicates. The silicates themselves form thick, viscous liquids that would gum up the inside of the furnace. But the heat of the furnace decomposes the $CaCO_3$ into CaO and CO_2, with the CaO forming a fluid slag with the silicates. This slag floats on top of the molten iron in the bottom of the furnace. Disposal of the slag can be a major problem. You may well have seen the enormous piles of slag that accumulate around most steel mills. Some slag can be converted to cement, and some can be made into gravel for concrete or

EXERCISE 6.10
The sum of the equations for the mechanistic, individual steps in any reaction must equal that for the net reaction. Show that this is true for equations 6.8 and 6.9, which should add to give equation 6.7.

roadbeds, but some just accumulates in the slag piles.

A serious pollution problem can be caused by the gases that come out of the top of the furnaces and out of the ovens in which coal is heated to produce coke. Coke-oven gases can contain large quantities of sulfur dioxide and other noxious volatile materials from the coal. Modern coke ovens have tightly fitting doors that minimize the escape of these products; in some cases the additional captured products are worth enough to pay for the better oven design.

A bigger source of air pollution and water pollution in many steel mills is the open hearth furnace where iron is converted to steel, and the mill itself where the steel is formed into shapes such as pipes, railroad rails, bars, and rods.

The *open hearth furnace* consists of a very large shallow tray into which the iron from the bottom of the blast furnace is run to have its carbon content adjusted and to have other alloying materials added. The carbon is burned out by bubbling air or oxygen through the molten pool. This lowers the carbon content from the solution produced in the blast furnace (remember, coke was used there). The open hearth furnace also produces vast quantities of fumes and smoke, including oxides of sulfur and nitrogen as well as carbon. Some calcium fluoride (CaF_2) is often added to help in slag formation, so that fluoride fumes are sometimes an additional problem.

Much of the air pollution from a steel mill comes from the open hearth, and it is difficult (that means expensive) to control. The managers point out that doing so increases the price of steel appreciably. They also point out that if only their mill does this, it will be driven out of business. Other mills (say, foreign ones) will be able to undersell them by paying no attention to the pollution problem. Should we support air pollution in a foreign country any more than at home? The air pollution can be serious, but it can be controlled, as Figure 6.11 shows.

Water pollution is also a serious problem for a steel mill, because it is necessary to remove oxide coats from the steel at various stages in the mill. This is normally done by "pickling" the steel in acid. The oxide dissolves, producing a solution of Fe(II) in water. The resulting solution is still acid and a serious contaminant if poured into the local waterway.

The acid solution can be neutralized by pouring it over old tin cans. They dissolve to produce more Fe(II) from the iron cans. Blowing air through the resulting solution converts the Fe(II) to insoluble Fe_2O_3, which can be settled out, dried, and run into the blast furnace for conversion back to iron. But the process does not pay for itself, and it further increases the cost of steel.

The potential magnitude of the pollution problem is shown in Table 6.4, which lists the materials required to produce 1000 kg of steel. Remember, most of the materials become wastes of one kind or another.

EXERCISE 6.11
Write a chemical equation (including phase designations) for the net reaction in which limestone decomposes in the blast furnace.

Figure 6.11
A steel mill before and after emissions were controlled.

Table 6.4 Materials and Electricity Needed to Produce 1000 Kilograms of Steel[a]

Material	Amount (kg)
Iron ore	2260
Coke	730
Limestone	270
Air	2500
Water	8000
Calcium fluoride	10
Total	13,770
Electricity	487 kwh

[a]The ratio of wastes to product is about 14 to 1. (Kaiser Steel Fontana Works)

Processes are known that can reduce the pollutant output from a steel-making complex to a low level. The final decision is based on how much *you* are willing to pay for steel and on how *you* view the problem of competition from abroad. It is also essential to remember that no process can reduce the pollution to zero and that the cost goes up very rapidly as the pollution level is lowered to very small values. Humans have always produced and tolerated wastes that are more polluted than the original air, water, and other raw materials. They will always have to do so. The hard decisions involve the question, what are tolerable levels?

Summary

Very sensitive qualitative and quantitative tests for the elements are now available—often more sensitive than 1 part in a billion. Some of the most sensitive tests depend on the absorption or emission of radiation by the element, because the energies of these radiations are often dependent only on the nuclear charge of the element, not on the other atoms in the vicinity. Mass spectra and chemical tests are also available in almost every case.

Some of the elements occur free in nature, but most are present as positive ions in sulfides or oxides. The metallic elements can be formed by reducing these positive ions to neutral atoms with a source of electrons. Coal, or coke, is the cheapest natural reducing agent, but the elements with the least tendency to hold electrons must be reduced electrolytically. For example, iron oxide ore is reduced with coke, whereas aluminum oxide is reduced by electrolysis of a molten solution.

All human activities produce pollutants, and large-scale industrial processes can produce pollution on a large scale. Methods are known for minimizing the pollution, but they are bound to cost money—and the cost rises rapidly as the level of pollution is lowered more and more. Many of the questions of siting industries and handling their waste products have as many political implications as economic or scientific ones, so all need careful consideration.

HINTS TO EXERCISES
6.1 Too soft and rare.
6.2 6.0×10^{11} impurity atoms.
6.3 Other compounds will precipitate. 6.4 Ca^{2+}, CaO, Sc^{3+}, Sc_2O_3. 6.5 Hg and SO_2 form. 6.6 Al_2O_3 and Cr form. 6.7 Chromium(II) chloride, chromium(III) chloride, arsenic(III) sulfide, lead(IV) oxide. 6.8 The relative change in radius on ionization differs. 6.9 Negatively charged dye. 6.10 Multiply the first by three, the second by two, and they will add. 6.11 CaO, CO_2 produced.

Problems

6.1 Identify or define: blast furnace, chemical symbols, electrolysis, flame tests, neutron activation analysis, open hearth furnace, ores, oxidation, oxidizing agent, qualitative and quantitative analysis, reducing agent, reduction, strip-mining.

6.2 Suppose the doubling rate of scientific information stays constant throughout your 40-year professional career. How much more will be known when you retire than is known now? Will all that is "known" now still be known then?

6.3 The chemical formula of sulfuric acid is H_2SO_4. Deduce as much information as you can from this formula about particles present, weight relations, and relative numbers of moles.

6.4 The compound $HgCl_2$ is a poison called *corrosive sublimate,* but Hg_2Cl_2 is *calomel,* often taken internally for liver ailments. If you were a nurse or pharmacist, what systematic names would you expect to find on these two bottles?

6.5 Write the chemical formula of manganese (III) oxide.

*6.6 Suggest possible industrial methods for the following conversions:

a. $KCl \rightarrow K$ d. $PbS \rightarrow Pb$
b. $ZnO \rightarrow Zn$ e. $NaBr \rightarrow Br_2$
c. $MnO \rightarrow Mn$ f. $Ca_3(PO_4)_2 \rightarrow P_4$

Do not look up possibilities.

6.7 Classify the following as reductions, oxidations, or neither:

a. $Na^+ \rightarrow Na$ d. $MnO \rightarrow Mn_2O_3$
b. $Cl_2 \rightarrow 2\,Cl^-$ e. $H_2O \rightarrow OH^- + H^+$
c. $O^{2-} \rightarrow \frac{1}{2}O_2$ f. $NO_2^- \rightarrow NO_2$

6.8 Comment on possible advantages and disadvantages of using only carbon steels in building automobiles.

6.9 Careful heating of lead sulfide in air produces metallic lead. Write a chemical equation for the probable change. What is apt to happen if the heating is not done carefully?

6.10 Which, in each of the following pairs, would emit the more energetic K x-rays: K–Ca, Ca–Sr, Ti–Al, Ti–Sc?

*6.11 Suggest why, in terms of equation 6.7 and the temperatures shown in Figure 6.10, the blast furnace has the double taper shown in the figure.

6.12 Would you lead a more satisfactory life today if humans had not learned to modify natural raw materials? Be specific. Would you probably be alive?

*6.13 Present practice is to add steam to the air entering a blast furnace. It is said to produce more reducing gas per unit of carbon, yielding 1 unit of carbon monoxide and 1 unit of hydrogen instead of just 1 unit of carbon monoxide in the carbon–oxygen reaction. Is it also true that more iron is reduced per unit of carbon, or that reduction is more rapid?

6.14 The rapid expansion of photography has played an appreciable part in the disappearance of silver as a coinage metal. What is the relationship?

*6.15 In 1850 aluminum cost more than gold. (Napoleon III of France ate with aluminum utensils while his guests had to make do with gold ones.) Now the cost of aluminum is less than 1% that of gold for the same weight. Discuss some factors in this shift.

*6.16 Mars seems to have large amounts of iron oxide on its surface. Will this ensure plenty of steel for a human colony on Mars?

ELECTRON STRUCTURE

Atoms can be ionized.

Ionization energies vary as the population of the orbital electrons changes.

Each atom has a pattern of electron energy levels.

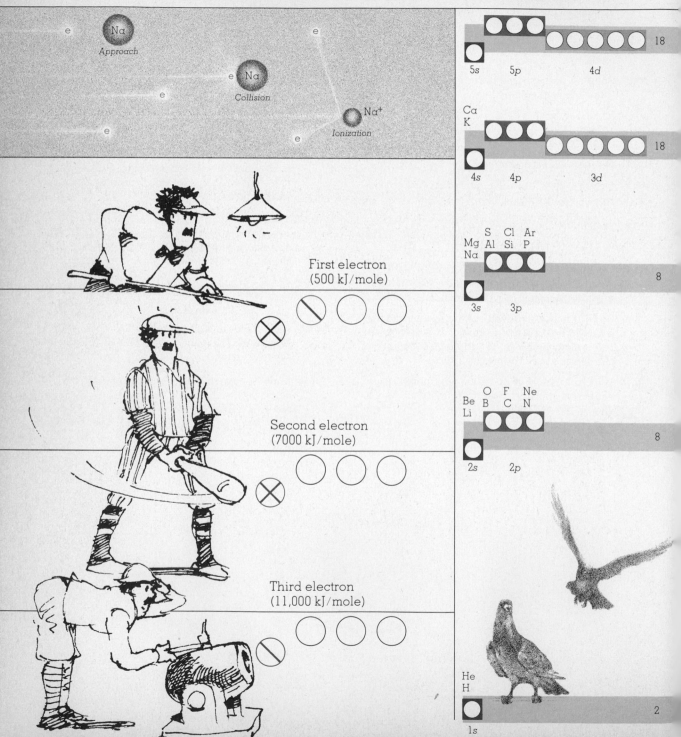

Na Approach

e Na Collision

Na⁺ Ionization

First electron (500 kJ/mole)

Second electron (7000 kJ/mole)

Third electron (11,000 kJ/mole)

18

5s   5p   4d

Ca
K

18

4s   4p   3d

S   Cl   Ar
Mg Al Si P
Na

8

3s   3p

O   F   Ne
Be B   C   N
Li

8

2s   2p

He
H

2

1s

The periodic table reflects the elements' electron structures.

Atoms of similar electron structure have similar properties…

and tend to be found together in nature.

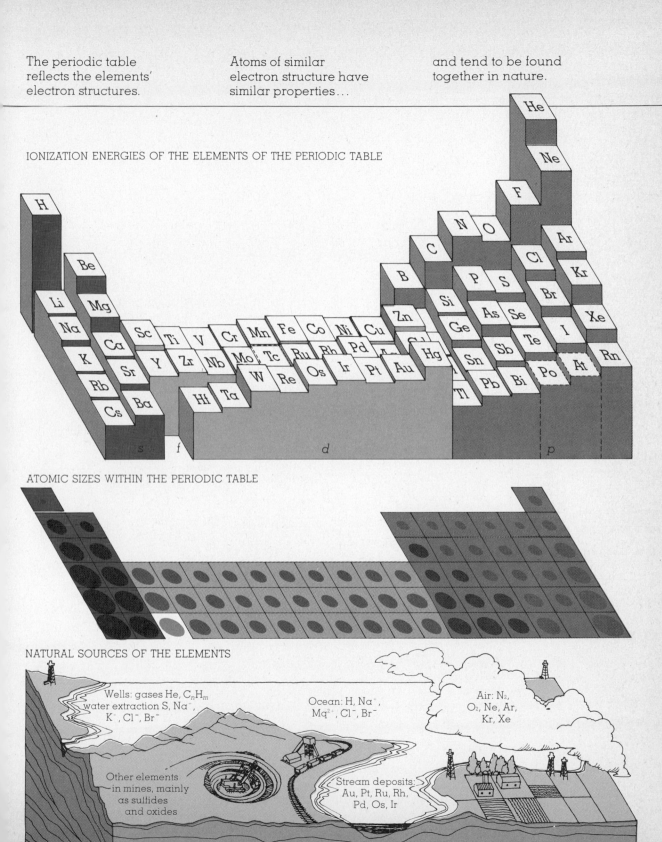

IONIZATION ENERGIES OF THE ELEMENTS OF THE PERIODIC TABLE

ATOMIC SIZES WITHIN THE PERIODIC TABLE

NATURAL SOURCES OF THE ELEMENTS

Wells: gases He, C_nH_m
water extraction S, Na^+,
K^+, Cl^-, Br^-

Ocean: H, Na^+,
Mg^{2+}, Cl^-, Br^-

Air: N_2,
O_2, Ne, Ar,
Kr, Xe

Other elements
in mines, mainly
as sulfides
and oxides

Stream deposits:
Au, Pt, Ru, Rh,
Pd, Os, Ir

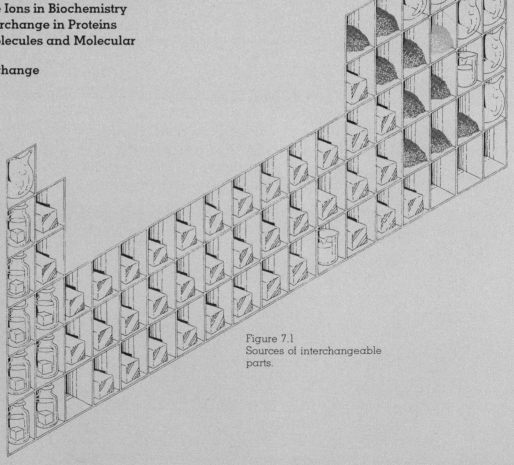

Figure 7.1
Sources of interchangeable
parts.

7 INTERCHANGEABLE ATOMS: FROM JEWELRY TO ENZYMES

The invention of mass production is usually credited to Henry Ford, with one of his great contributions the idea of interchangeable parts. But, as with many other human "inventions," the idea had been in practice for several billion years in nature. Humans just hadn't recognized it.

You have already seen that atoms of different elements can have similar electron structures, similar sizes, and identical ionic charges. It should not surprise you to find that such similar atoms can be substituted for one another with little change in the properties of the parent substance. We shall explore some of the large range of known substitutions, beginning with atoms of the elements, then moving to monatomic ions, polyatomic ions, and finally groups of atoms incorporated in electrically neutral molecules.

Some of the most interesting examples come from the beautiful materials we call *jewelry,* both metals and gems. Others involve some of the molecules, called *enzymes,* that make possible the rapid chemical changes required in living systems. Most of the examples I shall discuss have only been clarified in the last 30 years. One result has been a great increase in the ability of chemists to design and carry out syntheses of substances to meet special needs, even substances that can cure individual diseases.

Jewelry Metals

Gold jewelry is known from most of the earliest civilizations. Perhaps the oldest existing piece dates from about 2500 BC in Minoa. Gold shares with other items of jewelry three valued traits. It is beautiful, it is rare, and it is lasting. But it is weakest in the last trait, for pure gold is soft and wears away readily.

When gold coins were much used, it was not uncommon to put them into a leather bag and shake it vigorously. When the coins were removed they were still clearly coins, but further emptying of the bag might give up to 5% additional gold—a quick return on investment. It became common to weigh gold rather than to accept it at the face value of the coin.

Then it was discovered that impure gold is harder than pure gold

The most incomprehensible thing about the world is that it is comprehensible.
—*Albert Einstein.*

Table 7.1 Carats and Gold Content

| Carat | Content (in percentages) | |
	Gold	Other elements
22	92	4.16 Ag 4.16 Cu
18	75	10–20 Ag 5–15 Cu
15	62	15 Ag 23 Cu
10	42	12–20 Ag 38–46 Cu

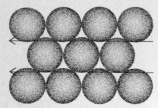

Smooth glide planes

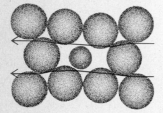

"Bumpy" glide planes

Figure 7.2
Glide planes are deformed and
metals hardened by substitutional alloying.

Table 7.2 Some Common Alloys

Alloy	Composition (percentage by weight)
Gold coin	90 Au 10 Cu
Silver coin	90 Ag 10 Cu
Hardware bronze	89 Cu 9 Zn 2 Pb
German silver	60 Cu 25 Zn 15 Ni
Nickel coin	75 Cu 25 Ni
Solder	67 Sn 33 Pb
Stainless steel	74 Fe 18 Cr 8 Ni
Bell metal	78 Cu 22 Sn

and that the hardness can be controlled by deliberately adding other metals (especially copper and/or silver) to melted gold before casting the coins. Pure gold came to be identified as "24 carat," with its alloys identified by smaller carat numbers (Table 7.1). Gold jewelry usually ranges from 10 to 22 carats, with 18 carats a common alloy. Can you see from the table approximately what *carat* means?

I discussed in Chapter 5 the mechanism by which hardness results in alloys. The glide planes are deformed when atoms of different size are introduced either *interstitially* or by *substitution* into the original crystal structure. Hardness in alloys depends on the extent of deformation of the glide planes—that is, the degree of lumpiness introduced in them (Figure 7.2).

The alloys may be just as beautiful as the original gold, but they are seldom as rare (silver is more common than gold), and they are often more reactive. This increase in reactivity can result from the presence of more reactive elements (silver is more reactive than gold, copper still more so). Or it can result from the strains caused in the crystal by the presence of atoms of different size. These strains can serve as points of chemical attack. Thus, gold alloys with a good deal of copper may tarnish and turn green because of the formation of copper compounds.

Silver, platinum, and copper are also used in jewelry, ordinarily as alloys. The advantages and problems are similar to those mentioned for gold. Some common alloys are listed in Table 7.2. Coins made from precious metals are not used much anymore, because the value of the metal in a coin of reasonable size can exceed the nominal (face) value of the coin. So coins now are more commonly made of aluminum or nickel alloys. There is no attempt to equate the nominal value of the coin with the value of the metal in it.

This same technique was used by many of the great civilizations of the past when inflation became a problem. Lead became a common diluent for gold because its presence did not noticeably change the weight of the coin. The difference between then and now is that governments then didn't admit they were demeaning the coinage. Now it is a published fact.

Not all elements dissolve in one another, especially if they differ greatly in atomic size and/or in electron structure. Elements that differ greatly often form compounds, but the properties of the compounds usually differ considerably from those of any of the constituent elements. Similar elements, however, may be interchangeable, even in compounds, as you will soon see.

Diamonds (A Girl's Best Friend?)

Diamonds (known as early as 400 BC) are, to most people, the outstanding jewel. They are the hardest naturally occurring substance, they have great flash and fire, and there are some magnificent great diamonds in royal jewelry. The Cullinan diamond from South Africa, the largest ever found, weighed 3106 carats, or 621.2 g (a carat of jewels is 0.2 g exactly). Prior to the eighteenth century, most diamonds were discovered in India, but 90% of current diamond production is in southern Africa.

Jewelry advertising sometimes states that "diamonds are forever." Yet diamond is the only gemstone that burns. Diamonds are pure carbon. One way to test an unknown stone is to put it in a very hot flame. If it burns, it *was* a diamond. A much preferred way is the scratch test, usually on a sapphire plate. Diamond is the only naturally occurring stone that will scratch sapphire (chemical name, aluminum oxide). A common scale of hardness, called the *Mohs scale,* is shown in Table 7.3. A gemstone must have a Mohs hardness of 7 or more if it is to withstand normal wear unchanged.

The flash and fire of diamonds are due to the fact that light entering a diamond is greatly bent from a straight-line path, and the degree of bending varies with the color of the light. Thus, cut diamonds serve as very effective prisms, separating white light into its colors and reflecting the colors. Scientifically, the diamond is described as having a high index of refraction that changes rapidly with variation in light color. Figure 7.3 indicates the flash and fire effects.

We saw in Chapter 5 that each atom of most metallic elements has 12 near neighbors surrounding it. In contrast, the number of near atomic neighbors of any one atom in diamond is low—only 4. Furthermore, diamond is a very poor conductor of electricity—in fact, it is an excellent insulator. All the electrons are tightly held and localized. The great hardness of diamond is consistent with this picture of tightly held electrons. The atoms are not free to move because the nuclei are strongly restrained by the rigid electron structure around them.

Figure 7.4 shows how the six electrons in a monatomic gaseous atom of carbon (nuclear charge of 6) would be arranged. There are two electrons in the $2p$ orbital, which can hold a maximum of six electrons. There are also two electrons in the $2s$ orbital and two in the $1s$. The $1s$ electrons will be very strongly attracted to their own nucleus and thus little influenced by neighboring nuclei. The $2s$ and $2p$ electrons are much further from their own nucleus and more readily attracted to neighboring nuclei. Thus there are *four* electrons, and vacancies for *four* more, in the $2s,2p$ orbital set of carbon. It is not merely a coincidence that *four* is the number of near atomic neighbors in diamond.

Notice that, if each atom shares a pair of electrons with each of its four neighbors, then each atom ends up with a share in its own original four plus four more (one from each of four neighbors), or a total of eight. This gives a full set of $2s$ and $2p$ electrons, which we have already seen is the structure of the noble gas neon (Figure 5.10). The same type of

EXERCISE 7.1
Why is it that lead dissolved in a gold coin does not make a noticeable difference in the weight of the coin (if the size remains the same), but silver does? (Hint: See the chart inside the back cover of this book.)

Table 7.3 Mohs Hardness Scale[a]

Hardness number	Substance
10	Diamond (C)
9	Sapphire (Al_2O_3)
8	Topaz[b]
7	Quartz (SiO_2)
	(6.5 steel file)
6	Feldspar
	($K_2O \cdot Al_2O_3 \cdot 6\,SiO_2$)
	(6 knife blade)
	(5.5 window glass)
5	Apatite
	($CaF_2 \cdot 3\,Ca_3P_2O_8$)
4	Fluorspar (CaF_2)
3	Calcite ($CaCO_3$)
	(2.5 fingernail)
2	Gypsum ($CaSO_4 \cdot 2\,H_2O$)
1	Talc
	($3\,MgO \cdot 4\,SiO_2 \cdot H_2O$)

[a] Each substance scratches all those below it in the scale.
[b] See p. 118.

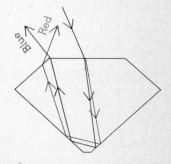

Figure 7.3
The internal reflection and prism effects in a diamond that lead to the diamond's flash and fire.

C (g)

Figure 7.4
Electron structure of gaseous C.

EXERCISE 7.2
Silicon, Si, and carbon, C, form
a compound, SiC, called *silicon
carbide*. It is very hard, but not
quite as hard as diamond.
Suggest a possible arrangement
of the atoms in SiC. (Hint: How
are the electron structures of
carbon and silicon related?)

Figure 7.5
Structure of diamond, a form
of pure carbon. All bonds are
identical. See Figure 8.3 also.
One unit cell is outlined here.
It is cubic in shape.

full set of eight *s* and *p* electrons is found in all the noble gases—and
also in many ions and other atoms, some of which I have already discussed. It is to the full set of sp^3 electrons that we attribute both the
hardness and the low reactivity of diamond.

Figure 7.5 shows a representation of the crystal structure of diamond, each atom having four near neighbors arranged symmetrically
about it. This arrangement is called *tetrahedral*. We shall see it again
and again in other structures. Thus, the diamond structure is rather
open compared to the tightly packed metals with 12 near neighbors.
Yet the interstitial spaces are small, because the carbon atoms themselves are small. In fact, the spaces are too small to accept interstitial
atoms. Furthermore, the carbon atoms themselves are much smaller
than other atoms with a similar electron structure (say, silicon), so
diamond cannot accept substitutional atoms either. Diamonds tend to
be quite pure. There are no interchangeable atoms for diamond except
carbon atoms.

The fact that the structure of diamond can be represented by a
figure showing only a few atoms is important. It can be done because
crystals in general consist of a regularly repeating pattern. Because
the pattern does repeat, it can be represented by a small section known
as the *unit cell*. The total structure consists of many unit cells side by
side.

You probably know that diamonds vary in color. There are two
main causes: displaced electrons and inclusions of foreign matter.

If colorless diamonds are irradiated with ultraviolet light or x-rays,
they may become green. When these diamonds are heated, they lose the
color. The radiation caused local ionization, and the free electrons moved

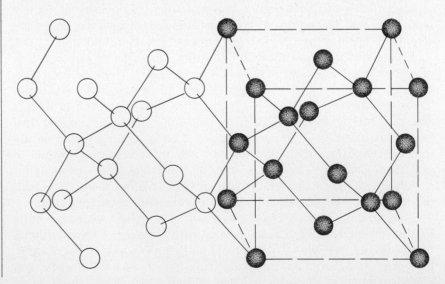

through the crystal to the interstitial spaces. Only at high temperatures can they move through the crystal to their original site. Electrons in interstitial spaces can absorb visible light and so give color. You have probably seen old glass bottles that have become colored from long exposure to sunlight. These bottles lose their color on heating just as diamond does, and for the same reasons.

An *inclusion* in a crystal occupies a hole in the crystal; it is not dissolved in it. The color of the crystal may change, but usually does not disappear, when the substance is heated. The actual color depends on the contaminant. I shall not discuss here the variation in diamond color in terms of inclusions.

One common inclusion in diamond is *graphite,* another form of pure carbon. In fact, if diamonds are heated long enough and hot enough in the absence of air, they will turn to graphite, the stable form of carbon at low pressures. Figure 7.6 shows the crystal structure of graphite. The layer structure of graphite is quite different from that of diamond, and the number of near neighbors has dropped to three. Graphite, like diamond, does not accept interstitial or substitutional atoms in the layers, and for the same reasons—the unique combination of atomic size and electron structure. But graphite will accept other substances *between* the layers. The layers are rather weakly held together, as shown by the large interlayer distance compared to the short internuclear distance in each layer. You probably know that graphite layers separate easily, hence its use in pencil lead and as a lubricant.

Why, if graphite is the stable form of carbon at usual conditions, does diamond exist at all? Actually there is no totally accepted theory

And they are both pure Carbon!

Figure 7.6
Structure of graphite, a form of pure carbon. There are some short (strong) bonds, and some long (weak) bonds.

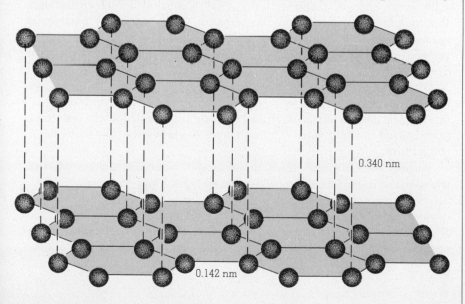

0.340 nm

0.142 nm

of the formation of natural diamonds. But it has been possible to apply basic chemical principles and to synthesize diamonds. Synthesis of this gem was one of the earliest syntheses to be attempted and one of the last to be accomplished. Let's see what was done.

Synthetic Diamonds

Two of the earliest attempts (around 1890) to synthesize diamonds were made by a famous French chemist, Henri Moissan, and by an otherwise unknown Scot, J. B. Hannay. Moissan apparently heard of the discovery in Arizona of iron meteorites that contained diamonds. He therefore dissolved graphite in molten iron, rapidly chilled the liquid, dissolved the iron away, and claimed to find diamonds. These were not seen by others, and they have been lost. Hannay said he heated fish oil, a complicated set of carbon compounds, with metallic lithium in a steel bomb. He deposited with the British Museum the stones he claimed to have made. They are indeed diamonds. No one has been able to produce diamonds by repeating either of these, or many other, early experiments, and it is too late to decide whether chicanery, self-delusion, or most unusual conditions accounted for the outcome of these early experiments.

The first successful synthesis was by Tracy Hall and his colleagues at the General Electric Company in 1955. It was based on the knowledge that diamond has a density of 3.51 g/cm^3 and graphite one of 2.25 g/cm^3. Clearly, the atoms in diamond must be much more tightly packed than those in graphite. High pressure squeezes atoms together and would be expected to convert graphite to diamond. The rate of such conversion should increase with rise in temperature. Decision: Heat carbon in various forms to a high temperature (up to 3000°C) in a high-pressure vessel (up to 10^5 atm, atmospheres). Result (eventually): synthetic diamonds.

Most theories of the formation of natural diamonds assume they were formed when carbon was subjected to high pressures and temperatures in the depths of the earth and then rapidly chilled as the resulting mass moved to the surface. The rapid chilling prevented the conversion to graphite.

Almost 100% of synthetic diamonds and 80% of natural diamonds are too small or imperfect to be used as gemstones. They are used industrially in cutting wheels, drills, dies for drawing small wires, and similar devices where high stresses would rapidly wear away most materials. Almost half the industrial diamonds now used are synthetics. But once again—the unending frontier. If we can make synthetic small diamonds, why not synthetic large ones?

When Is a Rock a Gem?

We have already seen that jewels must have the qualities of beauty, rarity, and permanence. All but pearl, opal, and turquoise of the birthstones listed in Table 7.4 have an additional characteristic. They are large crystals. "Rocks," on the other hand, are composed of small crystals. We have already looked at the crystal structure of diamond and the reasons diamonds tend to be so pure and reproducible. We shall see that the other gemstones can exist in an almost infinite variety of compositions and colors because of the ready interchangeability of atoms.

An outstanding example of variation in appearance with a small change in chemical composition is the ruby–sapphire pair. Rubies are red, sapphires are blue—yet both are almost pure aluminum oxide, Al_2O_3. *Ruby* has 1–2% of chromium—$Cr(III)$—substituted for some of the $Al(III)$. *Sapphire* has a small amount of iron—$Fe(III)$—and titanium—$Ti(III)$; a third gem, alexandrite, forms if vanadium—$V(III)$—is the substituting ion. Part of the color of some sapphires seems to be due to free electrons and is lost if the gem is heated.

It is simplest to think of Al_2O_3 as closely packed oxide ions, with $Al(III)$ in the interstitial spaces between the oxides, balancing the negative charges of the oxide ions (Figure 7.7). The situation is comparable to a set of closely packed negatively charged marbles with smaller, positively charged marbles in the interstitial holes. When chromium substitutes for some of the $Al(III)$, which it matches in size and charge, the characteristic red of ruby results as the chromium ions absorb part of the visible light. Substitution by iron and titanium gives

Table 7.4 Birthstones

January	Garnet
February	Amethyst
March	Aquamarine
April	Diamond
May	Emerald
June	Pearl
July	Ruby
August	Peridot
September	Sapphire
October	Opal
November	Topaz
December	Turquoise

EXERCISE 7.3
Rubies and sapphires are seldom found in the same underground mine, though they may occur in the same streambed. Why?

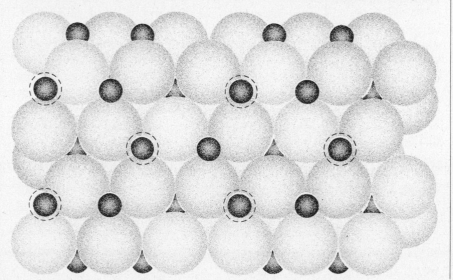

Figure 7.7
Structure of Al_2O_3 as in ruby and sapphire. The smaller spheres represent Al^{3+}; the larger spheres, O^{2-}. Each O^{2-} is surrounded by four Al^{3+}, each Al^{3+} by six O^{2-}. "Hidden" Al^{3+} are shown by dashed circles.

POINT TO PONDER
There is an ancient Burmese legend, recorded by Marco Polo around 1300, that rubies were mined by throwing sticky pieces of meat into an uninhabitable fever-ridden valley, and letting trained birds retrieve the meat—and the jewels that stuck to it. (This is the origin of the "roc" story told by Sinbad the Sailor in the *Arabian Nights' Entertainment*.) It is true that more than one mining expedition in Burma has foundered in fevers.

blue. The depth of the color is related to the concentration of the substituting ions, the hue to the kind of ion. Hence the great variety of hue and depth of color.

Silicates of Infinite Variety

Zircon and peridot are a second pair of closely related gemstones. They are silicates. The most common atom at the earth's surface is oxygen, and the next most common is silicon. The range of silicon–oxygen compounds is very wide, and you should not be surprised that the majority of gems are silicates—that is, compounds of silicon, oxygen, and other metals and nonmetals.

If only silicon and oxygen are present, the formula of the compound is SiO_2, of which quartz is one of the most common forms (Figure 7.8). *Quartz* consists of a three-dimensional structure in which each silicon atom is surrounded by four oxygen atoms, with each oxygen atom between two silicon atoms. In another form of SiO_2 (cristobalite), the silicon atoms are in positions similar to those of the carbon atoms in diamond, with an oxygen atom in between every pair of silicons. As with diamond, the silicon dioxide structure here is quite open—but the holes are big enough for other atoms to enter the interstices. A third form of SiO_2, tridymite, also has an open structure. Thus, interstitial forms of SiO_2 are common and give the varieties you can find on most beaches—white, rosy, black, brown, and other colors, depending on the interstitial elements present. *Amethyst* is SiO_2 that contains some interstitial iron, titanium, and oxide ions.

Window glass is made by melting beach sand (a crude SiO_2) with calcium carbonate, $CaCO_3$, and sodium carbonate, Na_2CO_3. Carbon dioxide, CO_2, escapes. The result is an interstitial solution of Na^+, Ca^{2+}, and O^{2-} in a highly disarranged quartzlike network. Pyrex glass contains boron oxide, B_2O_3, which gives a low thermal expansion to the glass. Glasses are not crystalline. They do not contain repeating unit cells but are instead random liquids resistant to flow.

Most silicates, as with window glass, can be represented as mixtures of various oxides with SiO_2. In fact, silicate analyses formerly were expressed in terms of possible oxide mixes. One such is ZrO_2SiO_2, written as an equimolar (50/50) mixture of zirconium oxide and silicon oxide—or $ZrSiO_4$, written in condensed form. Here, as with many silicates, the condensed form is closer to the truth, for the fundamental units in the crystal are SiO_4^{4-} and Zr^{4+} ions (Figure 7.9). This crystal is called *zircon*.

Peridot has the same SiO_4^{4-} ions, but contains Mg^{2+} and Fe^{2+} ions rather than Zr^{4+} to balance the electrical charge. The ratio of Mg(II) to Fe(II) varies from sample to sample, so that the formula is commonly written $(Mg,Fe)_2SiO_4$. The best peridot has an iron to magnesium ratio

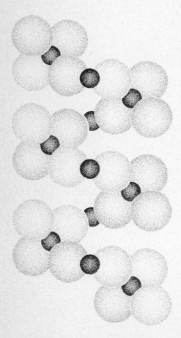

Figure 7.8
Structure of quartz (SiO_2): continuous, linked spiral chains of alternating Si—O—Si—O—Si—O. Each silicon (smaller spheres) is surrounded by four oxygens (larger spheres); each oxygen by two silicons.

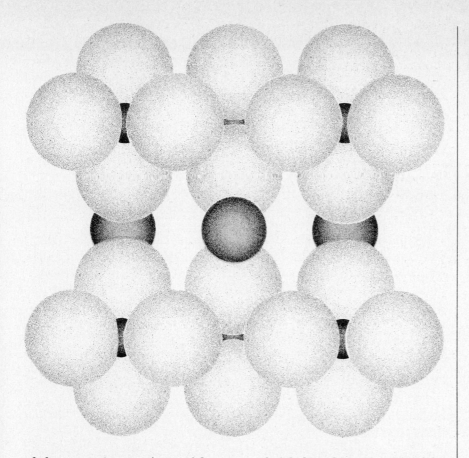

Figure 7.9
The zircon structure, alternating Zr^{4+} and SiO_4^{4-} ions. Smallest atoms are Si, largest O, and intermediate Zr.

of about 1 to 8, sometimes with a trace of nickel, and is green. Note the similar ionic radii (0.066, 0.074, and 0.069 nm) and identical charges of Mg(II), Fe(II), and Ni(II), which account for their interchangeability.

Zircons always contain some *hafnium,* Hf (from 0.5 to 4%), consistent with the very great similarity of Zr and Hf. They also typically contain traces of iron, thorium, and uranium, which affect the color. Over geologic time, the radiations from the radioactive thorium and uranium displace electrons. The colors of these zircons may fade to a blue-white when the gems are heated to 1000°C. If the uranium and thorium concentrations were high enough originally, the continuous bombardment actually will have led to separation of some of the $ZrSiO_4$ into ZrO_2 and SiO_2 and a frosty appearance in the stone. Heating these stones results in a recombination into $ZrSiO_4$.

Garnet is a complicated silicate that can form from many ionic combinations, as its rather unusual formula suggests: $(Ca,Mg,Fe,Mn)_3$ $(Al,Fe,Cr)_2(SiO_4)_3$. The three silicate ions have a total charge of $12-$. The negative charge of the silicates must be balanced by some combination of interchangeable positive ions. In garnet this means three (II)

EXERCISE 7.4
Look up the ionic size of Zr^{4+} and of Hf^{4+}. Are the sizes consistent with the observed interchangeability?

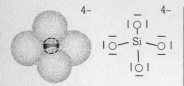

Figure 7.10
Tetrahedral structure of the orthosilicate ion, SiO_4^{4-}. See Figure 7.9 also.

ions totaling 6+ and two (III) ions totaling 6+. The commonness and great variety of garnets reflect this wide variety of possible combinations. Wouldn't Dalton and Berthollet have a fine debate over garnet?

The silicon–oxygen ratio of 1 to 4 (as found in zircon, peridot, and garnet) is the highest that exists in silicates. The net charge of 4− per SiO_4 group requires the presence of four units of positive charge. The crystals are hard, consistent with the large attractive forces between the oppositely charged ions of high charge.

The technical name for a compound containing the SiO_4^{4-} ion is *orthosilicate* (Figure 7.10). Zircon, peridot, and garnet are all orthosilicates, but the positive ions vary. Many other orthosilicates are known, with the number of positive ions always just sufficient to balance the charges on the silicates.

Topaz is an orthosilicate containing additional hydroxyl (OH^-) and fluoride (F^-) ions. Its formula is $Al_2(F,OH)_2SiO_4$; the F^- to OH^- ratio is variable. Note the presence of $2\,Al(III)$ per SiO_4^{4-} requires additional negative ions—here the similarly sized and identically charged F^- and OH^- ions. Topaz color is usually due to $Cr(III)$.

The basic formula of *emerald* and *aquamarine* is often written $Be_3Al_2(SiO_3)_6$. Actually, we now know that it might better be written $Be_3Al_2Si_6O_{18}$ because the $Si_6O_{18}^{12-}$ ion is present in the crystal as a large ring of alternating —Si—O—Si—O—, and so on. Each Si holds two more oxygens. They project somewhat like thorns from the ring of the Si—O—Si—O— chain. The aluminum and beryllium ions are positioned within and between the rings, balancing the charge and holding the crystal together.

The green color in emerald is due to $Cr(III)$ and some vanadium, $V(III)$; the color in aquamarine is due to $Fe(III)$. Emerald, which was described by the Babylonians in 4000 BC, appears to be the oldest known gemstone.

Radius Ratio

As far as size effects are concerned, atoms may substitute for one another if they are of similar size. But if atoms are to enter the interstitial positions in a crystal, they must match the size of the interstitial holes—not the sizes of the other atoms already present. It is a simple problem in geometry to calculate how big these holes are, if we assume the atoms act as spheres.

The larger the atoms already present, the larger the interstitial holes between them. But the ratio of the radius of the holes to that of the initial atoms remains constant. Table 7.5 lists the *radius ratio* (radius of interstitial hole/radius of initial atom) as a function of the number of initial atoms surrounding the hole. The smallest hole, with three atoms surrounding it, is a trigonal hole. It is said to have a *coordi-*

EXERCISE 7.5
Sketch a picture of the $Si_6O_{18}^{12-}$ ring. Describe its relation to the tetrahedron found in the orthosilicate ion.

Table 7.5   Some Common Atomic Packings and Their Radius Ratios

Geometry of packing	Trigonal plane	Tetrahedral	Octahedral	Simple cubic	Closest packed
Structure[a]					
Coordination number	3	4	6	8	12
Radius ratio[b]	0.155	0.225	0.414	0.732	1.000

[a]Note that the size of the hole increases as the coordination number around it increases.
[b]Radius ratio = radius of the interstitial hole/radius of initial atom.

nation number (CN) of 3. Other types of holes are listed in Table 7.5. Table 7.6 gives some data on actual crystals. Note the agreement between predicted and observed CNs.

Atoms having a radius smaller than the interstitial hole can, of course, fit into it. This is not common in ionic crystals because such packing allows the large negative ions to contact one another. Introduction of atoms larger than the hole is more common. Then the whole crystal must swell as the original atoms are pushed slightly apart.

Remember that positive interstitial atoms (positive ions) can fit only into holes surrounded by negative ions. Negative ions can fit only into holes surrounded by positive ions. Many crystals can be considered as made up of small positive ions in interstitial holes between the larger negative ions. Look at Figures 7.7, 7.8, 7.9, and 8.14 for examples.

Chromium(III): Both Green and Red?

I realize that you have been exposed to many colors and structures in this discussion of gems, but you may remember that Cr(III) gives ruby

Table 7.6   Coordination Number (CN) of Oxide Ions Around Various Cations as a Function of the Radius Ratio

Cation[a]	Radius of cation (nm)	Radius ratio[b]	Predicted CN	Observed CN
B^{3+}	0.020	0.15	3 or 4	3, 4
Be^{2+}	0.035	0.27	4	4
Al^{3+}	0.051	0.39	4 or 6	4, 5, 6
Ti^{4+}	0.068	0.52	6	6
Zr^{4+}	0.079	0.60	6 or 8	6, 8
Na^+	0.097	0.73	6 or 8	6, 7, 8
K^+	0.133	1.0	8 or 12	6, 8
Cs^+	0.167	1.3	12	12

[a]Ions with a positive charge are called cations; those with a negative charge are called anions.
[b]Based on the radius of O^{2-} ion of 0.132 nm.

POINT TO PONDER
The general composition of gems
has been known for a long time,
but silicate analyses have been
among the hardest to do using
aqueous-solution chemistry.
Now x-ray analysis and neutron
activation analysis may allow
one chemist to do as many good
analyses in a year as all
chemists had done in any single
year before 1940. Knowledge
of silicate chemistry is increasing
rapidly.

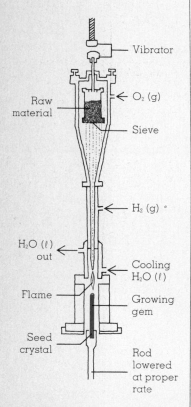

Figure 7.11
Verneuil synthesis of gems.

EXERCISE 7.6
Should the Verneuil flame have
excess O_2 or excess H_2 when
used to synthesize rubies?
Cr(III) is easy to oxidize.

its red color, yet also gives emerald its green color. How is this possible?
The answer is that color is due to the absorption of visible light and the
resulting movement of electrons from one orbital to another. But the
energy of the chromium orbitals is not determined solely by the central
nucleus. The oxygens surround the chromium very similarly in ruby
and in emerald, but the silicon(IV) atoms in emerald attract the oxygen
electrons more than do the aluminum(III) atoms in ruby. The differently charged oxygens interact differently with the chromium ions, and
so different light is absorbed in the two crystals.

Synthetic Gems and Artificial Gems

Most of the crystalline gemstones can now be grown synthetically.
There are two main methods. The first is called the *Verneuil method*
after its inventor. It is simple in concept and difficult in practice. The
idea is to use an oxygen–hydrogen flame to melt the ingredients in the
right proportion and then to let the molten drops fall onto a "seed"
crystal at just such a rate that a single crystal will grow, taking care not
to melt the growing crystal or to let the drops run off (Figure 7.11). A
single seed crystal is used for two reasons:

1. Most of these gem crystals are complicated structures, difficult to
   form.
2. If more than one seed crystal were used, many small crystals might
   form rather than the one large crystal desired.

The Verneuil method has now been perfected so that large crystals
(several centimeters in diameter and many centimeters long) may be
synthesized. Large ruby crystals, for example, are used to make very
intense light sources called *lasers*.

The second general method relies on the fact that water is a good
solvent for reactions involving ions. Clearly, all gems are almost totally
insoluble in water at room temperature. But, at temperatures of 500°C
and above, the ions are often sufficiently soluble to migrate to a seed
crystal from the appropriate raw materials, so that the seed will grow.
High pressures are always required because the water, at these temperatures, would otherwise be a dilute gas unable to transport the ions.
Crystallization from water at high temperature and pressure is known
as *hydrothermal growth*. It probably comes close to the conditions
under which many of the gems were formed in nature.

Other methods involve drawing a crystal slowly from a liquid melt,
or melting powdered material into a single crystal. The gems made by
any of these methods can be chemically identical with natural gems.
Often they have fewer flaws and are consistently larger than the majority of natural gems. It is interesting that they often cost less at the
jeweler's.

The term *artificial gem* should really be used for those that purport to be something they are not chemically. Thus, *strontium titanate* and *yttrium aluminum garnet* are often used as diamond substitutes in jewelry. Both have high flash and fire, but are softer than diamond. Or diamonds may be backed with dye or colored materials to enhance their color. Or thin sections of diamond may be glued onto a base of lower cost to give *diamond doublets*. There are many artificial ways to produce most of the properties of most gemstones.

A particularly interesting application of human-controlled synthesis and artificial material is the *cultured pearl*. The general technique has been known in the Orient for hundreds of years, during which tiny Buddha images have been introduced into appropriate shellfish, which coat the images with pearl and convert them into votive offerings. This rough process was finally reduced to an exact one by K. Mikimoto in Japan, and cultured pearls are now produced by the thousands in almost any size.

Quite large round grains are introduced into the shellfish, so that a thin layer of pearl forms. (The process takes several years.) External examination cannot distinguish natural from cultured pearls, but tests of density and/or x-rays can (Figure 7.12). Pearls consist of calcium carbonate ($CaCO_3$) and the protein conchiolin in an 8 to 1 heterogeneous mix. They normally are produced to cover sand or some other irritating particle that has entered the oyster.

Interchangeable Ions in Biochemistry

Just as it is possible to interchange ions in minerals, making only minor alterations in overall properties, so it is sometimes possible to do the same thing in living systems. Some especially simple examples are found in enzyme systems, though the enzymes themselves are complicated molecules often containing thousands of atoms. Enzymes are synthesized by living systems and are used to control the speeds of chemical reactions in the systems. The most common atoms in enzymes are hydrogen, carbon, oxygen, nitrogen, sulfur, and phosphorus. But many enzymes require one to a few ions of a metal per enzyme molecule before they can become reactive. (The magnesium necessary to activate chlorophyll is a case you may have heard about.)

An enzyme called *phospholipase* normally contains two zinc ions (Zn^{2+}) per enzyme molecule. One of these ions can be rather readily removed by adding ethylene diamine tetraacetate (EDTA) to the system. EDTA has a high affinity for zinc ions, so one Zn(II) leaves the enzyme; the ability of the enzyme to control is destroyed. The ability will be fully recovered if zinc ions are again added to the system, as you might have guessed. It will also be recovered if cobalt(II) ions (Co^{2+}) are placed in the solution. Addition of manganese(II) ions (Mn^{2+}), or

EXERCISE 7.7
The formula for strontium titanate can be written as an equimolar mix of the common oxides of strontium and titanium, or it can be written as a condensed formula. Write the two formulas.

Figure 7.12
Cross sections of natural and cultured pearls.

POINT TO PONDER
Annoy an oyster and get a pearl.

EXERCISE 7.8
Look up the ionic radii of Mg^{2+}, Mn^{2+}, Co^{2+}, and Zn^{2+}. Do they correlate with the effects on phospholipase mentioned in the text? Compare with the radii of Ca^{2+}, Cd^{2+}, Cu^{2+}, Fe^{2+}, Hg^{2+}, and Ni^{2+}.

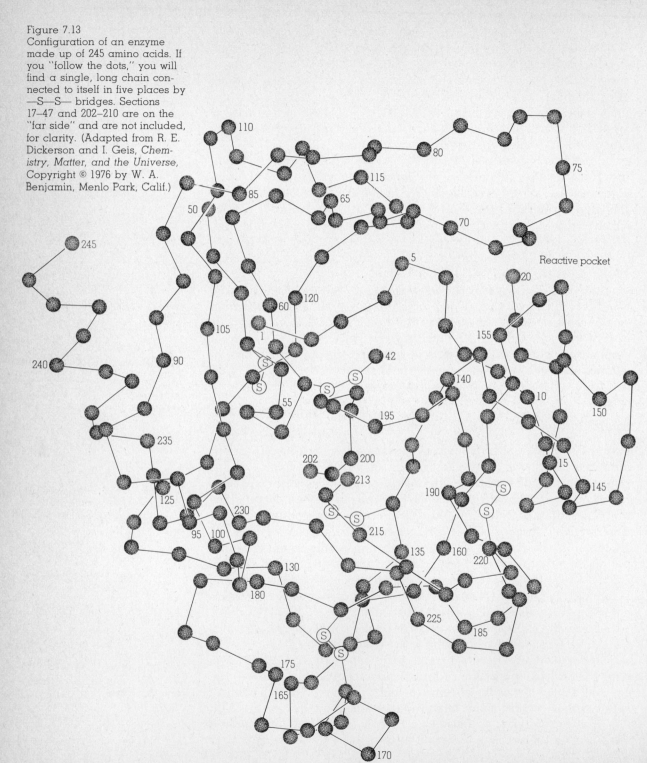

Figure 7.13
Configuration of an enzyme
made up of 245 amino acids. If
you "follow the dots," you will
find a single, long chain connected to itself in five places by
—S—S— bridges. Sections
17–47 and 202–210 are on the
"far side" and are not included,
for clarity. (Adapted from R. E.
Dickerson and I. Geis, *Chemistry, Matter, and the Universe*,
Copyright © 1976 by W. A.
Benjamin, Menlo Park, Calif.)

Reactive pocket

magnesium(II) ions (Mg^{2+}) leads to partial recovery of activity. This means that each molecule of the enzyme will react if one zinc is replaced by manganese or magnesium, but it will react more slowly. None of the following will restore activity: Cd^{2+}, Ca^{2+}, Ni^{2+}, Cu^{2+}, Fe^{2+}, or Hg^{2+}. Orthophenanthroline has an even greater attraction for zinc(II) than does EDTA. If it is used, more zinc is removed. The activity of the enzyme is destroyed and can only be renewed by the presence of zinc ions. None of the other ions is effective.

The most commonly interchangeable ions in biochemical systems are Mg^{2+} and Mn^{2+}; K^+, $NH_4{}^+$, and Tl^+; and Co^{2+} and Zn^{2+}. Note the identity of charge in each set, and use the chart inside the back cover of this book to confirm that the ionic radii are similar. You might be interested in working out the electron structures of some of these sets. In no case do all members of the set have the electron structure of a noble gas—in fact, only Mg^{2+} and K^+ do. In these sets, at least, ionic charge and size are more critical in determining function than are the details of the electron structure. However, the fact that the sets are small and do not include all ions of similar charge and size shows that only certain electron structures can meet the functional requirements.

Amino Acid Interchange in Proteins

Enzymes provide additional examples of interchanging parts with little change in function within the main part of the complicated molecule. Each enzyme is a protein consisting of one or more long chains of smaller units called *amino acids*. You have almost certainly seen figures like Figure 7.13 that represent these protein chains. Twenty different amino acids are found in various sequences in the chains. Each amino acid has the general formula $RCHNH_2COOH$, with the atomic composition of R differing from acid to acid. The amino acids can be represented in some detail, as in Figure 7.14, or by single letters, as in Figure 7.15, which gives the sequences found in five different species

EXERCISE 7.9
Using ionic radii alone as a criterion, which pair of ions should be most interchangeable: Mg^{2+} and Mn^{2+}, K^+ and Tl^+, or Co^{2+} and Zn^{2+}?

Glycine (G)   Alanine (A)   Asparagine (N)

Figure 7.14
A short sequence of protein showing variation in amino acid R groups (colored). This sequence contains three amino acid residues: glycine, alanine, and asparagine.

Figure 7.15
Hormone (βMSH) structures that stimulate pigment cells. The amino acids are identified by the letters listed in Table 1.2.

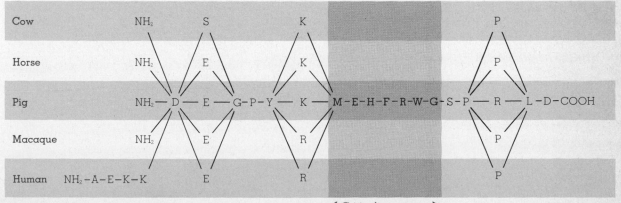

Cow	NH_2		S			K										P			
Horse	NH_2		E			K										P			
Pig	NH_2—D	—E —G-P-Y	—K	—M-E-H-F-R-W-G	-S-P	—R	—L-D-COOH													
Macaque	NH_2		E			R										P			
Human	NH_2-A-E-K-K		E			R										P			

←Critical sequence→

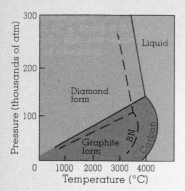

Figure 7.16
Regions of pressure and
temperature in which liquid,
diamond, and graphite forms
of carbon (colored line) and
boron nitride, BN (dashed line)
are stable.

for the hormone (βMSH) that stimulates the pigment cells (in forming
suntan, for example).

Don't worry about the full names for the amino acids represented
in Figure 7.15. But do note that there is one stretch of the molecules
where all species of animals have the same sequence (the *critical
sequence*), and other sections where considerable interchangeability
is found, though each species shows only one sequence. Sometimes the
amino acids in proteins are interchangeable, but those in the critical
regions are not.

Isoelectronic Molecules and Molecular Fragments

As disparities in size, electric-charge distribution, and electron structure
get bigger and bigger, the degree of interchangeability of two sets of
atoms decreases. One type of group that shows enough similarities to be
useful is an isoelectronic set. An *isoelectronic set* is a group of molecules
or molecular fragments that have the same number of nuclei, not counting hydrogen, and the same valence-electron structure. Table 7.7 lists
several such sets and some properties that are similar in each set.

Figure 7.16 shows the conditions of pressure and temperature for
which two isoelectronic substances, boron nitride (BN) and carbon (C),
are stable. Each, boron nitride and carbon, exists in two forms—a
graphite and a diamond form, with the latter stable only at high pres-

POINT TO PONDER
The fact that many species of
animals synthesize the same, or
nearly the same, molecules to
do the same jobs suggests a
common origin for these
syntheses. In fact, the small
changes found from one species
to another are consistent with
an evolutionary sequence
stretching back to a common
set of single-cell ancestors.

Table 7.7   Some Isoelectronic Sets

Molecule	Number of nuclei (not H)	Number of valence electrons	Bond length (nm)	Solubility in H_2O[a]	General chemical reactivity
HF	1	8	0.0917	High	High
H_2O	1	8	0.0960		High
NH_3	1	8	0.101	High	High
NN	2	10	0.1098	23.3	Very low
CO	2	10	0.1128	35	Moderate[b]
CN^-	2	10	0.115	High	Moderate[b]
HCCH	2	10	0.1201	Very low	Very high
NO_3^-	4	24	0.124	High	Moderate
CO_3^{2-}	4	24	0.129	High	Moderate
BO_3^{3-}	4	24	0.136	High	Moderate

The following 12 molecular fragments are isoelectronic:

—F	>CH_2	—NH_2	—Br	>O	—SH
—CH_3	—Cl	>NH	—OH	—I	>S

[a]In cubic centimeters per liter H_2O at 25°C.
[b]Note that both CO and CN^- are poisons.

sure, as you see. Note how similar the behaviors are. The diagram gives conditions at which the graphite form can be expected to transform into the diamond form, and vice versa.

Molecular Interchange

There is another type of molecular interchange that is very common yet causes no change in properties. Molecules can move around and interchange positions in the gas, liquid, or crystalline phase. These motions are at a minimum in crystals because of the tight packing and strong forces. Motion in gases is rapid because of the large amount of open space and the very small intermolecular forces. Gases interdiffuse at a rate of about 1 m (meter) per hour.

Motion in liquids is intermediate and is slow over long distances. It may take a week for any appreciable number of molecules to move as much as 1 cm from their original positions if no stirring occurs. But motion at the molecular level is rapid and interchange frequent, as Figure 7.17 shows through a computer simulation of molecular motions over a period of about 10^{-9} seconds in a crystal and in a liquid. You see that the position of the center of the molecule is in violent and rapid motion. This motion is centered around a single spot in the crystal.

Figure 7.17
Computer-simulated motion of molecular centers in an interval of about 10^{-9} seconds in (a) a crystal and (b) a liquid. (Courtesy B. Alder, Lawrence Radiation Laboratory, University of California)

Interchanges occur rarely in a crystal. But the molecules in a liquid move with considerable freedom and commonly interchange positions.

Summary

Sets of atoms, of ions, or of molecules that have similar electric charges, similar sizes, and similar electron structures often show similar properties. In some cases the similarities are so great that members of the set can be interchanged in a substance with no appreciable change in the properties of the substance. Thus, many alloys and minerals vary greatly in composition, but not in properties or structure, when various atoms are replaced by others. Some enzymes can function with any of several ions in their functional group, and with various sequences of amino acids in the noncritical sections of their chains. Isoelectronic molecules and molecular fragments are often similar enough in many properties to make correlations between them worthwhile.

All these observations are consistent with the idea that atoms and molecules consist of nuclei surrounded by electrons. It is the size, shape, electric-charge distribution, and electron structure that determine the properties of the substance.

HINTS TO EXERCISES
7.1 Similar atomic size and mass. 7.2 Like diamond. 7.3 Compositions are too different. 7.4 0.079 nm, 0.078 nm, interchangeable. 7.5 Place silicon symbols at corners of a regular hexagon, six oxygens at corners of a smaller hexagon within the silicons and with each oxygen corner halfway between two silicon corners. Add pairs of oxygens to outside of each Si. 7.6 Slight excess of H_2. 7.7 $SrO \cdot TiO_2$, $SrTiO_3$. 7.8 Correlation is good. 7.9 Co^{2+} and Zn^{2+} are very similar.

Problems

7.1 Identify or define: alloys (interstitial and substitutional), coordination number, diamond, displaced electrons, gemstone, glass, glide plane, graphite, hydrothermal growth, interchangeability (atoms, ions, molecular fragments), isoelectronic sets, quartz, radius ratio, ruby–sapphire, silicate, unit cell.

7.2 Look up the ionic sizes of Cr(III), Fe(III), V(III), and Al(III) and discuss their relation to rubies, sapphires, and alexandrite.

7.3 Would zinc be more likely to form an interstitial or a substitutional alloy with cadmium?

*7.4 What element would you suggest adding to metallic copper to harden it? Reasons? Addition of beryllium (Be) gives spring metal. Why so?

7.5 What change in composition must occur when Cr(III) enters topaz?

7.6 What abrasive material would you recommend for grinding and polishing diamonds?

7.7 You wish to synthesize red zircons. Is the addition of Cr(III) a likely possibility?

7.8 Show that the formula of turquoise, $CuAl_6(PO_4)_4(OH)_8 \cdot 5H_2O$, is consistent with the charges on the ions present.

7.9 United States coin silver was 90% Ag, 10% Cu; jewelry silver 80% Ag, 20% Cu. Why not use the same alloy for both coins and jewelry?

*7.10 Many platinum pen-points are a 50% Pt, 38% Ag, 12% Cu alloy by weight. What fraction of the atoms are of each element? (Hint: Calculate the moles of each element in 100 g of alloy.) [Atomic ratio $\cong$ 7 Pt/9 Ag/5 Cu]

7.11 Cementite, Fe_3C, is a compound. What is the weight of 1 mole of Fe_3C? How much iron and how much graphite would you melt together in an attempt to make 180 g of Fe_3C? [$\sim$10 g C]

*7.12 How can the external faces and glide planes of a crystal be used to deduce the arrangement of molecules in the crystal?

*7.13 What do you predict would be the physiological activity of a "βMSH" in which the histidine (H) and phenylalanine (F) were interchanged? How about the alanine (A) and glutamic acid (E) at at the NH_2 end for humans?

7.14 The manufacture of linear alkyl benzene sulfonates requires the isolation of straight-chain hydrocarbons from petroleum. This can be done by passing the hydrocarbon mixture through specially synthesized silicates called *molecular sieves*. They have structures full of long tunnels, like those in quartz but more open. Which chains, linear or branched, are passed by the sieves?

*7.15 The sodium ions in the surface of glass may be exchanged with lithium ions, producing compression forces and hardening the surface of the glass. Why does the compression occur? Automobile and airplane windshields are often treated in this way.

7.16 Why do synthetic gems ordinarily sell for less than the same gems found in nature?

*7.17 Why are hydrogen nuclei not counted in determining which molecular fragments are isoelectronic?

*7.18 Why do you suppose the practice originated of putting little ridges around the edge of a coin?

7.19 Why can't diamonds be produced by either the Verneuil or the hydrothermal method?

*7.20 Radioactive strontium, Sr ($Z = 38$) is often said to be one of the most dangerous products of nuclear fission. Note its place in the periodic table and, in light of the composition of humans given in Table 1.1, comment on the danger.

7.21 Relate the discussions in this chapter to previous discussions of the similarity of the chemistry of isotopes of the same element.

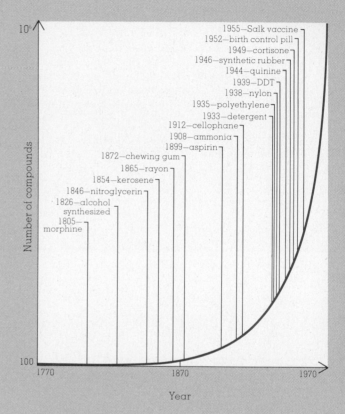

Figure 8.1
Fewer than 100 compounds
had been identified at "the
beginning of modern chemistry"
in 1770. Several million are now
known. This is an annual
growth rate of almost 5%, or a
doubling time of about 15
years.

8 THE FORMATION OF CHEMICAL BONDS

There are about 90 chemical elements common enough in the known universe to affect appreciably our observations of nature. All are present in the earth's crust or atmosphere and in the stars as well. Scientists have synthesized measurable (and sometimes very large) quantities of 15 additional elements. We have seen that only a few of the elements commonly occur on earth in their elemental state. Most are present in compounds. The great variety of materials we find in nature and synthesize in laboratory and factory is a result of the many ways in which the elements can combine.

Several million compounds have been identified already, and new ones are prepared or identified at a rate of more than 1000 per working day (Figure 8.1). The potential number is known to be indefinitely large—say, millions of millions and more. The present industrial scale is illustrated by the synthesis of more than 2×10^{11} kg of a total of almost 10,000 different substances annually. In addition to asking how we identify, characterize, and synthesize the presently known millions, we need to explore whether it seems worthwhile to add to the list. It will not surprise you if the answer seems to be yes.

One, Two, Three, . . . , Infinity: Counting the Chemical Compounds

You can be quite sure that the millions of known compounds are a tiny fraction of the possible ones.

How can I write this with such confidence?

Consider first the number of compounds containing only 2 elements; we call them *binary* compounds. Each of the 90 elements could combine with each of the 89 others to give a total of 90×89, or about 8000 compounds if each pair formed only 1 compound. However, this simple arithmetic counts each compound twice (NaCl and ClNa are not different compounds), and it does not allow for the fact that many pairs of elements can form more than 1 binary compound. Furthermore, not all binary combinations can form stable compounds. Simple arithmetic does not give us much help in counting possible compounds.

You read in Chapter 7 about proteins being made up of strings of amino acids connected in long sequence. There are 20 different amino acids found in these sequences, and every amino acid can be connected in any sequence with the others. Thus, for each position in the sequence

It has now been recognized that advances in theoretical or "pure" science eventually carry in their train changes in practice of the most far-reaching nature—changes which are usually far more radical than those caused by progress in the applied sciences directly concerned. —*R. J. Muller.*

POINT TO PONDER
The unlimited number of chemical compounds formed by 90 elements is similar to the unlimited number of books written with some 50 symbols, or of musical compositions based on a few octaves of notes. Even after following the laws of grammar, syntax, and spelling in writing, or the canons of composition in music, the number of possible products is without limit. As with chemical compounds, we might well ask whether more should be synthesized.

EXERCISE 8.1
Cytochrome c, an iron-
containing protein, has a protein
chain of 104 amino acid links.
How many different proteins this
long could be made from the
20 common amino acids?

—— Known
    tank

—— Spill

Figure 8.2
Comparison of oil recovered
from an oil spill and from
the storage tank presumed
to be the source of the oil.
The curves are obtained by
a special technique (similar to
a distillation) called *vapor
phase chromatography.*
(Courtesy the U.S. Coast
Guard, Miami, Florida)

there are 20 possible amino acids. There will be 20×20 simple pairs of acids, because each first position could be occupied by any one of the 20 acids, as could the second. The sequence alanine-glycine, for example, is not the same as the sequence glycine-alanine. (See Figure 7.13.) For a chain of three the number of possibilities is $20 \times 20 \times 20 = 20^3$. For any sequence containing a total of n amino acids—n standing for the number of amino acids—the number of possibilities is 20^n.

Some sequences longer than 100 have been completely identified. This means that any particular sequence is one of at least 20^{100} or 10^{130} possible sequences of 100 amino acids. Now 10^{130} is a very large number, but the number of possible sequences longer than 100 units is even greater. There appears to be no limit to the possible length or complexity of the sequences in proteins, so there is no limit to the number of possible molecules. And we have discussed only sequences of amino acids!

Units other than amino acids can also link in long sequences. The process is called *catenation,* from a Greek word meaning "chain formation." The simplest unit that can form very long chains is the carbon atom. You will recall that carbon forms very large planar structures in graphite and large three-dimensional structures in diamond. Carbon can also form long one-dimensional structures, as in paraffin waxes and polyethylene plastics, and it can form structures that combine the long linear structures with interspersed planar and three-dimensional structures.

The outside of each of the carbon structures can be covered with hydrogen atoms. Such compounds of carbon and hydrogen are called *hydrocarbons,* and their number is indefinitely large. Yet this infinitely large array of compounds contains only carbon and hydrogen; each is a binary compound. But each of these in turn can retain the same carbon skeleton and intersperse some oxygen, nitrogen, chlorine, or other atoms around the molecule. Note that $-CH_3$ is isoelectronic with $-NH_2$, $-OH$, and the Group VIIA elements, and that $=CH_2$ is isoelectronic with $=NH$ and $=O$. Such interchanges can be made in various places—so that the number of possible compounds, even with a rather small carbon skeleton, becomes very large indeed. It is the great variety of possible catenations of carbon that accounts for most of the millions of known compounds. In other words, most known compounds are based on carbon skeletons interspersed with other elements and with hydrogen atoms on the periphery.

Natural gas and petroleum are composed primarily of hydrocarbons, and a great variety of hydrocarbons at that. This mixture varies with the source, and samples can often be identified as to source in terms of the hydrocarbons they contain. For example, it is sometimes possible to identify the source of an oil spill by comparing the composition of the oil spill with the composition of the oil in the suspected source (Figure 8.2).

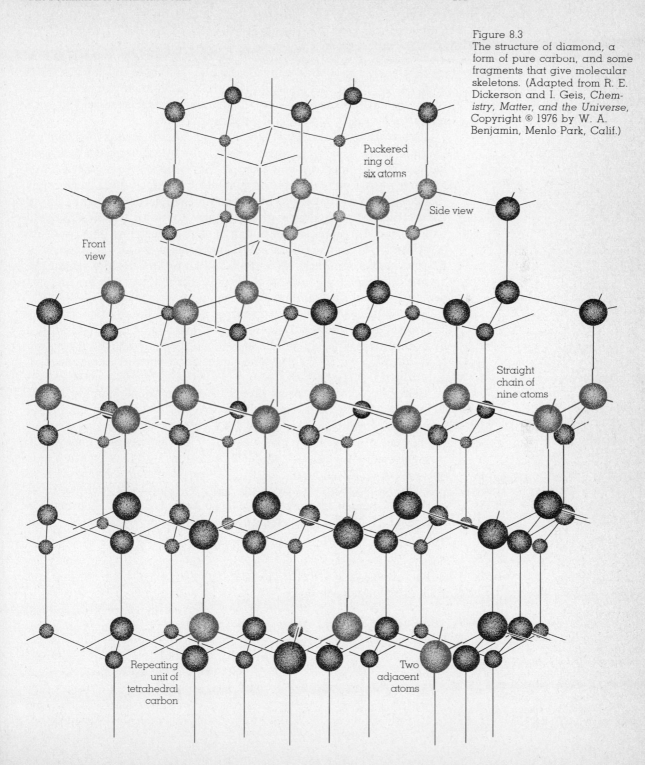

Figure 8.3
The structure of diamond, a form of pure carbon, and some fragments that give molecular skeletons. (Adapted from R. E. Dickerson and I. Geis, *Chemistry, Matter, and the Universe*, Copyright © 1976 by W. A. Benjamin, Menlo Park, Calif.)

Puckered ring of six atoms

Side view

Front view

Straight chain of nine atoms

Repeating unit of tetrahedral carbon

Two adjacent atoms

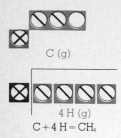

C (g)

4 H (g)

C + 4 H = CH₄

Figure 8.4
Electron structures of gaseous
atomic carbon and hydrogen
needed to make methane, CH₄.

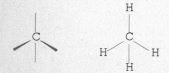

Figure 8.5
Valence electron structure of
methane, CH₄. (In the left
figure a solid line indicates a
bond in the plane of the page;
a dashed line, a bond extending backward; a wedge, a
bond extending forward; all to
hydrogens.)

Figure 8.6
Valence-electron and
molecular-model structures of
ethane, C₂H₆, and propane,
C₃H₈. (See Figure 8.5 for an
explanation of bond symbols.)

Catenations and Combinations

The ultimate in three-dimensional catenation of carbon atoms is diamond, the repeating unit being a carbon atom that is sharing electrons
with four adjacent carbon atoms (Figure 8.3). Suppose you extract one
such tetrahedral carbon atom from diamond and surround it with
hydrogen atoms. What molecule will result?

Figure 8.4 will help you recall that a carbon atom has four valence
electrons and room in the valence orbitals for four more. Each hydrogen
has one valence electron and room in that orbital for one more. Thus
four hydrogens can provide the four electrons for which carbon has
room, and one carbon can provide the four electrons needed to fill up
the orbitals in four hydrogen atoms. Figure 8.5 indicates how this could
happen, each line representing a pair of shared electrons. The resulting
compound (formula CH_4) is called *methane,* a major component of
natural gas. Experiments show that the four hydrogens in methane are
arranged tetrahedrally around the central carbon, just as were the four
carbon atoms in the diamond.

Removing two adjacent atoms from diamond involves breaking six
carbon-to-carbon bonds. Each of the broken bonds can unite with a
hydrogen atom—just as in the case of methane—with the formation
of $H_3C{-}CH_3$ (or C_2H_6), *ethane,* another important component of
natural gas (see Figure 8.6).

Removing three adjacent atoms of carbon from diamond breaks
eight bonds. The three-atom fragment can hold eight peripheral
hydrogens—forming $H_3CCH_2CH_3$ (or C_3H_8), called *propane,* a minor
component in natural gas and a common bottled gas (see Figure 8.6).

Removing four atoms from diamond brings in a new factor. There
are two possible fragments: a string of four atoms, or a string of three
with the fourth atom joined to the central atom of the string of three
(Figure 8.7). Each fragment can add 10 hydrogens and each has the
formula C_4H_{10}, but the structural arrangement is different. This
difference is sometimes represented by writing the two formulas as

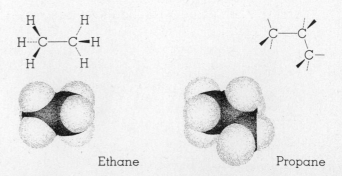

Ethane   Propane

(a)

(b)

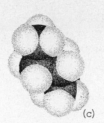
(c)

(d)

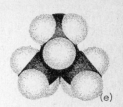
(e)

$H_3CCH_2CH_2CH_3$ (butane) and $(H_3C)_3CH$ (2-methyl propane). Molecules having the same molecular formula but differing structures are called *isomers*. Both C_4H_{10} isomers are common in bottled gas, often mixed with propane.

The $(H_3C)_3CH$ isomer can also be written $H_3CCH(CH_3)CH_3$ to indicate that the $—CH_3$ group inside the parentheses is a branch on a straight chain of three carbon atoms. You just saw that a straight chain of three carbon atoms is called *propane*. The $—CH_3$ group (a methane molecule minus one hydrogen atom) is called a *methyl* group. So the name for $H_3CCH(CH_3)CH_3$ is 2-methyl propane; the *2* indicates that the side methyl group is on the second atom of the propane chain. (See Figure 8.7 and Table 8.1.)

It should be clear that there are several ways to extract five connected carbon atoms from diamond, more ways to extract six, still more for seven, and so forth. All these fragments can be surrounded with hydrogen atoms to give molecules with formulas C_5H_{12}, C_6H_{14}, and C_7H_{16}. The isomeric molecules with eight carbons, C_8H_{18}, are the main components of gasoline. They are called *octanes* (as in "octane rating"). Later on I will go into more detail on the properties of gasoline in terms of its molecular makeup.

All the possible hydrocarbons up through the C_8H_{18} molecules have been purified and characterized, as have many of the more complicated

Figure 8.7
Molecular models of the two C_4H_{10} molecules. Butane can twist and turn into many forms (four are shown, a–d) because the single bonds allow rotation within the molecule. The carbon skeleton of 2-methyl propane (e) is fixed, but the three methyl groups rotate (much like propellors) so there are many possible patterns for the hydrogens.

EXERCISE 8.2
There are three structurally different molecules of formula C_5H_{12}. Write formulas for each one, and name the substances in terms of the longest "straight" chain of carbons in each one.

Table 8.1   **Names of Some Hydrocarbon Molecules and Fragments**

Molecules		Fragments	
Formula	Name	Formula	Name
CH_4	Methane	$—CH_3$	Methyl
C_2H_6	Ethane	$—C_2H_5$	Ethyl
C_3H_8	Propane	$—C_3H_7$	Propyl
C_4H_{10}	Butane	$—C_4H_9$	Butyl
$H_3CCH(CH_3)CH_3$	2-Methyl propane		
C_5H_{12}	Pentane		

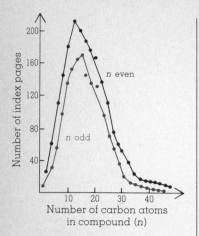

Figure 8.8 (above)
Sample count of index pages of reported research on hydrocarbons, versus n in C_nH_{2n+2}.

Figure 8.9 (right)
Structure of graphite, a form of pure carbon.

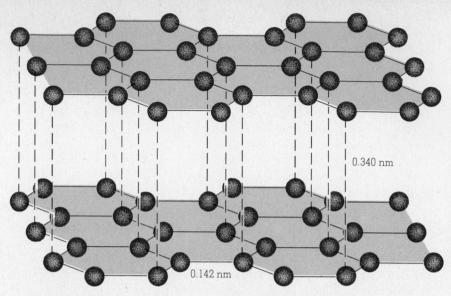

Benzene

Naphthalene
Anthracene

Phenanthrene

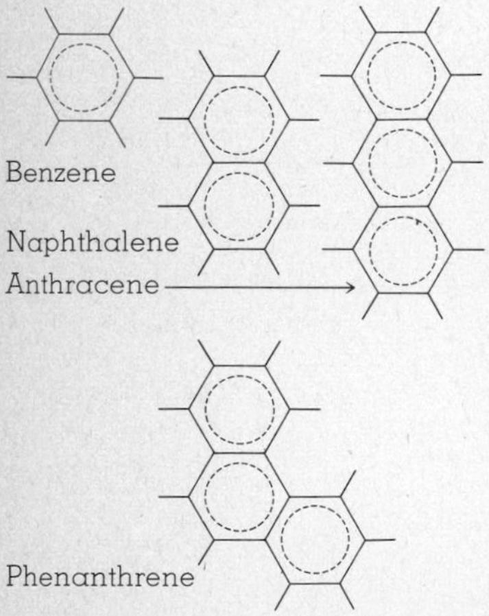

Figure 8.10
Some atomic arrangements found in ring structures. Hydrogen atoms are arranged around the edges. When rings of atoms share one or more edges they are called *condensed rings*. All the molecules here, except benzene, have condensed-ring structures.

hydrocarbons. But it should not surprise you that there are many complicated hydrocarbons that have not been isolated—even though we are sure that most of them could be made if it were worth the effort.

Because the number of fragments that could be "extracted" from diamond is without limit, the number of possible hydrocarbons is without limit.

Figure 8.8 demonstrates, however, that the extent of interest in individual long-chain hydrocarbons is limited. Partly this is due to the difficulty of synthesizing and purifying them. But there also seems little more to learn about or make use of in individual large hydrocarbons.

Graphite is also made up of linked carbon atoms. (Figure 7.6 is reproduced here as Figure 8.9 for easy review.) The most obvious unit here is the hexagon of six carbon atoms. If one of these hexagons is removed from graphite, and hydrogen atoms are added to the periphery of that hexagon, the molecule produced is called *benzene,* C_6H_6. An indefinitely large number of complicated units can be "extracted" from graphite and covered with hydrogen to produce complicated condensed-ring structures. Figure 8.10 gives some examples. There are many other structures with rings of more than or fewer than six atoms.

Many condensed-ring structures cause cancer in humans. The first evidence of this was the high incidence of cancer in chimney sweeps, who were continually exposed to the soot from coal fires. There are many condensed-ring compounds in soot, and also in cigarette smoke. They are also common by-products from the conversion of coal to coke.

A smaller possible unit in graphite is

$$C{=}C$$

This can hold four hydrogen atoms to give $H_2C{=}CH_2$ (or C_2H_4), called *ethene* (also *ethylene*). Ethene can combine with itself to produce the common plastic polyethylene, as we shall see later. In the same way the fragment C—C—C can add hydrogens to give $H_3C{-}CH{=}CH_2$, *propene* (formula C_3H_6). Propene can combine with itself to give the plastic polypropylene, much used in making transparent plastic bottles.

Just as the single dash symbolizes a *single bond,* the double dash between two carbon atoms indicates the presence of four electrons shared between the two atoms. It is the common symbol for a *double bond.* The suffix *-ene* in names like *ethene* indicates the presence of a double bond. Carbon compounds containing double bonds (or triple bonds—three pairs of shared electrons) are said to be *unsaturated* (Figure 8.11). They can add further atoms—for example, hydrogen. There are many compounds having long chains of carbon atoms, some of which are connected by double bonds and some by single bonds. We shall look at several such compounds in later chapters. Note that in all the formulas you have seen each carbon atom has four bonds (eight electrons) connected to it, consistent with its four valence orbitals (one *s* and three *p* orbitals).

Some of the reasons for so many carbon compounds should now be apparent to you—and as yet we have examined only hydrocarbons. Remember that oxygen, nitrogen, and other atoms can enter the chains and that chlorine and the other Group VIIA elements, as well as other isoelectronic groups, can replace hydrogen on the periphery. All these variations further increase the number and variety of compounds. We shall run into many of them but shall not concern ourselves with their systematic names at present. Rather, let's turn to the interpretation of these shared electron bonds that can lead to such a rich variety of substances.

Share the Wealth? Well, Sometimes

Chemical bonds are always due to the simultaneous attraction of one or more electrons by more than one nucleus. The early examples you studied involved atoms that differ greatly in ionization energy. In these cases, one or more electrons leave one atom and move to another—producing charged atoms called *ions.* Sodium chloride (NaCl) and the gemstones provided examples. In general, the positive ions come from atoms of metals, the negative ions from nonmetals—consistent with the relative abilities of these two types of elements to hold electrons.

But carbon and hydrogen are both more nonmetallic than metallic. Thus carbon and hydrogen do not differ greatly in their attraction for electrons and, of course, two carbon atoms do not differ at all. Let's consider just the carbon atoms first.

You have already seen (in the discussion of diamond) that the four

A graphite fragment

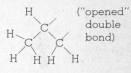

when "covered" with
H gives propane,

which can "open"
a double bond,

("opened"
double
bond)

and add H_2
to give propene

Figure 8.11
Unsaturation involving a double bond, and the addition of hydrogen to give all single bonds.

EXERCISE 8.3
The most common alcohol has the formula C_2H_5OH and is called *ethanol.* Another alcohol, CH_3OH is called *methanol.* What is the name for C_3H_7OH? Is there only one isomer of C_3H_7OH?

The simplest chemical bond.

bonds are consistent with the orbital electron structure of atomic carbon. When two carbon atoms do share electrons, the two nuclei are identical. Therefore, the electrons will experience equal forces from both nuclei. The electrons will be shared equally. The hardness and very low electrical conductivity of diamond are consistent with this idea. The bond formed by the sharing of electrons between two atoms is called a *covalent bond.*

It was G. N. Lewis in 1916 who first suggested that chemical bonds could be interpreted in terms of *electron pairs.* The diagrams we have been using, with dashes representing pairs of bonding or nonbonding valence electrons, are called *Lewis electron structures.*

But the two atoms bonded by a covalent bond need not be identical. Methane (CH_4) has four covalent bonds, one between each of the four hydrogens and the carbon. Because the carbon and hydrogen nuclei are different, you would expect the sharing to be unequal. The greater the difference in the attraction of the two bonded nuclei for the electrons, the more unequal the sharing. Where there is a big difference in attraction, the electrons move entirely from one atom to the other and form electrically charged ions.

When the two ends of the bond are the same, the electrons are shared equally, and the bond is said to be nonpolar. Diamond is an example. When the two ends of a bond differ, the sharing is unequal; if the bond is still covalent, the bond is said to be polar. The C—H bond is slightly polar. When the polarity becomes so large that the electrons actually spend essentially all their time on one atom and make it negative, the other being positive, the bond is said to be ionic. Sodium chlo-

POINT TO PONDER
Most of the several million known chemical compounds are held together by covalent bonds, especially between C, H, O, N, and S. Thus, 96% of all compounds contain H, 95% contain C, and 64% contain O. Most of these compounds are covalent. Perhaps it is even more revealing to list the abundances of the known compounds containing only certain elements: 16% contain only C, H, and O; 27% only C, H, O, and N; 25% only C, H, O, N, and S; and 19% only C, H, O, N, S, and halogens (chlorine, Cl; fluorine, F; iodine, I; and bromine, Br). Compounds containing the other 90 or so elements are only 13% of the total.

Table 8.2   The Electronegativity Scale Arranged in Periodic Table Form

H 2.1																	
Li 1.0	Be 1.5											B 2.0	C 2.5	N 3.0	O 3.5	F 4.0
Na 0.9	Mg 1.2											Al 1.5	Si 1.8	P 2.1	S 2.5	Cl 3.0
K 0.8	Ca 1.0	Sc 1.3	Ti 1.5	V 1.6	Cr 1.6	Mn 1.5	Fe 1.8	Co 1.8	Ni 1.8	Cu 1.9	Zn 1.6	Ga 1.6	Ge 1.8	As 1.9	Se 2.4	Br 2.8
Rb 0.8	Sr 1.0	Y 1.2	Zr 1.4	Nb 1.6	Mo 1.8	Tc 1.9	Ru 2.2	Rh 2.2	Pd 2.2	Ag 1.9	Cd 1.7	In 1.7	Sn 1.8	Sb 1.9	Te 2.1	I 2.5
Cs 0.7	Ba 0.9	La–Lu 1.1–1.2	Hf 1.3	Ta 1.5	W 1.7	Re 1.9	Os 2.2	Ir 2.2	Pt 2.2	Au 2.4	Hg 1.9	Tl 1.8	Pb 1.8	Bi 1.9	Po 2.0	At 2.2
Fr 0.7	Ra 0.9	Ac 1.1	Th 1.3	Pa 1.5	U 1.7	Np–No 1.3										

NOTE: Reprinted from Linus Pauling, *The Nature of the Chemical Bond.* Copyright 1939, 1940, third edition © 1960 by Cornell University. Used by permission of Cornell University Press.

ride, NaCl, is a good example. Figure 8.12 shows some Lewis electron structures with polar and nonpolar bonds.

As with all classifications, so here—there is a gradual transition from nonpolar to polar to ionic, with many intermediate steps. Most of the bonds discussed in an elementary treatment like this belong near one of the extremes. They will either be strongly ionic, or else they will be nonpolar or only weakly polar. But the molecules with intermediate bond types do exist.

An easy way to assess bond polarity is to use Table 8.2, which tabulates a *relative electronegativity* (often abbreviated to *electronegativity*) value for each element. This value is a measure of the tendency for an element to hold electrons. Note the regular trends, both in the rows and in the groups of the periodic table. If the difference in relative electronegativity of two bonded elements is greater than 2.0, the bond is primarily ionic. If the difference is less than 1.5 (but not zero), the bond is primarily polar.

Let it be said once more that we shall not deal much with bonds involving an intermediate difference in electronegativity. However, the most common example of a bond of intermediate difference in electronegativity is that between silicon and oxygen. These elements differ in electronegativity by 1.7 units. The bond may be considered as halfway between a completely nonpolar and a completely ionic bond. Thus, the tetrahedral orthosilicate ion, SiO_4^{4-} (as found in zircon, peridot, and garnet), can be considered either as a silicon ion with a charge of $4+$ surrounded by four oxide ions each of charge $2-$, or as a tetrahedral set of atoms held together by four covalent bonds. Note that when SiO_4^{4-} is viewed as made up of ions each atom has the electron structure of a neon atom.

In either case, it is easy to rationalize the great stability and lack of dissociation of the orthosilicate ion. It may be interpreted as due to the large ionic forces found between a $4+$ ion and four surrounding $2-$ ions. Or it may be interpreted as due to a silicon atom sharing electrons in its available s and p orbitals with four oxygens. Figure 8.13 shows the two extreme presentations. The nonbonding pairs of valence electrons on oxygen, three pairs per atom, are represented by the three peripheral dashes around each oxygen.

Bond Angles and Molecular Shapes

If we view the orthosilicate ion as held together by ionic bonds, it is easy to rationalize its tetrahedral shape. The four negative oxide ions mutually repel each other, but they are strongly attracted to the $4+$ silicon. Thus the oxides take up tetrahedral positions, which puts them as close to the silicon but as far from each other as possible. And it turns out that

Nonpolar

H:H   or   H—H

:N:::N: or   N≡N|

:Cl̈:Cl̈:   or |Cl̄—Cl̄|

Polar

H:Ï: or   H—Ï|

:C:::O: or |C≡O|

:Br̈:Cl̈:   or |Br̄—Cl̄|

Figure 8.12
Some Lewis electron structures. In polar structures, positive ends are to the *left*.

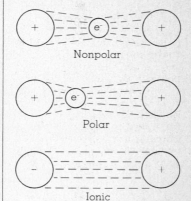

EXERCISE 8.4
Calculate differences in relative electronegativity for several pairs of elements and arrange the bonds you have chosen in order of increasing polarity.

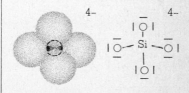

Figure 8.13
Four doubly negative oxygen ions packed around a $4+$ silicon in a tetrahedral orthosilicate structure. (Recall the structure of zircon.)

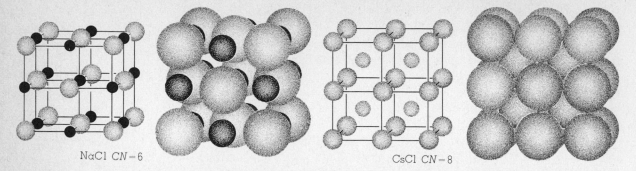

NaCl $CN=6$ CsCl $CN=8$

Figure 8.14
Expanded picture of packing
of smaller Na^+ and larger Cs^+
in chloride structures. The
coordination number, CN, is 6
or 8, depending on the radius
ratio. Ions are in contact in
actual crystals.

EXERCISE 8.5
What ions, in addition to
aluminum, would probably fit
readily into an octahedral hole
in an ionic oxide structure?
(As an example, Ca^{2+} will fit
into such a hole.)

Figure 8.15
All bond angles = 109.5°,
giving a tetrahedral arrangement in silicates.

the hole left between four tetrahedrally packed spheres the size of oxide
ions is just big enough to hold a sphere the size of a silicon ion. The
radius ratio, 0.042/0.132, favors a tetrahedral structure. The silicon is
said to be in a *tetrahedral hole*.

Aluminum ions are bigger than silicon ions and fit better into a
larger hole. In most aluminum ores and minerals, the aluminum ions
are found in holes bordered by six oxide ions. These are *octahedral
holes*. They have eight sides, each an equilateral triangle. They accom-
modate a larger radius ratio than do tetrahedral holes.

In general, in ionic crystals the packing and the bond angles are
determined by the tendency of the large, negative ions to get as close to
the positive ions as possible. The small, positive ions fit into the holes
between the negative ions. The radius ratios help determine which
holes are occupied. The net effect is to bring oppositely charged ions as
close together as possible, and to keep ions of like charge as far apart as
possible. Figure 8.14 shows how this works out in two and three dimensions for some simple ionic substances when the ions vary in radius
ratio. The gemstones in Chapter 7 provide other examples.

The same factors of charge repulsion and attraction can correlate
the shapes of covalent substances. If the orthosilicate ion is held together covalently by sets of electrons shared between the silicon and
the four oxygens, there will be four such sets. The four sets of electrons
will try to get as far apart from one another as possible so they will
assume tetrahedral positions around the silicon. The electrons will still
be between each oxygen and the silicon, and so will be attracted strongly
to the two nuclei at the end of each bond. The tetrahedron will be the
most stable form for the SiO_4^{4-} ion (Figure 8.15).

In diamond there are four sets of electrons (one pair in each of the
four valence orbitals), so they, too, take up tetrahedral positions. In
graphite there are only three near neighbors to each carbon, therefore
only three sets of valence electrons. The electrons get as far from one
another as possible (at the corners of a planar equilateral triangle), and
bond angles of 120° result. The same bond angle will be found in ethene

and benzene, and around the doubly bonded carbons in propene, for the same reason—three sets of bonding electrons will be at a maximum distance from one another if they form bond angles of 120° (Figure 8.16).

Those Repulsive Electrons: The Nyholm–Gillespie Rule

The great majority of chemical compounds follow the simple rule put forward in the preceding section. The rule was developed in some detail by Ronald Nyholm and Ronald Gillespie: Bond angles reflect the mutual repulsion of the sets of electrons around the bonded atoms. To make the best use of the Nyholm–Gillespie rule, we must consider one additional factor. Ammonia, NH_3, will provide the initial example.

Ammonia has a total of eight valence electrons, five from the nitrogen (a member of Group V in the periodic table) and one from each of the three hydrogens. Nitrogen has four valence orbitals, each hydrogen only one. Thus the three bonds from the nitrogen to the three hydrogens will use up the hydrogen orbitals and will also use up three of the four nitrogen orbitals. There is a pair of electrons in the fourth nitrogen orbital, but they are not involved in the bonding of the hydrogens. But the *nonbonding electrons* still repel the three sets of bonding electrons. The experimentally observed bond angle in ammonia is 105°, slightly smaller than the 109.5° found in a tetrahedron (Figure 8.17). This is close to prediction, but the Nyholm–Gillespie theory goes a step further.

Because the fourth pair of electrons is not bonding, this fourth pair repels the others more strongly than the bonding pairs repel each other, forcing the bond angle to be less than the tetrahedral one. In most molecules that have nonbonding pairs of electrons, these pairs repel other electrons more than do bonding pairs. Water is a good example (Figure 8.18). The fact that the "nonbonding dashes" take up more room than the "bonding dashes" may help you remember their effect.

Water has a bond angle of about 105°. The oxygen has two nonbonding pairs of electrons, and we can attribute the bond angle being smaller than tetrahedral to the greater repulsion of these nonbonding pairs for each other and for the bonding pairs. In the molecule CH_3F, where all the valence pairs are involved in bonding, the angles are all about 109° (Figure 8.19). Figure 8.20 gives further examples of molecular shapes, all consistent with the Nyholm–Gillespie rules.

Molecules From Atoms: Molecular Electronic Energy Levels

I hope that you find it reasonable that sets of bonding electrons try to get as far from one another as possible, just as do similarly charged ions.

Figure 8.16
All bond angles = 120°, giving a planar molecule in ethylene, C_2H_4.

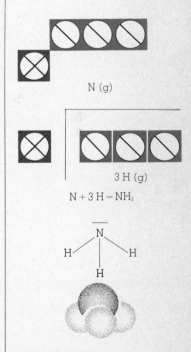

Figure 8.17
Formation of Lewis electron structure and Nyholm–Gillespie predicted shape in ammonia, NH_3, from N + 3H = NH_3.

Figure 8.18
Water, H_2O, forms a flat, bent
molecule with two unshared
pairs of electrons.

Figure 8.19
CH_3F forms a tetrahedral molecule with no unshared pairs
around the central carbon
atom.

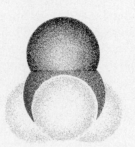

Figure 8.20
Some molecular shapes (with
examples), in terms of distributions of mutually repulsive
sets of electrons. Atoms
(A and X) are indicated by
circles. "Clouds" indicate
unshared pairs of electrons, E;
lines indicate covalent bonds
(single, double, or triple).

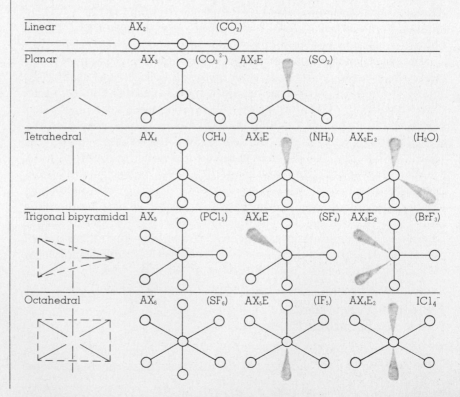

But I also hope that at least one idea introduced in the last two sections made you wonder what was going on. How did the *s* orbital and the three *p* orbitals of carbon suddenly produce four identical tetrahedral bonds, when these orbitals have quite different energies in the gaseous atom? And a very good question that is! But you might even have guessed an answer based on, of all things, the difference in color of Cr(III) in red ruby and green emerald.

The orbital patterns in Figure 5.6 (reproduced here as Figure 8.21), and the portions of the figure we have used from time to time, are for gaseous atoms. There the electrons are attracted to a single, central nucleus. In covalent bonds, every electron is being attracted by at least two nuclei simultaneously. It should not be surprising that this shift from one to two nuclei changes the patterns of the energy levels. It does, just as changing the environment of Cr(III) changes its electron structure and leads to different colors of light being absorbed.

A common effect is that the *s* and three *p* orbitals become identical in energy and allow the formation of four identical bonds, as in methane, CH_4. They may become nearly identical, as in NH_3 and H_2O, when only three or two orbitals of the sp^3 set are engaged in bonding and the other oribitals hold nonbonding electrons. (Note the abbreviation here, sp^3, for a set of *s* and three *p* orbitals all with the same energy level number—for example, the 2*s* and 2*p* orbitals in carbon. Recall that the noble gases all have outer sp^3 sets consistent with their generally low reactivity.)

The technical name for the equalizing of orbital energies in a molecule (as can happen with an atomic sp^3 set) is *hybridization* of the atomic energy levels. Sometimes the electrical forces in molecules are so strong that much more complicated shifts occur in the atomic energy levels, and atomic orbitals become almost useless in interpreting molecular properties. It is necessary then to go to a description in terms of *molecular energy levels* derived from, but having a different distribution and description than, the sets of atomic energy levels outlined in Figure 8.21. Fortunately, the force fields in most of the molecules we shall consider cause relatively small shifts in the atomic energy levels, and the molecules can be adequately represented in terms of Figure 8.21, using the idea of hybridization.

Summary: A Handy Guide to Compound Formation

Let's use the groups in the periodic table to summarize the behavior of elements when they form compounds:

1. Metals tend to lose one, two, three, or four electrons when combining with nonmetals. The number lost never exceeds the group number and (especially with Groups VB, VIB, VIIB, and VIII) may be less.

EXERCISE 8.6
What shapes do you predict for the molecules $SiCl_4$ and SeF_6?

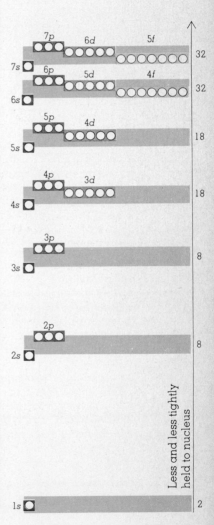

Figure 8.21
Available electron energy levels.

EXERCISE 8.7
What formula do you predict
for the compound of titanium
($Z = 22$) and oxygen? Would
the compound be ionic or
covalent?

HINTS TO EXERCISES
8.1 20^{104}. 8.2 C—C—C—C—C,
$(C-)_2$C—C—C, $(C-)_4$C. Last
is 2,2 dimethylpropane.
8.3 C—C—C—OH,
C—C(—OH)—C. Latter is
2-propanol. 8.4 Largest difference in electronegativity gives
most polar bonds. 8.5 A radius
ratio with oxide (0.132 nm)
between 0.4 and 0.7. 8.6 Tetrahedral, octahedral. 8.7 Ionic
TiO_2.

2. Nonmetals tend to gain one, two, three, or four electrons to form the electron structure of the nearest noble gas element. The number of electrons gained usually equals the quantity obtained by subtracting the group number from 8. Thus sulfur in Group VI would normally gain $8 - 6$, or 2 electrons.

3. Electron gain from metals by monatomic nonmetals leads to the formation of negative ions of charge $1-$, $2-$, or $3-$. Examples are Cl^-, O^{2-}, P^{3-}.

4. Nonmetals often gain electrons from other nonmetals by sharing to form polar covalent bonds. Examples are HF, H_2O, NH_3, CH_4, PCl_3, BrO_3^-, PO_4^{3-}, and hydrocarbons in general.

   There are exceptions and additions to this set of rules, just as there are with all rules, but these four will account for almost all the substances we shall investigate. You should be able to interpret the formulas of all the examples given in terms of Figure 8.21. It is the large number of ways in which atoms can exchange and share electrons that leads to the unlimited number of chemical compounds.

Problems

8.1 Identify or define: binary compound, bond angle, bonds (covalent, double, ionic, nonpolar, polar, single, triple), catenation, hydrocarbons, isomers, Lewis electron structures, Nyholm–Gillespie rules, nonbonding electron pairs, relative electronegativity.

8.2 Write probable formulas for the following stable compounds:
   a. calcium oxide (an ingredient of wall plaster);
   b. sodium bromide (a sedative);
   c. tungsten(VI) oxide (forms when a light bulb cracks);
   d. mercury(II) sulfide (an important ore, cinnabar);
   e. sulfur trioxide (a component of acid smog);
   f. the products that form when hydrogen sulfide reacts with sulfur dioxide at the mouth of a volcano;
   g. the products that form on heating $CaCO_3$ (an important source of building material) to 400°C.

8.3 Arrange the following molecules in order of increasingly polar bonds: N_2, HF, NaCl, CaO, HCl, BN, TiS.

8.4 Butyl mercaptan is the principal chemical in the odor of skunks. *Mercaptan* means an —SH group. Write a possible structural formula for butyl mercaptan. Are there any isomers?

8.5 Both potassium (a very reactive element) and copper (a rather unreactive one) have a single *s* valence electron, actually $4s$ in both cases. How do you account for the great difference in reactivity?

8.6 How many different monochloropropanes (formula C_3H_7Cl) are there?

8.7 Mothballs are dichlorobenzene, $C_6H_4Cl_2$. How many isomers of this structure are there based on a benzene ring? Mothballs have the chlorines as far apart as possible. Which isomer is that? (It is called *paradichlorobenzene*.)

*8.8 Transistors and other solid-state devices are often made by "doping" silicon with gallium or arsenic. The doped atoms fit into the silicon crystal lattice as substitutional atoms. Would an arsenic atom so introduced tend to be positive or negative in the silicon crystal? How about gallium?

8.9   Name, and predict likely shapes for the following: H_2S, CCl_4, HCl, C_2H_3Cl, PH_3, N_2H_4.

*8.10 Discuss the probable reliability in identifying oil spills with the originating source by methods such as that used in Figure 8.2. (Hint: What happens to an oil slick as it contacts the air?)

8.11 Use the radius ratio to account for the different structures of NaCl (c) and CsCl (c) shown in Figure 8.14.

*8.12 Rotation occurs around single bonds, but not around double bonds. How many isomers are there of the following: propane (C_3H_8), propene ($CH_3CH=CH_2$), n-butane ($CH_3CH_2CH_2CH_3$), and 2-butene ($CH_3CH=CHCH_3$)?

*8.13 Figure 8.8 shows a single maximum. I suggested in Chapter 4 that such curves can often be interpreted in terms of two effects—one for the increase, one for the decrease. Suggest two possible such effects here.

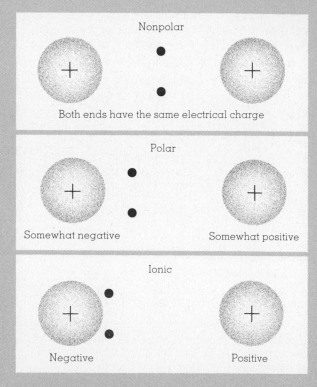

Figure 9.1
Bond polarity varies according
to how electrons are shared.
Nuclei are indicated by ⊕;
electrons by dots.

9 BONDS AND PROPERTIES

Chemical bonds form when one or more electrons are attracted by more than one nucleus. As Figure 9.1 shows, if the electrons are transferred from one atom to another, charged molecules called *ions* are formed. When the two ends of the bond are identical in attraction, the electrons are shared equally and a nonpolar covalent bond results. When the ends of a bond differ (but not enough to transfer electrons completely), the electrons are shared unequally. The bond will be polar, with one end somewhat positive and the other somewhat negative.

These various distributions of electric charge affect the properties of the substance in many ways. This chapter deals with the interactions of charge distributions in molecules and their relationships to molecular and bulk properties. Metals, with their loosely held electrons, were discussed in Chapter 5 and will not be mentioned here.

How Long? (Bond Distances)

You have seen that bond angles in covalent compounds are determined largely by the mutual repulsions of the valence electrons around each atom. Bond angles in ionic compounds are determined by the repulsions between ions of like charge and by the ability of the smaller positive ions to fit into holes between the larger negative ions. In other words, charged particles take up positions that minimize their mutual repulsions and maximize their mutual attractions. At the equilibrium position, the forces of attraction and repulsion just balance.

Bond distance—that is, the distance between two adjacent nuclei—is also determined by electrical forces. Each nucleus repels the other, and each electron in the bond repels the other electrons. But there are counterbalancing forces of attraction between the nuclei and the electrons. These form the bond. (See Figure 9.2.) At equilibrium the two forces just balance.

Remember that every chemical bond is due to the simultaneous attraction of one or more electrons by more than one nucleus. The equilibrium bond distance is determined by a balancing of the electrical forces of attraction and repulsion.

Just as bond angles are delightfully constant for atoms of any one element, so also are bond distances quite constant between the same two kinds of atoms in many different compounds—if the same number of bonding electrons is involved in the bond. Remember that single

Science cannot discover truth, but it is an excellent means of discovering error. The residuum left over after errors are eliminated is usually called scientific truth.—*Kenneth Boulding*.

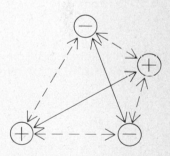

Figure 9.2
A schematic representation of the forces in molecular hydrogen, H_2. Solid lines indicate repulsions, dashed lines, attractions. The positions of the nuclei, $\oplus$, are relatively fixed, but the electrons, $\ominus$, move in an unknown fashion. You might view this representation as a snapshot of one of many possible arrangements.

bonds involve one pair of shared electrons, double bonds two pairs. Triple bonds involve three pairs of bonding electrons shared between two atoms.

Even more convenient for tabulation, the bond distances can be treated as the sum of atomic radii characteristic of each atom as shown in Figure 5.27 and on the inside back cover of this book. A radius of 0.030 nm is given for a covalently bonded hydrogen atom, and a radius of 0.077 nm for a single-bonded carbon atom. Adding these radii, we can predict an internuclear distance of $0.030 + 0.077 = 0.107$ nm in a carbon–hydrogen single bond. Observed carbon–hydrogen distances vary slightly in different compounds, but the most common observed value is 0.110 nm. The agreement between prediction and observation is within 3%, which is typical of results obtained by using the values in the figure. For example, the observed C—Cl distance in carbon tetrachloride, CCl_4, is 0.176 nm; adding the radii from the figure gives a predicted bond distance of $0.077 + 0.099 = 0.176$ nm.

Of course, the values given in the figure were derived from observed bond distances in the first place. The significant point is that the measurements are consistent with the concept of covalent radii; the values given can be used successfully to predict bond distances not used in constructing the figure.

Both covalent radii and ionic radii are tabulated in the figure inside the back cover. We shall find these distances useful in correlating chemical properties. They should also provide you with an experimental basis for confidence in the models of molecular size and shape that are used so frequently to correlate chemical information.

How Strong? (Bond Energies)

Just as bond angles and bond distances describe the geometric arrangement of atoms, so bond energies describe the difficulty of separating atoms. In any chemical reaction, some bonds break and others form. The weakest bonds in the reactants are the ones that usually break, to form the strongest possible bonds in the products. Because bond energies measure the attractive forces between atoms, it should not surprise you that there is a tendency for systems to move in the direction of forming the strongest possible bonds.

However, even the strongest bonds can be broken—by heating, for example. So it is not correct to say that reactions always lead to a strengthening of chemical bonds. You might note that at very high temperatures reactions occur in the direction of forming no bonds at all!

Table 9.1 shows that average values can be assigned for the strengths of bonds between each pair of elements. It is *not* possible to find a "partial strength" for each element, such that the strength of a bond

EXERCISE 9.1
What bond distance do you expect in sodium fluoride, NaF (used in fluoridating water)? In table salt, NaCl? In gaseous Cl_2? Are these substances ionic or covalent?

could be found by adding the partial strengths of the two elements involved. Instead, we must specify a bond strength for each pair of atoms.

If you compute bond distances from covalent radii for each of the bonds listed in Table 9.1, you will find that in most cases longer bonds are weaker. Consider this sequence: H and F (0.094 nm, 565 kJ/mole), H and Cl (0.129 nm, 431 kJ/mole), H and Br (0.144 nm, 364 kJ/mole), H and I (0.163 nm, 297 kJ/mole). Or this sequence: C and C (0.154 nm, 347 kJ/mole), Si and C (0.194 nm, 289 kJ/mole), Si and Si (0.234 nm, 176 kJ/mole).

Table 9.1   Some Average Bond Energies (measured in kilojoules per mole)

I. Single Bonds[a]

	Si	H	C	I	Br	Cl	N	O	F
F	540	565	440	—	260	260	270	180	160
O	370	460	350	—	240	210	220	140	
N	—	390	290	—	—	200	170		
Cl	360	430	330	210	220	240			
Br	290	350	280	180	190				
I	210	300	240	150					
C	290	410	360						
H	290	440							
Si	230								

II. Multiple Bonds

C=C	830	C≡N	890
C=C	610	C=O	710
		N≡N	950

NOTE: Many of the values in this table were originally suggested by Linus Pauling in his pioneering study of the chemical bond: *The Nature of the Chemical Bond*, copyright 1939, 1940, third edition © 1960 by Cornell University. Used by permission of Cornell University Press.
[a]Each bond energy in Part I is that of a single bond between the element listed at the left of the row and the element listed at the top of the column.

The inverse relationship between bond length and bond strength seems reasonable. After all, the stronger the bond, the more the atoms should be pulled together—and therefore the shorter the bond should be. Between any given pair of atoms, a double bond is stronger than a single bond—and a triple bond is stronger than a double bond. Compare the strengths of C—C, C=C, and C≡C, for example. As we would expect, the triple bond is the shortest and the single bond the longest in such a sequence of bonds.

The bond strength is measured by the energy needed to break the bond, so these values are often called *bond energies*. We shall find these bond–energy values useful in interpreting the mechanisms of chemical reactions. It is common to find that a reaction proceeds through a series of steps, in which the weakest bonds of the reacting substances are broken and new bonds established to form products that contain the strongest possible bonds. This sequence of events should strike you as eminently reasonable, because it involves the minimum expenditure of energy.

Water: An Interesting Compound

The most common small molecule on earth is undoubtedly water, H_2O. It is the only compound present at high concentrations as all three phases: crystal, liquid, and gas. It is also essential to all living systems. In most living systems, water constitutes most of the weight and most of the volume and is the most numerous molecule. Its formula, H_2O, is probably the first chemical formula learned by most people. But the formula implies a great deal more information than merely the atomic composition (two atoms of hydrogen with one atom of oxygen) or the weight ratio of oxygen to hydrogen (16 to 2, or 8 to 1).

The formula also tells us that the water molecule has eight valence electrons, because we know that two atoms of hydrogen have one valence electron each and that oxygen has six. The valence electrons are distributed in the $2s$ and $2p$ orbitals of the oxygen as an sp^3 set, and in the $1s$ hydrogen orbitals to give each hydrogen a share in the electron pair. The overall electron distribution can be written H—$\overline{O}$—H. The four pairs of valence electrons take up roughly tetrahedral positions around the oxygen. The two nonbonding pairs of electrons around the oxygen repel the two bonding pairs slightly more than they repel each other. The observed H—O—H angle is 105°.

Hydrogen and oxygen differ considerably in their tendency to hold electrons. This makes the H—O bond a highly *polar* one, consistent with the difference in electronegativity of 1.4. Thus the hydrogen "side" of the molecule is positively charged, the oxygen "side" negatively

POINT TO PONDER
Changes usually occur by paths that accomplish the most effective expenditure of the available energy. By most effective, I mean that the maximum possible change is accomplished with the energy that is available. The more energy that is supplied to a system, the more change that occurs. The normal human desire is to accomplish particular changes, not necessarily maximum ones, and therein lie many difficulties.

EXERCISE 9.2
If the tetrahedral compound $CHFCl_2$ is struck by a very energetic molecule, which bond will probably break?

charged. The molecule as a whole, though electrically neutral overall, is polar. Its electrical charge is not distributed symmetrically over the molecule. So, just as there are polar chemical bonds, there are also polar molecules—and water is a good example (Figure 9.3).

A great deal of information other than the actual bond angle and the actual difference in electronegativity is implicit in chemical formulas and symbols. Similar information can be deduced for molecules other than water from their structural formulas.

The spelling out of some of the meanings of the formula H_2O is not of itself very useful. But the ideas deduced allow a correlation of many of the properties of water in terms of molecular behavior. Water is colorless because its electronic structure has no energy level gaps that correspond to the energy of visible light. But it does absorb high-energy ultraviolet light, which moves the valence electrons into higher levels. It can be ionized by even higher energy light into free electrons, H^+ and OH. You have already seen that the OH can produce highly reactive molecules damaging to living systems.

The water molecule is so polar that it ionizes to a small degree at room temperature into H^+ and OH^- ions.

$$H_2O\ (\ell) = H^+\ (aq) + OH^-\ (aq) \qquad (9.1)$$

Each of these ions, once formed, will be surrounded by polar water molecules, indicated by (aq) for aqueous. This will stabilize the ions and increase the likelihood of their existing in appreciable concentration. The actual concentration of each ion in pure water is small, 10^{-7} moles/liter. Thus, at room temperature only about 10 out of every billion H_2O molecules ionize, but the presence of these ions leads to a small electrical conductivity in pure water.

The polar water molecules can similarly surround other ions, accounting for the high solubility of many ionic substances in water (Figure 9.4). Of course, if the interior forces in a crystal are very high (as they are in gemstones), the stability offered by interaction with polar water molecules is much smaller than that given by the strong ionic interaction in the crystal. Then the ionic substance does not dissolve appreciably.

Other polar molecules will dissolve in water because the polar water molecules can orient around the new molecules to enhance the interaction of unlike charges. In fact, the polar water molecules orient with respect to each other. This gives water a much higher melting and boiling point than other molecules with such small surface area for attractive interactions.

The strong polarity coupled with the bond angle in water leads to the formation of a loosely packed structure in ice. Each water molecule

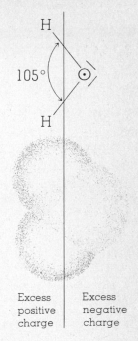

Excess | Excess
positive | negative
charge | charge

Figure 9.3
The water molecule is bent and polar; the oxygen end is negative; the hydrogen ends, positive.

EXERCISE 9.3
Cattle on a cattle drive are often able to detect water supplies before they see them. Do you believe they *smell* water? Discuss in terms of molecular interactions.

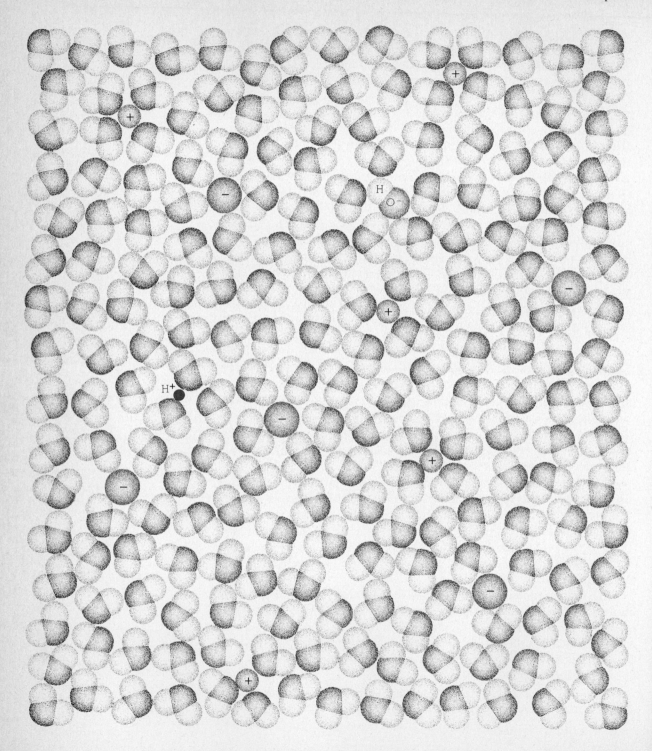

is surrounded by only four other molecules, with the positive hydrogen atoms located between the negative oxygens (Figure 9.5). The resulting open structure (with only four near neighbors) gives a lower density to ice than the somewhat collapsed (and denser) structure of liquid water. So ice floats in liquid water—with very great consequences in nature. Floating ice insulates the water below from cooling by evaporation or contact with cold air. If ice were to sink, bodies of water would tend to be frozen at the bottom, causing greatly different currents than those we see—and very different weather.

Like diamond, but unlike quartz (SiO_2), ice has a four-near-neighbor structure that does not leave any large holes between units (water molecules, in this case). Thus ice, if formed slowly to minimize large-scale inclusions, is always quite pure. One effective way to desalinate seawater is to freeze it slowly.

The tendency of water molecules to interact strongly with one another (plus the open structure of ice) accounts for the maximum density found in water at about 4°C. Below this temperature, the water becomes less dense because of the formation of icelike structures (less dense than the usual liquid). Heating disrupts the orientation of the water molecules so that, at temperatures above 4°C, the water molecules become less tightly packed.

The strong interaction between water molecules (coupled with their ability to ionize) produces an interesting effect during conduction of electricity. Neither the hydrogen ions (H^+) nor the hydroxide ions (OH^-) need to move freely through the solution. An ion merely jumps from one molecule to the next, forming an unstable intermediate (H_3O^+ or $H_3O_2^-$), which quickly releases a similar ion on its other side. That ion leaps to the next site, releasing a third. Thus, the net conductivity is much higher than it would be if each ion actually had to push its way through the liquid, as most ions must.

Figure 9.4 (opposite) Positive ions in water attract the negative oxygens of the surrounding water, and negative ions attract the positive hydrogens. The ions become insulated and separated by the water. Water is sufficiently polar that at any one time about 10 molecules per billion are ionized into H^+ and OH^-. (Adapted from R. E. Dickerson and I. Geis, *Chemistry, Matter, and the Universe*, copyright 1976 by W. A. Benjamin, Menlo Park, Calif.)

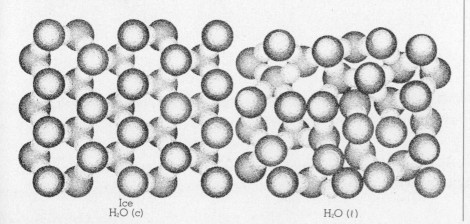

Ice
H_2O (c) H_2O (ℓ)

Figure 9.5
Structures of ice and liquid water. The ice structure collapses on melting, so at 0°C the density of liquid H_2O is greater than that of ice by about 10%.

The most common role of water is as a solvent. It dissolves a great many different substances. Water makes it easy for many substances to ionize, and it allows ready escape of gases and precipitation of crystals. It often provides an easy separation of soluble substances by partial evaporation of the water.

In one common chemical reaction of water, the oxygen retains all eight valence electrons and the hydrogens each gain one electron from other atoms—water is an oxidizing agent. For example, water reacts with most metals to give a positive ion of the metal, either oxide ions or hydroxide ions, and hydrogen gas.

The table of electronegativities (Table 8.2) is useful at this point. Note that, except for fluorine, oxygen is the most electronegative element. This is consistent with what was just said about reactions of water—the electrons tend to stay with the oxygen, freeing hydrogen to accept electrons from other atoms and become hydrogen gas. Only in reaction with elemental fluorine (or in electrolysis with an outside electrical generator) is it common for water to give up electrons and form oxygen gas or hydrogen peroxide (H_2O_2).

Let's look at equations for some typical reactions involving water. First, consider the common case where *water is an oxidizing agent*. Water will react with almost any metal (M) from the left half of the periodic table to form hydrogen gas (H_2), the positive ion of the metal, and hydroxide (or oxide) ions:

EXERCISE 9.4
Why is water such a good
extinguisher of fires? Base your
discussion on the nature of
water molecules. Would water
extinguish an aluminum fire—
say, a burning aircraft?

$$H_2O\,(l) + Na\,(c) = \tfrac{1}{2}H_2\,(g) + Na^+\,(aq) + OH^-\,(aq) \qquad (9.2)$$

$$3\,H_2O\,(l) + 2\,Al\,(c) = 3\,H_2\,(g) + Al_2O_3\,(c) \qquad (9.3)$$

$$2\,H_2O\,(l) + Cr\,(c) = H_2\,(g) + Cr(OH)_2\,(c) \qquad (9.4)$$

In a few reactions, *water is a reducing agent*. Its reaction with fluorine gas is one example of this rare role:

$$H_2O\,(l) + F_2\,(g) = \tfrac{1}{2}O_2\,(g) + 2\,HF\,(aq) \qquad (9.5)$$

When *water is electrolyzed* (by passing electric current through it), the following reaction occurs:

$$H_2O\,(l) = H_2\,(g) + \tfrac{1}{2}O_2\,(g) \qquad (9.6)$$

Under certain special conditions, a different reaction can occur to produce hydrogen peroxide (H_2O_2):

$$2\,H_2O\,(l) = H_2\,(g) + H_2O_2\,(aq) \qquad (9.7)$$

Note the importance of indicating phases in these equations. The symbols (c) for crystal, (l) for liquid, (g) for gas, and (aq) for aqueous help you visualize the actual experiment.

Graphite and Its "Fragments": Delocalized Electrons

Earlier, we looked at the properties of diamond in terms of its bonding patterns. Now let's explore graphite. Table 9.2 compares some properties of the two forms of carbon. The major differences are readily interpretable in terms of the layer structure of graphite, compared to the three-dimensional framework in diamond. Graphite is soft (as in pencil lead) and it is a lubricant, in spite of its very high bond strength, because the tightly bonded atomic layers can slide with respect to one another. It has very flat glide planes, corresponding to the interplanar spaces. The high electrical conductivity parallels these planes; electrical conductivity from one plane to another is poor, however. The bond energy in the planes (477 kJ) is intermediate between the carbon–carbon single bond (347 kJ) and the carbon–carbon double bond (611 kJ). All three of the bonds to every carbon atom in a graphite plane have the same length, energy, and bond angles, though the representation in Figure 8.9 suggests that two are single bonds and one is a double bond. The actual equality is not properly represented.

The hydrogen-surrounded graphite "fragment" benzene (C_6H_6) shows similar properties. The simple bond structure (Figure 9.6a) suggests alternating single and double bonds, but the real structure has all the bonds of the same length (0.1397 nm) and the same energy (506 kJ). All the bond angles are the same (120°), as in Figure 9.6b.

The theoretical picture used to describe electron structure is the same in both cases (benzene and graphite). It applies to many ring molecules (and to some linear ones) that in the simple picture would contain both single and double bonds. It is especially useful if the single and double bonds appear to alternate in a ring structure, as in Figure 9.6a.

In structures like graphite or benzene, in which single and double bonds appear to alternate through the structure, there is a very great tendency for the electrons to become delocalized, much as do those in a

Table 9.2 Properties of Diamond and of Graphite

Property	Diamond	Graphite
Mohs hardness (see p. 111)	10	1
Bond strength (kJ/mole)	347	477
Electrical conductivity	Low	High
Main use	Abrasive	Lubricant
Structural unit	$\overset{\shortmid}{\underset{\shortmid}{C}}$	$-C\!\!\!\diagup^{\diagup}$
Shape of unit	Tetrahedral	Planar

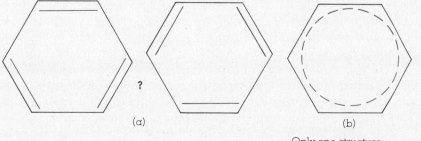

(a)

?

(b)

Only one structure, completely symmetrical, is found

Figure 9.6
(a) Alternating single- and double-bond picture of benzene. But experimentally we do *not* find different kinds of carbon–carbon bonds in benzene. (b) All bond distances, energies, and angles in benzene are the same. The dashed circle represents six delocalized electrons in benzene.

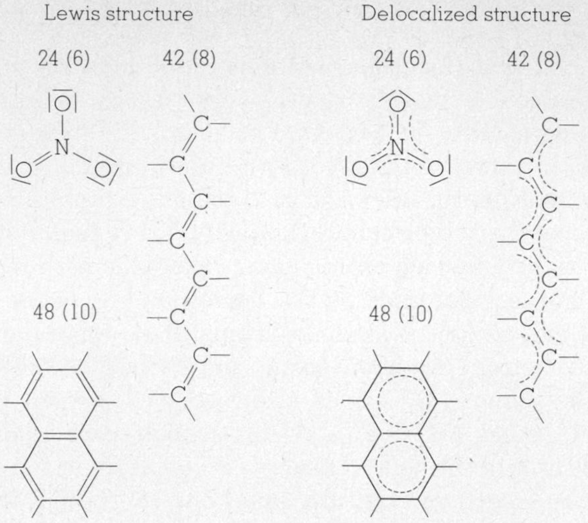

Figure 9.7
Some molecules having delocalized electrons (represented by dashed lines). The numbers refer to the number of localized electrons, and (in parentheses), the number of delocalized electrons. The sum represents the total number of valence electrons.

metal. You can think of this as a situation where there are insufficient electrons to fill the available atomic orbitals completely, so they fill all of them equally but partially. Or you can think of these cases as having such strong nucleus–electron forces contributed simultaneously by so many atoms that the atomic orbitals become greatly changed. This leads to delocalized orbitals characteristic of the molecule. Note that delocalization occurs only when there are many nuclei and many electrons close together—as in metals, graphite, or benzene.

Electrons in delocalized systems, such as benzene, are attracted to several nuclei, not just two. One result is that delocalized structures are more stable than localized structures with the same number of electrons. Figure 9.7 gives some further examples. In each case the fundamental bonding structure is made of single bonds, with the rest of the electrons delocalized. The degree of delocalization depends on the symmetry of the molecule. Benzene is highly symmetrical and has six highly delocalized electrons. The condensed rings (naphthalene) of Figure 9.7 have 10 highly delocalized electrons. Electrons in the long chain will be less delocalized, and this structure is much less stabilized than the ring structures. I shall indicate delocalized electrons by dashed lines (dashed circles in ring structures), but you will have to count valence electrons to see how many are delocalized (total valence electrons minus single bonding and nonbonding electrons).

Molecules that contain delocalized electrons in a ring structure (benzene, or several benzene rings condensed together) are said to be *aromatic*. Aromatic molecules do have odors, but as used here *aromatic*

POINT TO PONDER
The term *resonance* has often been used, instead of *electron delocalization*. Resonance is the physicist's word for the alternating flow of energy between two systems. But molecules like benzene have only *one* state. They do not have several states that alternate, and the term resonance often creates a poor understanding. The field of science, as with all fields, has a large collection of terms based on historical precedent that may have outlived their usefulness. It is interesting to speculate on how many of the terms you learn in this course will be replaced in your lifetime.

means that the rings are less reactive than rings with fewer delocalized electrons (*nonaromatic* rings).

What Makes Molecules Sticky?

Covalent bonds (that is, electron sharing between atoms) can involve highly localized sharing, as in diamond, or delocalized sharing, as in graphite. In each molecule it is, however, possible to count the valence electrons and to assume they are located in that molecule. This is true whether the molecule is electrically neutral (as is C_6H_6) or ionic (as is NO_3^-, pictured in Figure 9.7). Ions are held together by the attractions between oppositely charged ions packed in alternating fashion. What makes electrically neutral molecules stick to one another in the crystals and liquids they form?

The most obvious intermolecular stickiness, or force, traces to the existence of polar covalent bonds. These bonds can cause unsymmetrical distributions of electrical charges in molecules, thus forming polar molecules that can attract one another, much as would ions of less than unit charge (see Figures 9.8 and 9.9).

But even spherical monatomic substances like the noble gases can be liquefied and crystallized, so they must also have intermolecular forces. These forces exist because the electrons in atoms are not stationary. As the electrons fluctuate in position, the charge distribution in the molecule fluctuates. Even neutral, spherical molecules without polar bonds are not continuously symmetrical. The charge at any given spot on the surface is sometimes slightly negative, sometimes slightly positive. These small charges attract similar small, but opposite, charges in neighboring atoms and hold the crystal or liquid together. Such attractions are called *van der Waals forces*. Their magnitude will depend on the ease with which the electron distribution can fluctuate (called its *polarizability*) and the total area of the molecule in contact with neighboring molecules. High polarizability (which goes with low ionization energies) gives large attractions. Even with molecules of low polarizability, such as the hydrocarbons, the intermolecular attractions become large as the surface area of the molecule increases. Table 9.3 gives some data on boiling points to illustrate this. (Remember that boiling involves separating the molecules from one another.)

Sometimes (as in the bottom compound in Table 9.3) the polarity is so high in a molecule that it pays to identify the fact with a special name. This is so whenever a hydrogen atom is bonded to a fluorine, an oxygen, or a nitrogen atom; less so when hydrogens are bonded to atoms of lower electronegativity.

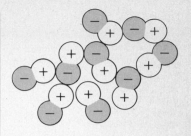

Figure 9.8
Representation of polar molecules orienting to cause maximum intermolecular attraction in a liquid.

EXERCISE 9.5
Classify the bonds in calcium carbonate, $CaCO_3$, as ionic or covalent. How about carbon dioxide, CO_2?

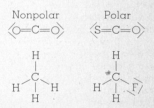

Figure 9.9
Some polar and nonpolar molecules. Negative region is colored. The bonds in O=C=O are polar but the molecule is not since its ends are equally negative; carbon-hydrogen bonds have little polarity.

EXERCISE 9.6
Use Table 9.3 to predict the
boiling point of hexane, C_6H_{14}.

Table 9.3   Boiling Point and (A) Molecular Surface Areas, (B) Molecular Polarity[a]

Molecule	Boiling point (°C)	Relative areas
A. H_3CH	−161	1.0
H_3CCH_3	−81	1.4
$H_3CCH_2CH_3$	−42	1.8
$H_3CCH_2CH_2CH_3$	0	2.2
C_5H_{12} (pentane)	36	2.6
C_5H_{12} (2-methyl butane)	28	2.4
C_5H_{12} (2,2-dimethyl propane)	10	2.3
B. $H_3CCH_2CH_2CH_3$	0	nonpolar
$H_3CCH_2OCH_3$	7	slightly polar, —O—
H_3CCH_2CHO	49	polar, =O
$H_3CCH_2CH_2OH$	118	highly polar, —O—H

[a]See also Figure 14.8.

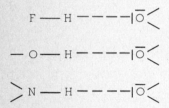

Figure 9.10
Some possible hydrogen-bond
structures. The right-hand
atoms could also be F or N.

A hydrogen atom can hold only two electrons. When hydrogens are bonded to a highly electronegative atom (fluorine, oxygen, nitrogen), the electrons are strongly drawn toward that atom, thus exposing the hydrogen nucleus to the surroundings. Because hydrogen atoms are located on the periphery of molecules, these positively charged hydrogens will be exposed to neighboring molecules. But the neighboring molecules may contain highly electronegative atoms. Their negative charge will interact with the positive charge of the exposed hydrogen. The interaction will be greatest if the hydrogen is on the straight line between the two electronegative nuclei (Figure 9.10). Such a bridging bond by a hydrogen between two highly electronegative atoms is called a *hydrogen bond.* You should now recognize that hydrogen bonds hold ice together in its tetrahedral structure. They are also highly important in holding *you* together.

You learned in Chapter 7 that proteins contain long chains of carbons interspersed with nitrogen and oxygen atoms. The nitrogens and oxygens, in turn, often are covalently bonded to hydrogens. The situation is ideal for hydrogen-bond formation. In fact, hydrogen bonds are a principal force causing protein chains to take the intricately folded structures required for their role in living systems. In the same way, hydrogen bonds are vital in establishing the structure of the genes that carry the messages of heredity from individuals to offspring (and from one cell to the daughter cells into which it splits). Figure 9.11 gives some examples of hydrogen bonds as intermolecular forces.

There is one more type of widely occurring intermolecular force in living and nonliving systems. It is found in genes and helps determine

EXERCISE 9.7
Which would be stronger, an
F--H—F or an O--H—O
hydrogen bond?

Table 9.4   Intermolecular Forces[a]

Name	Forms between	Examples	Usual range of bond strengths (kJ/mole)
Ionic	Electrically charged molecules	Na^+Cl^- $Ba^{2+}O^{2-}$	500–2000
Hydrogen bond	Highly electronegative atoms, and hydrogen atoms covalently bonded to other highly electronegative atoms	Ice, proteins R—O—H--O—R′	2–10
$\pi-\pi$ (pi–pi)	Delocalized electrons	Interplanar forces in graphite	2–6
Dipole–dipole	Polar molecules	H_3CCl, F_2CCl_2,	1–50
van der Waals	Every set of molecules due to fluctuations in instantaneous distributions of electrons	All pairs of molecules	0.1–200

[a]Covalent bonds range between 100 and 1000 kJ/mole. See Table 9.1.

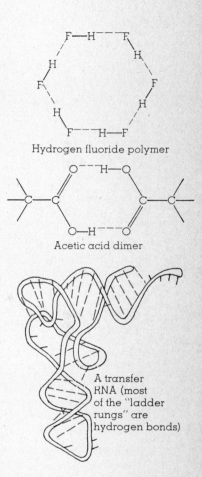

Hydrogen fluoride polymer

Acetic acid dimer

A transfer RNA (most of the "ladder rungs" are hydrogen bonds)

Figure 9.11
Some additional hydrogen-bonded structures.

heredity. You have already met this force in graphite. It is the same as that found between the nonpolar graphite layers, and it is stronger than van der Waals forces. We attribute this force to the interaction of the delocalized electrons in one graphite layer with the delocalized electrons in adjacent layers. Delocalized electrons are often in atomic *p* orbitals (as you can verify in graphite). But they are delocalized *p* orbitals. Thus they are given the special name of *pi* (π) orbitals, and the forces such as those between the graphite planes are called *pi–pi* ($\pi-\pi$) interactions. Table 9.4 summarizes the intermolecular forces in order of decreasing magnitude per molecule (or per bond for $\pi-\pi$ and hydrogen bonds).

The relative weakness of the $\pi-\pi$ bonds (10–30 kJ per six-carbon ring as in graphite) is consistent with the ability of alkaline metal atoms (Li, Na, K, Rb, and Cs) and many other substances to insert themselves between the planes of carbon atoms to form *intercalated* compounds of formula MC_{12} or MC_8. Intercalation is the basis of a most promising new type of electric cell, as you will see in Chapter 17.

Summary

The millions of known chemical compounds exist because the chemical elements can share and exchange electrons in so many ways. Additional millions could, and probably will, be identified and/or synthesized. One thing that contributes significantly to the wealth of compounds is the ability of some elements, especially carbon, to form long chains whose periphery is often covered by hydrogen atoms. Substitution of isoelectronic groups ($-NH_2$, $-OH$, and the members of Group VIIA for $-CH_3$, and of $>NH$ and $>O$ for $>CH_2$) provides many additional compounds. The general structure of many of these compounds may be interpreted as fractions of diamond and/or of graphite to which hydrogen and other atoms have become attached.

Bond angles, energies, and distances may be correlated with the fact that all chemical bonds are due to the simultaneous attraction of one or more electrons by more than one nucleus. For example, short bonds accompany large attractive energies. Most bonds involve two electrons between two nuclei; this we call a *single* bond. If four electrons or six electrons are bound between two nuclei, we call these *double* and *triple* bonds, respectively. Multiple bonds often are better described in terms of delocalized electrons. A very large number of molecules, both neutral and electrically charged (ions), contain just enough electrons so that every atom has an electron structure corresponding to a noble gas.

Intermolecular forces are interpreted in terms of van der Waals, dipolar, pi–pi, hydrogen bond, and ionic forces. All are electrical in origin.

It is possible to correlate many of the properties of a substance with its molecular formula (coupled with information that the formula implies about valence electrons, atomic size, bond polarity, bond distance, and bond energy and their effects on molecular polarity and reactivity).

HINTS TO EXERCISES
9.1 NaF, 0.242 nm; NaCl, 0.277 nm; Cl_2, 0.198 nm. 9.2 C—Cl bond will break. 9.3 Water is always in the nose and is odorless. 9.4 Water is incombustible, it cools, it readily forms a gas excluding air; and it is cheap, nontoxic, low in corrosivity, available. 9.5 $Ca^{2+}-CO_3^{2-}$, ionic; C—O, covalent. 9.6 Boiling point of C_6H_{14} = 71–76°C (pub. = 68°C). 9.7 F--H—F.

Problems

9.1 Identify or define: aromatic (molecules), chemical bond (distances, energies), delocalized electrons, hydrogen bonds, pi–pi ($\pi-\pi$) bonding, polarizability, stickiness, van der Waals forces.

9.2 Show the probable orientation of the water molecules that surround a positive ion, say K^+; a negative ion, say Cl^-.

9.3 Dimethyl ether, $(CH_3)_2O$, and ethanol, C_2H_5OH, are isomers, yet they differ considerably in boiling point. Which has the higher boiling point, and why? Which of the two isomers of C_4H_{10} is higher boiling? Check your predictions with a handbook.

9.4 Comment on the truth or falsity of the following statements:

a. Negative monatomic ions are always larger than the same atoms when neutral.

b. Compounds containing only nonmetals are nonpolar.

c. Molecules of elements cannot be polar. (Hint: O_3, ozone, has a structure like that of SO_2, sulfur dioxide.)

d. The gaseous molecule Na_2 should be stable.

e. A molecule with polar bonds is polar.

f. Ionic forces are the strongest forces known between polyatomic molecules.

*9.5 Draw possible electron distributions (Lewis electron structures) for (a) P_4, a tetrahedral molecule with a P atom at each corner of a regular tetrahedron; (b) CO; (c) S_8, a ring of sulfur atoms, each having two near neighbors; (d) P_2.

(The bond distance in P_2 is 0.189 nm, in P_4 0.221 nm. Are these distances consistent with your suggested structures?)

*9.6 Are there any delocalized electrons in acetic acid, CH_3COOH? How about in CH_3COO^-? What effect would delocalization have on the tendency of CH_3COO^- to add H^+ and form CH_3COOH? (Hint: See Figure 9.11 for some help on the molecular structure.)

*9.7 *Hypo* (photographers' "fixer") is $Na_2S_2O_3 \cdot 5 H_2O$. Suggest a structural formula for the $S_2O_3^{2-}$ ion. (Hint: In what periodic table group is sulfur?) If hypo is acidified, sulfur precipitates, and an odorous gas forms. How does your formula account for these changes?

*9.8 (a) Frying and broiling produce kitchen "smog," as any cook knows. One of the irritants, also found in environmental smog, is acrolein. Its structure is $H_2CCHCHO$. Draw the Lewis electron structure and estimate all the bond angles. (b) Do the same for PAN, peroxyacetyl nitrate, another irritant molecule in smog, formula H_3COONO_2.

9.9 Draw a Lewis electron structure for ozone, O_3. Are the bonds in O_3 polar? Is O_3 a polar molecule?

9.10 Why can table salt, NaCl, and sugar, $C_{12}H_{22}O_{11}$, be easily ground to powders, whereas copper, Cu, and polyethylene, $-(CH_2CH_2)-_n$, cannot.

*9.11 Silica gel (sponge-like SiO_2) is an excellent adsorbent for gaseous water. The world's largest air dryer for the U.S. Air Force's wind tunnels contains 600 tons of gel. Why is silica gel so effective?

*9.12 The plastic polymer Saran, $-(CH_2CCl_2)-_n$, is sticky to itself, to glass, and to many substances. Give a reason for the effect.

9.13 Comment on the problems that may result when living systems contact newly synthesized molecules never found in nature.

*9.14 The statement is made in this chapter that $\pi-\pi$ bonds are stronger than van der Waals bonds. Yet Table 9.4 seems to indicate the opposite. Relate these two apparently disparate claims.

9.15 There are seven crystal forms of ice in addition to the one you know. Suppose the one with a melting point of 20°C were the stable form at atmospheric pressure (instead of, as is true, at 21,000 atm and above). Describe some differences this would cause in the environment.

9.16 Diagram the mechanism by which H^+ (and/or OH^-) conducts electricity through water. (It is called a Grötthus chain reaction.)

9.17 Which of the following would you expect to be crystalline, which liquid, and which gaseous at ordinary room conditions: KI, C_9H_{20}, Xe, HBr, Fe_3C, NH_3? Give reasons, of course.

*9.18 The theoretical strength of a silicate glass fiber is about $2 \times 10^5 \text{ kg/cm}^2$. This is greater than any other common material of construction. How can it be greater than that of iron, for example, which melts at a higher temperature? (Unfortunately only 10% of this potential has been achieved, and then only as a very special case.)

*9.19 Monolithic transistor circuits are completed by evaporating gold film to make the interconnections. Gold does not adhere well to the oxide layer of the base but titanium does, and gold adheres well to titanium (though an intermediate palladium layer is usually used). Is the adherence of titanium and the nonadherence of gold to an oxide consistent with other things you know of the chemistry of gold and of titanium?

9.20 At high pressures (10^5–10^6 atm) most substances become metallic. For example, the interior of the planet Jupiter is thought to be a collection of metallic H (hydrogen) atoms. Why should electrons become delocalized to metallic arrangements at high pressures?

9.21 A Helena Rubinstein ad for "Radiant Action" cream says, "Actually *feel* your dull, tired-looking skin respond in seconds. It's incredible." Are you asked in the second sentence to believe or disbelieve the first sentence?

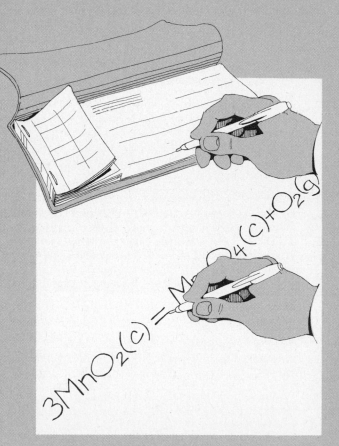

Figure 10.1
Producing a balanced
statement.

10 BALANCING THE BOOKS IN CHEMISTRY

A chemical equation is a convenient, quick, shorthand way of representing chemical changes and of understanding some of the factors that must be controlled to produce desired results. You learned early in this book that atoms and electrical charge, as well as mass and energy, are conserved in chemical reactions. You have seen many chemical equations—all of them consistent with the conservation laws. We also used conservation of mass and electrical charge to write equations for nuclear reactions, even those where atoms are not conserved.

Writing chemical equations is similar to balancing a checkbook—often easier. The methods are the same, but the penalty for a mistake may not be so severe—at least as long as you don't take an inaccurate equation into the laboratory and try to carry it out as a reaction. If you do try that, you may get a very severe penalty indeed!

Inspecting the Balance

After writing a check, you compute a new balance for your checking account. The dollars and cents of the old balance must be equal to the sum of the dollars and cents of the check plus the new balance. In a chemical equation, the sum of atoms and electrical charges in the products must be equal to the sum of atoms and electrical charges in the reactants.

If you enter a wrong number in the checkbook, you may find it impossible to balance the account—or you may be misled about how much money you have left in the bank. In writing a chemical equation, you must get the formulas of the reactants and of the products correct before you can balance the equation. In fact, it is incorrect chemical formulas that are responsible for most incorrect chemical equations. Once you have correct formulas for the substances involved, you write the formulas of the reactants on the left of the arrow and those of the products on the right.

Let's try to describe the combustion of gasoline in your car engine. Gasoline is a mixture of many compounds, but octanes (the eight-carbon hydrocarbons) are the most abundant, so we will consider the burning of octanes, C_8H_{18}. The products of the reaction are carbon dioxide (CO_2) and water (H_2O). One other reactant is necessary for the

There is only one good, namely knowledge. There is only one evil, namely ignorance. —*Diogenes.*

161

combustion; it is oxygen, formula O_2. (Note that oxygen gas has two atoms in each molecule; it is a *diatomic* gas. The proper modern name for O_2 is *dioxygen*. We shall use this name from now on, and shall follow similar modern nomenclature for other elements in their molecular forms.)

Now we write the formulas for the reactants and products:

$$C_8H_{18} + O_2 \rightarrow CO_2 + H_2O \qquad \textit{not an equation}$$

A quick check shows that the preceding listing is *not* balanced. For example, there are eight carbon atoms among the reactants, but only one among the products. How can we obtain a balance?

First, select the most complicated formula and put a 1 in front of it; this establishes a basis for the whole equation. We could put the 1 anywhere, but the job is usually simplest if it is placed before the most complicated formula, here C_8H_{18}. If we have one molecule of C_8H_{18} as a reactant, we have eight carbon atoms that must appear among the products. Only the CO_2 contains carbon, so we must put an 8 in front of the CO_2 to give the proper amount of carbon products. Similarly, the 18 hydrogen atoms from the octane must all appear in water molecules among the products, so we put a 9 in front of H_2O. We now have

EXERCISE 10.1
Gasoline contains both C_8H_{18} and C_7H_{16} molecules. Will C_7H_{16} or C_8H_{18} require more moles of O_2 to burn 1 mole of hydrocarbon?

$$1\,C_8H_{18} + O_2 \rightarrow 8\,CO_2 + 9\,H_2O \qquad \textit{not an equation}$$

We now have eight carbon atoms on each side of the arrow and 18 hydrogen atoms on each side. The check marks indicate these atoms are balanced. But oxygen is not balanced. We find $(8 \times 2) + (9 \times 1)$ or 25 atoms of oxygen among the products. Dioxygen contains two atoms per molecule, so we need $12\frac{1}{2}$ dioxygen molecules among the reactants.

$$1\,C_8H_{18} + 12\tfrac{1}{2}O_2 = 8\,CO_2 + 9\,H_2O \qquad \textit{an equation}$$

Double-checking our work, we find that all the atoms are now balanced. We use the equals sign instead of the arrow to emphasize this fact. We can now write the equation in the usual form (dropping the 1 at the beginning because it is understood):

$$2\,C_8H_{18}\,(\textit{l}) + 12\tfrac{1}{2}O_2\,(g) = 8\,CO_2\,(g) + 9\,H_2O\,(g) \qquad (10.1)$$

If we want only whole-number coefficients, we can multiply the equation by two to get

$$2\,C_8H_{18}\,(\textit{l}) + 25\,O_2\,(g) = 16\,CO_2\,(g) + 18\,H_2O\,(g) \qquad (10.2)$$

All the reactants and products are electrically neutral, so the charges are also balanced. The equals sign is justified because atoms and charges are conserved.

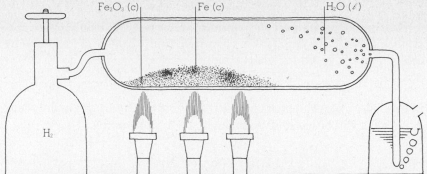

Figure 10.2
Conversion of Fe_2O_3 to Fe
by H_2:

$$3H_2 \text{ (g)} + Fe_2O_3 \text{ (c)} = 2Fe \text{ (c)} + 3H_2O \text{ (}l\text{)}$$

I have described the balancing process in terms of atoms and molecules, but I could just as well have talked about moles of reactants and products. The balanced equation can be read either "in molecules" or "in moles."

Most chemical equations can be written and balanced by simple inspection. The following examples show how to develop equations by this quick inspection method. (I use an arrow to indicate an unbalanced reaction, an equals sign to indicate a balanced equation.)

I. Hot iron(III) oxide reacts with dihydrogen; you see metallic iron form as a product (Figure 10.2).

$$Fe_2O_3 + H_2 \rightarrow Fe \qquad \textit{not an equation}$$

Some other product *must* form to account for the hydrogen and oxygen among the reactants; water is the only likely one. (It escapes as a gas.) Now let's balance the equation. (I use check marks above the symbols to indicate the elements that have been balanced at each step.)

$$1\overset{\checkmark}{Fe_2}O_3 + H_2 \rightarrow 2\overset{\checkmark}{Fe} + H_2O$$

$$1\overset{\checkmark}{Fe_2}\overset{\checkmark}{O_3} + H_2 \rightarrow 2\overset{\checkmark}{Fe} + 3\overset{\checkmark}{H_2}O$$

$$1\overset{\checkmark}{Fe_2}\overset{\checkmark}{O_3} + 3\overset{\checkmark}{H_2} = 2\overset{\checkmark}{Fe} + 3\overset{\checkmark}{H_2}\overset{\checkmark}{O}$$

or, $$Fe_2O_3 \text{ (c)} + 3H_2 \text{ (g)} = 2Fe \text{ (c)} + 3H_2O \text{ (g)} \qquad (10.3)$$

Atoms and charges are balanced, as they must be, in equation 10.3. The reverse of equation 10.3 sometimes occurs in the rusting of iron.

II. An aqueous solution of hydrogen peroxide (H_2O_2) gives off dioxygen when gently heated (Figure 10.3):

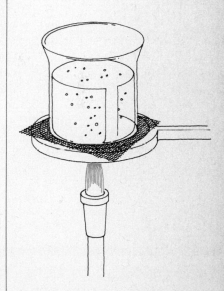

Figure 10.3
Decomposition of H_2O_2:
H_2O_2 (aq) $= H_2O$ (l) $+ \frac{1}{2}O_2$ (g).
This must be done very carefully to prevent an explosion.
What leads to the explosion?

$$H_2O_2 \rightarrow O_2$$

Again, there must be a second product, and water is the only likely possibility. (Hydrogen cannot be separated from oxygen completely by gentle heating; recall the great stability of H_2O.)

$$H_2O_2 \rightarrow O_2 + H_2O$$

In this case, oxygen appears in both products but hydrogen in only one, so our task will be easier if we balance the hydrogen first:

$$1\,H_2O_2 \rightarrow O_2 + 1\,H_2O$$

Now we can balance the oxygen:

$$1\,H_2O_2 = \tfrac{1}{2}O_2 + 1\,H_2O$$

or,

$$H_2O_2\,(aq) = \tfrac{1}{2}O_2\,(g) + H_2O\,(\ell) \tag{10.4}$$

Atoms and charges balance, as they must, in equation 10.4. You can consider this reaction a possible step in the action of hydrogen peroxide as a bleach.

III. Iron(II) ions react with permanganate ions (MnO_4^-) in dilute acid solution (H^+) to form iron(III) ions and manganese(II) ions (Figure 10.4):

$$Fe^{2+} + 1\,MnO_4^- + H^+ \rightarrow Fe^{3+} + 1\,Mn^{2+}$$

The standard fate of oxygen stripped from ions like MnO_4^- is to form H_2O, so we add water to the products and balance the oxygen:

$$Fe^{2+} + 1\,MnO_4^- + H^+ \rightarrow Fe^{3+} + 1\,Mn^{2+} + 4\,H_2O$$

$$Fe^{2+} + 1\,MnO_4^- + 8\,H^+ \rightarrow Fe^{3+} + 1\,Mn^{2+} + 4\,H_2O$$

Noting that there is one iron symbol on each side, we might think that our balancing job is done. But note that the charges are *not* balanced. Among the balanced reactants there is a net charge of $7+$ ($1-$ from the permanganate ion, and $8+$ from the hydrogen ions). Among the balanced products there is a net charge of only $2+$ (from the Mn^{2+} ion). Five positive charges have been lost somewhere. We note that each iron ion changes from $2+$ to $3+$, gaining one positive charge during the reaction. Therefore, we can account for the missing $5+$ by assuming that five iron ions participate:

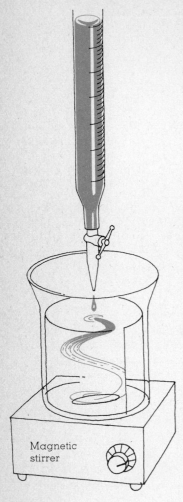

Figure 10.4
Reaction of Fe^{2+} with MnO_4^-.
The process is called a *titration*
if measured quantities are used
in the ratio in which they occur
in equation 10.5; that is,
$5Fe^{2+}$ (aq) to $1MnO_4^+$ (aq).

$$5\overset{\vee}{Fe}^{2+} + 1\,\overset{\vee}{Mn}\overset{\vee}{O_4^-} + 8\,\overset{\vee}{H}^+ = 5\,\overset{\vee}{Fe}^{3+} + 1\,\overset{\vee}{Mn}^{2+} + 4\,\overset{\vee}{H_2}\overset{\vee}{O}$$

$$5\,Fe^{2+}\,(aq) + MnO_4^-\,(aq) + 8\,H^+\,(aq)$$
$$= 5\,Fe^{3+}\,(aq) + Mn^{2+}\,(aq) + 4\,H_2O\,(\ell) \quad (10.5)$$

Verify that the atoms balance (5 Fe, 1 Mn, 4 O, and 8 H on each side), and that the charges balance (a net charge of $17+$ on each side). Note that we only need to *balance* the net charge on the two sides of the equation; this net charge need *not* be zero on each side (any more than your bank account balance must equal zero after each check).

Equation 10.5 gives a very common reaction used in analyzing for Fe(II). It is also the reaction that occurs when permanganate (MnO_4^-) is used as a disinfectant on vegetables and fruits. Permanganate in slightly acid solution is a powerful oxidizing agent—note that the iron has lost electrons (has been oxidized).

A Half-Equation Is Better Than None

Sometimes it is difficult to write an equation by inspection. Many methods are used in such cases, but one that works with minimal effort is the use of *half-equations*. This method involves explicit recognition that molecules contain electrons and that electrons as well as atoms may interchange in a chemical reaction. So you write two half-equations. One shows where the electrons came from, the other shows where they went. The two half-equations are then added in such a way that just as many electrons are given up as are used up, so that net charge is conserved.

Using half-equations is the method of choice only after balancing by inspection has proved difficult. But half-equations are very useful when the reaction gets complicated. Let's try one, even though inspection is usually sufficient.

Fool's gold (FeS_2) is commonly distinguished from gold by the action of nitric acid (HNO_3), which contains H^+ and NO_3^- ions, as mentioned in Chapter 6 (Figure 10.5). Fool's gold dissolves to give Fe(III) and brown nitrogen dioxide gas, NO_2. The sulfur forms sulfate ions, SO_4^{2-}.

$$FeS_2 + H^+ + NO_3^- \rightarrow Fe^{3+} + SO_4^{2-} + NO_2$$

Inspection is a difficult method of getting the equation for this reaction, as you can see by trying it.

However, it is clear that the FeS_2 forms Fe(III) and SO_4^{2-} and that

Figure 10.5
Gold or not gold? The brown gas says, "Not this time."

the NO_3^- forms NO_2. Let's set up separate half-equations for these changes, first the NO_3^- to NO_2 because it involves only two kinds of atoms and seems simpler:

$$1\,NO_3^- \rightarrow 1\,NO_2$$

Another product containing oxygen must be found. We shall assume, since the process is carried out in water, that water is the product, and that the required hydrogens come from the H^+ of nitric acid:

$$2\,H^+ + 1\,NO_3^- \rightarrow 1\,NO_2 + 1\,H_2O$$

The atoms are balanced, but not the charges. They become balanced if we add one electron to the reactants:

$$e^- + 2\,H^+ + NO_3^- = NO_2 + H_2O \qquad \textit{a half-equation} \quad (10.6)$$

This half-equation requires electrons, so the other half-equation must produce some. Let's see:

$$1\,FeS_2 \rightarrow 1\,Fe^{3+} + 2\,SO_4^{2-}$$

There must be some source of oxygen in the reactants. The oxygens can come from the H_2O in the solution, leaving the hydrogen as hydrogen ions.

$$4\,H_2O + 1\,FeS_2 \rightarrow 1\,Fe^{3+} + 2\,SO_4^{2-} + 8\,H^+$$

Atoms are now balanced, but to balance charges we must add seven electrons to the products (just where we thought they would appear):

$$4\,H_2O + FeS_2$$
$$= Fe^{3+} + 2\,SO_4^{2-} + 8\,H^+ + 7\,e^- \quad \textit{a half-equation} \quad (10.7)$$

Half-equation 10.6 requires one electron; half-equation 10.7 produces seven electrons. We now must add the two half-equations together in order to obtain a complete equation that does not include any "extra" electrons on either side. We can do this if we multiply half-equation 10.6 by a factor of 7 before adding it to half-equation 10.7. The electrons cancel when we add the half-equations.

$$\cancel{7e^-} + 14\,H^+ + 7\,NO_3^- = 7\,NO_2 + 7\,H_2O$$
$$4\,H_2O + FeS_2 = Fe^{3+} + 2\,SO_4^{2-} + 8\,H^+ + \cancel{7e^-}$$

$$14\,H^+ + 4\,H_2O + 7\,NO_3^- + FeS_2$$
$$= Fe^{3+} + 2\,SO_4^{2-} + 8\,H^+ + 7\,NO_2 + 7\,H_2O$$

Note that some formulas appear on both sides. Just as in an algebraic equation, we can subtract equal quantities from both sides. Let's eliminate $8\,H^+$ and $4\,H_2O$ from each side to simplify things:

$$6\,H^+ + 7\,NO_3^- + FeS_2 = Fe^{3+} + 2\,SO_4^{2-} + 7\,NO_2 + 3\,H_2O$$

Verify that we have simply dropped out those molecules and ions that appeared both among reactants and products. Then verify that we have obtained a reaction in which both atoms and charges are balanced.

$6\,H^+ \text{ (aq)} + 7\,NO_3^- \text{ (aq)} + FeS_2 \text{ (c)}$

$$= Fe^{3+} \text{ (aq)} + 2\,SO_4^{2-} \text{ (aq)} + 7\,NO_2 \text{ (g)} + 3\,H_2O \text{ (ℓ)} \quad (10.8)$$

We have obtained an equation for the reaction we set out to describe.

Relax! You will seldom find such complicated equations in this book. We went through the process to prove that even the use of half-equations to write equations involves only fairly simple arithmetic, not chemical information. All you have to do is this:

1. Write half-equations and balance atoms.
2. Balance charges, using electrons.
3. Multiply each half-equation by factors such that both half-equations contain the same number of electrons.
4. Add the resulting half-equations.

The result will always be a (balanced) chemical equation.

Perhaps you are convinced both that inspection should be tried first and that the half-equation method can be used to handle the difficult cases.

The Iffyness of Chemical Equations

You have seen that correct equations require correct formulas for reactants and products. You have also seen that balancing an equation involves only arithmetic, not chemistry; atoms and charges must be conserved. But the equation does suggest some chemical information. A chemical equation represents what *would* happen—

1. *if* the reactants are mixed;
2. *if* they actually react to form the listed products;
3. *if* the reaction goes to completion (changing all reactants to products);
4. *if* you wait long enough for the reaction to occur.

A chemical equation *never* tells you—

1. whether the formulas are correct;
2. whether the listed reactants will form only the listed products;

EXERCISE 10.2
Write the $Fe^{2+} + MnO_4^- + H^+$ equation of the preceding section using half-equations. Which method is easier in this case?

3. whether the reaction will go to completion;
4. whether the reaction will occur in a reasonable time.

The matters of correct formulas, other possible reactions between the reactants and products, extent of reaction, and rate of reaction can be determined only from experimental data, not from arithmetical balancing of atoms and charges.

We shall take up the subjects of determining correct formulas and producing a particular set of products in the rest of this chapter. Extent of reaction, discussed in terms of *chemical equilibrium,* will be handled in later chapters, as will the rate of reaction, discussed in terms of *rate and mechanism.*

Limiting Reagents, or the Law of the Minimum

The great Swedish chemist J. J. Berzelius suggested more than 150 years ago that the extent of a reaction might be controlled if one of the reactants was present in a *limiting amount.* For example, if 1.00 g of metallic aluminum is burned in air, the amount of aluminum oxide never exceeds 1.89 g, no matter how much air is present. On the other hand, the amount of aluminum oxide formed decreases regularly as the amount of air falls below the amount needed to supply 0.89 g of dioxygen. Under the latter conditions, dioxygen becomes the *limiting reagent,* whereas under the earlier conditions aluminum is the limiting reagent. Remember, a major way of extinguishing fires is to make air the limiting reagent. The fire goes out if the air is exhausted (Figure 10.6).

The equation for the aluminum reaction is

$$2\,Al\,(c) \quad + \quad 1.5\,O_2\,(g) \quad = \quad Al_2O_3\,(c) \qquad (10.9)$$

$$\underset{2\times 27 = 54\,g}{2\text{ moles}} \quad \underset{1.5\times 32 = 48\,g}{1.5\text{ moles}} \quad \underset{1\times 102 = 102\,g}{1\text{ mole}}$$

The equation shows that if the reaction proceeds completely, and exactly as written, 2 moles of Al (54 g) would combine with 1.5 moles of O_2 (48 g) to produce 1 mole of Al_2O_3 (102 g); 1.00 g of aluminum is 1/54 as much aluminum. It would require 1/54 as much O_2 (48 × 1/54 = 0.89 g), and produce 1/54 as much Al_2O_3 (102 × 1/54 = 1.89 g).

If either the Al or the O_2 is a limiting reagent, the amount of reactants used and of product formed will be determined by its amount. Half as much aluminum would use half as much dioxygen and produce half as much oxide. Similar statements could be made if dioxygen were the limiting reagent.

The formulas of many chemical compounds are analyzed by using the idea of limiting reagents. Not many years ago a new oxide of titanium (Ti) was prepared, starting with 0.382 g of titanium. A series of

Figure 10.6
Magnesium aerial flares. What is the limiting reagent? (Official U.S. Air Force photo)

EXERCISE 10.3
Many flashbulbs burn zirconium wire in oxygen to produce the flash. Which element would you make the limiting reagent? Why?

reactions led to the preparation of 0.574 g of the oxide, and great care was used to make sure all the original titanium was retained in each step. In other words, titanium was the limiting reagent. The other reagents were always in excess. The sample of oxide, then, must contain 0.382 g of titanium and 0.574 g − 0.382 g = 0.192 g of oxygen. Note that this implies an uncertainty of ± 0.001 in each weight, or about ± 0.002 in 0.192. The uncertainty in 0.192 is about 1%.

To compute the formula of the oxide, we must work with moles rather than grams. From the definition of a mole and the list of atomic weights given inside the front cover of this book, we know that 1 mole of titanium atoms weighs 47.90 g and 1 mole of oxygen atoms weighs 15.9994 g. Therefore (to the three significant figures in the data):

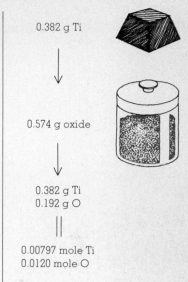

0.382 g Ti

0.574 g oxide

0.382 g Ti
0.192 g O

0.00797 mole Ti
0.0120 mole O

Ratio is 1 mole Ti
to 1.5 mole O
or Ti_2O_3

$$0.382 \text{ g Ti} = \frac{0.382 \text{ g}}{47.9 \text{ g/mole}} = 0.00797 \text{ mole Ti} \qquad (10.10)$$

$$0.192 \text{ g O} = \frac{0.192 \text{ g}}{16.0 \text{ g/mole}} = 0.0120 \text{ mole O} \qquad (10.11)$$

The ratio of O to Ti (in moles) is

$$\frac{0.0120 \text{ mole O}}{0.00797 \text{ mole Ti}} = 1.51 \qquad (10.12)$$

For each 1.00 mole Ti there are 1.51 moles O, so for each 2.00 moles Ti there are 3.02 moles O. Could this actually be a ratio of 2 to 3? A difference of 0.02 in 3.00 is 0.7% — less than our experimental uncertainty of 1%. Therefore, we may tentatively conclude that the formula of the oxide is Ti_2O_3. But we will want to do some more experiments to make sure that the discrepancy is indeed due to uncertainties in the measurements, rather than to a more complicated true formula of the oxide.

Note that keeping titanium as the limiting reagent throughout is central to the reasoning of the analysis. It was the amount of titanium present that determined the amounts of intermediates and the amount of final oxide. Similar techniques are used in most chemical analyses.

Limiting Reactants in Solution

Consider a determination of the amount of Fe(II) in a sample, using acidified permanganate as in equation 10.5, which we repeat here:

$$5 \text{ Fe}^{2+} \text{(aq)} + \text{MnO}_4^- \text{(aq)} + 8 \text{ H}^+ \text{(aq)}$$
$$= 5 \text{ Fe}^{3+} \text{(aq)} + \text{Mn}^{2+} \text{(aq)} + 4 \text{ H}_2\text{O} \text{(}l\text{)} \quad (10.5)$$

You see that 1 mole of permanganate reacts with 5 moles of Fe(II). This reaction goes rapidly and completely in the ratios shown. It has the additional great advantage that permanganate ion is intensely colored, so that any excess of it shows up as a purple color in the system.

A solution containing the unknown amount of Fe(II) is used as the limiting reagent, and a carefully metered solution of permanganate of known concentration is added (Figure 10.4). The flow is stopped when the first permanent purple shows that excess permanganate has been added. The limiting reagent Fe(II) is now known to be exhausted, and its amount is determined from the amount of permanganate added. Let's do it.

A solution of MnO_4^-, known to contain 0.1134 mole MnO_4^- per liter, is put into a buret calibrated in cubic centimeters (cm^3) and is slowly added to an acidified iron solution. The flow is stopped at the first permanent permanganate color (a light purple). The final buret reading is 24.82 cm^3; the original reading, 1.07 cm^3. So $24.82 - 1.07$ or 23.75 cm^3 of MnO_4^- solution was used.

Adding MnO_4^- (aq) to Fe^{2+} (aq)

One liter (1000 cm^3) contains 0.1134 mole of permanganate ions; we used 23.75 milliliters (ml) or 0.02375 liter of solution, so we used

$$0.02375 \text{ liter} \times 0.1134 \text{ (mole } MnO_4^-/\text{liter)}$$
$$= 0.002693 \text{ mole } MnO_4^- \quad (10.13)$$

Each mole of MnO_4^- reacts with 5 moles of Fe(II), so the unknown solution must have contained

$$0.002693 \text{ mole } MnO_4^- \times 5 \text{ moles Fe(II)/mole } MnO_4^-$$
$$= 0.01347 \text{ mole Fe(II)} \quad (10.14)$$

If we need the result in grams, we use the list of atomic weights to find that 1 mole of Fe atoms weighs 55.85 g, so

After one drop excess MnO_4^- (aq)

$$0.01347 \text{ mole} \times 55.85 \text{ g/mole} = 0.7523 \text{ g Fe(II)} \quad (10.15)$$

This is the weight of Fe(II) originally present in the unknown solution.

Suppose that we obtained this Fe(II) from the analysis of 2.1881 g of a pure iron chloride of unknown formula (carrying out the analysis in such a way that the iron was always the limiting reagent). We have just determined that the sample contained 0.7523 g Fe, so we can now compute the amount of chlorine in the sample as

$$2.1881 \text{ g} - 0.7523 \text{ g} = 1.4358 \text{ g chlorine, Cl} \quad (10.16)$$

We can now compute the formula of iron chloride as we did for titanium oxide. (We already know from equation 10.14 that the sample contained 0.01347 mole Fe.)

$$1.4358 \text{ g Cl} = \frac{1.4358 \text{ g Cl}}{35.453 \text{ g/mole}} = 0.04050 \text{ mole Cl} \quad (10.17)$$

$$\frac{0.04050 \text{ mole Cl}}{0.01347 \text{ mole Fe}} = 3.007 \text{ moles Cl/mole Fe} \quad (10.18)$$

This result differs by only 0.2% from a ratio of 3 to 1, which seems reasonably within the uncertainty of our experiment. So we conclude that the formula of the iron chloride is $FeCl_3$.

One Difficulty With Being Analytical

I have no wish to turn you into an analytical chemist. But the next time you read that a can of tuna contains 1.0 part per million (ppm) of mercury, you might think of the problem of keeping such tiny amounts of material constant as the limiting reagent in a complicated analysis. Clearly, any gain or loss of mercury would destroy its role as limiting reagent and lead to erroneous results.

A great tribute is due modern analysts. They work with such care that limiting reagents can sometimes be determined reproducibly at a concentration of 1 part per billion (ppb) or lower. It is these very high accuracies that allow us first to detect, and then to monitor, levels of pollution while they are only alarming—before they become lethal or toxic to the general public. The continually increasing life expectancy suggests that we are doing a better job than ever before of protecting humans from toxic and lethal agents, especially bacteria.

A current challenge to the analyst is to develop methods that will detect the many newly synthesized compounds that might be lethal or toxic. The challenge to the toxicologist is to discover the danger level for each compound. The analyst can then monitor concentrations in appropriate sites. (Remember: *Every* chemical is toxic and lethal to human beings if present at sufficiently high concentrations.)

Environmental Limiting Reagents

The concept of the limiting reagent has another highly useful aspect. The effects of potentially dangerous concentrations of reactants can often be held near zero by limiting the concentration of some other chemical also required to produce the dangerous effect. Phosphate pollution of natural waters provides good examples.

Phosphate is required for the growth of all organisms, but it is essentially never a limiting reagent in the ocean. There is more than enough phosphate almost everywhere in the ocean to ensure good growth. Usable nitrogen (as nitrate) is the most common limiting reagent.

The reverse situation exists in many freshwater lakes. There is plenty of nitrate from decomposing plant growth, but phosphate is low and limiting in concentration.

Thus the use of phosphate detergents (which I shall discuss later) in oceanside cities may cause no problems when the waste is dumped into the sea. However, seriously excessive green algae and other growth may occur in freshwater lakes if the same phosphate detergent waste enters

EXERCISE 10.4
A 200 g sample of tuna fish is found to contain 1 ppm mercury. How many atoms of mercury are in the sample?

EXERCISE 10.5
Suppose you have 1 kg of uranium whose impurities have been reduced to 1 part per 10^5. How many molecules of impurities do you have in your sample if their average molecular weight is 100?

POINT TO PONDER
A dramatic result of the discovery of a limiting reagent occurred in the British navy in the eighteenth century, when limes were added to the shipboard diet. This addition completely did away with scurvy, a degenerative disease that had been incapacitating up to half of some crews on long voyages. We now know even more about that limiting reagent, vitamin C. But we still are not sure what the best dosage level is, even though the minimum dose rate needed to prevent scurvy is well established.

Figure 10.7
Where do all these chemicals
come from?

them. Many lakes in the United States are being rapidly cleared of algae by removing most of the phosphate from municipal wastewaters before they enter the lakes.

Knowledge about the limiting reagent allows appropriate measures to be taken to control undesired changes. It should be clear also that much money and effort could be wasted in controlling an ingredient that is not the limiting reagent. With remarkable regularity, it turns out that money is not the limiting reagent in human affairs—knowledge or wisdom is.

Raw Materials in Chemistry

Nowadays if you want a chemical, you can often go to the store and select the one you wish from thousands of substances. But almost none of these compounds occurs pure in nature. Thus, few of them are *raw materials*. Let's explore where, in nature, materials do come from (Figure 10.7).

You might think that the most abundant raw materials in the world would be agricultural products. But, even if you lump all foods into a single category, they will be outdone on a weight basis both by silicates and by coal and petroleum. Table 10.1 lists approximate figures for world production of the most common raw materials—that is, substances found in nature that are then modified by man for his use. I shall discuss foods later, so they are not included here. The international nature of the supply picture is also indicated in Table 10.1. More than 90% of the supplies of some other, critical materials (columbium, sheet mica, strontium, manganese, cobalt, tantalum, and chromium) are obtained abroad by the United States.

The largest demands on raw materials are for construction materials for shelter and for transportation—silicates and iron ore principally. The next largest demand is for energy producers—mainly coal and petroleum (plus food of course). Much of the coal and petroleum is consumed in the form it is mined with little modification, though this is changing as pollution (especially from the sulfur in coal) becomes a bigger and bigger problem. Almost everything else in Table 10.1 must go through a chemical process before it reaches its end use. The chemical industry is widespread and highly varied. Because it deals with enormous weights of raw materials (much of them often destined for a quick consignment to the waste pile), it is easy to see why the chemical industry encounters so many pollution problems. We will return to these problems many times. For the moment, let's concentrate on some of the processes and products, and on the raw materials required.

I have already discussed the coal—limestone—iron—steel and the aluminum industries. Most of the other raw materials in Table 10.1 lead to

Table 10.1   The Principal Raw Materials of the World[a,b]

Raw material	World production (kg/yr)	Typical products (kg/yr)[c]	Percentage of U.S. supply obtained abroad
Silicates	1.3×10^{13}	Cement, 2×10^{12}; asbestos, 10^{10}; mica, 10^9	4
Coal and petroleum	1.1×10^{13}	Gasoline, 10^{12}; cheapest reducing agent (C)	0,40
Iron ore	1.3×10^{12}	Steel, 2×10^{12}	29
Mg^{2+} (ocean)	2.2×10^{11}	Magnesium, 4×10^9	3
Sodium chloride	2.2×10^{11}	Chlorine products (Cl_2, HCl)	6
Limestone	2.2×10^{11}	Steel-making materials, 2×10^{11}; cheapest base (CaO), 2×10^{10}	0
Phosphate rock [$Ca_3(PO_4)_2$]	1.3×10^{11}	Fertilizer	0
Gypsum ($CaSO_4 \cdot 2H_2O$)	1.1×10^{11}	Construction materials	39
Bauxite (Al_2O_3)	7×10^{10}	Aluminum, 3×10^{10}	85
Nitrogen	4×10^{10}	Fertilizer	0
Sulfur	2×10^{10}	Sulfuric acid (H_2SO_4), 10^{11} (cheapest acid)	0
Potassium salts	2×10^{10}	Fertilizer	49
Uranium ore	1.3×10^{10}	Nuclear fuel (U_3O_8), 10^8	30
Copper ore	1.1×10^{10}	Copper, 2×10^{10}	10
Zinc ore	1.1×10^{10}	Zinc, 2×10^{10}	64
Lead ore	7×10^9	Lead, 10^{10}	4
Ilmenite (TiO_2)	7×10^9	White pigment, 2×10^7	28
Fluorite (CaF_2)	7×10^9	Fluorite for processing steel and uranium; fluorocarbon refrigerants, 4×10^8	82

[a]The U.S. production ranges from 10 to 40% of the totals.
[b]Water (H_2O) and air (N_2 and O_2) are other major raw materials.
[c]Note that the most common categories of products are (1) construction materials (silicates, iron, magnesium, gypsum, metals), (2) energy producers (fossil fuels, uranium), and (3) fertilizers (phosphate, nitrogen, potassium); all these are indispensable to mankind at the present population level.

familiar end products—mainly the common metals of commerce, reduced from their ores in processes similar to those discussed for iron and aluminum.

We shall now briefly look at two processes that do not give metals: the conversion of atmospheric nitrogen to fertilizers, and the conversion of sulfur to sulfuric acid. In the process I shall mention something about the conversion of petroleum, coal, and natural gas to pure chemical compounds.

Needed: 100,000,000,000 Kilograms of Protein per Year; Catalyze!

One of the most common kinds of starvation is *kwashiorkor,* or protein deficiency. In mild forms, it disables millions of humans annually. In

POINT TO PONDER
Humans eat about 1 kg of solid food per day, or about 400 kg per year (a liberal estimate). Thus, the almost 4×10^9 people eat about 1×10^{12} kg annually. Note that the sum of the fertilizer items in Table 10.1 comes to about 10^{11} kg, or one-twentieth the weight of the food produced. This gives some idea of how dependent agriculture is on chemical fertilizers. Of course, there are many other chemical inputs to food production: insecticides, herbicides, fuel, vehicles. Without these inputs it would be quite impossible to feed the present human population.

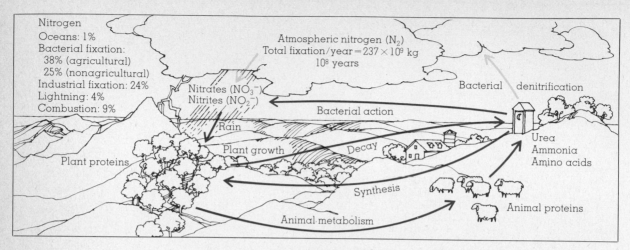

Nitrogen
Oceans: 1%
Bacterial fixation:
 38% (agricultural)
 25% (nonagricultural)
Industrial fixation: 24%
Lightning: 4%
Combustion: 9%

Atmospheric nitrogen (N$_2$)
Total fixation/year = 237×10^9 kg
10^8 years

Bacterial   denitrification

Nitrates (NO$_3^-$)
Nitrites (NO$_2^-$)

Bacterial action

Rain

Urea
Ammonia
Amino acids

Plant growth

Decay

Plant proteins

Synthesis

Animal proteins

Animal metabolism

Figure 10.8
The nitrogen cycle.

severe forms, it kills hundreds of thousands. Kwashiorkor is due to inadequate protein diet, and the limiting reagent from a chemical point of view is usually nitrogen, present in protein to about 16%. Why don't chemists solve this problem?

There is plenty of nitrogen in the atmosphere. About 80% of the molecules you breathe are nitrogen. But most animals and plants are totally incapable of converting gaseous nitrogen into protein. Instead, their proteins can decompose into gaseous nitrogen. For example, some intestinal gas is nitrogen, consistent with the effect on your intestinal tract of a meal of high-protein beans.

The main natural agents that convert atmospheric nitrogen into forms usable for protein synthesis are combustion (8%), lightning (4%), and bacteria and algae that live in the soil and sea (68%). Lightning or combustion can provide the energy required for the reaction. Energy is sometimes explicitly listed in a chemical equation:

EXERCISE 10.6
Obtain equation 10.20 by the method of half-equations. (Hint: NO clearly forms NO$_3^-$. So, let O$_2$ form H$_2$O in the other half-equation.)

$$N_2 \,(g) + O_2 \,(g) + 181 \text{ kJ} = 2 \, NO \,(g) \qquad (10.19)$$

The NO then reacts with O$_2$ and H$_2$O to give HNO$_3$ (ionized in H$_2$O), which plants can convert to proteins:

$$4 \, NO \,(g) + 3 \, O_2 \,(g) + 2 \, H_2O \,(l) = 4 \, H^+ \,(aq) + 4 \, NO_3^- \,(aq) \qquad (10.20)$$

The nitrogen cycle in nature is very slow (Figure 10.8). Man has appreciably speeded this cycle in agriculture by efficient recycling of plant and animal waste. But for the last 60 years, only the addition of synthesized nitrogen fertilizers has allowed world agriculture to feed the expanding population (which quadrupled in this interval). About 20% of all nitrogen fixation is now done by the chemical industry.

One of the earliest methods of converting nitrogen to fertilizer was similar to the action of lightning. But the costs of maintaining a high-intensity electric arc are now prohibitive in most parts of the world. The most-used modern process of nitrogen fixation converts the nitrogen into ammonia (NH_3) by direct reaction with hydrogen gas. This process is a good example of a chemical reaction that does not "go to completion." No matter how long the nitrogen and hydrogen are kept together, conversion to NH_3 is incomplete.

The extent of reaction is influenced by the temperature (less ammonia at high temperatures) and by the pressure (more ammonia at high pressures). Figure 10.9 shows the variation of completion with variation in temperature and pressure. We shall look into the molecular interpretation of the effects of pressure and temperature in Chapter 13.

There is still one more problem: The reaction is too slow under any of these conditions for commercial feasibility. As you might guess from your everyday experiences, the reaction rate increases if the temperature does. But such an increase also decreases the potential yield of NH_3. What to do?

The problem was initially solved by a self-educated physical chemist, Fritz Haber, in Germany about 1910. He discovered that the presence of certain substances greatly speeded up the reaction, but they were not themselves used up in the process. Similar substances had been discovered before. Berzelius (early in the nineteenth century) had named them *catalysts*. Catalysts are substances that increase the rate of a chemical reaction but are not themselves used up. A "good" catalyst not only speeds up reaction of the reactants but also leads to production of the *desired* products—not just any old products.

I shall discuss the nature of catalytic action in Chapter 14. Suffice it to say here that the great majority of industrial and laboratory chemical reactions, as well as those occurring in living systems, owe their success (that is, rapid yet controllable rates) to the discovery of suitable catalysts. (Our language recognizes a similar feature in human affairs when we say that an individual catalyzed a group into action.)

The chemical equation for the reaction of dihydrogen and dinitrogen to give ammonia could hardly be simpler:

$$N_2\ (g) + 3\,H_2\ (g) = 2\,NH_3\ (g) + 92\ kJ \qquad (10.21)$$

I indicate here the heat generated in the reaction as 92 kJ. Heat changes accompany most chemical reactions and, as in this case, often have important effects. Here, for example, the heat must be removed to prevent a rise in temperature, which would decrease the yield of ammonia. Figure 10.10 outlines schematically the flowchart for a plant manufacturing synthetic ammonia.

Even at very high pressures, and with catalysts and temperatures high enough to give a good rate, the initial yield of NH_3 is only 20–25%. Internal recycling is used to increase the conversion. In other words,

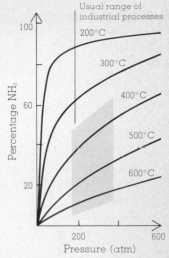

Figure 10.9
The effect of varying pressure and temperature on the percent conversion in the reaction

$$N_2\ (g) + 3H_2\ (g) \\ = 2\,NH_3\ (g) + 92\ kJ$$

(From J. T. Gallagher and F. M. Taylor, *Education in Chemistry*, **4**, 1976, p. 30)

EXERCISE 10.7
If 92 kJ is generated per mole of N_2 consumed in equation 10.21, how much heat would be generated by the reaction of 3×10^7 kg of N_2 (the amount of N_2 synthesized into NH_3 annually)?

Figure 10.10
Flowchart for the industrial
production of ammonia.

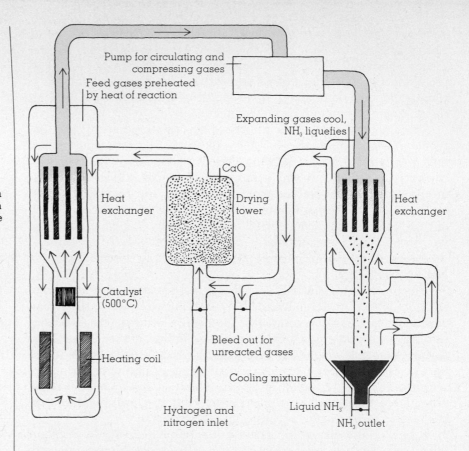

Pump for circulating and
compressing gases

Feed gases preheated
by heat of reaction

Expanding gases cool,
NH₃ liquefies

Heat
exchanger

CaO

Drying
tower

Heat
exchanger

Catalyst
(500°C)

Heating coil

Bleed out for
unreacted gases

Cooling mixture

Liquid NH₃

NH₃ outlet

Hydrogen and
nitrogen inlet

POINT TO PONDER

Haber was working on ammonia synthesis primarily as a problem in chemical synthesis—though he knew, of course, of the many uses of ammonia. One of the main reasons he persisted in the light of repeated failures was the acidulous criticism of Walther Nernst, one of the famous chemists of the day. Both Haber (in 1918) and Nernst (in 1920) won Nobel Prizes. Without Haber's work, world population would probably have leveled off by now.

But an interesting note is that the first big use of "his" ammonia was to provide high explosives to Germany during the 1914–1918 war, when Germany was cut off from overseas supplies. The war would probably have ended much sooner if Germany had not developed the Haber process.

Again we find a single chemical (here NH_3) essential for both war and peace. Should a scientist control the use of his discoveries, and if so, how?

the readily liquefied ammonia is removed from the product gases and the unreacted hydrogen and nitrogen are run around the cycle again and again, with fresh nitrogen and hydrogen added. (Some of the exit gas must be bled off continuously. Otherwise the argon and other noble gases in the reactants would build up to large concentrations in the plant and would interfere seriously with the reaction.)

Catalyze, Catalyze, Catalyze— Then Catalyze Some More

It is easy enough to see where the nitrogen comes from—the atmosphere. But where does one get hydrogen, and what happens to the oxygen mixed with the original nitrogen? Many methods have been used, but I shall discuss only one, which involves using natural gas (primarily methane, CH_4) as a single solution to both problems.

First, excess methane and steam are passed over a catalyst to give a mixture of H_2, CO_2, and CO called *synthesis gas*. This gas can be used for many further syntheses.

$$\text{catalyst I}$$
$$CH_4 \,(g) + H_2O \,(g) \rightarrow H_2 \,(g) + CO_2 \,(g) + CO \,(g) \qquad (10.22)$$

For ammonia synthesis, air and more methane are mixed with the synthesis gas and passed over a second catalyst to give the further reaction

$$\text{catalyst II}$$
$$4\,N_2 \,(g) + CH_4 \,(g) + O_2 \,(g) = 2\,H_2 \,(g) + CO_2 \,(g) + 4\,N_2 \,(g) \qquad (10.23)$$

The gases resulting from this reaction are mixed with steam (gaseous water), and passed over a third catalyst to convert most of the CO to CO_2:

$$\text{catalyst III}$$
$$CO \,(g) + H_2O \,(g) = H_2(g) + CO_2 \,(g) \qquad (10.24)$$

The carbon dioxide is then removed, leaving mainly dihydrogen and dinitrogen.

There will still be about 1% carbon monoxide in the gas. Carbon monoxide is isoelectronic with N_2 and it "poisons" the "ammonia" catalyst (reduces its reactivity). See Figure 10.11. So the gas is passed over catalyst IV under conditions where the H_2 present converts any CO and residual O_2 or CO_2 to CH_4 and H_2O. This gas is then dried and sent on to NH_3 synthesis, which requires catalyst V. Note that each catalyst and set of reaction conditions are chosen to produce a particular set of desired products.

The relative amounts of air, steam, and natural gas are adjusted so that the overall reaction is approximately

$$7\,CH_4 \,(g) + 10\,H_2O \,(g) + 2\,O_2 \,(g) + 8\,N_2 \,(g)$$
$$= 7\,CO_2 \,(g) + 24\,H_2 \,(g) + 8\,N_2 \,(g) \qquad (10.25)$$

Note that if this were the exact reaction it would produce 24 moles of H_2 mixed with 8 moles of N_2. This is the 3 to 1 ratio that equation 10.21 requires for the synthesis of ammonia. Actually, the mixture is made richer in N_2 by admitting slightly more air and less steam. This minimizes the overall cost because nitrogen is cheaper than hydrogen. Making hydrogen the limiting reagent ensures that it will be used up as completely as possible, rather than being bled back to the atmosphere with the noble gases and the surplus nitrogen.

From a Few Raw Materials—Many Products

The commercial synthesis of ammonia, so important to the health of humans, is an excellent example of the folly of using natural gas and

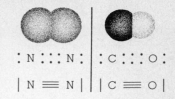

Figure 10.11
Isoelectronic N_2 and CO.

EXERCISE 10.8
Show that a 5 to 1 molar ratio of steam to oxygen corresponds closely to a 1 to 1 molar ratio of steam to air for the reactions summed in equation 10.25.

petroleum for fuels. These raw materials are potentially of far greater worth as sources of industrial chemicals. In fact, petroleum and natural gas are already the raw materials from which stem the great majority (probably over 90%) of all the chemicals currently synthesized.

Societies would be well advised to conserve their oil and gas for chemical synthesis by developing nuclear and solar energy as rapidly as possible. Of course, given enough energy, one can synthesize any chemical from any source of its atoms. If oil and gas become depleted, it will be possible to use agricultural products (or even water and carbon dioxide) as raw materials for the wealth of carbon compounds we use and need. But the syntheses from oil and gas are more direct, cheaper, and less polluting with one of the most threatening pollutants of all— heat. But more on this later.

Ammonia is a gas at room temperature and pressure. Its boiling point is $-33°C$. Liquid ammonia is sometimes injected directly into fields as a fertilizer (Figure 10.12), or it can be converted to solid fertilizers for ease of shipping and handling. A common method is to oxidize the ammonia with air. A 90/10 (by weight) platinum/rhodium catalyst is used (Figure 10.13), and the final product is nitric acid (HNO_3), a strong and highly corrosive acid much used by chemists. The corrosiveness of the nitric acid can be neutralized by reaction with ammonia to form ammonium nitrate, a good fertilizer. (Again the problem of usage comes up. Ammonium nitrate is not only an excellent fertilizer; it is also a very powerful explosive for either war or peace. About 75% of manufactured NH_3 goes into fertilizers; 10% goes into explosives.)

The overall equations for these synthetic steps are

$$2\,NH_3\,(g) + \tfrac{7}{2}O_2\,(g) = 2\,NO_2\,(g) + 3\,H_2O\,(g) \qquad (10.26)$$

$$2\,NO_2\,(g) + \tfrac{1}{2}O_2\,(g) + H_2O\,(\ell) = 2\,HNO_3\,(aq) \qquad (10.27)$$

$$HNO_3\,(aq) + NH_3\,(g) = NH_4NO_3\,(c) \qquad (10.28)$$

Figure 10.12
Fertilizing with liquid ammonia.
(Courtesy the McGregor
Company)

Still another common nitrogen fertilizer is urea, H_2NCONH_2, which can be synthesized cheaply by catalysts. The overall reaction is

$$2\,NH_3\,(g) + CO_2\,(g) = H_2N\overset{\overset{\displaystyle O}{\|}}{C}NH_2\,(c) + H_2O\,(l) \qquad (10.29)$$

As long as cheap energy and a cheap source of hydrogen are available, kwashiorkor and the other protein deficiencies cannot be blamed on nitrogen being the limiting reagent worldwide. Figure 10.14 shows the recent history and a projection of world fertilizer production.

Note that at least one of the reactants has been a gas in every synthetic step for converting atmospheric nitrogen to other compounds. This should remind you of the discussion of the reduction of iron in the blast furnace. There are many advantages to reactions in which at least some reactants and products are gases.

Figure 10.13
A catalyst for the oxidation of ammonia.

Fire and Brimstone

The brimstone of the ancients is the sulfur of today, and burning sulfur smelled just as bad to them as it does to us. But we know (and they did not) that the sulfur dioxide (SO_2) that is produced assails more than our nostrils. It corrodes our lungs, has similar bad effects on plants, and is a great enhancer of smog formation. In fact, very great efforts are expended in many cities (and in many industries, even in remote places) to minimize discharge of sulfur dioxide into the atmosphere. Yet sulfur dioxide is intentionally produced in amounts greater than 10 billion kg per year. Most of this sulfur dioxide goes into sulfuric acid, one of the most widely used of all chemicals.

As you saw in Figure 6.5, sulfur is found free in nature. Its occurrence around volcanoes supported the idea that the hell beneath us must have fires of sulfur. Much larger deposits are found in coastal Louisiana and Texas in underground salt domes. The sulfur is mined by pumping down hot water, which melts the sulfur, brings it to the surface, and then separates from it, leaving very pure solid sulfur upon cooling.

Conversion to sulfuric acid by way of a four-step process is straightforward, thanks once more to catalysts—mainly vanadium oxide—which speed the conversion of SO_2 to SO_3. (See Figure 10.15.)

$$S\,(c) + O_2\,(g) = SO_2\,(g) \qquad (10.30)$$

$$SO_2\,(g) + \tfrac{1}{2}O_2\,(g) \overset{\text{catalyst}}{=} SO_3\,(g) \qquad (10.31)$$

$$SO_3\,(g) + H_2SO_4\,(l) = H_2SO_4 \cdot SO_3\,(\text{oleum}) \qquad (10.32)$$

$$H_2SO_4 \cdot SO_3\,(\text{oleum}) + H_2O\,(l) = 2\,H_2SO_4\,(l) \qquad (10.33)$$

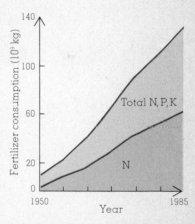

Figure 10.14
World nitrogen (N) and total NPK fertilizer consumption, 1950–1985 (projected). It is common to mix sources of phosphorus (P) and potassium (K) with fertilizer containing nitrogen. These are called NPK fertilizers. (From *A Hungry World: The Challenge to Agriculture*, University of California Task Force, 1974)

Figure 10.15
A sulfuric acid plant.

The net equation is

$$S(c) + 1.5\,O_2(g) + H_2O(\ell) = H_2SO_4(\ell) \qquad (10.34)$$

Pure sulfuric acid (H_2SO_4), like pure nitric acid (HNO_3), is a mainly covalent molecule. Each contains at least one OH group, so that the structural formulas may be written

$$\begin{array}{c} H \quad |\overline{O}| \\ |\overline{O}-S-\overline{O}| \\ |\underline{O}| \quad H \end{array} \qquad \text{and} \qquad \begin{array}{c} |\overline{O} \quad |\overline{O}-H \\ \diagdown N \diagup \\ |\underline{O}| \end{array}$$

or $(HO)_2SO_2$ and $HONO_2$

Note the delocalized electrons indicated in nitric acid.

   Just as water ionizes slightly, so do these acids. In fact, when dissolved in lots of water, HNO_3 and H_2SO_4 ionize almost completely to give hydrated ions. We shall soon look into this more fully. I mention it here because sulfuric acid is the cheapest highly ionized acid, and it is these hydrogen ions that account for many of its uses. Nitric acid is

EXERCISE 10.9
How many kilograms of sulfur are required for the U.S. annual production of 3×10^{10} kg of sulfuric acid? This sulfuric acid contains 90% (by weight) H_2SO_4 and 10% H_2O. Compare with Table 10.1.

much more expensive. It is used when the nitrate as well as the hydrogen ions can react. Nitric acid is seldom used for its acidity alone, as sulfuric acid is.

$$HNO_3 \text{ (aq)} = H^+ \text{ (aq)} + NO_3^- \text{ (aq)} \qquad (10.35)$$

$$H_2SO_4 \text{ (aq)} = 2H^+ \text{ (aq)} + SO_4^{2-} \text{ (aq)} \qquad (10.36)$$

In the last 10 years, the primacy of the underground sulfur deposits as a source of sulfur has been threatened, and all because of a pollution problem. Increasing demands for more and more energy have led to the use of coals and oils with a high sulfur content. Early users burned them anyway, in spite of the undesirable by-products formed. But then anti-pollution laws appeared, and a pleasant discovery was made. The sulfur could be extracted from the oil before the oil was burned, and often could be sold at a profit—if not at a financial profit, at least at a profit in terms of air pollution and its effects on the neighborhood. As a result, almost half the sulfur now used to make sulfuric acid is by-product sulfur. By-product sulfur is that which has been reclaimed from oil or coal or mineral ores.

Summary

Chemical equations and chemical formulas are shorthand devices for expressing the compositions and reacting ratios of chemical compounds. They are based on the laws of conservation of atoms, charge, and mass. The idea of the limiting reagent allows formulas and equations to be used to correlate the results of chemical analyses and to direct chemical syntheses.

All chemical formulas contain uncertainties. Few substances have compositions identical with their simple chemical formulas. And few chemical reactions proceed exactly as represented by the simple chemical equations. The reaction may not proceed "to completion." It may produce other products in addition to those listed, and/or it may proceed so slowly as to be of negligible interest (unless a catalyst can be found). All these questions should be faced and answered when one attempts to reduce to practice the predictions implicit in the equations. But the equations provide valuable guidance and limits to the questions that must be answered and are central to all work in chemistry.

Every one of the millions of processed substances you see traces back to a few raw materials. There are probably fewer than 1000 common ones, with fewer than 100 providing the great bulk of products you use. Silicates and iron ore provide most of the structures, petroleum and coal most of the energy. But petroleum and coal are also the richest known sources of chemical raw materials. It behooves us to use them more and more to balance the books in chemistry, and to find other sources of energy.

EXERCISE 10.10
Heat is always produced when chemical bonds are formed. Why? The reaction that occurs when sulfuric acid is poured into water produces a great deal of heat. What bonds are forming?

POINT TO PONDER
Balancing in chemistry is not limited to equations. All changes conserve mass. Furthermore, all materials eventually become waste. Thus the societies that have the largest gross national product (in terms of matter processed) are the societies that have the greatest gross national waste.

HINTS TO EXERCISES
10.1 C_8H_{18}. 10.2 $5[Fe \text{ (c)} = Fe^{2+}(aq) + 2e^-]$, plus $2[5e^- + 8H^+ \text{ (aq)} + MnO_4^- \text{ (aq)} = Mn^{2+} \text{ (aq)} + 4H_2O \text{ (l)}]$ gives net equation. 10.3 O_2 (g). 10.4 6×10^{17} atoms. 10.5 10^{20} molecules. 10.6 $4[NO \text{ (g)} + 2H_2O \text{ (l)} = NO_3^- \text{ (aq)} + 4H^+ \text{ (aq)} + 3e^-]$, plus $3[4e^- + O_2 \text{ (g)} + 4H^+ \text{ (aq)} = 2H_2O \text{ (l)}]$. 10.7 10^{11} kg. 10.8 To get 1 mole of O_2 requires 5 moles of air. 10.9 9×10^9 kg S. 10.10 Hydrogen bonds to the water.

Problems

10.1 Identify or define: catalyst, chemical equation (writing by inspection and by half-equations), chemical formulas, half-equations, limiting reagent, raw material.

10.2 An early Flemish experimenter, Jan Baptista van Helmont, planted a willow tree in a pot of earth. He then kept it watered. Five years later the tree had gained 75 kg, the soil had lost 0.057 kg. Evaluate his conclusion that the gain in weight, almost 75 kg, must have come from the water (the only thing he added).

10.3 Nicotine has an atomic carbon–hydrogen–nitrogen ratio of 5 to 7 to 1. Its molecular weight is 162. What is its molecular formula?

10.4 Two common ores are CuS and $CuFeS_2$. Which ore has the higher weight of sulfur per ton of ore? Do the problem in your head without written calculations.

10.5 Much of the putrid odor of human waste is due to skatole: 7% H, 11% N, with the rest carbon. The molecular weight is 131.2. What is the molecular formula? Many of the odorous excretions of animals are nitrogen compounds. (Note the common stem with *scatological,* also derived from the Greek word for excrement.) [C_9H_9N]

10.6 Passage through a bed of activated (finely ground) charcoal can be used to remove chlorine from water. The net equation is

$$2\,H_2O\,(aq) + C\,(c) + 2\,Cl_2\,(aq)$$
$$= CO_2\,(aq) + 4\,H^+\,(aq) + 4\,Cl^-\,(aq)$$

Home dechlorinators contain about 100 g of carbon, as charcoal. Estimate the time between rechargings if the water contains 1 ppm Cl_2. [$\sim$30 years]

10.7 Write chemical equations for—

a. The dissolving of shells in acid,

$$CaCO_3 + H^+ \rightarrow Ca^{2+} + CO_2 + H_2O$$

b. The fermentation of glucose to alcohol in wine making,

$$C_6H_{12}O_6 \rightarrow C_2H_5OH + CO_2$$

c. The precipitation of "bathtub ring,"

$$Ca^{2+}\,(aq) + C_{17}H_{35}CO_2^-\,(aq)$$
$$\rightarrow Ca(C_{17}H_{35}CO_2)_2\,(c)$$

d. The rusting of iron in water and air,

$$Fe + H_2O + O_2 \rightarrow Fe(OH)_3\,(s)$$

e. The combustion of natural gas (CH_4) in your stove.

f. A rocket motor burning hydrazine (H_2NNH_2) in fuming nitric acid (HNO_3) to produce N_2 and H_2O. (In what molecular ratios should the fuels be mixed?)

g. The burning of charcoal in your backyard grill. (Is only one product actually formed?)

h. A spacecraft rebreather in which KO_2 reacts with CO_2 to give K_2CO_3 plus "fresh air."

*10.8 The compound $Pb(C_2H_5)_4$ precipitates lead, Pb, in a car engine unless $C_2H_4Br_2$ is added to form PbClBr. (The Cl comes from other substances.) What should be the molar ratio of $Pb(C_2H_5)_4$ and $C_2H_4Br_2$ in gasoline? What weight of $C_2H_4Br_2$ should be added per liter, if 1 g/liter of $Pb(C_2H_5)_4$ is present? [$\sim$0.3 g]

10.9 Tuna fish averages about 0.25 ppm mercury (Hg). How much tuna can you eat per day without exceeding the suggested safe dose of 0.01 mg Hg/day? [$\sim$40 g]

*10.10 The pollution standards for the United States in 1980 require cars to emit less than 0.15 g of unburned hydrocarbons per kilometer driven. Estimate (to one significant figure) what fraction this is of the weight of gasoline entering the motor. [$\sim$0.1%]

*10.11 (a) Platinum is insoluble in most chemicals, but it does dissolve in aqua regia (royal water) to give $PtCl_6^{2-}$ and NO_2. Aqua regia is made up of concentrated, but ionized, HCl and HNO_3. Write an equation for the reaction, and specify the ratio in which the acids should be mixed to prepare aqua regia. (b) Aqua regia dissolves gold (to give $AuCl_4^-$), as well as all the Group VIII metals. Write an equation for dissolving gold in aqua regia.

10.12 Many farm and suburban houses discharge their sewage into septic tanks, in which bacteria digest the sewage to form innocuous products. What advice concerning using high- or low-phosphate detergent would you give the owner of a septic tank?

10.13 Zinc (Zn) and cadmium (Cd) occur together as sulfide ores. Refining the ores reduces both elements to the metals. A mine produces an ore 95% ZnS/5% CdS; its annual production is 10^5 kg ore. What weight of zinc and what weight of cadmium are available for sale each year, assuming 100% recovery? [$\sim 6 \times 10^7$ g Zn, $\sim 4 \times 10^6$ g Cd]

10.14 Aspirin, $C_9H_8O_4$, is made into pills by mixing and pressing with $CaSO_4$. The aspirin is assayed by burning 0.200 g of ground pill in O_2 and trapping the carbon dioxide, 0.120 g, that forms. What percentage (by weight) of the pill is aspirin? [$\sim 30\%$]

10.15 Borax, $Na_2B_4O_7 \cdot 10 H_2O$, can also be prepared in the anhydrous form, $Na_2B_4O_7$. What fraction of the shipping costs would be saved if the $Na_2B_4O_7$ were shipped, then later hydrated back to borax? [$\sim 50\%$]

10.16 Estimate the weight of a cigarette. Assume that this weight constitutes half the weight of the materials (of average molecular weight 100) produced when a cigarette burns. If you inhale half these molecules from one pack of cigarettes a day, how many molecules of combustion products do you inhale? [$\sim 10^{24}$ molecules]

10.17 The use of water-based paints minimizes air pollution. Why?

*10.18 A recent book made this statement: "Substances that are not consumed in use create waste-disposal problems." What would you say about substances that *are* consumed in use?

10.19 An oxide of nickel (Ni) is 78.75% nickel. Determine the formula and comment on it. [$\sim NiO$]

10.20 An average chemical "formula" for algae is sometimes given as $C_{106}H_{181}O_{45}N_{16}P$. About 10^9 kg per year of phosphorus is used in detergents in the United States. Assuming P is the limiting reagent for algae growth, what mass of algae could result from these detergents? Estimate the volume of this mass of algae in cubic kilometers (km^3). [~ 0.08 km^3]

10.21 Oxygen difluoride, OF_2, is a yellow gas that explodes when mixed with steam. Write a probable equation for the reaction. Write a Lewis electron structure and predict the bond angle and bond distance in OF_2.

10.22 A mixture of NH_4NO_3 and fuel oil, $(CH_2)_n$, is the explosive most used in mining today. What weight of fuel oil should be mixed with each mole of NH_4NO_3 to use up all the oxygen, yet ensure that little CO (poisonous to miners) would form in the explosion? [Hint: The nitrogen is emitted as N_2.) [~ 5 g oil/mole NH_4NO_3]

*10.23 The lowest melting alloy of sodium and potassium melts at $-12°C$. It is known as NaK. Because it is a liquid metal, NaK is a good heat transfer agent and is used in the valves of racing cars. It is made up of 23% Na and 77% K by weight. How many potassium atoms are present for each sodium atom in the alloy? [~ 2 atoms K]

*10.24 Antlers contain about 50% calcium, yet horned animals are almost entirely herbivorous. Males normally grow a new set of antlers each year. Estimate the daily foliage requirement for an elk with 20 lb antlers if the foliage contains an average of 0.02% (by weight) calcium. Is there a possible correlation between the large antlers of moose and their use of aquatic plants as food?

Figure 11.1
Chemicals for treating surfaces.

11 STICKINESS AND CLEANLINESS: THE CHEMISTRY OF SURFACES

You have seen that two substances can react only if they come into contact. (Recall the problem of getting coke and iron ore to react in the blast furnace.) Reactions occur only at surfaces. To minimize reaction, you use minimum surface areas and inert surface materials. To maximize reaction, you use maximum surface areas and reactive surface materials.

One of the reasons that chemistry seems so remote from everyday life is that most of the chemicals surrounding us have minimum reactivity. Chemical changes in our surroundings are not encouraged. Floors, walls, ceilings, roofs, household utensils, clothing, cars, furniture, roads, even humans themselves—all have surface coatings selected to give long life to the object.

Paints, tiles, and roofing materials are designed to protect surfaces from wear and weather; tin, zinc, nickel, and chromium are used to plate surfaces to minimize corrosion; polishes and waxes minimize surface scratches; and each is also intended to beautify as it protects. Nor is this the end of surface treatment. Toothpastes, soaps, and detergents are designed to clean surfaces; lubricants, to minimize surface wear; enamels, Teflon coatings, and waterproofing, to minimize adherence of other materials to surfaces; dyes, inks, glues, and adhesives, to maximize adherence of other materials to surfaces. And, of course, there are the surface coatings that can be the most expensive of all—cosmetics. (Sometimes beauty isn't even skin deep.)

Because all reactions, at the molecular level, involve collisions between the outer electrons of atoms, surface chemistry can be interpreted ultimately in terms of the behavior of valence electrons.

Some Surface Effects

You are so familiar with some surface effects that they usually go unnoticed. Without the surface interaction between your fingers and the paper, you could not turn the pages of this book. Without the "friction"

Men who have excessive faith in their theories or ideas are not only ill-prepared for making discoveries; they also make very poor observations.
—*Claude Bernard.*

EXERCISE 11.1
One indication that friction results from the formation and breaking of chemical bonds is the transfer of radioactive material from one lubricated surface to another when the surfaces are rubbed together. Outline what probably happens at the atomic level.

between foot and floor, you could not walk. And without "friction" in the shoelace, you could not tie your shoes. Practically all frictional forces appear to be due to chemical bonding between surfaces.

You know that surfaces often become electrically charged. If you rub a balloon vigorously, it will adhere to a wall. If you pull back bed-sheets or pull off a shirt rapidly, you can hear the spark discharges—and, if it is dark, even see them. You also know that many things stick together in the absence of an adhesive. Newspaper pages are hard to turn. Freshly mimeographed paper is difficult to separate. Book pages often fly back just after you have turned them. And a piece of lint often goes directly back to your clothing if you drop it close by. All these effects result from the tendency of separated electrical charges to reunite.

Soap bubbles (and many insect, reptile, and fish eggs) cluster into "rafts" and minimize their exposed surface. Insects that land *on* the water can take off again without difficulty, but if they get *into* the water they do not have the strength to escape. There are too many chemical bonds to the liquid that they would have to break. In fact, were humans suddenly reduced to the size of insects (as happens in some science fiction), they would be unable to get their clothing off. They would be unable to turn the pages of an ant-sized book, and unable to swim through water at its surface. All these problems would be due to surface forces too strong to be overcome by the muscles of the insect-sized man.

The Nature of Surfaces

Only crystal and liquid phases have surfaces, but these surfaces can be exposed to one another or to gases. Thus, crystals can form *interfaces* with crystals, liquids, or gases; and liquids can also form interfaces with crystals, liquids, or gases. Before exploring these interfaces, let's look briefly at the surfaces of pure substances, starting with liquids.

A liquid suspended in a vacuum, say in a spaceship, quickly takes on a spherical shape, no matter how big the volume of the liquid. If you think for a moment about the differences between a molecule at the surface and a molecule in the interior of a drop of liquid, you can see why. Each molecule in the interior of the liquid is completely sur-rounded by other molecules to which it is attracted. A molecule on the surface has fewer neighbors, hence fewer bonds to other molecules. Thus, bonds must be broken for a molecule to move from the interior to the surface. This requires energy. So, in the absence of an external source of energy, as few molecules as possible will move into the surface. A drop of liquid forms a sphere because a sphere has minimum area for its volume. Every surface molecule will be much like every other one,

EXERCISE 11.2
Discuss at the molecular level a possible interpretation of a fly's ability to walk upside-down on a ceiling.

EXERCISE 11.3
Interpret the lack of interfaces between gases.

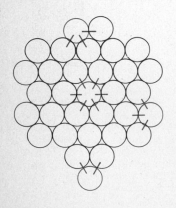

and the spherical liquid surface will be uniform.

An object that lands on a liquid first dents the surface. This requires energy to move more molecules into the new surface (Figure 11.2). Note that a surface molecule has about half as many molecules around it as does a molecule within the liquid. Thus the energy required to break half the intermolecular bonds and move a molecule into the surface is about half that required to vaporize one molecule.

If an object does not form bonds to water, it may only dent the surface of liquid water and not break through. We say water does not "wet" it. This is what happens when an insect alights on water, or when a duck "floats" on water. The energy of creating new surface area offsets the gravitational energy, and the animal stays on the surface rather than breaking through. You can see the effect if you carefully place a steel needle or piece of wire gauze flat on a clean water surface. Each will dent but not penetrate the surface.

Crystals are characterized by flat surfaces, edges, and corners. Molecules at corners are less tightly bound than those in edges, and these in turn are less tightly bound than the ones in faces. The ones in the interior of the crystal are the most tightly bound of all. Furthermore, if the exposed crystalline faces are not the same, the atoms in each face will have different properties. This can readily be demonstrated by machining a single crystal into the shape of a sphere and then exposing it to a chemical reagent, as was done to the copper crystal in Figure 11.3. You can see from the figure that our expectation is borne out. Differently exposed atoms have reacted differently; atoms that are the same have reacted in the same way. The high symmetry of the picture matches the known symmetry of the copper crystal. At the atomic level the reactive atoms are still edge and corner atoms, even though the overall shape appears spherical.

Because of the greater reactivity of corner atoms and edge atoms compared to face atoms (and because of the varying reactivity of atoms in most faces), crystals normally grow so they are bounded by only a few relatively unreactive faces. Many solid catalysts, on the other hand, are synthesized in such a way as to maximize the number of reactive face, edge, and corner atoms.

The higher reactivity of edge and corner atoms explains the phenomenon of *sintering*. In many cases, if crystals melting at high temperatures are finely powdered, packed together, and heated to 100–200°C below their melting points (but not to the melting point), they become a single piece. The individual particles of powder weld together at their corners and edges of contact to produce an open structure like a rigid sponge. Clearly the loosely held atoms have moved to more stable positions by forming the welds.

Crystals, being more rigid than liquids, are more apt to have imper-

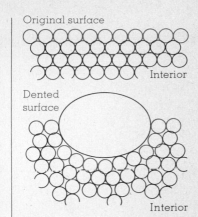

Original surface

Interior

Dented surface

Interior

Figure 11.2
Denting a liquid surface requires energy to break intermolecular bonds and bring more molecules from the interior into the surface.

EXERCISE 11.4
The text says that crystals are normally bounded by atoms in flat faces because of the greater reactivity of edge and corner atoms. Explain this more fully. In other words, what happens to reactive faces during normal crystal growth?

Figure 11.3
Different chemical reactivities on different exposed faces of a crystal of copper machined into the shape of a sphere. The most reactive areas on the sphere become the most patterned. (Courtesy A. T. G. Wathamey)

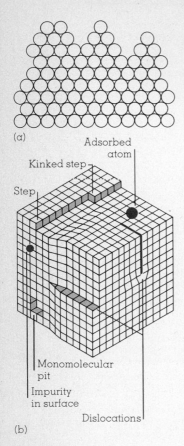

Figure 11.4
(a) A smooth surface with three rough edges. (b) Types of surface imperfections in a crystal.

POINT TO PONDER
It is common in industry to use expressions like *five-nines purity,* or *three-nines purity.* These refer, respectively, to 99.999% and 99.9% pure bulk material. The second figure is uncommon, the first very rare indeed. But these figures do not give the surface compositions. *All* surfaces, even ones called clean (unless very specially treated), are covered by adsorbed films of the substances in air, most commonly oxides.

fections both internally and in their surfaces. Thus even the boundary faces may not be flat. They may have pits, cracks, and protuberances, which are not found in liquids. (Figure 11.4 shows some common surface effects in crystals.) As a result, study of surface chemistry in crystals is much more difficult than in liquids, with surfaces between two crystals the most difficult class of all. (Here is that blast-furnace effect again!)

Figures 11.2 and 11.4 assume that bulk materials are pure substances and that the surface atoms are all of the same kind. Clean surfaces are achieved only with great difficulty. The materials must be specially treated to remove all interior impurities, and then they must be treated to remove all surface impurities. Once the clean state is achieved it can only be maintained in a perfect vacuum. Only in space can really good (but still not perfect) vacua be achieved. So some space experiments have been done on *ultrapure substances* and *ultraclean surfaces.* The results so far are interesting, but not extensive enough to indicate their potential usefulness. In our further discussion of surfaces you must remember that completely clean surfaces and completely pure materials are not attainable.

Surfaces of Metals

The tendency to form surface coatings is great in metals exposed to air. Essentially all metal surfaces quickly become covered with oxide. Oxide coats stick to the metal if there is neither swelling nor shrinking in converting metal to oxide. Such is the case with chromium, nickel, black iron oxide, and aluminum oxide. Porcelain enamels (such as the coatings on stoves) are mixtures of oxides. The oxides are carefully selected to adhere tightly to iron and expand and contract just as the metal does. Once the scientific requirements are met, colored oxides can solve the esthetic problems. Rust (red iron oxide) is much more voluminous than the iron from which it comes—so it swells, forms cracks, and exposes fresh surface to further attack.

One way to prevent the formation of oxide coats is to cover the metal with something else. Grease, paint, and coats of other metals (tin plate, galvanizing, chrome plate) are examples. Most commonly, one metal is merely covered with a thin film of another, the coating metal being chosen so that its oxide coat adheres. But not all metals stick well to one another. For example, the chrome metal trim on a car is usually made by first plating the iron metalwork with copper and/or nickel, and then plating with chromium.

Before plating it is normal to start with a *stripping operation,* in which a very thin layer of metal is dissolved from the surface of the object to be plated, to give a really clean surface. Stripping is commonly done by using sulfuric acid, which dissolves most metallic oxides and some of the metal itself. Or stripping may be done by using an electric current to oxidize some of the base metal to ions that are soluble in the plating solution. The new surfaces formed by stripping can then accept the plating material in a uniform way.

Often large pieces of metals must be joined. Many metals, because they have similar crystal structures, will stick together if their surfaces are clean. *Soldering* and *welding* are two techniques you have probably heard of. In each, the metals to be joined are heated; then a *flux* (almost always a weakly acid material) is applied to remove the oxide coats. Molten metal is poured over the fresh surfaces and is then allowed to cool, joining the two pieces of metal together. As with plating, so here—the key to a successful joint (in addition to matching crystal structure) is the cleaning of the surfaces. The most common soldering flux is *rosin,* the active chemical in which is abietic acid, $C_{20}H_{30}O_2$. A flux has two functions: It must first clean off the surface, and it must then cover the fresh surface to prevent further contamination before the metal joint is made. Borax, $Na_2B_4O_7 \cdot 10\,H_2O$, is a common high-temperature flux because it dissolves oxides and forms an adherent coat stable at high temperatures. Aqueous zinc chloride containing some free HC¹ is the usual "acid flux" used at lower temperatures.

When surfaces are made very flat (for instance, flat to 10 nm or less) they can sometimes be joined by simple contact and pressure. Two pieces of gold, one radioactive, the other not, trade atoms when clamped tightly together for several years. Very flat metal blocks may be stuck together by wringing them (twisting them together while the surfaces are made to overlap). The surfaces are slightly lubricated first with a liquid. And it is not unknown for a person working with *optical flats* (very flat objects, often of glass) to place two together accidentally and then find them impossible to separate without breaking.

A particularly simple experiment showing the effects of various adsorbed layers is outlined in Figure 11.5. It consists in measuring the angle at which a nickel collar just begins to slide down a nickel wire as

POINT TO PONDER
The beauty of many sculptures and architectural details is enhanced by surface effects. But sculptors and architects must be careful to use only metals that give pleasant and long-lasting coatings—for example, steels that "rust" to attractive surfaces rather than to growing piles of dark brown sludge.

EXERCISE 11.5
Aluminum can be *cold-welded* merely by squeezing two pieces together at high pressure. Discuss the processes that must occur at the weld in terms of atomic behavior.

EXERCISE 11.6
If two adjacent copper surfaces are each flat to 10 nm, what is the maximum distance apart of the atoms in the two surfaces, measured in atomic diameters?

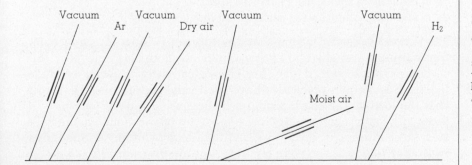

Figure 11.5
The angle at which a nickel collar begins to slide down a nickel wire varies as the surrounding atmosphere is varied. (From R. Holm and B. Kirchstein, *Wiss. Veröff. a.d. Siemenswerken,* 15, 1936, p. 122)

the nature of the adsorbed gas changes. Note the variations in angle. So, two similar surfaces brought together will stick if

1. they are clean so they are really similar;
2. they have complementary shapes, preferably flat, so that the surface atoms can make contact.

It is claimed, for example, that if a metal rod is broken under mercury and the pieces repositioned, the rod will regain a large fraction of its original strength—but only if the metal itself does not react with the mercury.

Making the Unbearable Bearable

The mutual stickiness of flat, similar surfaces is a major problem in mechanical bearings such as axles, roller bearings, and sliding parts. The continual wear tends to flatten the surface. The metals, if similar, tend to *seize*. This means that the metals tend to weld together to form a tight bond. Squeaking in a bearing is the sound of the breaking of these intersurface welds. It means "Help me!" Seizure can be minimized in two main ways: (1) The two bearing surfaces can be made of dissimilar metals so they cannot form matching bonds, and/or (2) lubricants can be applied between the metal surfaces.

A common kind of bearing has the axle or moving part made of a single hard-surfaced metal (usually an alloy). The fixed surface is then made of pieces of a different hard-surfaced metal (alloy) surrounded by a softer metal. The breaking-in of a car or a new piece of machinery gives the bearings a chance to adjust their mutual surfaces at low loads. The surfaces come to fit one another rather than tear one another up.

It is estimated that, even in well-fitted surfaces, only 1/10,000 of the surface areas are in actual contact.

Almost all metal bearings are lubricated. A good lubricant is much more than a "slippery liquid"—in fact, some of the best are solids. A good lubricant must have two properties:

1. It must keep the bearing surfaces apart.
2. It must not offer much resistance to movement itself.

The second criterion means that the lubricant's own intermolecular forces must be small. The hydrocarbons are nonpolar and meet this second criterion well, if they are chosen carefully in terms of chain length. Short-chain molecules have such small intermolecular forces that the load on the bearing will push them aside and the bearing surfaces will come together. Molecules made up of very long chains have such large intermolecular forces, and tend to get so tangled, that their resistance to flow is too high. Hydrocarbon chains from 19 to 25 atoms long are usually light oils (good for light loads); from 25 to 30 atoms

EXERCISE 11.7
Why is it common to make bearings out of two different iron alloys (steels), or out of brass and steel, but not out of two pieces of brass?

POINT TO PONDER
Lubricants as we know them were unknown to the ancients, yet they moved massive objects weighing tens of tons many miles (as at Stonehenge, on Easter Island, in Egypt, or in Peru). Such feats would be quite impossible without some lubricant. But no one knows what was used. Probably animal fat and water.

long, medium oils, as for automobile crankcases; and from 30 to 35, heavy oils, for very heavy loads. But hydrocarbons alone are not good lubricants if the loads become appreciable. Reason? The hydrocarbons do not adhere well either to metals or to their oxide surfaces, so they tend to squeeze out from the bearing, especially when it is stationary. This causes excessive wear, especially as motion starts or at low speeds.

Special Jobs, Special Lubricants

Most oils, and all heavy-duty lubricants, contain special additives that adhere to the bearing surfaces. Designing these additives raises interesting problems in surface chemistry because it is desirable for a bond to form between the additive molecules and the surfaces. This gives a new surface that will be lubricated by the bulk of the lubricant.

One of the most common type of additives to oils and greases is a heavy-metal soap, often calcium stearate. Calcium stearate consists of calcium ions, Ca^{2+}, bound ionically to stearate ions. Each stearate ion consists of a long hydrocarbon chain terminating in a negatively charged COO^- group:

$$CH_3CH_2CH_2CH_2CH_2CH_2CH_2CH_2CH_2CH_2CH_2CH_2CH_2CH_2CH_2CH_2CH_2COO^-$$

There are twice as many stearate ions as calcium ions in order to balance the charge: Ca (stearate)$_2$. The calcium ions, with their positive charge, will be strongly attracted to any oxide surface and will form a strong ionic linkage between the oxide surface and the stearate ions. This leaves the stearate ions extending out from the surface. At the same time other negative stearate ions will adhere strongly to any exposed metal surface. (Recall that metals act as positive ions immersed in a sea of electrons.) Thus, greases coat both oxide and metal surfaces with a strongly adherent "fuzz" of hydrocarbon chains. When the bearing moves, these chains tend to orient and form a smooth, furry layer that slides easily against the hydrocarbon oil between the two surfaces (Figure 11.6). The oil may squeeze out when the bearing stops moving, but the attached stearates will keep the metal surfaces out of contact and lubricate the starting motion until hydrocarbon is drawn into the gap.

The lubrication of aluminum surfaces was a difficult problem for years. The positive ions in the lubricants stuck to the oxide all right, but the oxide did not adhere very tightly to the metal and the negative ions in the lubricants did not adhere well to the metal. Then it was discovered that aluminum metal is attracted to π electron systems—that is, to double bonds between carbon atoms. Use of vinyl stearate, $CH_3(CH_2)_{16}COOCH=CH_2$, now gives a good lubrication between aluminum surfaces.

Sintered metals can be impregnated with lubricants to give *self-*

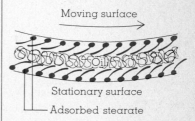

Moving surface

Stationary surface

Adsorbed stearate

Figure 11.6
Some effects in a lubricated moving bearing.

EXERCISE 11.8
A greased bearing may have been moving readily in one direction, but it may be difficult to reverse after it has stopped. Yet, once started in the reverse direction, it moves readily. Interpret at the molecular level.

Figure 11.7
Structure of graphite.

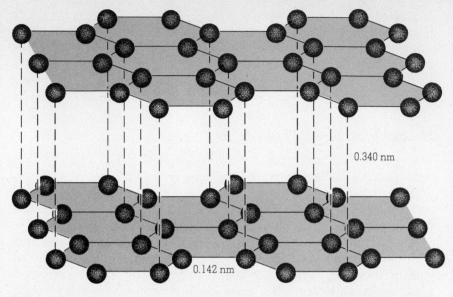

0.340 nm

0.142 nm

EXERCISE 11.9
Why is graphite less likely to
squeeze out of a stationary
bearing than is a $C_{20}H_{42}$ hydrocarbon chain?

lubrication. The spongy matrix retains the lubricant and supplies fresh
lubricant whenever fresh surface forms. *"Permanent" lubrication* can
be attained by filling the voids with *Teflon,* $CF_3(CF_2)_n CF_3$, a very good
lubricant.

Several solid lubricants have been introduced recently. (One solid
lubricant, graphite, has been used for a very long time.) All these solid
lubricants have layered structures similar to graphite's, which is shown
in Figure 11.7. Solid lubricants do not squeeze out in a stationary bearing, so there is no starting problem. They are also unusually stable at
high temperatures and pressures. Molybdenum disulfide, MoS_2, is one
such layered solid. Its cost is high, and its own resistance to flow is so
high that it is usually mixed with hydrocarbon or other liquid lubricants. But the load-bearing properties of these solid lubricants makes
them the choice for some high-load applications.

How to Come Clean (Soaps and Detergents)

Soap and detergent sales are big business. After all, they not only help
keep a nation clean—they also support a large fraction of actors, writers,
and media personnel.

Calcium stearate, used in heavy-duty greases and oils, is a calcium
soap. In fact it is the same chemical as the one that causes the "ring
around the bathtub," or the "luster-dulling film on your hair," when
you use soap while bathing in hard water. (Hard water contains calcium
ions; soaps contain stearate ions. Their compound, calcium stearate, is
insoluble in water and precipitates on hair and tub as a greasy, white
substance, colored gray by dirt.)

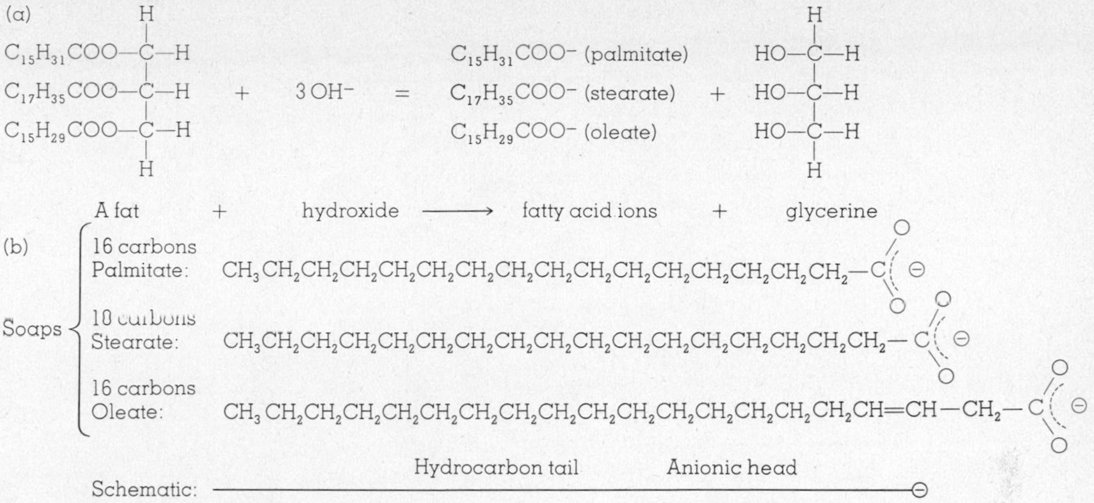

(a)

$$C_{15}H_{31}COO-\underset{\underset{H}{|}}{\overset{\overset{H}{|}}{C}}-H$$

$$C_{17}H_{35}COO-\overset{|}{C}-H \quad + \quad 3\,OH^- \quad = \quad$$

$$C_{15}H_{29}COO-\underset{H}{\overset{|}{C}}-H$$

$C_{15}H_{31}COO^-$ (palmitate)

$C_{17}H_{35}COO^-$ (stearate)   +

$C_{15}H_{29}COO^-$ (oleate)

$$HO-\underset{\underset{H}{|}}{\overset{\overset{H}{|}}{C}}-H$$

$$HO-\overset{|}{C}-H$$

$$HO-\underset{H}{\overset{|}{C}}-H$$

A fat   +   hydroxide $\longrightarrow$ fatty acid ions   +   glycerine

(b)

Soaps

16 carbons
Palmitate: $CH_3CH_2CH_2CH_2CH_2CH_2CH_2CH_2CH_2CH_2CH_2CH_2CH_2CH_2CH_2CH_2-C\overset{O}{\underset{O}{\big\langle}}\ominus$

10 carbons
Stearate: $CH_3CH_2CH_2CH_2CH_2CH_2CH_2CH_2CH_2CH_2CH_2CH_2CH_2CH_2CH_2CH_2CH_2-C\overset{O}{\underset{O}{\big\langle}}\ominus$

16 carbons
Oleate: $CH_3\,CH_2CH_2CH_2CH_2CH_2CH_2CH_2CH_2CH_2CH_2CH_2CH_2CH_2CH_2CH_2CH=CH-CH_2-C\overset{O}{\underset{O}{\big\langle}}\ominus$

Hydrocarbon tail   Anionic head

Schematic: ————————————————————————————————————⊖

Most household soaps are sodium compounds, though liquid soaps are potassium compounds. The negative ions, such as stearate, are obtained by decomposing vegetable and animal fats (cottonseed oil, palm oil, beef fat) into glycerine and the "fatty acid" negative ions, mainly palmitate, stearate, and oleate. We will not concern ourselves with the details of the structures of these ions, except that each contains a long hydrocarbon chain (16 to 18 carbons long) with a negatively charged COO⁻ group at one end. (See Figure 11.8.)

We shall find that soaps and detergents (the collective noun is *tensides*) have similar structures with long hydrocarbon tails and polar (often electrically charged) heads. Sometimes the electrical charge is negative as in soaps; the molecules are then called *anionic tensides*. Sometimes the charge is positive; the molecules are then called *cationic tensides*. (This is because *anion* is the collective noun for negative ions; *cation* is the collective noun for positive ions.) *Neutral tensides* have no net charge. It is easy to use the schematic formula in Figure 11.8 to represent all tensides, adjusting the charge as appropriate, depending on the net charge on the tenside.

The function of a soap or detergent is to help remove soil and leave a clean surface. Usually the surface is on a solid and the cleaning agent is an aqueous solution. For many specialty applications, however, other liquids may be used, such as the chlorinated hydrocarbons common in dry-cleaning fluids. The best cleaning agents have a threefold action:

1. They stick to the surface better than the dirt does.
2. They stick to the dirt better than the surface does.
3. They tend to concentrate at the surface of the liquid, so they come in contact with the dirty surface as soon as it is immersed in the liquid.

Figure 11.8
(a) Manufacture, and (b) structure, of soaps. The ratio of length to radius for the molecular "tail" is about 10 to 1.

POINT TO PONDER
Soap was known to the Phoenicians before 600 BC. Its earliest manufacture involved boiling goat tallow (a fat) and wood ashes (to provide K⁺ and OH⁻). It is interesting to speculate how this early technique developed. Did an observant shepherd, having dropped some tallow in the hot ashes of his fire while butchering a goat, note, in spreading the ashes later, that his hands were cleaned?

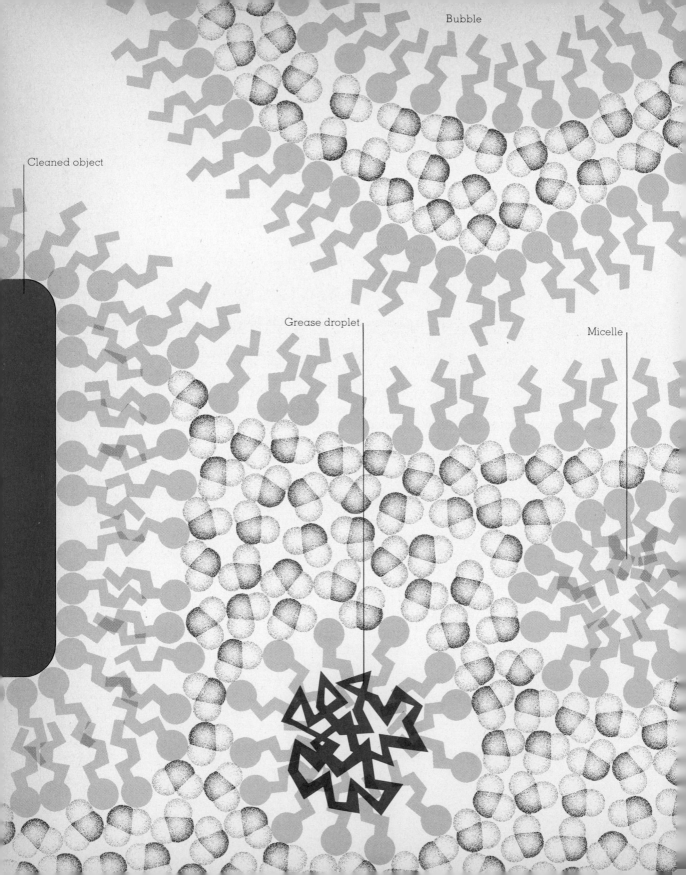

Bubble

Cleaned object

Grease droplet

Micelle

The aqueous tensides concentrate at the water surface because, though their polar heads are highly attracted to water, the nonpolar tails are not. So the nonpolar tails extend beyond the water surface. Once the surface is saturated with tenside, additional tenside enters the bulk water—but as clusters of molecules, known as *micelles*. The nonpolar tails intermingle in the interior of the micelles. The polar heads cover the outsides of the micelles and interact with the water.

When the layer of tenside on the surface of the water contacts a grease spot, the nonpolar tails enter the grease. The polar heads stay in the water. This creates bridges that allow the water to surround droplets of the grease. At the same time, tenside will usually coat the surface, polar head toward the surface, nonpolar tail interacting with the nonpolar tenside tails on the water surface. Of course this requires more tenside than was originally in the water surface. These additional molecules come from the micelles, which become fewer as tenside action cleans, and covers, more and more surface. Figure 11.9 shows the sequence.

Layers and Double Layers

The stability of an ionic tenside surface film is greatly increased if molecules containing an electrically neutral polar group are also present. The many un-ionized $—CO_2H$ molecules mixed with the $—CO_2^-$ ions serve this function in soaps. However, the ionic groups in most tensides are $—OSO_3^-$, $—SO_3^-$, and similar groups that do not combine readily with hydrogen ions. The simultaneous presence of electrically neutral molecules ending in $—OH$, $—OCH_3$, and similar polar groups greatly improves the tenside's *lathering* abilities. The reason is simple. Concentrating the ionic heads in a surface leads to strong repulsions and holds down the surface concentration of tenside. The neutral molecules get in between the ionic heads and insulate them, thus greatly increasing the total tenside concentration in the surface.

A double-layer effect on the cleaned surface may be readily demonstrated by using a cationic detergent (for example, cetyl trimethyl ammonium bromide) to clean glass. A glass surface is negatively charged when in contact with water because some sodium ions present in glass enter the solution, leaving the silicate framework of the glass negatively charged. The cationic (positively charged) head of the detergent adheres firmly to the clean glass. Its nonpolar tail attracts a second layer of detergent tails, which then expose their cationic heads to the bulk water. Draining leaves a thin, continuous film of water on a sparkling clean surface. After three or four rinses, however, the water film becomes discontinuous and spotty because the second layer of detergent has been removed and a nonpolar surface is exposed. The surface now acts

Figure 11.9 (opposite)
The formation of micelles and surface films.

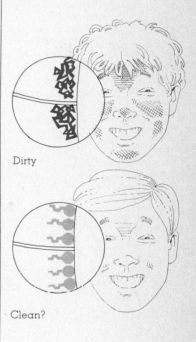

Dirty

Clean?

EXERCISE 11.10
Diagram the steps in the action of a cationic detergent in cleaning glassware when a final hot rinse with lots of clean water is used.

POINT TO PONDER
When soaps gave way to detergents for the family wash, more was changed than just the formula of the tenside. The washing machine changed from having a wringer dryer to having a centrifugal dryer. One result is that modern machines do not handle soaps well. Soaps form precipitates with calcium and magnesium ions in the water, and these are filtered onto the clothes in a centrifugal machine, whereas the wringer forced most off the clothes. So soaps tend to produce stiff laundry rather than the fluffy results of the detergents. Changing back to soaps would not only create an unfillable demand for natural fats—it would require equipping home laundries with new machines.

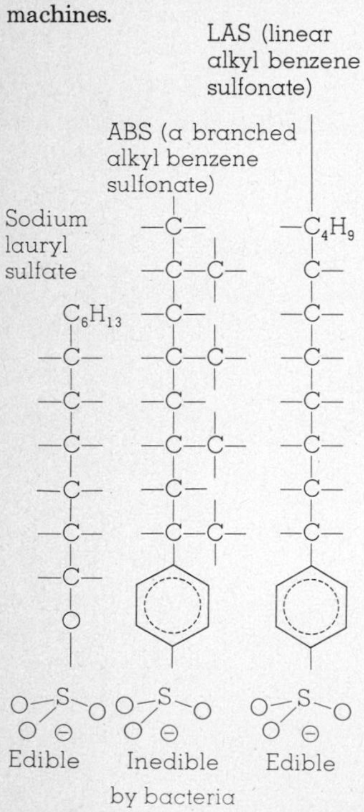

LAS (linear alkyl benzene sulfonate)

ABS (a branched alkyl benzene sulfonate)

Sodium lauryl sulfate

C_6H_{13}

Figure 11.10
One nonbiodegradable and two biodegradable tensides. Straight chains are needed for biodegradability.

as though it were greasy. It is not wet by water.

Thus, many clean surfaces are actually covered by a film, or even a double film, of tenside. The crispness of newly washed clothes and the silkiness of some fabrics after being washed in certain tensides are due to these surface layers and their stiffness, or their lubricating qualities. The layers may also minimize the rate of further soiling by presenting a less soilable surface than the fabric itself. In fact, it has become common to treat fabrics with surface coatings to minimize soiling rate. The trick is to make a surface lumpy at the molecular level to reduce contacts, with the lumps nonpolar to minimize intermolecular forces.

Most of the tensides you will meet work best against greasy dirt on fabrics. But some common dirt is made up of silicates (like mud) or protein food (like eggs). The tensides are often not very effective against such soils, so other chemicals are added to increase the range of cleaning action. Otherwise the tensides may merely disperse the soil uniformly, leaving the fabric looking dirtier than ever. Phosphate and borate extenders can help remove silicates, and enzymes and acidity control have been used to attack food stains. Enzymes are not used as much today because of effects (such as skin rash and nasal irritation) on those handling them.

And Always Those "Side" Effects

The activity of a tenside does not stop in the laundry or the plant. It continues in the sewers, creeks, and rivers unless the molecules are destroyed.

The fatty acids of soaps, with their linear arrays of carbon atoms, had been food for bacteria for years. Most animals can also digest them. But the first detergents, in the 1940s, were not digestible. They contained branched hydrocarbon chains that bacteria could not attack, and the sewage-disposal plants, creeks, and rivers foamed enough to cause problems.

The first step was to add foam killers, but the next was to produce digestible (biodegradable) molecules, made (as are soaps) of linear chains of carbons (Figure 11.10). This has now been done well enough so the principal pollution problem remaining is usually the phosphate that is present in the tenside to adjust acidity and increase effectiveness by tying up the Ca^{2+} and Mg^{2+} in hard water. Some communities that discharge sewage wastes into water where phosphate is often the limiting reagent to algal growth now forbid the use of detergents containing phosphate.

Tensides must be chosen carefully to avoid attack on skin and flesh. Ionic tensides, especially, interact strongly with skin proteins. This is one reason it has taken so long to produce a tenside other than soap that can be used in a bathtub. It is, however, possible to add substances that are more strongly adsorbed to skin than the ionic tenside and so protect the skin from attack. The oil obtained from the sperm whale is one such substance. Or should the whale have the oil for its own skin?

Waterproofing

Tensides increase the ability of water to wet surfaces. Waterproofing does the opposite. Manufacturers have become much more careful lately and refer to *water-repellent* fabrics rather than *waterproofed* ones. The fact is, of course, that the tiny, polar water molecules are potent penetrators and hard to exclude completely. Furthermore, a truly waterproof covering (say, a plastic raincoat) can be most uncomfortable, for it does not allow air to circulate and carry away evaporated perspiration.

Among the most common water-repellents are silicon compounds. Exposing a fabric to gaseous $(CH_3)_3SiCl$ or $(CH_3)_2SiCl_2$ leads to the formation of covalent bonds between the silicon and the fabric, with the nonpolar methyl groups pointed away from the surface. These are not wet by water. Water hitting the fabric forms into drops and runs off, rather than wetting the fabric and soaking in. Coating the fabric with stearate residues is the basis of another water-repellent. One of the most vital uses of waterproofing is on aircraft electric wiring, which otherwise shorts out when water condenses on the cold wires as a plane lands.

Making It Stick

Almost any two rigid surfaces will stick strongly if they are forced together tightly. The van der Waals forces are always present and are strong when the surface of contact is large. For example, a polyethylene molecule is a long-chain hydrocarbon with only van der Waals forces acting toward other molecules. Wood, on the other hand, contains long-chain cellulose molecules held together by polar forces and hydrogen bonding due to its many surface OH groups. An attempt to make plywood using polyethylene as the adhesive failed, even when pressure was applied for up to 10 minutes at 115°C. However, if the same technique is used but the joint is clamped overnight at 110°C, a tight bond occurs. Only on long standing is the adsorbed film of air and water squeezed out so the surfaces come into contact. When they do, the joint is a good one.

The most common reason for adhesive failures is that the surfaces were not clean and never came into good contact. Even adsorbed water and air can keep surfaces apart. For example, graphite is not a good lubricant in a vacuum. It is water and other small molecules adsorbed

POINT TO PONDER
Oil spills often coat wild birds. Tensides can be used to remove the oil but they also remove the natural oil that prevents water from wetting the feathers. It took conservationists a few sad experiences before they learned to keep the cleaned birds confined for a few days before release.

POINT TO PONDER
The effects of some earthquakes (that in 1964 in Alaska, for example) are much worse when the quake shakes up a clay soil surrounded by water. The shaking converts the firm soil to a soft jelly by separating the clay particles so water lubricates them. Buildings have been known to sink in clay during a quake while undergoing almost no structural damage.

EXERCISE 11.11
Cotton surfaces are negatively
charged in contact with water.
Other factors being the same,
would you recommend a
cationic or an anionic dye for
cotton?

between the graphite layers that make it easy for them to slide over one
another. Sliding is much more difficult when the rigid layers are in good
contact. Thus graphite is not nearly as good a lubricant in a spaceship
as in a vehicle on earth. Similarly, powders on earth are well lubricated
by adsorbed air and will not support weight. You cannot walk across a
bin full of dry flour. But moon dust is not lubricated, so it supports
astronauts without difficulty.

Though the prime requisite for adherence of one rigid surface to
another is good contact, the nature of the forces acting should also be
considered. You have already seen many examples of this in the discussions of lubricants and tensides. Similarly, dyes, glues, and paints are
chosen with the surfaces to be covered in mind. A simple reverse example
is this: How do you cover or close a container for glue so you can get it
open the next time you want some glue?

The cheapest adhesive is a water solution of sodium silicate, Na_2SiO_3,
called *water glass*. Its biggest use is in making corrugated cardboard for
packing and for cartons. The ionic silicate forms tight bonds to the polar
paper. The water provides good contact, then evaporates. The resulting
layer of adhesive is thin and it is not readily attacked by a humid atmosphere, though soaking in water will undo the bond. Most glues for
paper and fabrics form hydrogen bonds as well as polar and van der
Waals bonds.

Natural raw materials from which adhesives are commonly made include hides, horns, hoofs, blood, milk (casein), soybeans, starch, cellulose,
and rubber. The first six are proteins, the next two, carbohydrates. Both
can form strong hydrogen bonds. Rubber is an unsaturated hydrocarbon. A long list of synthetics could be added. But all have in common—

1. a degree of fluidity and wetness, so they can squeeze out adsorbed
   layers;
2. molecules with large areas to give large intermolecular attractions;
3. hydrogen bonding, pi bonding, polar, and even ionic forces, which
   interact with the surfaces to be bonded.

In general, the thinner the layer of adhesive the better; adhesives seldom
have much strength by themselves.

EXERCISE 11.12
Try erasing a Xerox copy,
a pencil mark, and a mimeographed or printed sheet. All
are legible because of finely
divided carbon. Account for
any differences in erasability.

Cell Membranes

Living systems contain a large number of surfaces of different kinds,
all of which total a large area. It is sometimes said that the basic unit of
life is the cell. Each cell is bounded by a cell membrane (Figure 11.11).
These membranes have some of the properties of the micelles in soap
solutions. The micelles form when the nonpolar tails cluster so that the
polar heads face out to the water. The double-layered cell membranes
(which have water solutions on both sides of them) are made of inter-

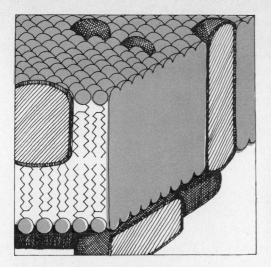 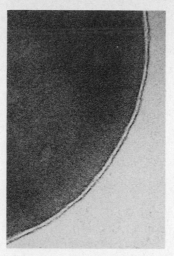

Figure 11.11
(left) A schematic diagram of a
cell membrane. A double layer
of phospholipid molecules provides the basic structure. The
large shapes represent protein
molecules. The proteins can be
in or at either surface, or can
extend clear through the membrane. (right) A photomicro-
graph of a cell membrane.
Note the double-layer structure.
(Courtesy J. David Robertson)

mingled nonpolar tails with polar heads faced outward. Cell membranes
are about 10 nm thick. The molecules that make up the membranes
have two long hydrocarbon ends fastened to a single molecule of glycerine. The third carbon of the glycerine holds a phosphate and sugar
composite that gives the polar head.

Embedded in and adhering to the cell membranes are protein molecules. Some of these control the rate of entry and departure of molecules. Some catalyze necessary metabolic reactions. Some synthesize
cell-wall structures that surround many cell membranes like a net and
so strengthen the cell. As a result some cells are rigid as in plants, others
highly flexible as are amoebas.

Perhaps one of the most remarkable things about human cells is
that all descend from the single cell formed when the sperm and egg
join. And from this single cell each human develops about 10^{14} cells of
at least a thousand different types, including different surfaces. How
this continual division and differentiation occurs is one of the most
actively researched areas of science, with study of effects at cell surfaces
one of the most exciting.

Summary

Because reactions occur when molecules or atoms or ions collide, it
could be argued that all of chemistry occurs at surfaces. As a minimum,
this does encourage us to interpret surface phenomena in terms of
molecular behavior. Just as every two molecules are attracted to each
other at close distances, so are every two surfaces. The intensity of the
attraction varies with the closeness of approach and the cleanliness and
molecular nature of the surfaces.

The design and behavior of welds, solders, platings, corrosion pre-

EXERCISE 11.13
Assume your average body
cell is a sphere 5000 nm in
diameter. Calculate the total
cell-surface area in your body.

HINTS TO EXERCISES
11.1 Different bonds form when
surfaces touch than break when
they separate. 11.2 A liquid
film provides good contact.
11.3 Gas molecules are
widely separated.
11.4 Exposed atoms are less
tightly bonded. 11.5 Al—Al
bonds form. 11.6 40 atomic
diameters. 11.7 Surfaces cannot
bond well if their atomic packings differ. 11.8 The orientation
of lubricant molecules must
reverse. 11.9 Graphite planes
are much larger in extent.
11.10 Layered cationic detergent is slowly removed.
11.11 Cationic dyes. 11.12 Pencil adheres, mimeo ink sinks in,
Xerox is bonded with plastic.
11.13 8×10^4 m².

venters, lubricants, soaps, detergents, water-repellents, paints, adhesives, inks, and many other items common in our society involve an understanding of surface effects at the molecular level. In fact, every living system depends for its life on the efficiency with which its own cell surfaces maintain themselves, transmit other molecules, synthesize needed ingredients, and interact with neighboring surfaces.

I did not specifically discuss smokes, dust, froths, films, or bubbles in this chapter, but all owe most of their properties to large ratios of surface area to volume. You should be able to interpret many of these properties in terms of the ideas that have been specifically discussed.

Problems

11.1 Identify and define: biodegradable, cell membrane, double layer, flux, friction, lubrication, micelle, "nines" purity, oxide coats, sintering, soap, soldering, surfaces (liquid and crystal), tenside (anionic, cationic, neutral), welding.

11.2 It is common advice to add no more tenside to wash water than is needed to give the first stable foam. Comment on the merits of this suggestion.

11.3 Why was rainwater once very popular for washing hair, and why do few bother to use it now?

11.4 Skiers use different waxes on their skis, depending on snow conditions. What molecular properties would you incorporate in a wax for warm "wet" snow? How about very cold snow?

11.5 Skis coated with Teflon, $(CF_2-CF_2)_n$, a long-chain molecule, are very fast, in fact, too fast for most skiers to handle. Why are they so fast?

11.6 Butter is slippery. Why not use it as a lubricant?

*11.7 Cetyl trimethyl ammonium bromide is mentioned as a tenside in this chapter. See if you can figure out its molecular structure from nomenclature you have seen in this book. (See also Problem 11.21.)

11.8 Soap micelles are spherical at low concentrations. The diameter of a micelle is about 5 nm. Estimate the number of $C_{17}H_{35}COO^-$ "ions" in a spherical micelle if the bulk density is 0.7 g/cm³. What would be the charge on the micelle

if all the molecules were ions? Do you think they are? [$\sim 10^2$ molecules/micelle]

*11.9 Concrete paving exposed to extreme cold fails more rapidly on bridges and overpasses than on roads. Epoxy toppings minimize the problem. What causes the problem, and why does epoxy minimize it? A recent advance is to mix in epoxy before casting the concrete.

11.10 What molecular properties would you incorporate in a "nonskid" floor wax? How about a wax for dance floors?

11.11 It is much easier to solder to copper than to a similar-sized piece of iron. Account for this difference at the molecular level.

*11.12 Foamed concrete consists of closely packed tiny bubbles whose thin shells are made of cement. Great savings in raw materials cost and in structural weight are made possible because of the low density of the foam. It is almost as strong in compression as solid concrete, though much weaker to bending forces or extension forces. Account at the molecular level for the fact that the bubbles are almost as strong to compression as they would be if they were solid. This is, incidentally, a universal property of rigid shells of material if the ratio of wall thickness to diameter is no less than about 1 to 10.

11.13 Cotton is negatively charged in contact with water. Account for the fact that soaps are strongly adsorbed to cotton in the presence of calcium ions, but not in the presence of sodium ions. Would magnesium ions cause adsorption?

11.14 Epoxy adhesives often come in two tubes: adhesive and catalyst. The catalyst is an acid in water. When the two tubes are mixed, the adhesive (some form of   O) opens up its oxygen

$$-\overset{|}{\underset{|}{C}}-\overset{|}{\underset{|}{C}}-$$

bridge, joins to other molecules, and forms a long polymer chain,

$$HOCH_2CH_2(OCH_2CH_2)_nOCH_2CH_2OH$$

where n may be up to 20 or so. Discuss why epoxy can glue metal to either metal or wood so effectively.

11.15 Discuss at the molecular level why a grease spot may be worse than a spot of silicate dirt in preventing a glue from working.

11.16 Polyethylene is made up of long chains, $CH_3(CH_2CH_2)_nCH_3$, and is a rather stiff plastic. If a few CH_3 groups are put along the chain instead of some hydrogens, the polymer is much more pliable. Suggest a molecular interpretation.

*11.17 If you sprinkle the whisker clippings from an electric razor on a large, clean water surface, they will rapidly spread out from the point of impact, then come to rest well separated from one another, yet not covering the whole area. What makes them spread?

11.18 If a drop of detergent is added to a glass of water "full to the brim" (that is, about to run over) the water does run over. A drop of water would not cause this. Explain the detergent action at the molecular level.

11.19 The colors of soap bubble films are interference effects due to the thickness (or thinness) of the film. On draining, films develop dark spots where the thickness is less than the wavelength of light. These thin black films can be preserved for years if ionic detergents are used, but films from nonionic detergents alone are seldom stable. Suggest why one film stabilizes whereas the other becomes so thin it breaks.

*11.20 Water and oil are almost insoluble but, when shaken together, can form either water-in-oil or oil-in-water emulsions. What happens depends on the tenside used. You are asked to recommend a tenside to clean up oil spills at sea. Which type of emulsion would you try to attain?

11.21 Cetyl alcohol, $C_{15}H_{31}CH_2OH$, has been sprayed on large lakes in Australia, resulting in the retention of as much as 6 feet of water depth per year otherwise lost by evaporation. Account for the effect. About how many dm^3 (liters) of cetyl alcohol should be used per square kilometer of lake? Assume the molecules are square rods $0.3 \times 0.3 \times 2$ nm. [~ 600 liters]

11.22 Most detergents are alkyl benzene sulfonates, like

$$CH_3(CH_2)_n - \left\langle \bigcirc \right\rangle - O - SO_3^-$$

Suggest a possible function for each of the three main parts of the molecule (alkyl, benzene, sulfonate) in the detergent action.

11.23 A Cascade detergent ad says, "It actually makes water rinse off in sheets." Suggest what is meant at the molecular level.

11.24 A Borden epoxy ad says, "The epoxy cement that holds a ton." No other description. Is this cement stronger than others?

11.25 Some scientists have suggested that the complex molecules of living systems first formed as micelles of simpler compounds in primitive oceans. What properties of micelles are consistent with such a suggestion?

11.26 Grinding and polishing the surfaces of two pieces made of the same metal first decreases, then increases, the friction between them. Suggest a mechanism at the molecular level. Friction between dissimilar surfaces also first decreases then increases upon polishing, but the increase is much less than for identical substances. Why? Be specific.

CHEMICAL BONDS

Bonds depend on
electron structures.

Bonds have constant
characteristic angles...

lengths, and strengths,
depending on the atoms.

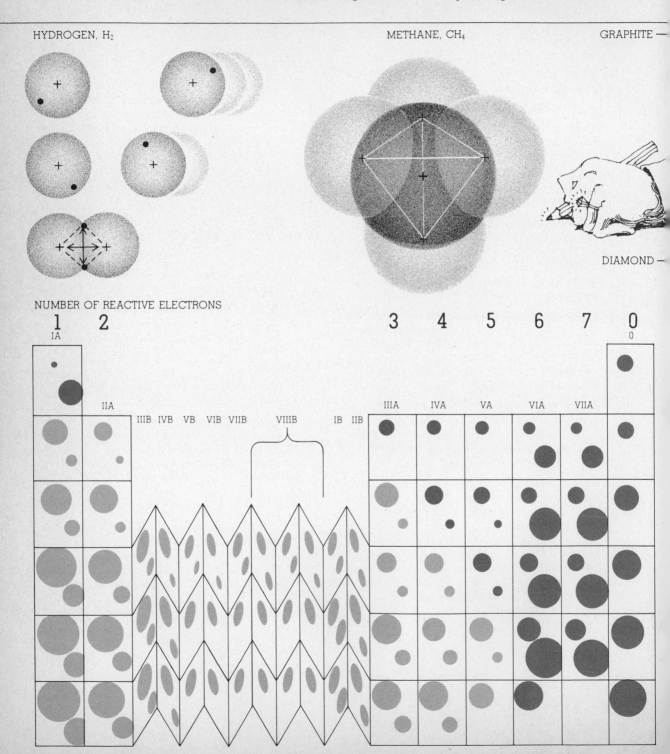

HYDROGEN, H_2

METHANE, CH_4

GRAPHITE —

DIAMOND —

NUMBER OF REACTIVE ELECTRONS

1   2   3   4   5   6   7   0

IA   0

IIA

IIIB IVB VB VIB VIIB   VIIIB   IB IIB   IIIA IVA VA VIA VIIA

Bonds can be *metallic* (with shared, delocalized electrons)...

covalent (with shared, localized electrons), or...

ionic (with transferred, localized electrons).

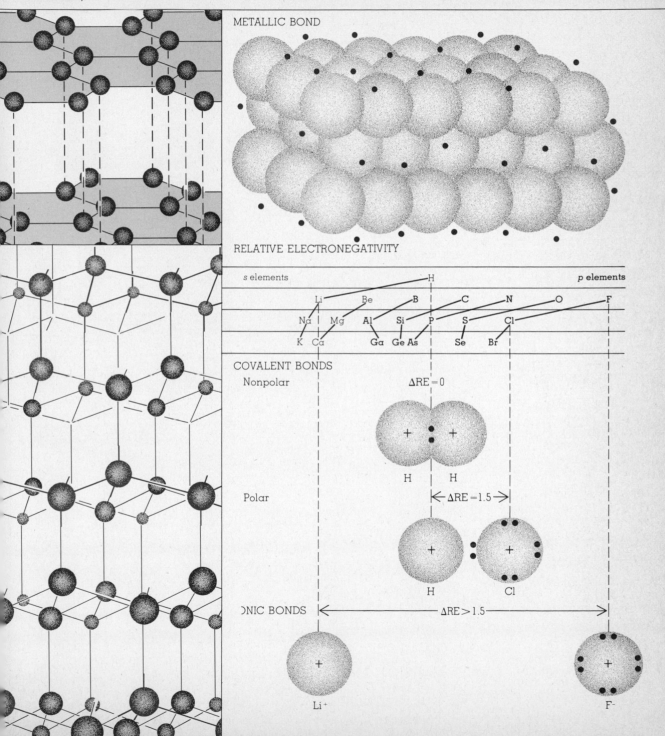

METALLIC BOND

RELATIVE ELECTRONEGATIVITY

s elements   H   *p* elements

Li   Be   B   C   N   O   F

Na | Mg   Al   Si   P   S   Cl

K Ca   Ga Ge As   Se   Br

COVALENT BONDS

Nonpolar $\Delta RE = 0$

H   H

Polar $\leftarrow \Delta RE = 1.5 \rightarrow$

H   Cl

ONIC BONDS $\leftarrow$————————$\Delta RE > 1.5$————————$\rightarrow$

Li⁺   F⁻

(Courtesy Culver Pictures)

12 THE SEA OF AIR, AND OTHER GASES

There is no chemical you are in better contact with or more dependent on than the atmosphere. Consciousness is lost after 3 or 4 minutes without air and is seldom restored without outside help. After 5 minutes without air, your brain will probably suffer permanent damage.

Humans have known these facts for a long time. But it was only with the work of Antoine Lavoisier, some 200 years ago, that serious efforts were made to understand the composition of the air and its detailed behavior. He was helped greatly by the studies of the preceding 100 years on the interrelationships between pressure, temperature, volume, and amount of gas.

But these four variables were not really tied together until around 1860, when the work of Stanislao Cannizzaro allowed the formulation of the

$$\text{IDEAL GAS LAW: } PV = nRT \qquad (12.1)$$

This expression, $PV = nRT$, has a great virtue. It expresses in a very direct and simple fashion the relationships between absolute pressure (P), volume (V), number of moles of gas (n), and absolute temperature (T). And it uses only a single arithmetical constant (R) called the *ideal gas constant*. The constant R is the *same* for *all* gases acting ideally. This equation and the information that may be correlated using it are primary topics of this chapter.

Even if you have never heard of the ideal gas equation, you have had a great deal of experience with the properties of gases. As a child you held balloons that floated in the air and "exploded" when you were rough with them. You probably popped bubble gum and learned that adding air increases the volume of the bubble. You may have had a toy sailboat, and you may have watched it being pushed around the pond by moving air. Your bicycle taught you that adding air to a tire increases the volume somewhat and the pressure a great deal. Your car added the knowledge that the pressure also increases when the tires get hot. You probably learned that the burning fuel in the car's cylinders produces the hot, expanding gas that pushes back the piston. And you may have seen birds or gliders soaring on updrafts of hot air.

Further thought will probably bring to mind many other examples of gaseous behavior, which you have come to expect, and about which you can now generalize with confidence. For example, it may have

A theory may be fruitful, leading to valuable research, even though it is eventually proved to be wrong—or, indeed, because it is formulated carefully enough so it can be disproved.—*Neal E. Miller.*

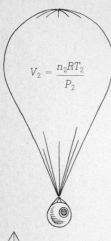

$$V_2 = \frac{n_2 R T_2}{P_2}$$

decreases.

pressure

atmospheric

mainly because

with height

balloon changes

flexible

of a sealed,

The volume

$$V_1 = \frac{n_1 R T_1}{P_1}$$

occurred to you that airplanes stay up because molecules hit the bottoms of wings more often and harder than they hit the tops. And why is it easier for a plane to fly faster than the velocity of sound at high altitudes than at low altitudes? Well, sound is transmitted by molecular collisions at nearly molecular velocities. The lower temperature at higher altitudes means lower molecular velocities, hence a lower velocity of sound. (Of course, there is also less resistance to the forward motion of the plane because there are fewer molecules present per unit of volume.)

I shall introduce here a molecular view of gas behavior, and outline some simple arithmetical methods of correlating properties of gases. This will give you a foundation for understanding, predicting, and exercising some control over the behavior of the gases with which you come in contact.

On the Need to Be Absolute

Many measurements are relative. They are made with respect to some arbitrary zero. A tire pressure of 30 lb/in.^{2} (pounds per square inch) is a pressure of 30 lb/in.^{2} above atmospheric. After all, if the tire is flat and has zero pressure according to the gage, it is still full of air at atmospheric pressure. Zero gage pressure does not mean "no pressure"; it means atmospheric pressure—usually about 14.5 (say 15) lb/in.^{2}, so that the absolute pressure inside an inflated tire is about 30 + 15, or 45 lb/in.^{2} (See Figure 12.1.)

Similarly, a temperature of 25°C means 25°C above the melting point of ice. A temperature of −10°C indicates a temperature 10°C below the melting point of ice. Zero degrees Celsius does not indicate an absence of temperature; 0°C is, rather, only an arbitrary point on a scale that extends both above and below that temperature.

In science it is often more useful to use absolute scales when possible. This avoids the mathematical problem of the quantity changing sign somewhere in its range of values. It also has the advantage that zero means the absence of the property. Thus, zero absolute pressure means no pressure. Zero absolute temperature means no temperature, just as zero volume or zero molecules or zero weight means that the property is absent in the sample. From now on you may assume that, unless otherwise specified, pressure and temperature data are given in absolute units.

Zero pressure is an easy concept; it suggests a perfect vacuum. But zero pressure is very difficult, indeed impossible, to achieve. I have already mentioned that outer space comes closest to it. But the fact that zero pressure cannot be attained does not keep us from using an absolute scale of pressure based on that zero. It just means we are a bit uncertain about all values for pressure, because we are not quite certain where the zero value is to which we should set our gages. But it is quite possible to attain a vacuum as low as 10^{-10} of an atmosphere, and to measure it to an accuracy of about 1 part in 10^{10}. Not a very big uncertainty!

Zero temperature is a much more difficult concept, mainly because most people have a poor idea of what temperature measures. Temperature is usually used to determine the direction in which heat will flow. If the outside temperature is low, you add clothing to decrease your heat loss. If the outside temperature is high, you take clothing off, go to areas having cool water, or turn on the air conditioner. All these increase your heat loss. But temperature is not heat; it merely determines the direction of heat flow.

It is easy to believe that absolute zero temperature will be cold, but it is difficult to see why there could not be a still colder temperature. The fact that the absolute zero of temperature is not attainable (compare zero pressure) does not help. The best efforts so far have reached about 0.00001°C above absolute zero. But measurements of temperature are difficult in that range, just as measures of pressure are difficult at very low values. Furthermore, it is more expensive to get close to absolute zero temperature than to absolute zero pressure. As a result we don't often try. Instead we define the standard of temperature as 273.1600 Kelvins (exactly) for a mixture of ice, water, and gaseous water. A temperature difference of 1 Kelvin is equivalent to a temperature difference of 1° Celsius (Figure 12.2). That is, the Kelvin and Celsius scales have units of the same "size," but different zero points.

The normal melting point of ice exposed to the atmosphere equals 273.15 Kelvins, the same temperature as 0°C. Celsius temperatures are converted into Kelvins by adding the Celsius temperature to 273.15:

$$K = {}^\circ C + 273.15 \qquad (12.2)$$

Thus 20.00°C is 293.15 K, and −10.00°C is 263.15 K. There are many other temperature scales but we shall not deal with them in this book.

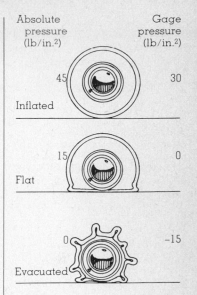

Figure 12.1
Comparison of gage pressure and absolute pressure for a tire.

Figure 12.2
Comparison of Kelvin and Celsius scales of temperature.

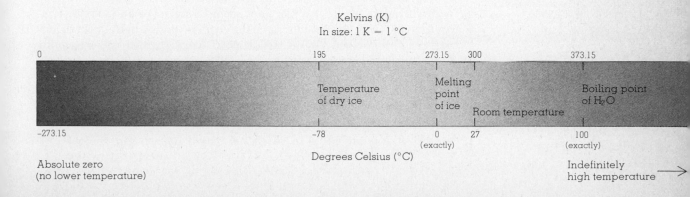

POINT TO PONDER
Australia used its postage
stamps to help with the change
in temperature scales. You may
have noted similar "helps"
appearing in the United States.

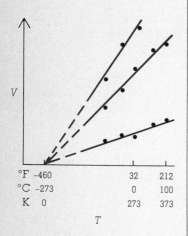

Figure 12.3
Charles' law: $V = k_C T$, where
k_C is a constant. Volume, V, is
proportional to absolute temperature, T, in Kelvins (when
pressure and amount are constant). Careful experimentation
shows $0 \text{ K} = -273.15°\text{C}$
$= -459.69°\text{F}$.

EXERCISE 12.1
Sketch an apparatus that would
keep the volume and amount
of gas constant, yet allow you
to measure the pressure as
temperature varies.

A common reaction to the preceding explanation is, "Picky scientists, why don't they leave well enough alone and stick to a sensible scale like Fahrenheit? Who needs to bother with unattainable standards and what the meaning of temperature is anyway? I can tell when to put on my overcoat." The logic is unanswerable, and most people will be annoyed with the coming change from Fahrenheit to Celsius in the United States. Because you are in a science course, you may be a bit more tolerant and think, "Okay, I can understand the desirability of absolute values, but where did those funny numbers 273.16 (exactly) and 273.15 come from? Couldn't you at least pick an integer, preferably something like 300?" (Perhaps you were somewhat relieved to see in Figure 12.2 that room temperature is about 300 K.) But let's look into the origin of 273.16. It will return us to the study of gases.

Hot Air Expands

Hot air balloons, soaring birds, and tire blowouts on hot roads are explained in popular language by saying, "Hot air expands." One of the great contributions of scientists, especially since the seventeenth century, has been to explore these popular qualitative observations in terms of numerical relationships—that is, *quantitatively*.

Figure 12.3 shows what happens if the volume of a gas is studied as a function of temperature—the pressure and amount of gas being kept constant. The relationship when the volume of a gas is plotted against temperature is a straight line over a range of both positive and negative Celsius temperatures. The relation is called Charles' law after its discoverer, J. A. C. Charles. The experiment is readily carried out by placing a small amount of gas in a capillary tube (say a thermometer with the bottom broken off). A drop of mercury is then introduced to trap the gas. The exterior atmosphere acting on the mercury applies a constant pressure on the trapped gas (Figure 12.4). The capillary tube is heated to various temperatures, and volume and temperature are recorded. The data give straight-line plots when plotted on the graph.

Repeating the experiment with different amounts of gas continues to give straight-line plots. The lines seem to converge at a single negative temperature. Can you imagine the surprise and delight of the scientists who discovered these regularities? Very careful measurement shows the convergence point to be $-273.15°\text{C}$ (or, in Fahrenheit, to mention it for the last time, $-460°\text{F}$). Actually, of course, all gases have turned to liquid well before this very low temperature is reached. But the extrapolated value (reached by the dotted lines in Figure 12.3) is $-273.15°\text{C}$. Furthermore, the same value, $-273.15°\text{C}$, is obtained regardless of what gas is in the apparatus if its pressure is kept low.

According to the plots, $-273.15°C$ is the temperature at which the volume of a gas would become zero if it did not condense—become liquid or crystal—first.

Suppose we keep the volume and the amount of gas (any gas, if pressure is low) constant, and study the effect of temperature on pressure. Figure 12.5 shows straight lines (originally discovered by G. Amontons). They again extrapolate to $-273.15°C$, this time at $P = 0$. Many other experiments on extrapolating temperature scales to low values give the same result: $-273.15°C$ is the lowest limit of temperature. So it is called the absolute zero of temperature—zero Kelvin. (Remember, a Kelvin is defined as a temperature change of the same "size" as the Celsius degree.)

Details, Details

For many years the normal melting point of ice was a basic standard of temperature. But this temperature varies slightly as atmospheric pressure varies and as more or less air is dissolved in the water. As temperature measurements began to require greater and greater accuracy, these variations became a nuisance. Hence a new standard, now the accepted standard temperature in all scientific laboratories, was defined as the temperature of a mixture of ice and water in the presence of gaseous water only. Because there are three phases present this is called the *triple point* of water. No air (or other soluble substance) can be present.

The triple point of water has many virtues as a standard temperature. One is that the temperature is constant to at least ± 0.000001 K. Another is that the system is very easy to set up. (See Figure 12.6.) Just put some very pure water in a very clean container, evacuate all gases, seal the container, freeze some of the water, and insulate the flask from the surroundings. The temperature inside the container will quickly come to 273.1600 K (exactly), and will remain there as long as all three phases of water are present.

The triple point of water, 273.16 K, is the basis of all present temperature measurements, though the normal melting point of ice (the *ice point*) is used in most routine work. The temperature of the ice point is 273.15 K. So much for why the standards of temperature have the values they do. (For a further discussion of standards see Appendix A.)

$PV = nRT$: The Ideal Gas Law

Late in the seventeenth century, Robert Boyle performed an experiment even simpler than those involving temperature variations. He used a glass U-tube (with one end sealed) to study the relation between

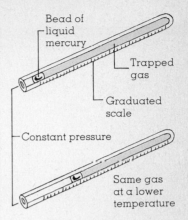

Figure 12.4
A simple method of checking Charles' law. Note the variations in length (volume) of the fixed amount of trapped gas as the temperature varies. The pressure is unchanged.

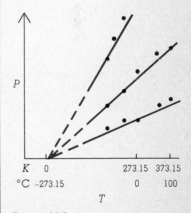

Figure 12.5
Amontons' law: $P = k_A T$, where k_A is a constant. Pressure, P, is proportional to absolute temperature, T, in Kelvins (when volume and amount are constant).

Figure 12.6
Determining the triple point
of H$_2$O.

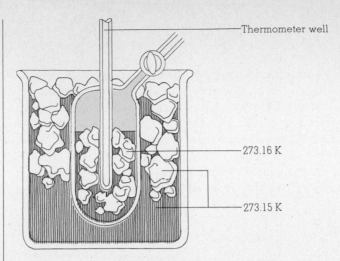

Thermometer well

273.16 K

273.15 K

Figure 12.6
Determining the triple point
of H$_2$O.

pressure and volume of a gas, keeping constant both the temperature
and the amount of gas. Boyle discovered that, under these conditions,

$$P \times V = PV = k_B \tag{12.3}$$

where k_B is a constant. Because the diameter of his tubing was almost
constant along the length of the tube, he actually used heights to
measure P and V, as shown in Figure 12.7.

There were no accurate thermometers in Boyle's time, nor any
devices to weigh gases and so measure their amounts. Hence the quantitative studies on temperature and amount effects were not done until
100 to 150 years later. About 1811, however, Amadeo Avogadro suggested a relationship that is now known as Avogadro's law: Equal
volumes of all gases contain the same number of molecules if their
pressures and temperatures are the same. Most prominent scientists
ignored or rejected this hypothesis until 1860. In that year, Stanislao
Cannizzaro presented an effective argument in favor of Avogadro's
law at the world's first international congress of scientists (held at
Karlsruhe, Germany). Impressed by Cannizzaro's experimental evidence and logic, scientists began a thorough investigation of the quantitative relationships between pressure, volume, temperature, and
numbers of moles. One result was the establishment of the ideal gas
law, which we saw earlier:

$$PV = nRT \tag{12.1}$$

where P = absolute pressure
         V = volume
         n = number of moles of gas
         R = a constant (the same for all systems)
         T = absolute temperature

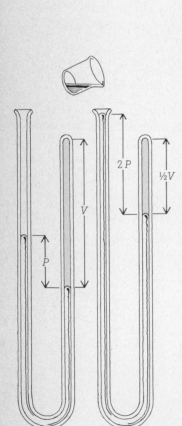

$2P$

$\frac{1}{2}V$

V

P

$PV = 2P \times \frac{1}{2}V$

Figure 12.7
Boyle's experiment: $PV = k_B$,
where k_B is a constant. Temperature and amount are constant.

Note that Boyle's law (PV = "constant"), Charles' law (V = "constant" $\times T$), Amontons' law (P = "constant" $\times T$), and Avogadro's law (n is constant if P, V, and T are constant) are all simplifications of the ideal gas law that follow when the appropriate properties are held constant. The substitution of a more inclusive law for several earlier ones is a constant aim of scientists. The more general law minimizes the amount of memory and work required to correlate observations and make predictions.

There are many units in which pressure, volume, and temperature can be expressed. Even n could be expressed in various ways—say, number of moles or number of molecules. For each set of units there is a corresponding value of R. In this book, for simplicity, we shall stick with pressure in atmospheres (atm), volume in liters (same as cubic decimeters, dm^3), temperature in Kelvins, and n in moles. Then

$$R = 0.08206 \text{ (liters} \times \text{atm})/(\text{mole} \times \text{K}) \qquad (12.4)$$

But you should see, for example, that if the volume were expressed in cubic centimeters (cm^3), the value of R would be expressed as

$$R = 82.06 \text{ (cm}^3 \times \text{atm})/(\text{mole} \times \text{K}) \qquad (12.5)$$

One word of caution before we try to use the ideal gas equation: No truly ideal gases exist. But all common gases are close enough to ideal that the ideal gas equation correlates their properties to within a few percent (at pressures less than a few atmospheres and temperatures greater than their boiling points). The lower the pressure and the higher the temperature, the closer to ideal every gas becomes.

Some Practical Problems

The ideal gas equation is much used, especially to calculate effects having to do with changing amounts of gas. Given values for any three of the four variables (P, V, n, T) the value for the fourth can be calculated, assuming only that the gas behaves ideally. Let's try.

Sample Problem 1
Some high-pressure cylinders, each with a volume of 40 liters, are to be stored in the tropics. Normal practice is to fill each with 200 moles of gas. Would the safe pressure limit of 200 atm be exceeded if normal fillings were made?

Answer: Assume that the ideal gas law is applicable. We know the safe limit is P = 200 atm, V = 40 liters, and n = 200 moles. We want to calculate the corresponding temperature (T). We use one significant figure (see Appendix A). From $PV = nRT$, we know that $T = PV/nR$. Let's plug in the values:

EXERCISE 12.2
Express the "constant" in Boyle's, Charles', and Amontons' laws in terms of the ideal gas law, and show that each constant *is* constant when conditions meet the limitations set by the originator of that law. Thus, Boyle's constant = nRT.

POINT TO PONDER
One of the great problems that comes up when using scientific equations is deciding in what units to express the numbers. If you remember that the units of the two sides of every equation must be the same, you will make few errors. The way to make sure the units are the same is always to write the units immediately after the number, then compare to see that the two sides of the equation are expressed in the same units. The worked-out problems on gas behavior provide examples. See Appendix A also.

SAFE?

POINT TO PONDER
Cylinders of high-pressure gas are potentially dangerous objects. Their protective caps should be kept on when the cylinders are not in use, and the cylinders should be firmly lashed in an upright position. When one falls over unprotected, the top valve may break off, allowing the high-pressure gases to rush out and create a high-velocity ground rocket. If the gas is combustible, as is hydrogen, it usually catches fire, adding to the "rocket effect." Such rampaging cylinders have been known to penetrate concrete walls.

$$T\,(K) = \frac{200\ \text{atm} \times 40\ \text{liters}}{200\ \text{moles} \times 0.08\ (\text{liters} \times \text{atm})/(\text{moles} \times K)}$$

$$= \frac{8{,}000\ (\text{liters} \times \text{atm} \times \text{moles} \times K)}{16\ (\text{liters} \times \text{atm} \times \text{moles})}$$

$$= 500\ K$$

The units come out the same on both sides, so we are okay there. The cylinders should be safe up to a temperature of 500 K, and $500 - 273 = 200°C$. It will never get this hot (unless there is a fire), so the usual filling seems quite safe.

Sample Problem 2

A helium-filled blimp has a volume of 10^6 liters. How many cylinders of gas, each containing 200 moles of helium, are needed to fill the blimp?

Answer: The pressure in the blimp is 1 atm, the temperature about 27°C or 300 K (always assume nice round numbers of reasonable value when given the freedom to do so). The volume is 10^6 liters. In this case, we need to solve for n, the number of moles needed: $n = PV/RT$. Again we have only one significant figure.

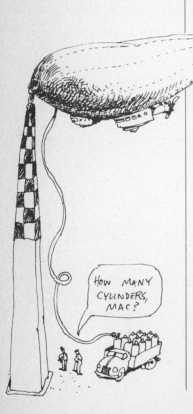

$$n\,(\text{moles}) = \frac{1\ \text{atm} \times 10^6\ \text{liters}}{0.08\ [(\text{liters} \times \text{atm})/(\text{moles} \times K)] \times 300\ K}$$

$$= \frac{10^6\ (\text{atm} \times \text{liters} \times \text{moles} \times K)}{24\ (\text{atm} \times \text{liters} \times K)}$$

$$= 4 \times 10^4\ \text{moles}$$

The units are okay. But, rereading the problem, we see that we aren't through yet. We need 4×10^4 moles of gas, but we were asked how many 200-mole cylinders we need.

$$\text{Number of cylinders} = \frac{4 \times 10^4\ \text{moles}}{2 \times 10^2\ \text{moles/cylinder}}$$

$$= 2 \times 10^2\ \text{cylinders}$$

Thus, 200 cylinders are needed.

Sample Problem 3

In its suggestion box, an airline finds the following note: "Heavier planes use more fuel. Each jet has 10 big tires. Fill the tires with helium instead of air to save weight." How much weight would this practice save? Would you recommend the change? (Tire pressure on the ground is about 10 atm.) The solution is lengthy, but we use only one significant figure. Good luck!

Answer: Assume that the ideal gas equation holds for both air and helium at 10 atm pressure and at ground temperature (assume 300 K). If we calculate the moles of gas in the tires, then we can compute the weight of gas. But we don't know V.

Airplane tires are large. If we imagine the donut-shaped tire straightened out into a cylinder, it would be about 0.5 m in diameter and 3 m high. From a math handbook we find a formula for the volume of a cylinder: $V = \pi d^2 h/4$ (d = diameter, and h = height). So, to one significant figure:

$$V\,(\mathrm{m}^3) = \frac{3 \times (0.5)^2\,\mathrm{m}^2 \times 3\,\mathrm{m}}{4} = 0.5\,\mathrm{m}^3$$

One liter equals 1 dm³, and 1 m = 10 dm. So, 1 m³ = 1000 dm³ = 1000 liters.

$$V\,(\text{liters}) = 0.5\,\mathrm{m}^3 \times 1000\,(\text{liters/m}^3) = 500\ \text{liters}$$

We now have an estimate of the volume of one tire as 500 liters, so we estimate the total volume of the 10 tires on a plane as 10 × 500 liters = 5000 liters. This is the value of V we need, so we return to the ideal gas law and calculate $n = PV/RT$.

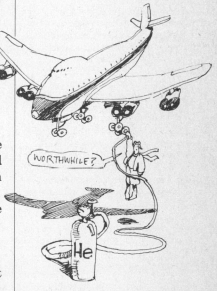

$$n\,(\text{moles}) = \frac{10\ \text{atm} \times 5000\ \text{liters}}{0.08\,[(\text{liters} \times \text{atm})/(\text{moles} \times \text{K})] \times 300\ \text{K}}$$

$$= \frac{50{,}000\,(\text{atm} \times \text{liters} \times \text{moles} \times \text{K})}{24\,(\text{atm} \times \text{liters} \times \text{K})}$$

$$= 2000\ \text{moles}$$

Note that we keep only one significant figure in the answer. The plane tires contain about 2000 moles of gas (either air or helium). Atomic and molecular weights are expressed in grams per mole (g/mole), so we can now find the weight of the gas.

Helium is a monatomic gas, so we find its molecular weight from the table of atomic weights to be 4 g/mole.

$$\text{Weight of He} = 2000\ \text{moles} \times 4\ \text{g/mole} = 8000\ \text{g, or 8 kg}$$

But what is the molecular weight of air? We know (see Table 6.1) that the ratio of O_2 (molecular weight, 32 g/mole) to N_2 (molecular weight, 28 g/mole) in air is about 1 to 4. So we won't be far wrong if we estimate the average molecular weight of air (to one significant figure) as 30 g/mole. Therefore,

$$\text{Weight of air} = 2000\ \text{moles} \times 30\ \text{g/mole} = 60{,}000\ \text{g or 60 kg}$$

The amount of weight saved by using helium in the tires will be 60 kg − 8 kg = 50 kg, which is about the weight of one small person.

EXERCISE 12.3
A container is weighed when evacuated, then weighed filled with dioxygen gas. The gain in weight is 0.83 g. The container is then filled with an unknown gas and weighed again. The difference from the weight when evacuated is now 1.66 g. What is the molecular weight of the unknown gas? (Hint: Remember the molecular weight of dioxygen is 32.) Could the gas be sulfur dioxide?

POINT TO PONDER
Using arguments based on his hypothesis, Avogadro computed atomic weights of the elements. At that time, Dalton and the Swedish chemist Jöns Berzelius had each proposed a different set of atomic-weight values, and there was strong rivalry between the two. In general, Avogadro's results confirmed the values given by Berzelius. Dalton refused to accept the validity of Avogadro's hypothesis. We don't know why Berzelius didn't accept the idea, because he simply ignored it. After half a century more of confusion about atomic weights, chemists were much more ready in 1860 to consider any theory that would offer a way out of the maze. Besides, the personal influences of Dalton and Berzelius were no longer overpowering. Human personalities and interactions are major factors in the history of science.

EXERCISE 12.4
A container of 0.30 liter volume was weighed empty, then filled with an unknown gas at 1.0 atm and 300 K. The gain in weight was 1.0 g. What is the molecular weight of the unknown? Could the unknown gas be sulfur trioxide?

We conclude that the weight saved would not be worth the trouble and expense of filling the tires with helium.

Note that this problem involves a lot of computation, but it's all simple arithmetic and logical reasoning. We've worked through it here to show you how much you can do by making some simple assumptions and just working carefully and logically with simple techniques. And, with one significant figure, you can usually do the arithmetic in your head.

And Now a Word From John Dalton

The ideal gas equation describes not only the behavior of pure ideal gases but also the behavior of their gaseous solutions. It was John Dalton who proposed that the total pressure of a gas is the sum of the separate pressures (called *partial pressures*) of the constituent gases:

$$P_{\text{total}} = P_1 + P_2 + P_3 + \ldots \qquad (12.6)$$

We can now go further. Each gas in a solution of ideal gases behaves independently of the rest. The gas law can be applied separately to each gas, as well as to the whole solution. Thus, if 10% of the molecules are hydrogen, 10% of the total pressure will be due to hydrogen. Or, if the gases are separated such that each separated gas has the same pressure as the original solution, the volume of the separated hydrogen will be 10% of the original volume. Similarly, 80% of the molecules in air are N_2. So 80% of the atmospheric pressure is due to nitrogen, and separation of pure N_2 would give a volume of dinitrogen gas equal to 80% of that of the original air at the same temperature and pressure.

So You Don't Have a Mass Spectrometer!

Nowadays, molecular and atomic weights are usually determined with a mass spectrometer. But these spectrometers were invented only 60 years ago, and thousands of molecular weights had been determined before then. In fact, thousands have been determined since then with no help from the mass spectrometer. All that is really needed for the determination of many molecular weights is a container and an analytical balance.

The simplest key to molecular weights is Avogadro's law: Equal volumes of gases contain equal numbers of molecules if their temperatures and pressures are the same.

If you (1) weigh an evacuated container, (2) fill it with a gas of known molecular weight and weigh it again, and (3) fill it with a gas of unknown molecular weight and weigh it a third time, then you can calculate the molecular weight of the unknown. The *difference* between the first and second weights is the weight of the known gas at some temperature and pressure in that volume. Neither the volume nor the temperature nor the pressure needs to be known if each remains constant. The *difference* between the first weight and the third weight is the weight of the unknown under the same conditions of V, T, and P. The ratio of these two weight differences must be the same as the ratio of the two molecular weights if Avogadro's law is followed, as it is by all ideal gases. The simplest known gas to use is dioxygen of molecular weight 32.00.

Sometimes it is easier to use an apparatus for determining gas densities. Density is weight per unit volume. Or,

$$\text{Density} = \frac{w}{V} \tag{12.7}$$

The ratio, r, of two gas densities, at the same T and P, is also the same as the ratio of their molecular weights. Do you see why?

If P, V, and T are known, you can calculate the number of moles (n) from the ideal gas equation, $n = PV/RT$. You can find the weight of n by calculating the difference between the weight of the container empty and the weight of the container full. Dividing the result, w, by n gives the molecular weight, M. The equation is $M = w/n$. Or the ideal gas equation may be written

$$PV = (w/M)RT \tag{12.8}$$

which gives

$$M = wRT/PV \tag{12.9}$$

It was Cannizzaro who invented a very clever method of getting atomic weights from molecular weights. He pointed out that the weight of an element in one molecular weight of a compound is given by multiplying the molecular weight by the weight percentage of the element. Furthermore, the weight of the element in a molecular weight must be some integral multiple of the atomic weight. Thus, examination of the weight of an element in a series of molecular weights of its compounds indicates the atomic weight. The atomic weight is the smallest weight of the element in any molecular weight. Table 12.1 illustrates the Cannizzaro method.

And What Are the Molecules Doing All This Time? The Kinetic Theory

The ideal gas law has a minimum of theory in it. It was discovered, and can still be formulated, directly from measurements of the bulk prop-

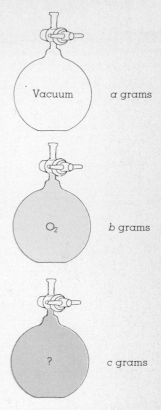

Vacuum a grams

O_2 b grams

? c grams

Unknown molecular weight =
$$\frac{(c - a)}{(b - a)} \times 32.00$$

EXERCISE 12.5
Analysis of the two gases of Exercises 12.3 and 12.4 showed both were oxides of sulfur, with molecular weights of 64 and 80, and percentages of sulfur of 50% and 40%, respectively. Calculate the atomic weight of sulfur, and the two molecular formulas.

Table 12.1   Determining Atomic Weights Following the Cannizzaro Method[a]

Compound	$r = \dfrac{\text{density gas}}{\text{density } O_2}$	Molecular weight (M) of gas $M = \dfrac{\text{g compound}}{\text{mole compound}}$ $M = 32 \times r$	Weight fraction (f) of N in compound $f = \dfrac{\text{g N}}{\text{g compound}}$ (from weight analysis)	Weight of N per mole of compound $M \times f = \dfrac{\text{g N}}{\text{mole compound}}$
Ammonia	0.53	17	0.82	14
Nitric oxide	0.94	30	0.47	14
Nitrous oxide	1.3	44	0.64	28
Ethyl amine	1.4	45	0.31	14
Ethylene diamine	1.8	60	0.47	28
Nitrobenzene	3.8	123	0.11	14
Nitrogen	0.87	28	1.00	28

[a]The atomic weight of nitrogen, N, is either 14 (most likely) or some simple fraction of 14.

erties of gases. There need be no reference to the existence of molecules and their behavior, except in terms of the number of molecules present. But chemists like to interpret properties in terms of molecular behavior and, pleasantly enough, this can be readily done with gases.

Our present model of molecular behavior in gases has been in use for about 100 years. It is called the *kinetic theory of gases,* developed since the middle of the nineteenth century.

The *kinetic-theory model of gases* is as follows:

1. Gases are made up of molecules separated by distances that are large compared with the size of the molecule.
2. The molecules are all in rapid, random motion with a range of molecular velocities varying from almost zero to very high values.
3. The average molecular velocity (v) decreases as the molecular weight increases, and increases as the temperature increases.

$$v = \sqrt{3RT/M} = 1.58 \times 10^4 \sqrt{T/M} \qquad (12.10)$$

where v = average velocity of the molecules in centimeters per second (cm/s)
R = the gas constant = 8.31×10^7 g cm^2/(s^2 K mole)
T = temperature in Kelvins
M = molecular weight in grams per mole.

This is the same R as in the ideal gas law, but in different units.
4. Pressure is due to the collisions of the molecules with one another and with the walls of the container.
5. The molecules of an ideal gas act as though they have no volume and no intermolecular forces except at the instant of collision (then they repel one another).

Let's look at a small part of the experimental support for these ideas, taking them up one by one, using for reference the numbers in the list just given.

1. Gaseous molecules are separated by distances that are large compared to molecular size. When liquids boil to gases, the volume increases by a factor of from 100 to 10,000, depending on the boiling points. If the molecules themselves do not change in size, they must be much further apart in the gas than in the liquid. Furthermore, it is far easier to walk through a gas than through a liquid, and gases are much easier to compress into a small volume than are liquids or crystals.

2. Molecules are in rapid, random motion. Dust particles in a sunbeam dance around in a random fashion called *Brownian motion*. This random dance is caused by collisions with the randomly moving molecules of gas. Sometimes a dust particle undergoes a particularly violent gyration when hit by some especially energetic (high-velocity) molecules. The range of velocities can be measured in an apparatus like that in Figure 12.8, where a beam of molecules is chopped into small bursts. The variation in arrival time at the rotating receiver gives a variation in density of the deposited molecules, from which we get the distribution of molecular velocities. The actual velocities of the molecules in air average about 500 m/s.

3. The average molecular velocity is given by the equation $v = 1.58 \times 10^4 \sqrt{T/M}$. See equation 12.10.

Aquanauts in a deep-sea habitat breathe a mixture of helium and oxygen rather than nitrogen and oxygen. Helium does not dissolve in the blood as much as nitrogen does, so it does not form gas bubbles in the blood vessels as pressure diminishes when the divers come toward the surface. The divers can decompress faster without getting the "bends."

Breathing helium has two undesirable side effects. Body heat is lost rapidly in a helium atmosphere, and the pitch of the voice rises. Both effects are due to the higher average velocity of helium molecules (of mass 4) compared to dinitrogen molecules (of mass 28). The greater number of collisions of helium with the skin removes heat more rapidly than in air, but the cooling can be offset if the helium is warm to begin with. The higher velocity of the helium carries the sound more rapidly in the larynx and so raises the frequency of the sound. This effect is not easily offset. Divers must try to get used to it.

The experiment shown in Figure 12.9 demonstrates that the average velocities of HCl and NH_3 molecules differ. Equal partial pressures of the two gases are released at the ends of a glass tube, say 1 m long. About half an hour later a ring of NH_4Cl will form where the diffusing molecules first meet. Their random collisions with the air in the tube have

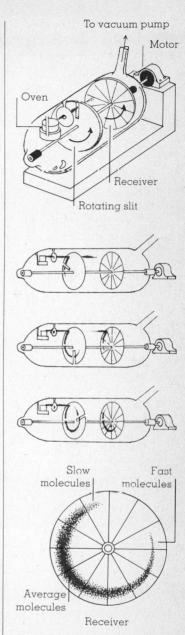

Figure 12.8
A simple method of determining that molecular velocities vary. The density of the molecular deposition on the receiver indicates the distribution of molecular velocities.

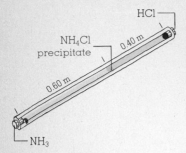

Figure 12.9
Gaseous NH_3 diffuses more
rapidly than gaseous HCl.

EXERCISE 12.6
If the odor of perfume does not
spread by random molecular
motion, how does it spread?

EXERCISE 12.7
If no molecules hit the top of
a stationary airplane wing, the
only pressure will be on the
bottom of the wing and will be
atmospheric. I have said that
atmospheric pressure is about
15 lb/in.[2] One pound equals
about half a kilogram, and 1 m^2
equals about 1500 in.[2] Show
that atmospheric pressure gives
a lift of about 10^4 kg/m^2.

allowed the gases to diffuse along the tube, but at different rates. The ammonia molecules (NH_3, molecular weight 17) move more rapidly on the average than do the hydrogen chloride molecules (HCl, molecular weight 36), and so the precipitate forms closer to the HCl side. The distances (0.60 m for NH_3, 0.40 m for HCl) can, of course, be measured.

The ratio is found to be consistent with a modern statement of a law originally suggested by Thomas Graham: The ratio of rates of diffusion of two gases is inversely proportional to the square roots of the molecular weights. Or,

$$\text{Rate}_1/\text{rate}_2 = \sqrt{M_2/M_1} = v_1/v_2 \qquad (12.11)$$

Here $\sqrt{36/17} = \sqrt{2.2} = 1.48$. The precipitate should form $1.48/(1 + 1.48)$ or 0.60 of the way from the NH_3 end to the HCl end of the tube. The experimental values agree with equation 12.11. Note that equation 12.11 follows from equation 12.10 for two different gases at the same temperature. But equation 12.11 was discovered first.

The fact that it would take these molecules almost an hour to move a meter when their average molecular velocities are about 500 m/s should reinforce the idea that their motion is random. It should also show the fallacy of the argument that smelling perfume in the rear of a room shortly after it is released at the front has much to do with molecular velocities. If users of perfume had to rely on the random diffusion of molecules to waft their scent away from them, the sale of perfume would drop dramatically!

4. Pressure is due to molecular collisions. The dancing dust particles mentioned earlier support the idea that pressure is due to molecular collisions. So does the flight of a glider or an airplane. The shapes of the wings are designed to ensure that more molecules hit the bottom than the top, and/or hit the bottom harder, to keep the plane airborne. If it were possible, the top of the wing would be designed so no molecules hit it. This would give a lift of 10^4 kg/m^2 even when the plane was stationary. This is 10 metric tons/m^2. A well-designed wing can actually give a lift that is a few percent of this when the airplane's air speed is 1000 kilometers (km)/hr.

5. Ideal gas molecules act as though they have no volume and no intermolecular forces. The linearity of the curves in Figures 12.3 and 12.5 shows that both the volume and the pressure extrapolate to zero at absolute zero temperature. This suggests that if the molecules condensed they would have no volume and no interactions. Neither of course is true. But the approximations must be quite good, judging from the high degree of linearity actually found in experimental curves.

Perhaps you can now see why all gases become more ideal as pressure is lowered and temperature is increased. Lowering the pressure increases the ratio of unoccupied space to that occupied by the molecules, so the effect of their volume becomes less. Raising the temperature increases the kinetic energy of the molecules, so the energy of their electronic interactions becomes relatively less important.

Conversely, at high pressures and low temperatures all gases deviate appreciably from ideality. The fact that all condense to liquids is ample proof of this.

For our purposes in this book, it will be adequate to treat all gases as ideal. If you ever need to do an accurate calculation on gases, you will have to learn how to treat deviations from ideality.

Chaos Without Change

The idea of random molecular collisions in systems where the actual molecular velocities vary over a wide range is a very useful one. I shall refer to it often in interpreting behavior at the molecular level. Some accurate experiments similar to that outlined in Figure 12.8 are summarized in Figure 12.10. These curves are known as *Maxwell–Boltzmann* curves after the two men who provided the theoretical framework for their interpretation. The curves are experimental. But the Maxwell–Boltzmann equation (which is too complicated to cite here) describes them in full detail, using only temperature and molecular weight to correlate the distribution of molecular velocities. Thus the general shapes of the Maxwell–Boltzmannn curves are similar for all gases.

Chemists are not just interested in the increase in average molecular velocity as temperature increases. They are also interested in the even more rapid increase in number of molecules with high velocities and energies. The molecules with low energy merely rebound when they collide. It is the high-energy molecules that are able to react with one another upon collision. The rapid increase in the number of high-energy

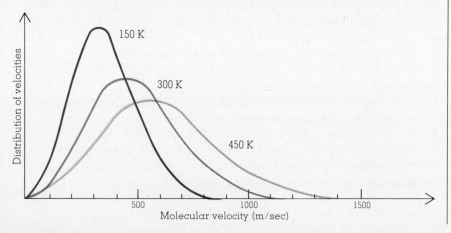

Figure 12.10
Maxwell–Boltzmann curves showing the distribution of molecular velocities at various temperatures: a most important and useful set of curves.

Figure 12.11
Random motion in liquids and
crystals. (Courtesy B. Alder,
Lawrence Radiation
Laboratory, University of
California)

collisions accounts for the well-known fact that the rates of most chemical reactions increase rapidly as the temperature increases.

It is interesting to note that the wild chaos of molecular motion in gases leads to such smooth curves. Similarly, gas pressure gages do not fluctuate wildly, but give steady readings. Nor do glass windows break under the incessant molecular bombardment, because the bombardment is equal on both sides of the pane (except in a windstorm or explosion). We shall see that most systems that give the outward, bulk appearance of calm and no change are actually tempestuously active at the molecular level. It is not just the molecules in the air that are moving with average velocities of some 500 m/s. So are the molecules in liquids and crystals (Figure 12.11). But the motions are truly random. For each "forward" there is soon a "reverse." Complete chaos leads to no net change at all. The successful application of statistical averaging to molecular properties such as velocities (as in Figure 12.10) is one of the greatest scientific contributions of the last 100 years.

"This Most Excellent Canopy"

The main topic of conversation involving the atmosphere used to be weather. But weather discussions are now frequently replaced by discussions of pollution. That this is not a new subject is shown by William Shakespeare in *Hamlet:* "This most excellent canopy, the air, look you this brave o'erhanging roof fretted with golden fire—why, it appears no other thing to me than a foul and pestilent congregation of vapors." Nor was Shakespeare alone. Other writers in seventeenth-century London, and elsewhere, comment on the stink of the cities, and it is easy to find similar comments even 2000 years earlier. The Black Death, which killed perhaps one-fourth of the people in western Europe in the fourteenth century, was due to pollution, partly airborne. This makes London's "killer smog" of December 1952 seem trivial, though some 5000 deaths in a period of about 1 week were attributed to it. (I was there and was among the millions unaffected.)

There are several reasons why air pollution is a bigger topic of conversation than either liquid or solid pollution (which we shall find to be much more serious problems).

1. Air moves rapidly from place to place. Pollution produced in one place quickly spreads elsewhere.
2. Most living systems on the surface of the earth take in large quantities of air continuously and so notice and are affected by its contents.
3. Millions of people have already assaulted their lungs with contaminated air, especially cigarette smoke. It is they who are at highest risk from lung cancer, and the risk for them is enhanced by general pollution. Most of the people killed by the

POINT TO PONDER
Complete chaos leads to no net change at all. Yet you observe change all about you. The changes are due to energy flow through the systems you see. Without a continuous supply of energy from such sources as the sun, chaos would reign, and all change would cease.

London smog of 1952 already had serious lung problems.

4. Air pollution is readily detectable long before it causes any physiological effect. Instruments easily do this, but so can the unaided eye. The view gets hazy (not romantically misty, but grayly grimy), there may be a brown tinge to the normally blue sky, and it becomes harder to get a nice tan! Air pollution is far from a trivial matter, but much of the complaining is based on trivia compared to other threats to a pleasant life.

Perhaps the most hopeful thing to remember about air pollution is that turning off the sources of pollutants would allow the atmosphere to clean itself of major pollutants worldwide in less than a month. This is because even the worst polluted air does not have much material suspended in it. The natural processes of rain, oxidation, and contact with ocean and ground can and do remove these materials quite rapidly and effectively. It is also encouraging that methods are available to detect most contaminants at concentrations far below toxic levels. On the other hand, the most discouraging thing about air pollution is the fact that most of the polluting in industrial societies is done by people quite aware of what they are doing.

One of the reasons the air (except for the very high altitudes) cleans itself so rapidly is its massive worldwide circulation. The principal driving force is the greater amount of solar energy at the equator compared to that at the poles. So equatorial air gets hot, becomes less dense (n/V decreases), and rises. Polar air is more dense and it sinks. The rotation of the earth gives a twirling effect and the atmosphere circulates. It flows, in a spiral fashion, from high-pressure regions to low-pressure regions, as a weather map will show. Another reason air cleans itself is the high reactivity of oxygen plus the washing effect of rains.

The problems at the end of this chapter will give you a chance to use the gas laws and kinetic theory to gain some insight into air pollution. And in the next chapter I shall discuss some of the chemistry that leads to the pollutants actually found.

Summary

We view a gas as composed of tiny molecules in endless, random motion — colliding with one another and rebounding with a wide range of velocities. A rise in temperature leads to an increase in the average molecular velocity. The interrelationships between temperature, pressure, volume, and amount (moles) of gas are given (under most common conditions) by the ideal gas law: $PV = nRT = (w/M)RT$. Though gas molecules in the atmosphere have an average velocity of about 0.5 km/s, they interdiffuse only at a rate of about 1 m/hr, consistent with their random motion.

Molecular weights, atomic weights, and molecular formulas may all

EXERCISE 12.8
The average private car is driven about 16,000 km/yr, consuming about 1 liter of gas for every 8 km driven. One liter of gas contains about 1 g of lead, and will yield about 2 g of carbon monoxide, 0.3 g of nitrogen oxides, and 0.5 g of sulfur oxides. How much of each contaminant (lead, CO, NO_x, and SO_x) did your family car produce last year? How much of this was necessary?

POINT TO PONDER
Atmospheric oxygen is probably responsible for the death of more species than any other air "pollutant." The earth's early atmosphere had no oxygen, and the early life forms did not need it. Most of these forms became extinct when the oxygen content rose after the beginning of photosynthesis. They were killed when the plants "polluted" the atmosphere with oxygen. If the amount of oxygen in the atmosphere were to drop sharply, humans (and many other species) would probably become extinct, but many simpler forms of life would live on. And, presumably, some of the now extinct forms might be re-formed.

Table 12.2   Solving Problems in Chemistry Using Conservation of Atoms (Mass) in a Closed System (solved in terms of moles)

1. Write the *net equation*. It must be *balanced* to be an equation.

2. Given:

Mass (g)	Solution volume (liters)	Number of molecules (N)	Gas volume (V, P, T)
↓	↓	↓	↓

3. Operation:

Divide mass by formula weight	Multiply solution volume by concentration	Divide by N_0, 6.02×10^{23}	Calculate $n = PV/RT$

Units:

$\dfrac{g}{(g/mole)}$	liters $\times$ (moles/liter)	$\dfrac{molecules}{(molecules/mole)}$	mole

4. To Get: Number of moles

↓

5. Then: Use net equation to determine numbers of moles of other reactants and/or products.

↓

6. Finally: Convert moles to units required, by reversing one of the four operations in step 3.

be obtained by the use of the ideal gas law and chemical analyses. These data, plus sensitive detection devices, provide means for us to monitor the atmosphere, to learn how it becomes polluted and how pollutants are transferred, and to invent control methods. But the application of these control methods is strongly influenced by political and social factors.

The gas law provides a means of rounding out chemical calculations in terms of moles. Table 12.2 summarizes the uses we have explored for moles in chemical calculations.

HINTS TO EXERCISES
12.1 A rigid container plus a pressure gage. 12.2 $k_C = nR/P$; $k_A = nR/V$. 12.3 64 g. 12.4 82 g; yes. 12.5 32 g S/mole; SO_2, SO_3. 12.6 Transported by currents of air. 12.7 10^4 kg/m^2. 12.8 2 kg Pb/year, 4 kg CO/year, 0.6 kg NO/year, 1 kg SO_x/year.

Problems

12.1   Identify or define: absolute pressure and temperature, air pollution, average molecular velocity ($v = 1.58 \times 10^4 \sqrt{T/M}$), Avogadro's law, Cannizzaro method to get atomic weights, Celsius temperature ($0°C = 273.15$ K), diffusion in gases ($v_1/v_2 = \sqrt{M_2/M_1}$), gas density (w/V), ideal gas law [$PV = nRT = (w/M)RT$], Kelvin temperature (T), kinetic theory, molecular weight determination for gases, Maxwell–Boltzmann curve, partial pressure ($P_{total} = P_1 + P_2 + \ldots$), triple point of water, velocity (molecular, sound).

12.2 To warn miners of impending danger, a tube of ethyl mercaptan (C_2H_5SH, which has an odor similar to that of skunks) may be broken in the air intake to the mine. Do the molecules get to the miners by simple diffusion through the intervening air?

12.3 Why do gases become more "ideal" (follow the ideal gas law better) at low pressures than they do at high pressures? Why better at high temperatures than at low temperatures?

*12.4 Which system of each of the following pairs would have the higher velocity of sound?
a. air or natural gas (CH_4);
b. air at sea level or at high altitude;
c. air at the South Pole or air at the equator;
d. atmosphere at the surface of earth or moon;
e. steel girder or air.

12.5 You find a thermometer that is complete, except it has no markings. How would you experimentally mark it so it would read degrees Celsius correctly over a common range?

12.6 Use Avogadro's law to show that the ammonia-synthesis reaction (1 volume of nitrogen gas plus 3 volumes of hydrogen gas gives 2 volumes of ammonia gas) proves that nitrogen molecules contain an even number of atoms. (Hint: Would the volume ratios be possible if nitrogen were monatomic? Diatomic? Triatomic?)

12.7 The molecular ratio of dinitrogen to dioxygen in air is about 4 to 1. What is the pressure ratio for the two?

12.8 You feel cooler when you enter a humid cave, come out of a swimming pool, or are fanned by a breeze, even though the air temperature in each case is the same 23°C. Interpret at the molecular level.

12.9 An underwater habitat of volume 100 m^3 is to be filled to 3 atm pressure with 0.1 atm of O_2, the balance He. How many moles of each gas are required? [$\sim 10^4$ moles He]

12.10 On a hot, humid day, the pressure of gaseous water in the air may be 0.05 atm. How big a bucket would be needed to contain the water that would be condensed if a room 10 m $\times$ 8 m $\times$ 3 m were dehumidified to a pressure of 0.02 atm of gaseous H_2O? [~ 10 liters H_2O]

*12.11 Estimate P_{CO_2} and P_{H_2O} in the exhaust gases from an automobile. (Hint: First write the equation for C_8H_{18} (gasoline) burning in *air*.) [$P_{CO_2} \cong 0.1$ atm]

12.12 It is fun to drop dry ice (crystalline CO_2, which vaporizes to gaseous CO_2) in water, but some children put it in bottles and drive in a cork. Calculate the probable result if 10 g are dropped into a 1-liter bottle, then sealed. Recommendation? [$P_{CO_2} \cong 6$ atm. DANGER]

12.13 How many grams of gasoline (C_8H_{18}) should enter a car's cylinder per cycle if the vaporized gas is just to fill the 600 cm^3 cylinder at 80°C and 0.10 atm? [~ 0.2 g C_8H_{18}]

12.14 The boiling point of liquid nitrogen is about 77 K. Helium is still an ideal gas at this temperature. What would be the pressure inside a 0.50-liter container holding 1.0 g of He and immersed in liquid N_2? If the bursting strength of the container were 10 atm, could it be safely warmed to room temperature? [~ 3 atm He]

*12.15 Automobile engines have been commonly adjusted to an air-to-fuel gaseous ratio of 60 to 1 on a volume (or mole) basis. What is the limiting reagent at this setting? (Assume gasoline is C_8H_{18}.) [air in excess]

12.16 One suggestion for reducing air pollution is to burn gasoline with pure oxygen, carried in a high-pressure cylinder in the car. What would the kilometer range of such a car be if the volume of the gas cylinder were 50 liters and the pressure 130 atm? A mole of gasoline will take a small car about 1 km. [~ 20 km]

12.17 To minimize the rate of evaporation of the tungsten filament, 10^{-5} mole of argon is placed in a 200-cm^3 light bulb. What is the pressure in the light bulb? [$\sim 10^{-3}$ atm]

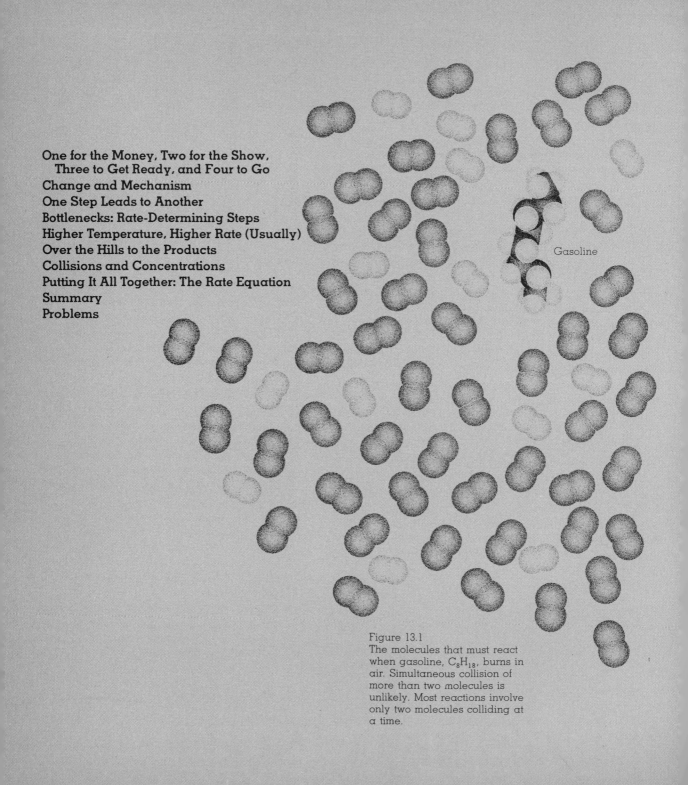

Gasoline

Figure 13.1
The molecules that must react
when gasoline, C_8H_{18}, burns in
air. Simultaneous collision of
more than two molecules is
unlikely. Most reactions involve
only two molecules colliding at
a time.

13 CHANGE, CHANGE, CHANGE: REACTION RATES

You have already seen that most of the chemicals around you have been designed and put in place with the deliberate intent of minimizing change. In fact, you are irritated when paint flakes off, carpets wear out, the car rusts, or you skin your knee. On the other hand, you are even more annoyed if fresh paint doesn't dry, the car doesn't start, or your knee doesn't heal. Sometimes you wish a given object to change, and at other times you wish the same object to resist change. Humans, more than any other species, are able to control changes. We explore here some of the ideas and methods we use to do so.

One for the Money, Two for the Show, Three to Get Ready, and Four to Go

There are three factors that must be satisfied for a change to occur. In terms of the preceding nursery rhyme they are—

1. *a collision* between the reacting substances;
2. *an appropriate geometric orientation* of the reacting substances;
3. *enough energy* to carry out the change.

Then change can occur. If any one of the three is missing, change does not occur.

Consider, for example, what must occur before your car can get a dented fender. (1) The car must collide with something; (2) the orientation must be such that the fender is involved; and (3) there must be sufficient energy in the collision to leave a dent. The same three factors account for when you do and when you do not skin your knee—and whether your car rusts, the carpet wears out, or the paint flakes. They also control whether the paint dries, the car starts, or your knee heals. They apply at the molecular level as well. In fact, why don't you pause in your reading and try to outline some change that does not involve all three factors? (If you find one, consider it further after you finish this chapter.)

So you see that at least some of the factors involved in change are easy to comprehend at the qualitative level. They are not difficult to understand at the quantitative level either. And it is quantitative understanding that leads to effective control.

Any system that resists change also resists investigation
—*William S. Bullough.*

POINT TO PONDER
One of the great advantages of
studying molecular behavior is
that molecules are ignorant;
their average behavior is delightfully repetitious and consistent,
so that you do not have to
"outguess" them. You have seen
examples of this already in the
wide applicability of the
Maxwell–Boltzmann equation
and in the behavior of *all* ideal
gases.

Change and Mechanism

Most changes seem too complex and difficult to understand because
they are, indeed, complex. Many things are happening during the
change, and too often we try to concentrate on the whole complexity.
Small wonder we may fail to comprehend.

Recall Chapter 2: "Simplify, Simplify." So, to simplify, let's concentrate on studying changes at the molecular level. Later we shall see
whether what we learn can be applied to larger systems of unknown
molecular composition.

First, one thing leads to another—every change can be considered
as a series of steps. The overall change may be highly complex, but the
individual steps may not be. Only after biologists and chemists made
this assumption did biochemists (scientists studying living systems at
the molecular level) begin to make the discoveries that have so rapidly
improved medical treatment as well as our understanding of health.

Consider a particular example with which you are quite familiar—the
combustion of gasoline in your car. The equation for the net reaction is

$$C_8H_{18}\,(\ell) + 12.5\,O_2\,(g) = 8\,CO_2\,(g) + 9\,H_2O\,(g) \qquad (13.1)$$

But what is the chance that enough randomly moving molecules of
dioxygen will simultaneously strike a single molecule of octane (C_8H_{18})
with enough energy and the right orientation to form 8 molecules of
carbon dioxide and 9 molecules of water? As you might guess, the probability of such a collision is almost zero. (And don't forget the 50 nonreactive, but interfering, N_2 molecules in the vicinity also.) Yet this
reaction is fast enough to drive your car down the road at a very good
speed. If you need further convincing, look at Figure 13.1. Note that the
low probability has nothing to do with the fact that the net equation
involves 12.5 molecules of oxygen. The low likelihood of collision, orientation, and energy requirements being met in a one-step process is due
to the *complexity* of the net equation.

Notice how much simpler it is to consider the net reaction as the sum
of a whole series of mechanistic steps—each step a simple *bimolecular
reaction.*

The initial reactive collision must be between one molecule of C_8H_{18}
and one of O_2—a highly likely event because both C_8H_{18} and O_2 are
present in high concentration. Because the periphery of C_8H_{18} is covered
with hydrogen, the oxygen will hit H atoms. The spark in the cylinder
heats the gas locally (as well as partly fragmenting it), so there are
plenty of high-energy molecules present. A likely partial set of mechanistic steps is the following (all are gases, so we leave out phase designations):

$$O_2 + C_8H_{18} = C_8H_{17} + HO_2 \qquad (13.2)$$

$$HO_2 + C_nH_m = H_2O + O + C_nH_{m-1} \qquad (13.3)$$

$$O + C_nH_m = HO + C_nH_{m-1} \qquad (13.4)$$

$$HO + C_nH_m = H_2O + C_nH_{m-1} \qquad (13.5)$$

$$O_2 + C_nH_m = CO + O + C_{n-1}H_m \qquad (13.6)$$

Here n and m refer to numbers from 8 to 1 for n and from 18 to 1 for m, as continuous reaction strips hydrogens and carbons from the initial C_8H_{18} molecules. Some of the equations shown contain formulas for molecules you have not seen, but otherwise the equations are simple:

1. They involve collisions between only two molecules at a time.
2. Every reaction requires a particular orientation.
3. In most of the equations, one old bond breaks as one new bond forms, thus requiring minimum energy.

So all three of the factors necessary for reaction are met—and in such a way that the overall reaction can be expected to be fast, as it is.

I have shown only some of the possible steps, so it is easy to see why the overall reaction may be difficult to understand, even though the individual steps are simple. In fact, this reaction (the burning of gasoline) is still not fully understood in spite of the thousands of man-years that have been spent studying it. But many of the individual steps are well understood. And enough is known about the overall change to begin to design more effective ways of ensuring complete reaction—and thus minimize unburned hydrocarbon, one of the causes of smog.

One important relation between mechanistic steps and the equation for the net reaction cannot be illustrated by the gasoline example. This is the requirement that the sum of the mechanistic steps must equal the overall net equation. We cannot do it here because we do not know all the steps and their relative importance. But we can be sure that they do add up in the required manner. Otherwise atoms would not be conserved, whereas experiments show they are. The law of conservation of atoms describes chemical reactions very well.

One Step Leads to Another

Let's proceed to a simpler system—the rapid reaction between Fe^{3+} and I^- ions, whose net equation is

$$2\,Fe^{3+}\,(aq) + 2\,I^-\,(aq) = 2\,Fe^{2+}\,(aq) + I_2\,(aq) \qquad (13.7)$$

(This is an important equation in the chemical analysis of either iodine

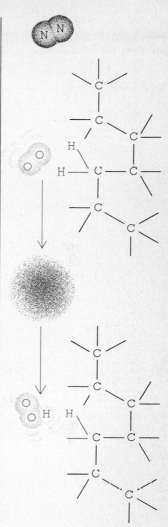

Possible reaction(s)

EXERCISE 13.1
Sketch a collision orientation for the reaction in equation 13.4 that would lead to the observed products. Assume the collision is between an oxygen atom and a molecule of C_8H_{18}. Would the collision orientation you sketched be a likely or unlikely one?

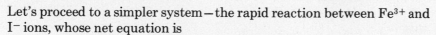

EXERCISE 13.2
What atoms must collide for the
reaction in equation 13.6 to
occur? Is this likely to be the
first step in the reaction of
gasoline with air?

or iron. It serves here as an example of a rather simple net equation.) It is unlikely to be a one-step reaction because that calls for the simultaneous collision of four molecules. But its reaction mechanism, or path, has been studied and is thought to follow these mechanistic steps:

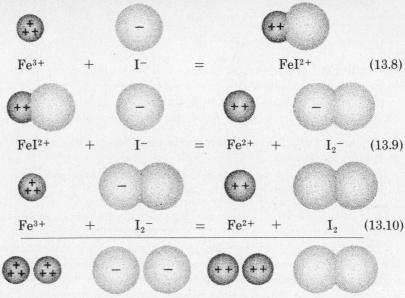

$$Fe^{3+} \quad + \quad I^- \quad = \quad FeI^{2+} \qquad (13.8)$$

$$FeI^{2+} \quad + \quad I^- \quad = \quad Fe^{2+} \quad + \quad I_2^- \qquad (13.9)$$

$$Fe^{3+} \quad + \quad I_2^- \quad = \quad Fe^{2+} \quad + \quad I_2 \qquad (13.10)$$

Net equation: $2\,Fe^{3+}\,(aq) + 2\,I^-\,(aq) = 2\,Fe^{2+}\,(aq) + I_2\,(aq)$ (13.7)

Again, each mechanistic step involves (1) the collision of only two molecules, (2) a simple geometric requirement, and (3) a low energy requirement. Note that one of the molecules in each collision is an original reactant and hence is present in high concentration, as the intermediate molecules might not be. Each collision is between ions of opposite charge, which increases the likelihood of collision and hence helps account for the observed high rate. One of each pair of reactants is spherical, which is the ideal of simplicity for meeting geometric requirements. And the sum of the mechanistic steps equals the equation for the net reaction, as must always be true. Overall complexity is reduced to sequential simplicity. Equally pleasing, the complexity is reduced to a simplicity that is easy to believe and understand.

As you may have noted, I said only that the reaction is *thought* to follow this mechanism. This is consistent with the general situation that there never is a way of proving that a reaction proceeds by only one mechanism. We can "prove" the equation for the net reaction by invoking experimental evidence and the well-established conservation laws. But it is unlikely that details of molecular mechanisms can ever be certain. For example, here is another mechanism that seems to meet the main criteria outlined (bimolecular, simple geometry, low energy, sums to the net equation):

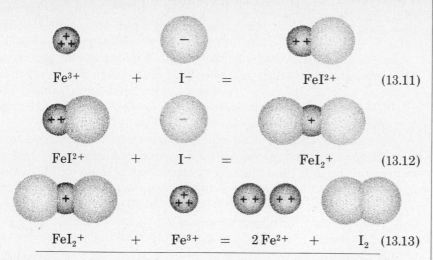

$$Fe^{3+} \quad + \quad I^- \quad = \quad FeI^{2+} \qquad (13.11)$$

$$FeI^{2+} \quad + \quad I^- \quad = \quad FeI_2^+ \qquad (13.12)$$

$$FeI_2^+ \quad + \quad Fe^{3+} \quad = \quad 2\,Fe^{2+} \quad + \quad I_2 \quad (13.13)$$

Net equation: $2\,Fe^{3+}\,(aq) + 2\,I^-\,(aq) = 2\,Fe^{2+}\,(aq) + I_2\,(aq)$ (13.7)

Do you see, however, that one of the steps is probably not a low-energy one? Note that one step requires a collision between two ions of the same sign of charge. That is unlikely.

There are many clever and useful experimental techniques for testing proposed mechanisms. Intermediate molecules may be trapped and identified. Isotopic tracers are used to follow the changes in different molecules. The experimentally observed energy requirements are compared with the computed energies required to accomplish the assumed mechanistic steps. Bonding theory is used to see whether the assumed changes are consistent with what is known about bond symmetry. But this is not a course in kinetics and mechanism, and we shall not study these methods. Rather we shall continue to concentrate on developing a simple, but adequate, picture of the factors that are important if comprehension and control of change are to be achieved.

Bottlenecks: Rate-Determining Steps

Many reaction mechanisms are composed of a single sequence of steps, one of which is slowest and therefore is the rate-determining step, or "bottleneck reaction." When there is a single mechanism, and it has only one slow step, then the system is easy to study. Understanding becomes more difficult if the mechanism consists of several parallel paths, possibly with branches in each.

POINT TO PONDER
The inability to be certain that a suggested mechanism is the only correct one is characteristic of so-called "positive" proofs—that is, proofs that a generalization is *always correct*. Negative proofs (that is, proofs that a generalization is *incorrect*) are common, because only one exception is needed. But no one can ever be sure that no exceptions will *ever* be found to positive generalizations concerning real systems, no matter how firmly they seem to be established at present.

Difficulty also arises when any one of several steps might become rate-determining with only a small change in conditions. It will not surprise you to learn that such systems exist, are even common; but we'll stick with simpler ones for the present. The others involve no new ideas, just greater difficulty in collecting and interpreting experimental data.

Just as there are three factors (number of collisions, geometric requirements, and energy requirements) involved when a reaction occurs, so each of these three factors may determine which step is the rate-determining step—that is, which single step in the mechanism determines the overall rate.

For example, the rate-determining step may be characterized by a low collision rate. The rate of rusting is most commonly diminished by preventing the iron from colliding with reactants that would corrode it. You keep from being poisoned by not exposing yourself to poisons. You drive your car at a high speed exposed on all sides to possible high-energy collisions, but you keep the collision rate low. Collisions between ions of like charge are unlikely and are often rate-determining mechanistic steps.

EXERCISE 13.3
Identify a rate-determining step in some change. You might consider the drying of laundry, the wearing out of a car tire, traveling 100 miles in the United States by air, or converting an infant into a reasonably good tennis player. Is the same step always the rate-determining step in any particular sequence?

Some changes proceed at rates determined by geometric factors. The rate of leakage of a gas from a container depends on the number of holes in the walls and their shapes. The chance of damage to your car in a slight collision is much greater if the contact occurs on the side rather than on the bumpers. The effectiveness of many enzyme catalysts is largely due to their ability to hold the reacting molecules in exactly the right position to ensure a rapid reaction.

Perhaps the most common factor in determining overall rate is the availability of energy. High-energy collisions tend to lead to reaction, because the original bonds can be broken. Molecular velocities increase as temperature rises, as summarized in the Maxwell–Boltzmann curves of Figure 12.10. The increased fraction of molecules with high velocity increases the number of violent collisions. Reactions requiring high energies become more and more likely, but those requiring the most energy will remain the most unlikely. Thus, the mechanistic step with the largest energy requirement is usually rate-determining.

To repeat, it is commonly found that one step in the mechanism is much slower than the others. That step is then the rate-determining step. Any attempt to control the system must concentrate on the rate-determining step and on whether its rate is primarily determined by collision rate or by an orientation or energy requirement. But note that identifying and speeding one rate-determining step will still leave some step in that rate-determining role.

The three factors, as they help determine the rate-determining step, may be more easily remembered if you recall the problems of the forty-niners crossing the United States in the nineteenth century. They could use the geometrically favorable direct route across the Rockies. But this faced them with the energy problem of crossing the Rockies and of facing the rare unfriendly Indians. Some chose a southern route, which avoided the Rockies and much of the fear of Indians, but was longer and drier. Some went by sea and the Isthmus of Panama, thus bypassing the Indians, but substituting a longer route and a high malarial energy barrier at the Isthmus. And some went around Cape Horn, a route that had no Indians, no malaria, and no mountains—but was the longest and provided a stiff energy barrier in the unfavorable winds around Cape Horn. (See Figure 13.2.) The dollar costs were comparable, so each party had to decide its route for itself in terms of hazards (real and imagined); the rate of migration was not the same for each route. Note that the rate-determining steps in all routes varied as time passed; today all the early ones have been removed. History, too, is affected (and directed?) by rate effects.

Figure 13.2
Four routes west.

EXERCISE 13.4
Suggest the probable rate-determining step for most people using each of the four routes mentioned to the American West. Which route would you have chosen? It has been said that if the Pilgrims had landed in California, the rest of the country would never have been settled!

Higher Temperature, Higher Rate (Usually)

The great majority of net reactions proceed more and more rapidly as the temperature is raised (Figure 13.3). You are familiar with many examples: the burning of a match, the cooking of food, the explosion of TNT, the wearing out of automobile tires, the evaporation of crystals and liquids, the dissolving of crystals in liquids, the fading of colors, the growing of plants, the chirping of crickets, the speed of movement of cold-blooded animals, and many more.

Svante Arrhenius (near the end of the nineteenth century) correlated the effect of temperature on rate in terms of an *activation energy,* ΔE_a, for each reaction. The expression ΔE_a is read "delta E sub a," and means the *change* in energy due to *activation*. The activation energy is the average amount of energy necessary to cause reaction of 1 mole of material.

In the simplest conceivable case, where reaction involves breaking a chemical bond completely, the activation energy is equal to the strength of that bond. In most cases, however, a new bond forms as the old one breaks, so the activation energy is smaller than the bond energy of the breaking bond. Checking this relationship is one way of testing suggested mechanisms.

Over the Hills to the Products

Figure 13.4 presents the concept of the activation energy in a particularly simple and effective way. It is assumed that reactions proceed through an intermediate *activated complex* in going from reactants to

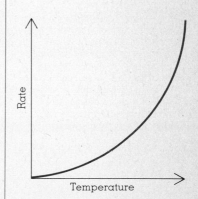

Figure 13.3
Rates of chemical reactions increase faster and faster as the temperature rises.

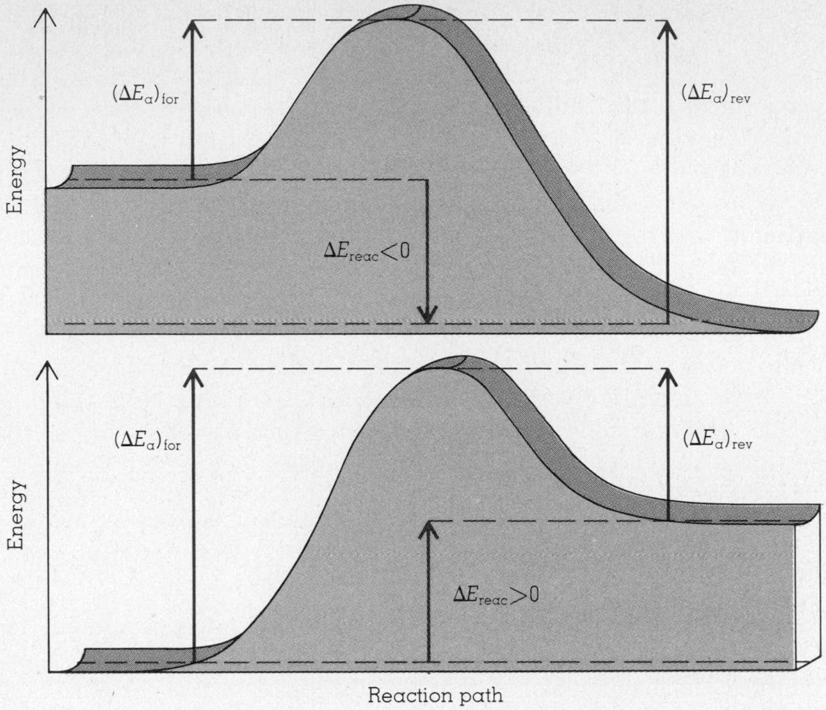

Figure 13.4
Reactants climb an energy hill
before they convert to products.
The path is reversible, so the
overall energy change equals
the difference between the two
activation energies: ΔE_{reac}
$= (\Delta E_a)_{\text{for}} - (\Delta E_a)_{\text{rev}}$.

POINT TO PONDER
This book uses the symbol ΔE_a
for activation energy. Other
books use other symbols. Scientists are not so single-minded in
their devotion to agreement as
is sometimes pictured. I actually
prefer the more complicated
symbol $\Delta E^{\ddagger}$ put forward by
Henry Eyring in his further
development of the Arrhenius
idea. The symbol Δ means
change, and the symbol $\ddagger$ was
chosen by Eyring because he
needed some symbol and the
tuberculosis seal came to mind!
But ΔE_a is simpler and we shall
use it. Perhaps the subscript a
(for *activation*) will also help
you remember Arrhenius.

products. The activation energy, ΔE_a, is the difference in energy between reactants and the activated complex. The reaction proceeds to
products when the activated complex disintegrates into the products.
It could, of course, disintegrate back into the reactants instead. Figure
13.4 shows this change in terms of "climbing an energy hill" as the molecules proceed along the reaction path.

Two additional important points are presented in Figure 13.4. The
difference in energy of reactants and products, ΔE_{reac}, is indicated. It
is called the *energy of reaction*. As Figure 13.4 shows, the products can
have less energy than the reactants, in which case ΔE_{reac} is negative. Or
they can have more energy. Thus, ΔE_{reac} can be either negative or positive.

More important, for the moment at least, is that there seems to be
no reason why the products could not react to return to the original
reactants. This conclusion is correct. When the products do react, they
can form the same activated complex, which can decompose into the
reactants. The activation energy for products doing this we shall call
$(\Delta E_a)_{\text{rev}}$, the activation energy for the reverse reaction. The activation
energy for the forward reaction would then be $(\Delta E_a)_{\text{for}}$. (Normally we
shall be talking of a particular reaction and ΔE_a will be sufficient.)

Note the simple relationship between $(\Delta E_a)_{\text{for}}$, ΔE_{reac}, and $(\Delta E_a)_{\text{rev}}$.

It is

$$\Delta E_{\text{reac}} = (\Delta E_{\text{a}})_{\text{for}} - (\Delta E_{\text{a}})_{\text{rev}} \qquad (13.14)$$

Both $(\Delta E_{\text{a}})_{\text{for}}$ and $(\Delta E_{\text{a}})_{\text{rev}}$ are always positive. But ΔE_{reac} can be either positive or negative (as was pointed out) depending on the relative magnitudes of the activation energies for the forward and for the reverse reactions.

I have drawn Figure 13.4 with only one "hump." This indicates that only one activated state forms between these reactants and these products. Actually there can be many. In fact, there must be as many activated states as there are steps in the mechanism. Figure 13.5 gives some data for a four-step mechanism that occurs in the heart of beef cattle. Here the first mechanistic step has a rather low activation energy $(\Delta E_{\text{a}})_1$, the second step a higher one $(\Delta E_{\text{a}})_2$, and the third a much higher one again $(\Delta E_{\text{a}})_3$. The activation energy for the fourth step, $(\Delta E_{\text{a}})_4$, is intermediate in value, but achieves the highest activated state. The experimentally observed overall activation energy is none of the individual values; it is ΔE_{a}, the difference between the reactants and the *highest* activated state. Step 4 will be rate-determining in this system unless collision probabilities and/or orientation factors are even more limiting than ΔE_{a}.

Now, there are three very important points to understand if you are to handle rate problems:

1. The activation energy, as experimentally determined, is for the overall net reaction, not for the individual steps; ΔE_{a} corresponds to the difference between the energy of the reactants and the *highest*-energy activated state in the process.

Figure 13.5
Energy profile for the reaction path catalyzed by the enzyme lactate dehydrogenase in beef heart. (After W. Borgman et al., *Canadian J. Biochem.*, **53**, 1975. p. 1196; reproduced by permission of the National Research Council of Canada)

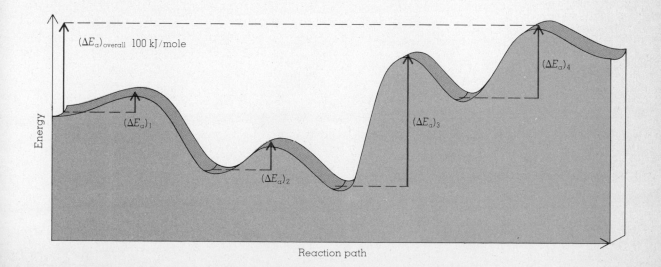

$(\Delta E_a)_{\text{overall}}$ 100 kJ/mole

$(\Delta E_a)_1$

$(\Delta E_a)_2$

$(\Delta E_a)_3$

$(\Delta E_a)_4$

Energy

Reaction path

2. The activation energy for each individual step is *always positive*.
3. Consistent with the Maxwell—Boltzmann distribution of molecular energies, *raising* the temperature *increases* the rates of reaction because more and more collisions have energies *exceeding* ΔE_a.

In other words *every* actual reaction—that is, *every mechanistic* step—has a positive activation energy, and its rate always increases with an increase in temperature because ΔE_a is more available at higher temperatures. (Rarely does ΔE_a equal zero; then the rate does not change with temperature.)

The energy requirements for chemical reactions are similar to those met in driving a car through hilly country where several roads lead to the destination. Minimum energy input is required if the route over the lowest passes is taken. This minimum energy equals that required to get from the start to the highest pass in the lowest route, because the car (if frictionless) can coast as far up each hill as it came down the preceding one.

Some net reactions appear to slow down at higher temperatures. This is because some competing net reaction accelerates faster with rise in temperature, and so slows the net rate of formation of the original products. Thus, if the reactants begin to form an alternative product more and more rapidly, the experimental rate of appearance of the original product may decrease. This will happen even though the original product is forming more and more rapidly from the smaller and smaller amounts of the original reactants that still can react. Enzyme reactions appear to slow down above some temperature because of alternative reactions, including those that decompose the enzyme. For humans the optimum temperature for body reactions is 37°C. Above

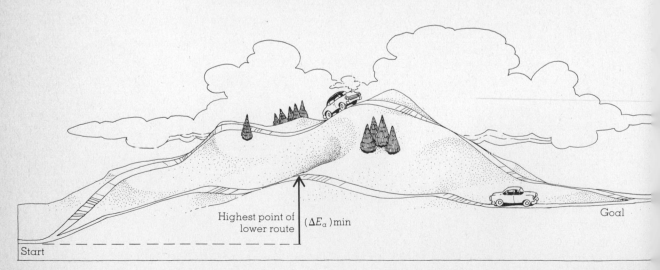

Highest point of lower route (ΔE_a)min

Start

Goal

41°C the rate of damage to enzymes is so great that death occurs unless the temperature is brought down quickly.

Collisions and Concentrations

Almost every mechanistic step, especially in gas-phase reactions, involves a collision between two and only two molecules. For reaction to occur, the collision must provide both the favorable orientation and the necessary activation energy. Of course, most collisions lead to no reaction at all. The molecules come together for a small fraction of a second, say 10^{-12} second, then rebound with all the original bonds intact.

The fraction of collisions that do lead to reaction will be determined (1) by the temperature and the activation energy on the one hand, and (2) by the chance of a favorable orientation on the other. But the total *number of collisions per second* varies with the concentrations of the molecules. In fact, the likelihood of collision is related to the product of the concentrations of the reacting molecules. Thus the likelihood of a bimolecular collision between B and C in a mixture of B and C is related to the quantity $[B] \times [C]$. The square brackets indicate a concentration, so the expression is read as "the concentration of B times the concentration of C."

This product should make sense to you from the following argument. Doubling the concentration of B, $[B]_1$, will make it twice as likely that a molecule of B will be present at a given time at any particular point in the system. Tripling the concentration of C, $[C]_1$, will make it three times as likely to find a C molecule at that point. The chance of finding both of them at the same point at the same time (so they will collide) will be related to the product of the two concentrations. It will have changed from $[B]_1 \times [C]_1$ for the first concentrations to $[B]_2 \times [C]_2 = 2[B]_1 \times 3[C]_1$, for a sixfold increase to $6 \times [B]_1 \times [C]_1$, for the second.

An analogy may help. The rate at which rabbits are shot on the first day of the hunting season is proportional to the concentration of rabbits that year. It is also proportional to the concentration of hunters. Actually it is proportional to the product of the two concentrations, consistent with the "gambler's law." A drop in either concentration lowers the rate proportionately, as the hunters discover during the later days of the season.

Now suppose two identical molecules react. The chance of a B molecule colliding with a B molecule is related to the product of their two concentrations. But the two concentrations are equal, and equal to the total concentration of B. Thus the probability of collision between two B molecules is related to $[B] \times [B] = [B]^2$.

In general, the likelihood of a particular molecular collision is related

POINT TO PONDER
There is a fundamental law of probability that is used with great confidence by gamblers, game players, scientists, and others interested in the chance of an event occurring. It is called the "gambler's law": If the chance of one event occurring is P_1 and the chance of another unrelated event occurring is P_2, then the chance of both occurring is equal to $P_1 \times P_2$. For example, the chance of throwing a coin "heads" is $\frac{1}{2}$. The chance of throwing two heads simultaneously (using two coins) is $\frac{1}{2} \times \frac{1}{2} = \frac{1}{4}$. The chance of throwing three heads (using three coins) is $\frac{1}{2} \times \frac{1}{2} \times \frac{1}{2} = \frac{1}{8}$, and so on.

EXERCISE 13.5
What concentration unit should
be used in equation 13.15?

to the product of the concentrations of all the molecules that collide. But each concentration must be raised to a power equal to the number of that kind of molecules involved in the collision. So we get

$$\text{Rate of reaction} = k \times [\text{B}]^b \times [\text{C}]^c \times [\text{D}]^d \qquad (13.15)$$

where k is a constant called the *rate constant* for the reaction at that temperature. The value of k usually increases rapidly as temperature increases—that is, as ΔE_a becomes more readily available.

The expression could contain additional concentration terms but there are few, if any, molecular reactions that require the presence of more than three different kinds of molecules. Remember that the bracketed terms are the concentrations of molecules B, C, and D, respectively. The superscripts (b, c, and d) are the numbers of each type of molecule taking part in the collision.

Putting It All Together: The Rate Equation

Equation 13.15, which describes the likelihood of collision, assumes a constant temperature. We know that ΔE_a is related to the effect of temperature on rate, and that orientation factors also affect rates. It would be nice if we could get a single equation relating rate, concentration, orientation, *and* temperature effects. And there is such an equation! This may remind you that the ideal gas law put into one equation the separate laws of Boyle, Amontons, Charles, and Avogadro. The situation here is similar. The equation involves a rather complicated exponent. Don't let it bother you if you are not familiar with exponents; I just thought you ought to see how all the factors are related.

Arrhenius, in his early formulation of a rate equation, summarized the effects of the variables we have outlined as follows. He wrote

$$\text{Rate of reaction} \quad = \quad \underset{\substack{\text{effect of} \\ \text{orientation}}}{A} \quad \times \quad \underset{\substack{\text{effect of} \\ \text{temperature and } \Delta E_a}}{10^{-(\Delta E_a/2.303RT)}}$$

$$\times \quad \underset{\text{effect of concentration}}{[\text{B}]^b \quad \times \quad [\text{C}]^c \quad \times \quad [\text{D}]^d} \qquad (13.16)$$

$$= \quad A \quad \times \quad 10^{-0.0522(\Delta E_a/T)}$$

$$\times \quad [\text{B}]^b \quad \times \quad [\text{C}]^c \quad \times \quad [\text{D}]^d \qquad (13.17)$$

POINT TO PONDER
The recurrence (sometimes
called *eruption*) of the same
constants in apparently
unrelated fields gives us confidence that there is a basic unity
to nature—a unity that is *really*
describable in mathematical
terms, not just accidentally so.
Thus R, the gas constant, occurs
in many descriptions of molecular behavior, suggesting that
the theory is well integrated.

where ΔE_a is in joules per mole and T is in Kelvins (0.0522 equals $1/2.303R$). You should recognize that the three concentration terms are exactly the same as the terms in equation 13.15 that describe the effects of concentrations on rate. Similarly,

$$k = A \times 10^{-0.0522(\Delta E_a/T)} \qquad (13.18)$$

So the rate constant, k, depends on three factors: A, ΔE_a, and T. Since ΔE_a and T appear in an exponent there will be a large change in k if ΔE_a and T vary. The rate of many reactions is doubled by a 10 K rise in temperature or a 5% decrease in ΔE_a.

The factor A is called the *Arrhenius collision factor*. It is related to the number of collisions per unit of time that will have a favorable orientation for reaction if all the concentrations are unity. You may wish to convince yourself this is reasonable by setting all the concentration terms equal to 1 and ΔE_a equal to 0. Only A remains. The value of A is essentially independent of temperature and of concentration. It is determined by the molecules involved in the reaction. If they are monatomic, hence can collide and react with almost every collision, A is large. If the reacting molecules are complicated and require a very special orientation upon collision in order to react, A is smaller.

The Arrhenius rate equation summarizes the effects of four factors affecting rate of reaction (A, ΔE_a, T, concentration), just as the ideal gas equation summarizes four factors affecting gas behavior (P, V, T, n). The actual rate is determined by three probability factors, which when multiplied give the rate of reaction.

Rapid rates accompany high temperatures, small values of ΔE_a, large values of A, and/or large concentrations in the rate-determining step. It is possible for ΔE_a to be 0, but ΔE_a is never negative. When $\Delta E_a = 0$, the rate is controlled by the rate of collisions. Such reactions have *diffusion-controlled rates* essentially independent of temperature.

Summary

Rates of reaction are dependent on the number of collisions occurring, the availability of the activation energy (ΔE_a), and the likelihood of a favorable orientation that can lead to products. Normally the overall rate is limited by that of a single rate-determining step. As with the "gambler's law" (p. 235), so here—the overall probability is the product of the individual probabilities. For example,

$$\text{Rate of reaction} = k \times [\text{B}]^b \times [\text{C}]^c \times [\text{D}]^d$$

where [B], [C], and [D] are concentrations involved in the rate-determining step, and k is called the *rate constant*. Each concentration term in the rate equation shows how the rate is affected by changes in the concentration of the corresponding reactant. The rate constant, k, is a function of the system and the temperature, but is independent of the concentration. The value of k is high if orientation requirements for reaction are easy to satisfy, if the temperature is high, and if the activation energy (ΔE_a) is small.

EXERCISE 13.6
Does the rate constant, k, increase or decrease with an increase in temperature?

HINTS TO EXERCISES
13.1 Highly likely. 13.2 An O atom must hit a C atom; cannot happen with C_8H_{18}. 13.3 Your choice. 13.4 Your choice: mountains, deserts, disease, storms, natives, etc. 13.5 Moles/liter. 13.6 The rate constant k increases as T increases.

Problems

13.1 Identify and/or define: activated complex, activation energy (ΔE_a), Arrhenius (collision factor and rate equation), molecular collision, energy of reaction (ΔE_{reac}), gambler's law, mechanistic step, orientation factor, rate constant (k), rate-determining step, reaction mechanism.

13.2 The sugar glucose ($C_6H_{12}O_6$) is sometimes advertised as a source of quick energy, presumable compared to sucrose ($C_{12}H_{22}O_{11}$), from which it can be made by the body. (Actually, the validity of the claim is doubtful.) Why might it be a "quicker" source?

*13.3 Glucose can polymerize in several ways into starches. Starches are either linear with two ends to the molecular chain (amylose) or branched with several ends (amylopectin). Both starches are slowly digestible, amylose more slowly. Glycogen (the form in which glucose is stored in the liver) is highly branched, has many end molecules, and is rapidly convertible to glucose. Suggest a mechanism to interpret the different rates at which these three substances convert to glucose.

13.4 Review the "mechanism" for the blast furnace suggested on page 102. Suggest a more believable set of steps.

*13.5 The most effective treatment for burns is immediate (within a minute or two) immersion in ice water. The ice-water treatment need last only 10 to 15 minutes for small burns. Suggest a possible "mechanism for burning" that would lead to this observed result. (Hint: Do all the bad effects occur during the actual "burning"?)

13.6 Bacillus spores are viable for two or three centuries, lotus seeds for 5000 years. What is a probable rate-determining step for "bringing them back to life"? It takes about 100 minutes in bacillus. Can you suggest why the later steps can then, and only then, occur? (From P. A. A. Sneath, *Nature,* **195**, 1962, pp. 643–646)

13.7 Interpret at the molecular level what happens when you blow out a candle.

*13.8 Polychlorobiphenyl (PCB) is now banned from most uses in the United States, because it is so slowly biodegradable, even though it is less toxic than DDT. It is made from $C_6H_5—C_6H_5$. There may or may not be a Cl on each of the 10 carbons not involved in the bridge. Calculate the number of possible isomeric PCBs. The structure of $C_6H_5—C_6H_5$ is

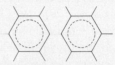

[~200 isomers of PCB]

13.9 J. A. Fraser-Roberts (in *An Introduction to Medical Genetics*) claims there are 10^6 people in East Cape (South Africa) descended from 40 couples in 300 years, a 12,000-fold increase. Is this possible? [~4 children per family would do it]

13.10 Bulk grain and flour are hard to burn, let alone explode. Yet on February 21, 1976, there was an enormous explosion in the concrete grain-storage bins of Goodpasture Inc., on the Houston Ship Channel in Texas. Three people were killed, and millions of dollars damage was done. The event was attributed to a "dust explosion." What is that? (Hint: At what molecular sites can reaction occur between gases and solids?)

13.11 How many of the four forward steps represented in Figure 13.5 have positive values for ΔE_a? How many have positive values for ΔE_{reac}? Answer both questions for the four reverse steps. Try to generalize about the sign of ΔE_a.

*13.12 Many rates of reaction double if the temperature rises 10 K. This is about a 3% rise if T is near 300 K. How much change in rate is caused by a 3% increase in concentration in the rate-determining step? Is a 1% change in temperature or in a limiting concentration the more effective way to change most rates of reaction?

13.13 The concentration of oxygen in natural bodies of water follows a daily cycle, called the *oxygen pulse,* maximizing at about 3 PM and minimizing at about 7 AM. Interpret these data.

13.14 Show why one log burns slowly (if at all), two placed next to each other don't do much better, but placing a third on top of the first two can lead to a steady fire.

13.15 Rewrite the set of equations 13.2 through 13.6, replacing the n's and m's with a set of possible actual numbers.

13.16 You have already noted that equation 13.13 involves the collision between two similarly charged ions. This is an unlikely occurrence.

Another unlikely occurrence is also involved in equation 13.13 (compared to equations 13.8 through 13.12). What is this unlikely feature in 13.13?

13.17 Analyze some simple change with which you are familiar and identify the probable rate-determining step. Suppose you found a means of tripling the speed of that step, would the overall rate of change triple? Why or why not?

Figure 14.1
More changes.
(Photos courtesy United Press
International; Naval Photo-
graphic Center; National
Oceanic and Atmospheric
Administration)

14 PYRES, POPS, AND PLAGUES: REACTION MECHANISMS

Other than all beginning with P's (which is one reason they were chosen) what do the three words titling this chapter have in common? *Answer:* They all refer to occurrences in which rapid change occurs: fires, explosions, and infectious diseases. Change, including rapid change, is the subject of this chapter. This includes, I hope, a change in your understanding of change.

A placid pond,
 a sitting frog.
 Splash!
—*Matsura Basho*
(translated by Larry McCombs).

Fire and Flames, Thermal Chains

Fires are among the most common fast reactions. You recognize at once that they are controlled by the three factors enumerated in Chapter 13: an initial activation energy (such as a match), access of the reactants to one another, and a favorable configuration. The first two are so familiar to you that I need not expand on them. And the orientation effects at the molecular level are also easy to understand in terms of previous examples. But there is an additional application here that we might call the "three-log effect." It is very difficult to keep a one-log fire burning. Even two logs will usually just smolder. But three logs, piled with one on top of the other two, burn along merrily.

Three logs burn because fires are *thermal chain reactions*. The initial heating of the logs and the ignition of escaping gases supply the activation energy to start the flame. But a fire gives out much more energy than was put into it originally. The reaction is said to be *exothermic* (from the Greek *exo,* "out," and *thermos,* "heat"). In terms of Figure 13.4, ΔE_{reac} is negative, or $(\Delta E_a)_{rev}$ is greater than $(\Delta E_a)_{for}$. Thus the reaction energy provides the activation energy for additional molecules to react. But the energy is generated in a flame as light, heat, and hot gaseous products. These tend to escape from the site of reaction quickly. The three-log setup, with its almost enclosed reaction volume, ensures that much of this product heat will be absorbed by further reactants. A thermal chain, in which the heat of the initial reaction activates further reaction, can then be maintained.

Most exothermic reactions can set up thermal chains if enough heat

EXERCISE 14.1
What do you think is the
principal reaction when charcoal burns as glowing coals?
And what is the principal
reaction in the flames over
the charcoal?

is retained in the system. In many systems, the amount of heat retained is more than enough to supply the continuing activation energy. The system begins to warm up. If the temperature rises, the rate of reaction may become unmanageably fast. Thus, exothermic reactions are usually provided with an external cooling system to assist in controlling them.

Fires are self-regulating. If there is a large quantity of fuel, they rather quickly come to a *steady state*. Then the rate of supply of the limiting fuel (usually oxygen) just balances the rate of heat escape. Some of the reaction occurs at surfaces and produces glowing coals, some occurs in the gaseous phase and produces flames. A *flame* is a gas-phase reaction that has achieved a steady state.

Pops and Booms, Branching Chains

A thermal chain that generates energy faster than the energy can escape causes the temperature to rise. In addition, if the rate of heat generation is high enough, more than one new reaction may be activated for each original one. This leads to a *branching chain,* a very rapid and accelerating reaction rate, and (usually) an *explosion*. Large amounts of hot gaseous products may cause a violent explosion from the rapid generation of high pressures. The explosions of dynamite and of trinitrotoluene (TNT) are good examples. Possible net equations are

POINT TO PONDER
Chain letters are a form of
branching chain. As with all
such chains they must peak and
then fade out as the "reactants"
are exhausted.

Dynamite: $CH_2ONO_2CHONO_2CH_2ONO_2$ (c)
$$= 1.5\,N_2\,(g) + 2.5\,H_2O\,(g) + 3\,CO_2\,(g) + 0.25\,O_2\,(g) + 1650\,kJ \quad (14.1)$$

TNT: $H_3CC_6H_2(NO_2)_3$ (c)
$$= 1.5\,N_2\,(g) + 2.5\,H_2O\,(g) + 3.5\,CO\,(g) + 3.5\,C\,(c) + 947\,kJ \quad (14.2)$$

Note that there is no second reagent; the explosive supplies all the fuel. Both reactions rapidly generate a great deal of hot gas; both substances are powerful explosives.

Dynamite contains enough oxygen to convert everything to colorless gases and can be smokeless; TNT does not have enough to do this and must be *smoky*. TNT can be mixed with ammonium nitrate, NH_4NO_3, to obtain a better *oxygen balance* and a less smoky explosion.

Many branching chain reactions are thermal chains, but you may recall from Figure 4.4 that the fission bomb and the nuclear pile work on *branching molecular chain* reactions. More product neutrons are produced than are used up, so the neutron chain reaction branches. In the bomb, this leads to an explosion. In the nuclear pile, control rods are inserted to soak up the extra neutrons so the pile does not explode. It is the possible, though unlikely, failure of such a control system that worries some people.

EXERCISE 14.2
Estimate very roughly (to one
significant figure) the pressure
that would be generated in a
cavity of 1 liter volume by the
explosion of 5 moles of dynamite
to give a final temperature of
1000 K. (See equation 14.1.)

Many chemical reactions are molecular branching chains. The explosion of an H_2-O_2 system is an example. So are the burning of gunpowder in a gun barrel and of gasoline in an internal combustion engine. Each reaction involves both thermal and molecular chains. With gunpowder and gasoline the increasing rate of reaction due to chain branching must match as closely as possible the increasing volume of the reaction chamber (and the increasing velocity of the piston or the projectile). If the fuel burns too rapidly, the push on the projectile or piston will be converted to a blow, great stress will be put on the system, and the pressure may actually rise so high that the system bursts. (See Figure 14.2.)

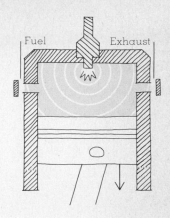

Figure 14.2
The rate of burning of fuel in a piston should increase as the piston accelerates, pushed by the expanding gases.

Plagues and Pestilences: Pass It On

Both simple and branching chain reactions are common in the propagation of disease. The mechanism is, of course, transmission of molecules and molecular systems. In most large cities there is a fairly constant number of people with chicken pox, mumps, colds, or similar infectious diseases. Most are branching chain effects, with one individual often infecting more than one contact. On the other hand, some individuals will infect no contacts. Perhaps their contacts were not intimate enough. The whole mechanism can be interpreted in terms of the three factors discussed in Chapter 13: activation energy, collision, and orientation (with appropriate interpretations at either the personal or the molecular level).

The situation often becomes dangerous when a population has been free of a particular infectious disease for a long time. Then one source often infects many contacts. Each of them infects more and the disease becomes an epidemic, comparable to an explosion. Eventually, immunity (or lowered rate of contact due to immobilization of the population) will cause the epidemic to peak and die out. But sometimes the death rates are worse than in war. About one-fourth of the population of western Europe was killed by the Black Death plague in the fourteenth century. Similar plagues have occurred in many populations whose health records have not been incorporated in written history.

On the other hand, population growth or maintenance is itself a chain reaction. Many of our present problems are due to the branching nature of this chain. Like all other branching chains, it *must* eventually peak and then (probably) fall off. The real question is whether the falling off will be controlled or catastrophic. The possibilities can be readily outlined in terms of the theory of reaction rates and limiting reagents. The only stable long-term possibility is a steady state of approximately zero rate of change in population. The first stages of an explosion can be exciting, but the peak pressures become unbearable.

Having explored some of the scope of rate theory, let's now return to simpler systems.

POINT TO PONDER
I have dealt with rate effects from the molecular point of view of a chemist in terms of collisions, activation energies, and orientation factors. Physicists also treat rates—often as velocities. Mathematicians may express rates as derivatives with respect to time; biologists, as changes in size or population, and so forth. The common factor is the measurement of changes in some quantity as elapsed time increases. Time has not appeared explicitly here because I have kept the time interval constant and talked of comparative rates—that is, amounts of change in some constant elapsed time.

EXERCISE 14.3
Enzyme reactions, like most
reactions, require a collision
between the enzyme (E) and
the other reacting molecule
(called the *substrate*, S). Show
that the rate equation for such a
reaction, rate = $k \times$ [E] $\times$ [S],
becomes equation 14.4 if the
substrate is present at concentrations much higher than that
of the enzyme.

And What Do the Experimental Data Look Like? From Rate Data to Rate Laws

You have already studied the rate of one simple type of reaction as a function of concentration—the decay of a radioactive isotope. In such a case the rate equation becomes

$$\text{Rate of decomposition} = k_1 \times [\text{isotope}] \quad (14.3)$$

Many reactions in living systems have a similar simple rate equation:

$$\text{Rate in the living system} = k_2 \times [\text{enzyme}] \quad (14.4)$$

In the first case the rate is determined solely by the concentration of the isotope; in the second, solely by the concentration of the reacting enzyme.Rates that depend on the concentration of only one substance (with its concentration present in the rate equation only to the first power) are called first-order reactions. The reactions described by equations 14.3 and 14.4 are said to be *first order* in isotope and enzyme concentrations, respectively.

Radioactive nuclei must contain all the ingredients—both matter and energy—to eject a nuclear particle, but they may take a very long time to do it. Half-lives of 10^{10} years and more are known. How does a nucleus store the required activation energy all that time, for it must have had it ever since the nucleus itself was formed? Almost certainly this indicates that nuclei are complicated—for example, made up of many particles, each of which has some variable fraction of the energy. Only occasionally does the energy concentrate on one particle and eject it. Even neutrons are radioactive ($^1_0 n = ^1_1 p + ^0_1 \beta$, half-life = 13 minutes), so apparently neutrons are not "simple" particles.

Note that enzyme control of reaction rates in living systems provides an excellent means of maintaining constant metabolic rates. No matter how much food you have just eaten, the rates of most body reactions are unchanged—even those using up products that come directly from the food. (The body has several mechanisms for maintaining constant concentrations of reacting enzymes.) Many biological reactions are *zero order* in one or more of the reactants. Changes in these concentrations do not affect the rate of reaction.

Consider the problems you would have if the rates of your metabolic reactions varied with the concentration of glucose (one of the products directly available from food). If the concentration of glucose were to double, your metabolic rate would double. As the concentration of glucose fell between meals (assuming no internal source of supply), your metabolic rate would fall. You might have trouble generating the energy to make it to the next meal!

Remember the exponents like b, c, and d in rate equations such as those developed in Chapter 13:

$$\text{Rate of reaction} = k \times [\text{B}]^b \times [\text{C}]^c \times [\text{D}]^d \qquad (14.5)$$

These exponents are known as the order of the reaction in the concentration of the substance (B, C, or D, respectively). The exponents usually have values of 0, 1, or 2. Doubling a concentration whose exponent is 1 (first order) doubles the rate. Doubling a concentration whose exponent is 2 (second order) quadruples the rate. Substances (like most foods in the body) whose concentrations have no effect on rates are said to be zero-order reactants. The rate is unaffected by a change in the concentration of a zero-order reactant.

Table 14.1 gives some rate and concentration data for several systems. Gaseous concentrations are expressed in atmospheres; solution concentrations, in moles per liter; and rates, as relative rates. These particular reactions are chosen not because they are important but because they are well studied and illustrate most of the simple types of rate equations. See if you can deduce the order of each reaction with respect to each reactant. Then fill in the exponents, e and f, in the rate equations.

Rate data such as those presented in Table 14.1 are not always so simply related, but let us be thankful for little things and see what can be determined from them. In System I, do you see that the concentration of iodide ions is constant in the first three lines, but the $S_2O_8^{2-}$ concentration decreases by half, then decreases by half again? Each time $[S_2O_8^{2-}]$ decreases by half, the rate also decreases by half. Thus the order with respect to this ion must be one. In the same way, comparing lines 1, 4, and 5 shows that the rate changes by a factor of 2 whenever the concentration of iodide ions changes by a factor of 2. The rate must be first order in iodide as well.

Net equation: $S_2O_8^{2-} \text{ (aq)} + 3\,I^- \text{ (aq)}$
$$= 2\,SO_4^{2-} \text{ (aq)} + I_3^- \text{ (aq)} \quad (14.6)$$

Rate equation: $\text{Rate} = k_I[S_2O_8^{2-}] \times [I^-] \qquad (14.7)$

The rate is first order in each reactant, or second order overall. *Second order overall* means the sum of the exponents is 2.

System II (in the same way and for the same reasons) gives a first-order rate in each reactant.

Net equation: $SO_2 \text{ (g)} + 2\,H_2 \text{ (g)} = S \text{ (c)} + 2\,H_2O \text{ (g)} \quad (14.8)$

Rate equation: $\text{Rate} = k_{II}[SO_2] \times [H_2] \qquad (14.9)$

The rate is first order in each reactant, and second order overall.

System III shows, in its first two lines, that doubling [NO] increases the rate by a factor of 4, suggesting that $[NO]^2$ (a second-order term

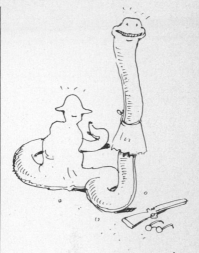

More active after a big meal?

Table 14.1   Rate and Concentration Data for Various Systems[a]

Determine the values for e and f for each system from the data given.

I.   Net equation

$$S_2O_8{}^{2-} (aq) + 3I^- (aq) = 2SO_4{}^{2-} (aq) + I_3{}^- (aq)$$

$[S_2O_8{}^{2-}]$	$[I^-]$	Relative rate
0.08	0.08	4
0.04	0.08	2
0.02	0.08	1
0.08	0.04	2
0.08	0.02	1

Rate $= k_I[S_2O_8{}^{2-}]^e \times [I^-]^f$

II.   Net equation:

$$SO_2 (g) + 2H_2 (g) = S (c) + 2H_2O (g)$$

$[SO_2]$	$[H_2]$	Relative rate
0.03	0.12	1
0.06	0.12	2
0.12	0.12	4
0.12	0.06	2
0.12	0.03	1

Rate $= k_{II}[SO_2]^e \times [H_2]^f$

III.   Net equation:

$$2NO (g) + 2H_2 (g) = N_2 (g) + 2H_2O (g)$$

$[H_2]$	$[NO]$	Relative rate
0.50	0.20	1.0
0.50	0.40	4.0
0.50	0.80	16.0
0.25	0.40	2.0
0.75	0.40	6.0
1.00	0.40	8.0

Rate $= k_{III}[NO]^e[H_2]^f$

IV.   Net equation:

$$CH_3CHO (g) = CH_4 (g) + CO (g)$$

$[CH_3CHO]$	Relative rate
0.146	2.5
0.231	6.3
0.093	1.0
0.121	1.6
0.189	4.1

Rate $= k_{IV}[CH_3CHO]^e$

V.   Net equation:

$$(CH_3)_2CO (aq) + I_2 (aq) = CH_3COCH_2I (aq) + H^+ (aq) + I^- (aq)$$

$[(CH_3)_2CO]$	$[I_2]$	Relative rate
0.0150	0.00234	1.0
0.0300	0.00234	2.0
0.0600	0.00234	4.0
0.0150	0.00489	1.0
0.0150	0.00672	1.0

Rate $= k_V[(CH_3)_2CO]^e \times [I_2]^f$

[a]Concentrations of gases are expressed in atmospheres; of solutions, in moles per liter.

in NO) is correct. Line 3 confirms this. Lines 4, 5, and 6 show that the reaction is first order in H_2:

$$\text{Net equation:} \quad 2\,NO\,(g) + 2\,H_2\,(g) = N_2\,(g) + 2\,H_2O\,(g) \quad (14.10)$$

$$\text{Rate equation:} \quad \text{Rate} = k_{III}[NO]^2 \times [H_2] \quad (14.11)$$

This is one of the few known examples of a gas-phase reaction that is third order overall. The majority of gas-phase reactions are second order overall (first order in each of two reactants).

The data in System IV are not well ordered. But if you rearrange them, say in order of increasing rate, you will see that the rate increases fourfold if $[CH_3CHO]$ doubles (see lines 3 and 5 in the table). All the other data are consistent with a second-order rate equation:

$$\text{Net equation:} \quad CH_3CHO\,(g) = CH_4\,(g) + CO\,(g) \quad (14.12)$$

$$\text{Rate equation:} \quad \text{Rate} = k_{IV}[CH_3CHO]^2 \quad (14.13)$$

This reaction is second order overall, and second order in one reactant. There are no gas reactions known that are higher than second order for any one reactant.

Finally, in System V, the first three lines of data clearly show that the reaction is first order in $[(CH_3)_2CO]$. The data in the last three lines show less obviously, but just as clearly, that the reaction rate is not influenced by varying $[I_2]$. The order with respect to iodine is zero.

$$\text{Net equation:} \quad (CH_3)_2CO\,(aq) + I_2\,(aq)$$
$$= CH_3COCH_2I\,(aq) + H^+\,(aq) + I^-\,(aq) \quad (14.14)$$

$$\text{Rate equation:} \quad \text{Rate} = k_V[(CH_3)_2CO] \quad (14.15)$$

Had System V been studied at varying acidity, you would have observed that the rate is first order in H^+. So a more complete statement is the following:

$$\text{Rate equation:} \quad \text{Rate} = k_V'[(CH_3)_2CO] \times [H^+] \quad (14.16)$$

I have said nothing of the mechanisms of these reactions. Rate data merely summarize overall effects. It is highly likely, however, that each mechanistic step is bimolecular, with no more than one bond breaking as one bond forms. Remember that steps tend to be simple. It is the overall mechanism that gives the complexity we find so often in our observations.

Rate Equations Do Not Come From Net Equations!

So what do we learn from these experimental net equations and rate equations? Remember that our goal is to be able to control rate pro-

EXERCISE 14.4
At what concentration of H_2 would the relative rate of System II (Table 14.1) become 3.0 if $[SO_2] = 0.12$?

EXERCISE 14.5
How do you suppose the data in Table 14.1 on System V were collected without discovering that the rate varied if $[H^+]$ varied?

EXERCISE 14.6
What can you deduce about
the rate equation for the gas
phase reaction from the net
equation $2H_2 + O_2 = 2H_2O$?

cesses. If so, perhaps the first and most important thing to notice is that there is no way of predicting the rate equation from the net equation. Sometimes the exponent in the rate equation is the same as the coefficient of that reactant in the net equation. But often it is not. Sometimes, as in System V, the concentration of one of the reactants does not even appear in the rate equation. And often the concentration of a substance not consumed (perhaps even produced) in the reaction does appear in the rate equation! Remember both the enzymes and System V.

Control of System V cannot be achieved by changing the iodine concentration. Iodine is a zero-order reactant here. But the rate can be varied by changing the concentration either of $(CH_3)_2CO$ or of the hydrogen ion. Note that hydrogen ions are produced in the reaction. Therefore, the rate will increase with time if the relative increase in $[H^+]$ is greater than the decrease in $[(CH_3)_2CO]$, as is apt to be the case. Only experimental data on rates could have revealed this.

Similarly, because System III is second order in NO and first order in H_2, the rate will be more sensitive to variation in $[NO]$ than to variation in $[H_2]$. It is clear which concentration should be monitored more carefully if a constant rate is desired—the $[NO]$.

Note that the exponents in the rate equation do not match the coefficients in the net equation for any of the systems we have studied. Systems are known, of course, where these numbers do match. But it is pure coincidence when they do. Net equations alone give *no* clear information on rate equations or reaction mechanisms.

Chemical Formulas From Rate Equations?

Yes, rate equations (derived from experimental data) do provide information about chemical formulas. The formula for the activated complex can be deduced from the rate equation. Remember that the activated complex is the collection of atoms formed briefly during the rate-determining step. (This is the step, as pointed out in Chapter 13, that has the highest energy as the reaction proceeds from reactants to products. See Figure 14.3.) The rate equation lists the molecules or ions involved in this critical collision. Thus, the formula of the activated complex may be obtained merely by summing the formulas that appear in the experimentally obtained rate equation for the reaction. From this molecular formula, we can guess at a likely structural formula. For the five systems of Table 14.1, we get the formulas shown in Table 14.2.

The molecular formulas for activated complexes obtained from rate equations (as in Table 14.2) are unambiguously correct. The structural formulas are guesses based on the following considerations:

1. Formation of the complex should involve minimal distortion of the

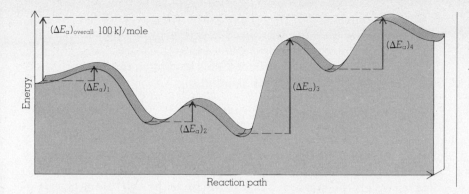

Figure 14.3
Schematic energy profile for a four-step reaction. (After W. Borgman et al., *Canadian J. Biochem.*, **53**, 1975, p. 1196; reproduced by permission of the National Research Council of Canada)

Table 14.2   Formulas of Activated Complexes, Derived from Experimental Rate Equations[a]

System	Experimental rate equation	Molecular formula	Possible structure[b]
I	Rate $= k_I[S_2O_8{}^{2-}] \times [I^-]$ (second order overall)	$S_2O_8I^{3-}$	
II	Rate $= k_{II}[SO_2] \times [H_2]$ (second order overall)	H_2SO_2	
III	Rate $= k_{III}[NO]^2 \times [H_2]$ (third order overall)	$H_2(NO)_2$	
IV	Rate $= k_{IV}[CH_3CHO]^2$ (second order overall)	$(CH_3CHO)_2$	
V	Rate $= k_V[(CH_3)_2CO] \times [H^+]$ (second order overall)	$(CH_3)_2COH^+$	

[a]In System I, five pairs of electrons surround the iodine atom, the O—O bond breaks.
In Systems II and III, the bond in H_2 breaks and the H atoms bond to the electron pairs of an oxygen atom to form H_2O.
In System IV, the collision breaks a C—H bond, and the H is pushed over to bond with the other C, breaking the C—C bond and forming CH_4 and CO.
In System V, the H^+ ion bonds to an electron pair of the electronegative oxygen atom.
Note the importance of proper orientation of the colliding molecules or ions in each case.

[b]Dotted lines represent bonds that form as the activated complex is created. Dashed lines represent bonds that break as the complex reacts further.

POINT TO PONDER
The configuration of a tennis racket and a tennis ball at the instant of contact provides a good analogy to an activated complex. Both racket and ball are deformed from their usual shapes, the potential energy is a maximum, the state lasts only a small fraction of the total reaction time, and orientation is important if an optimal result is to be achieved. A golf club meeting a golf ball and a player's foot meeting a football provide similar analogies.

concentration

activation

Orientation

Reaction!

reactant molecules.

2. Breakdown of the complex should lead with minimal distortion to the observed product molecules.

3. The proposed structure of the complex should be reasonable in terms of polarities and orbital populations.

By definition, an activated complex is unstable. It is located at a peak in the energy curve from reactants to products. Moreover, if two different paths from reactants to products are possible, we can expect that the reaction will proceed along the path with the lower energy peak. Therefore, if two different structures seem possible for the activated complex, the most probable structure is the one that is closest to being a stable molecule (that is, the one that has the lower energy).

Note that all the criteria used to guess at a structure for the activated complex are consistent with the three factors used throughout the discussion of rates—activation energy, orientation, and collision probability.

And So to the Rate-Determining Step!

Knowledge of the rate-determining step is the key to controlling the rate of any change. The activated complex shows the most energetic configuration and the most favorable orientation formed between reactants and products. It will be the most difficult thing to form and will be the result of the rate-determining step. Each of the formulas for activated complexes, and each of the presumed structural formulas, must have been formed in the slowest step of the reaction mechanism.

If you look at the structural formulas in Table 14.2, you will see that each of them "violates" some simple rule of chemical stability—hence these substances should be difficult to form. But they do not violate the rules *much,* so they are not totally unreasonable. Consider each one.

The formula for the activated complex in System I has an iodine atom (actually from a negative ion) attached to an electronegative oxygen. This also forces the iodine to have five pairs of valence electrons. Neither of these situations is unknown in other iodine compounds, but it should be clear that the likelihood of the collision of a negative I^- ion with a negative $S_2O_8^{2-}$ is not high. The fact that the $S_2O_8^{2-}$ is large helps. This distributes the doubly negative charge over eight oxygens, making the charge on each one rather small. So it is easy to believe that the formation of this unstable structure is possible but slow.

Both System II and System III show a hydrogen molecule attached to a single oxygen atom. The polarity is not unreasonable. The complexes can readily dissociate into the known product, water. These facts, plus the existence in each molecule of two different oxygens where the hydrogen could have attached, tend to make the structures reasonable, but the rate of formation will be slow.

The structural formula in System IV assumes that the rate-determining step is the transfer of the CHO hydrogen to the CH_3 carbon, and that the function of the second molecule is to drive it over. Because a rather special direction of approach would have to be used by the colliding molecules, it is easy to see why this step would be slow.

The structural formula in System V seems to cause no real difficulties. The exposed electronegative oxygen seems a natural site for a hydrogen ion. And once this hydrogen-to-oxygen bond has formed, it is easy to see why the iodine could attack the other end of the molecule (now made slightly positive by the presence of the hydrogen ion).

Catalysts—Reactants That Regenerate

You will recall that the success of the ammonia synthesis depended on the discovery of many catalysts. Without them, the world would be unable to support its present population. And you will also recall that every living system has thousands of internal catalysts called enzymes, without which metabolic rates would be far too low for survival. Catalysts are essential to life. They are also essential to the production of about one-sixth of all the manufactured goods in the United States—goods worth about $100 billion annually. How do they work?

One answer is, "Very well, indeed." Enzyme catalysts are known that increase the rate of a reaction by a factor of 1 billion. Averaging all known enzymes, we find that 1 molecule of enzyme can catalyze the formation of about 1000 molecules of product per second. The enzyme *catalase* is even more effective. Its *turnover* number (number of product molecules per molecule of catalyst) is 600,000 per second. No catalysts synthesized by humans (other than enzymes) can approach these turnover numbers. The best synthetic catalysts probably have turnover numbers of about 100 per second.

Catalysts were recognized as such more than 100 years ago, though they have been used by humans since prehistoric times. Lubricants, for example, are catalysts for motion. The lubricants participate in the motion, but are still present when the motion stops. It is this ability to increase a rate without being used up or permanently consumed that characterizes catalysts. The catalyst does this by providing a new mechanism for the reaction. In chemical reactions, the catalyst is one of the reactants in the new mechanism, but the catalyst is later regenerated into its original form and can go around the cycle again and again. Eventually the catalyst may be destroyed. For example, it may collide with a very high energy molecule and be decomposed itself. But good catalysts go through thousands and thousands of cycles without losing the ability to continue the cycle.

Catalysts cannot change the concentrations of reactants. They

EXERCISE 14.7
For the net equation in H_2O,

$$2Fe^{3+} + 2I^- = 2Fe^{2+} + I_2$$

the rate equation at high values of $[Fe^{3+}]$ is

$$\text{Rate} = k[Fe^{3+}] \times [I^-]^2$$

Show that only one step outlined in equations 13.8–13.10 (in Chapter 13) is consistent with the activated complex of the experimental rate equation.

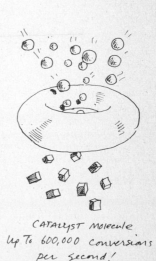

CATALYST Molecule
up to 600,000 conversions
per second!

cannot increase collision frequencies. But they can and do provide different activation energies and orientations. Hence the effects of catalysts can be measured best in terms of their effects on ΔE_a and A (the Arrhenius factor discussed in Chapter 13). Catalysts not only provide a new mechanism for the reaction; good catalysts provide a mechanism in which ΔE_a is smaller and/or A is larger than in the uncatalyzed mechanism.

The molecular effect of a catalyst is well illustrated in the common chemical reaction of hydrolyzing an ester (formula R—COOR′) into its constituent acid (RCOOH) and alcohol (R′OH). The hydrolysis of a fat into fatty acids and glycerine is an example you have seen (Figure 11.8).

Hydrolysis means reaction with water (from the Greek *hydro*, "water," and *lysis*, "take apart"). The rate equation for hydrolysis of an ester is

$$\text{Rate} = k_1\,[\text{ester}] + k_2\,[\text{ester}] \times [\text{H}^+] + k_3\,[\text{ester}] \times [\text{OH}^-] \quad (14.17)$$

EXERCISE 14.8
Write out chemical equations for a possible set of mechanistic steps for the OH⁻ catalyzed hydrolysis of RCOOR′ to the final products RCOO⁻ and R′OH.

The three additive terms in the rate equation indicate three different and independent mechanisms for the hydrolysis. The total rate is the sum of the rates of three separate mechanisms. The first mechanism depends only on concentration of ester to the first order. The second mechanism is first order in both ester and H^+. The third mechanism is first order in ester and OH^-. In each case, water is known to be a reactant, but it is present in such high concentrations that its own concentration does not vary measurably. So, no change in $[\text{H}_2\text{O}]$ occurs to change the rate of hydrolysis, and $[\text{H}_2\text{O}]$ does not appear in the experimental rate equation. The three activated complexes are shown in Figure 14.4. The bonds that break and form are indicated.

$$\text{Net equation:} \quad \text{RCOOR′} + \text{H}_2\text{O} = \text{RCOOH} + \text{R′OH} \quad (14.18)$$

The effect of the H^+ catalyst is to make the central carbon atom slightly more positive and so slightly more attractive to the negative oxygen of the incoming water molecule. The negative OH^- ion attacks the central carbon much more vigorously than a neutral water molecule can. This increases the rate of reaction at which the C—OR′ bond breaks.

In each case, the catalyst is regenerated. The H^+ will ionize, once the new bond has formed and the old bond has broken. The collision of

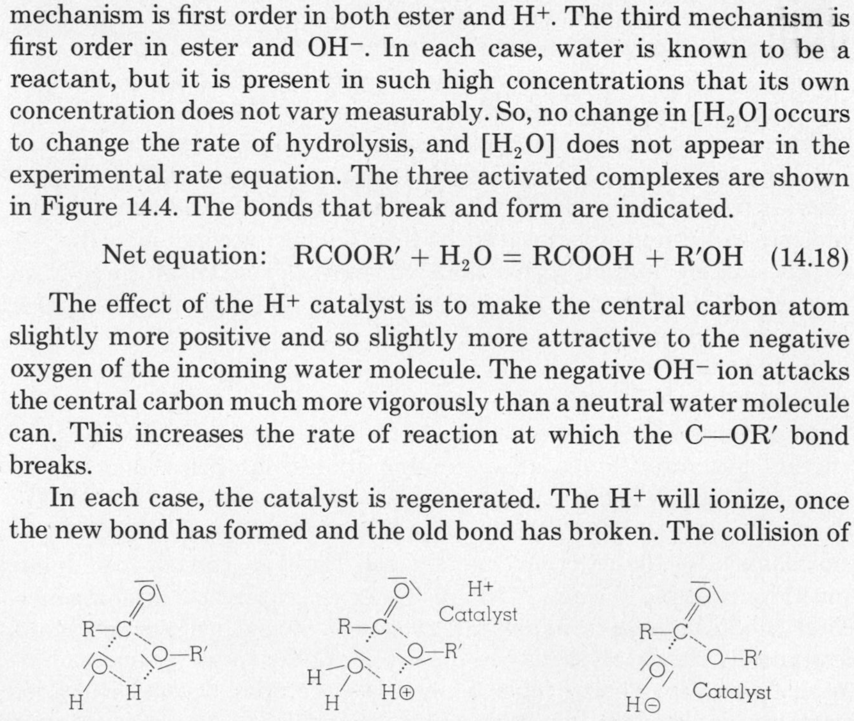

Figure 14.4
Three mechanisms—(a) uncatalyzed, (b) catalyzed by H^+, and (c) catalyzed by OH^-—for the hydrolysis of an ester, RC$\diagdown$$_{\text{OR′}}^{\text{O}}$ to give RCO$_2$H and R′OH. The dashed lines indicate bonds that are breaking; dotted lines, bonds that are forming.

(a) Uncatalyzed (slowest)   (b) Catalyzed by H^+ (faster)   (c) Catalyzed by OH^- fastest)

Corrosion of iron chains.

OH⁻ drives off an R′O⁻ ion, which reacts with water (H_2O) to form R′OH and OH⁻. The regenerated catalysts can then attack additional ester molecules and go around the cycle again and again.

Corrosion (Rust Leads to Bust)

If societies seem dedicated to increasing the rate of consumption of fossil fuels as rapidly as possible, they are also dedicated to the idea of minimizing the rate of corrosion of iron. Use of fuels is intended to decrease the cost of living. Corrosion of iron is known to increase it—by about $5 billion per year in the United States. It is hard to think of any way of making money quicker, or increasing the standard of living faster, than by finding a method of stopping corrosion. So let's look at some of what is known.

Rust is an ill-defined, complex, hydrous oxide of iron, $Fe_2O_3 \cdot n\,H_2O$. As you could guess from its formula, the formation of rust requires the simultaneous presence of iron, water, and dioxygen (or at least some oxidizing agent to remove electrons from the iron). A typical net equation would be

$$2\,Fe\,(c) + 1.5\,O_2\,(g) + n\,H_2\,(\ell) = Fe_2O_3 \cdot n\,H_2O \qquad (14.19)$$

Table 14.3 lists some methods of minimizing the rate of this corrosion reaction.

The corrosion of iron proceeds by many mechanisms, so there is no simple rate law. One common mechanism (outlined in Table 14.4) is well established.

Table 14.3 Some Methods of Reducing the Corrosion Rate of Iron

Treatment	Mechanism of protection	Faults
1. Controlling atmosphere	Lowers O_2 and/or H_2O concentration at the iron surface	Difficult to maintain with O_2 and H_2O so common
2. Painting	Covers iron surface	Paint develops cracks, weathers off
3. Coating with grease or oil	Covers iron surface	Messy, slowly permeable to both O_2 and H_2O
4. Galvanizing	Zinc covers iron surface and converts oxidized iron back to metal	Expensive. Zinc eventually oxidizes and iron corrodes
5. Plating with tin	Tin covers iron surface	Expensive. Iron oxidizes rapidly if there are pinholes in the tin coat, since oxidation is catalyzed by the tin
6. Plating with nickel and chromium	Covers iron surface	Very expensive. Corrosion occurs when pinholes develop
7. Sherardizing with PO_4^{3-}	Adsorbed PO_4^{3-} covers iron surface	Rubs off easily
8. Electrolyzing	Aluminum or magnesium bars fastened to surface corrode rather than iron	Only applicable to large surfaces (ships, pipes, vats) surrounded by an electrically conducting medium such as salt water or moist soil
9. Controlling acidity	Lowers concentration of H^+, which is both a good oxidizing agent and a good catalyst for corrosion	Difficult to do uniformly without thorough stirring of surroundings
10. Keeping other corrosive substances to a minimum	Cl^- and many other ions form soluble iron compounds and increase the corrosion rate	Atmosphere, water, and other surroundings are difficult to monitor

Table 14.4 A Mechanism of Rusting

1. Oxidation of iron	$Fe\,(c) = Fe^{2+}\,(aq) + 2e^-$	As iron contacts H_2O, electrons accumulate in metal
2. Reduction of dioxygen	$2e^- + \frac{1}{2}O_2\,(g) + H_2O\,(l)$ $= 2\,OH^-\,(aq)$	Involves several steps, but occurs when O_2 picks electrons off metal
3. Precipitation of iron(II) hydroxide	$Fe^{2+}\,(aq) + 2\,OH^-\,(aq)$ $= Fe(OH)_2\,(c)$	Involves two steps; occurs when Fe^{2+} and OH^- meet
4. Formation of rust	$Fe(OH)_2\,(c) + \frac{1}{4}O_2\,(g)$ $+ (m-1)H_2O\,(l)$ $= \frac{1}{2}Fe_2O_3 \cdot m\,H_2O$	Involves several steps leading to the actual deposit of rust

Net equation: $Fe\,(c) + \frac{3}{4}O_2\,(g) + m\,H_2O\,(l) = \frac{1}{2}Fe_2O_3 \cdot m\,H_2O$ *sum of steps 1–4*

Treatment 1 in Table 14.3 minimizes steps 2 and 4 in the proposed mechanism. Treatments 2, 3, 4, 5, 6, and 7 minimize step 1 by preventing electrons from leaving the iron. They also minimize steps 2 and 4 by helping to exclude H_2Q and O_2 from contact with the iron. Treatment 8 reverses step 1 each time it occurs by providing electrons from the aluminum or magnesium where they are less tightly held. Treatment 9 encourages the immediate occurrence of step 3 at the corroding surface rather than elsewhere in the system, and thus tends to cover the surface with a semiprotective coat of $Fe(OH)_2$. It also diminishes the rate at which H^+ can pick up electrons from the iron in a reaction parallel to step 2. Treatment 10 normally concentrates on excluding substances that would prevent the precipitation of $Fe(OH)_2$ or rust itself. After all, if these do not precipitate, the surface remains even more exposed to fresh attack than if they partially covered it. Such corrosive substances, which tie up the iron ions in solution, also increase the tendency of step 1 to occur.

In spite of all these measures, iron still rusts—often in places that lead to extensive damage. What is needed is a cheap, tightly adherent, electrically insulating, neat, self-healing coat, resistant to abrasion and weathering. The coat should prevent the escape of electrons from the iron and prevent either water or oxygen, preferably both, from getting to the iron surface. Such a coating probably will be developed sometime. After all, you are covered with a coating that meets many of these criteria. And mechanistic studies will help develop that "ultimate" rust preventive.

Smog Could Be Worse!

There are two main kinds of smog. One, sometimes called *London smog,* is a mixture of coal smoke and fog (hence the name *smog*). It has been a nuisance or worse for more than 800 years. The Royal Court of England had to leave the city of Nottingham in 1257 because the smoke-filled air was so foul. The other smog, known (without pride of ownership of the name) as *Los Angeles smog,* is due to a combination of industrial and traffic exhausts with the famous pure air and bright sunshine of Southern California (or similarly favored places). A more scientific name is *photochemical smog.* Let's look at it (ugh) for a bit (Figure 14.5).

As with rusting, so with photochemical smog, the mechanisms are varied and complicated—otherwise the control would probably be much simpler than it is. The ideal, of course, is a single, well-understood rate-determining step! Table 14.5 summarizes some of the ideas of the mechanism of Los Angeles smog.

You see that reaction 1, a key step to the formation of photochemical smog, is the absorption of light by NO_2, a brown gas. The resulting

Figure 14.5
(top) Los Angeles smog.
(bottom) Los Angeles on an unsmoggy day. (Courtesy UPI)

Table 14.5   A Possible Mechanism for the Formation of Photochemical Smog[a,b]

1. Light absorption	$NO_2 + h\nu = NO + O$ (SO_2 and SO_3 are other light-absorbers)

2. Atomic oxygen chain	**$O + R \rightarrow R + RCHO$** (an aldehyde)
	$R + O_2 \rightarrow RO_2$
	$RO_2 + NO \rightarrow RO + NO_2$ Net equation:
	$RO \rightarrow R + O$ $NO + O_2 = NO_2 + O$

3. Ozone chain	**$O + O_2 \rightarrow O_3$** (ozone)
	$O_3 + R \rightarrow RCO_2 + RCHO$
	$RCO_2 + NO \rightarrow RCO + NO_2$
	$$\overset{O}{\underset{\|}{}}$$
	$RCO + NO_2 + O_2 \rightarrow RC{-}O{-}O{-}NO_2$ (probably two steps)
	acyl peroxynitrate

4. Chain termination	$O + NO_2 \rightarrow NO + O_2$
	$O_3 + NO \rightarrow NO_2 + O_2$
	$RO + NO_2 \rightarrow RONO_2$

[a]R = unburned hydrocarbon residues.
[b]The four reactions set in boldface type produce ozone, aldehydes, and acyl peroxynitrate, the main irritants in photochemical smog. The overall process can be described as the photochemical initiation of the oxidation of gaseous hydrocarbon residues to noxious chemicals.

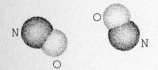

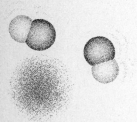

Fast step

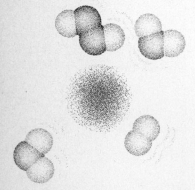

Rate-determining step

oxygen atoms, O, then initiate either the atomic oxygen or the ozone chain reactions (2 or 3) that involve unburned hydrocarbons and produce the irritants and active pollutants. But where does the NO_2 originate? Usually from high-temperature processes in which air is involved. Under these conditions, dinitrogen and dioxygen react to give NO, as outlined earlier in Chapter 10. But NO is not NO_2!

However, if you were to generate NO, which is colorless, in the laboratory and then mix it with air, brown NO_2 would form almost instantaneously. Yet the same reaction, which does indeed give the NO_2 necessary to form photochemical smog, requires hours in the air over a big city. Peak production of NO occurs from 7 to 9 AM and from 4 to 6 PM, but the peak smog hours are often from 2 to 4 PM. The rate equation for the conversion of NO to NO_2 provides the interpretation:

$$\text{Rate} = k[NO]^2 \times [O_2] \qquad (14.20)$$

The reaction of NO with O_2 has a third-order rate, second order in NO. But exhaust fumes do not contain much NO, and it is quickly diluted until its concentration may be only a few parts per million. Because [NO] appears in the rate equation as its square, any decrease in [NO] has a big effect on the rate. Decreasing [NO] from, say, the 0.1 atm you might have generated in the laboratory, to 10^{-6} atm (as over Los Angeles) changes the concentration by a factor of 10^{-5}, and decreases the rate by a factor of 10^{-10}! So bad smogs occur in the afternoon rather than all day long (unless the NO_2 hangs around from the day before).

Suppose the reaction were still third order, but first order in NO and second order in O_2. Then the rate would decrease only by a factor of 10^{-5} from the lab rate, and we'd have smog most of the day.

It is data such as those in Table 14.5, and the rate data from which they come, that are being used to seek better methods of smog control, including emission control. Unfortunately emission of NO is one of the hardest types to control. And most past methods of decreasing [NO] have increased unburned hydrocarbons, so that the smog level has not changed much.

Back to the Balky Car

Days, weeks, or even months ago, you read in Chapter 2 about some problems with starting a car—problems that were traced to the gasoline and solved by using lighter fluid in the air intake. And you were asked to consider *why* the molecules in lighter fluid boil more readily than those in gasoline, and how gasoline could be "winterized" for easier starting of cars in cold weather. You may recall that I indicated that vaporization of the gasoline was the rate-determining step on that cold morning.

Vaporization occurs when a molecule is hit vigorously enough and in the right direction so it achieves the activation energy necessary to burst through the surface and escape into the gas phase. The reverse process *(condensation)* merely involves a molecule striking the surface and sticking, and so may have no activation energy. Thus the activation energy of vaporization will be very close to the energy of vaporization. To a good approximation this is the same as the heat of vaporization. (See Figure 14.6.) Now vaporization involves breaking all the intermolecular bonds to the molecule, which vaporizes and enters the gas. So we can expect the tendency of a liquid to vaporize to be directly related to its heat of vaporization, which is in turn determined by the strength of the intermolecular bonds. This is so.

The molecules in gasoline are mostly C_8H_{18}, those in lighter fluid C_5H_{12}. The smaller molecules will have less surface area on which van der Waals forces can act. So they will have weaker intermolecular bonds and a smaller heat of vaporization. The smaller heat of vaporization will lead to higher vapor pressure. Because liquid gasoline does not burn well, but the vaporized gas does, a high vapor pressure is needed in the engine.

The chemical reasoning for designing a winterized gasoline is as follows:

1. Combustion in the engine requires gaseous gasoline mixed with air.
2. The rate-determining step in a cold engine is vaporization of the liquid gasoline.

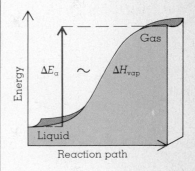

Figure 14.6
Reaction path for vaporization. Note that the activation energy for condensation (the reverse process) is almost zero.

3. The activation energy equals the heat of vaporization.
4. At lower temperatures, a fuel with lower heat of vaporization must be used.
5. The heat of vaporization of hydrocarbons of small molecular surface area is lower than that for those with larger area at the same temperature.

So we add to the gasoline some smaller molecules that can vaporize readily and initiate the combustion reaction.

Gasoline combustion provides a good example of orientation effects

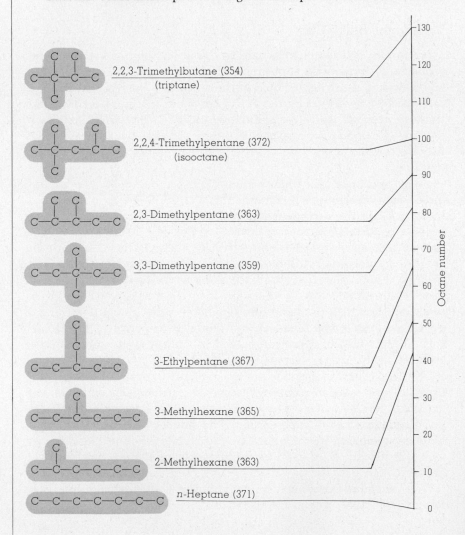

Figure 14.7
Structure and octane rating. The octane rating of normal heptane (n-heptane) is zero; of isooctane, 100. Other values are determined by relative performances in test engines. Boiling points (in Kelvins) are given in parentheses.

2,2,3-Trimethylbutane (354)
(triptane)

2,2,4-Trimethylpentane (372)
(isooctane)

2,3-Dimethylpentane (363)

3,3-Dimethylpentane (359)

3-Ethylpentane (367)

3-Methylhexane (365)

2-Methylhexane (363)

n-Heptane (371)

Octane number

if you consider the relation between octane rating and the molecular structure of the gasoline. Figure 14.7 shows the structures of some molecules common in gasoline, their boiling points (Kelvin), and the octane rating (0 to 130) of each pure isomer. Do you see any likely correlation? The differences in boiling points are too small to have much effect.

Octane rating measures the speed with which the isomer burns in the automobile cylinder. Thus, 2,2,4-trimethylpentane is given a rating of 100, heptane one of 0. Low octane number means a high rate of combustion, high octane number means a lower rate. High-octane fuels can be mixed with air, heated, and compressed more than those with lower octane numbers. All the low-octane fuels are extended molecules. All those with high octane numbers are compact molecules with a smaller surface area that can react with the oxygen. So their rate of reaction will be lower, the combustion will be slower and smoother, and the car will not knock because of premature ignition or too rapid combustion in the cylinder. Addition of $Pb(C_2H_5)_4$, lead tetraethyl, at a rate of about 1 g/liter also slows combustion and raises the octane rating about 10 units.

A Word to the Wise

One word of warning: Both at the molecular level and at the bulk level, mechanisms, slow steps, activated complexes, activation energies, and catalysts often change if the conditions change a great deal. Only experimental data taken under conditions close to those of maximum interest should be used in designing controls.

A simple example is the silicone called *Silly Putty*. The activation energy for slow flow is low, and the viscosity (resistance to flow) changes very slowly with temperature. But the activation energy for rapid flow is very high. Even large amounts of energy do not lead to rapid flow but to elastic bounce. Larger energy inputs lead to shattering of the sample rather than to flow.

If the main subject of this book were kinetics and mechanisms, we could continue with these and many more systems to explore total mechanisms from start to finish. But such work would be of doubtful value to those not going on in science.

So why should you, as a probable nonscientist (though I hope always scientific when that is helpful), have labored this far through the chaotic collisions that lead to the smooth reactions and the final order we observe? Well, see if the summary convinces you that the gain in understanding of changes and their control has justified the venture. Then try the problems. May you find them interesting and answerable.

EXERCISE 14.9
Are the differences in boiling points in Figure 14.7 consistent with the differences in rate of evaporation that you would predict for these molecules?

Summary

Even complicated changes proceed through a series of simple steps. These can be interpreted at the molecular level in terms of

1. the likelihood of a molecular collision;
2. the availability of an activation energy;
3. the occurrence of a reasonable orientation of the reactants so that the collision leads on toward the products.

Varying temperature allows us to determine the energy of activation and to predict rates at other temperatures. Determining the experimental rate equation allows us to determine the formula of the activated complex. The rate equation also allows us to make good guesses about the detailed nature of the collisions that have led to reaction. Catalysts supply additional mechanisms, usually with lower activation energies and/or more favorable orientation requirements.

The experimental determination of the rate equation reveals the effect of varying the concentration of each reacting substance. Often, varying the concentration of a reactant has no effect on the rate. Often, varying the concentration of something that does not appear in the equation for the net change does have an effect. Control is possible only if you find which concentrations are related to the rate, so you can direct the control efforts at them.

The same factors found at the molecular level are also useful in interpreting and controlling change in macroscopic systems: energy, orientation, and number of contacts.

HINTS TO EXERCISES
14.1 Forms CO at surface, CO_2 in gas. 14.2 3×10^3 atm. 14.3 All E's will be tied up with S's as ES's. 14.4 $[H_2] = 0.09$. 14.5 Initial $[H^+]$ must be large compared to that generated by the reaction. 14.6 Only the sum of the mechanistic steps. 14.7 Equation 13.9 involves $FeI_2{}^+$ as an activated complex. 14.8 $RCOOR' + OH^- \rightleftharpoons RCO(OH)OR'^- \rightleftharpoons RCOOH + OR'^-$, $OR^- + H_2O \rightleftharpoons ROH + OH^-$. 14.9 Boiling point falls as the molecules have less surface area.

Problems

14.1 Identify or define: catalysis, chain reaction (branching, molecular, thermal), corrosion, exothermic reaction, explosion, flame, hydrolysis, octane number, order of reaction, smog, steady state, vaporization.

14.2 The inside of your galvanized steel trash can has begun to rust. Which of the following would help minimize further rusting: keeping the inside dry, painting the inside, painting the outside, wire-brushing the rust and treating the spot with aqueous trisodium phosphate (Na_3PO_4), using plastic liners, or cleaning off the rust with rust remover?

14.3 Tin cans at the bottom of a swampy lake often stay unrusted for long periods. Why?

*14.4 A drop of water on a piece of iron leads to solution of iron under the center of the drop, but rust forms around the edge of the drop. Draw a diagram and interpret the observations.

*14.5 Corrosion (solution of metal) occurs much more rapidly at the head and at the sharp tip of an iron nail than along the cylindrical shank. Review Chapter 11 on surface effects; then consider how a nail is made by deforming a piece of wire, and interpret these observations.

14.6 An old rule was that quiet streams purified themselves after 7 miles, mountain brooks after a few hundred yards. Any justification for either statement? If so, what pollutants would and would not be removed? Why does the mountain brook require less distance?

14.7 Gunpowder for large guns is often made of cylindrical "grains" that may be 1 cm in diameter. These grains have several holes penetrating them from end to end, parallel to the axis of the cylinder. Gunpowder ignites and burns only at its surface. What is the function of these holes? (Hint: What would happen to the burn rate during burning if only solid cylinders were used?)

*14.8 Would you expect the fraction of hydrocarbons containing C=C bonds to be higher in gasoline or exhaust fumes?

*14.9 *Homeostasis* is the general term for the mechanism by which a human being maintains weight and other physiological properties at almost constant values. Suggest one or two types of level that could help maintain homeostasis.

14.10 A catalyst often first forms a bond and later breaks the same bond to be regenerated. Should the bond formed be very strong, say 400–500 kJ/mole? How about very weak, say 20–30 kJ/mole?

14.11 A certain extract from living cells proved to be a powerful catalyst for the hydrolysis of sucrose. One researcher argued that the catalyst was an enzyme. Another said no, it was protons. Design an experiment that should quickly settle the discussion.

14.12 Self-cleaning ovens (introduced by General Electric in 1963) burn off the oven residues at about 470°C and then pass the gases over the platinum and palladium oxide catalysts at 650–750°C. Write equations for some possible reactions the catalysts might affect.

14.13 High-energy light of wavelength from 297 to 315 nm causes sunburn and tan; lower-energy light of wavelength from 315 to 330 nm, only tan. The compound 2-ethoxy ethyl *p*-methoxy cinnamate, $CH_3CH_2OCH_2OCO$ ⬡ OCH_3, prevents burn but allows tan. How does it probably do this? What happens to the energy of the the 297–315 nm light?

14.14 Formaldehyde (H_2CO) is a constituent of smog and causes eye irritation at 1×10^{-8} mole/liter. It can be formed by the reaction

$$O_3 \text{ (g)} + C_2H_4 \text{ (g)} = 2\,CH_2O \text{ (g)} + O \text{ (g)};$$

Rate $= 2 \times 10^3$ [liters/(moles $\times$ s)]
$$\times [O_3] \times [C_2H_4]$$

If $[O_3] = 5 \times 10^{-8}$ mole/liter and $[C_2H_4]$

$= 2 \times 10^{-8}$ mole/liter, what is the rate of H_2CO formation in moles/(liters $\times$ s)? Would these conditions lead to eye irritation levels of $[H_2CO]$? [Smog forms in about 2 hours]

14.15 The morning rush hour in Los Angeles produces much more smog than does the evening one. Why so?

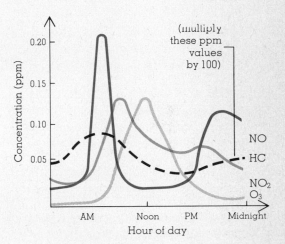

14.16 You are designing a machine to produce ice to cool drinks. Comment on the relative advantages of the following shapes: large cubes, small cubes, crushed ice, cubes with a hole through the center.

*14.17 According to Grand and Braarud (*J. Biol. Board of Canada,* **1,** 1935, p. 279), the temperature for the maximum growth rate of four plankton species of the same genus is 3°C, 6°C, 9°C, and 12°C (in April, May, June, and August, respectively). How would you account at the molecular level for (a) the fact that there is a maximum, and (b) the variation within a given genus?

*14.18 A gaseous mixture of hydrogen and air will not explode if the volume percentage of hydrogen is greater than 74% or less than 6%. Why will mixtures outside the 6–74% range not explode?

14.19 Fast-draw expert Bill Munden is said to have been clocked at 0.001 second for drawing a Colt 45, firing it, and jamming it back into its holster. Evaluate this claim from a molecular point of view.

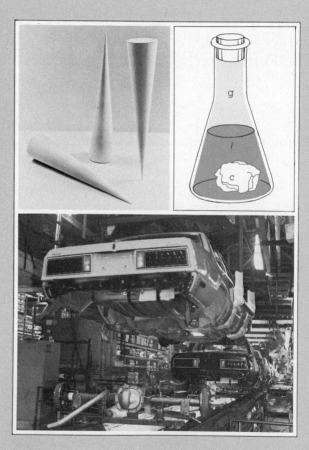

Figure 15.1
Constancy yet change—static,
dynamic, and steady states.

15 CONSTANCY YET CHANGE: EQUILIBRIUM STATES AND STEADY STATES

Many of the systems around you are continually undergoing change; yet at the same time they show remarkable constancy. Turn on an electric light. Obviously, changes are occurring because the bulb emits light and we know that electric current is flowing through it. However, the properties of the bulb itself remain constant, as do the rates of electrical input and light output. On a much larger scale, ocean currents behave similarly. So does the sun, or a gas flame, a human heart, an automobile assembly line, or a mountain lake with a constant water level. Or consider the contents of a sealed tin can, the concentration of dissolved calcium carbonate in the ocean, the gas pressure in a sealed soda bottle, or the position of an undisturbed pool ball on the table or in a pocket.

Such apparently constant systems are often said to be *at equilibrium*. However, we shall find it convenient to classify only the latter four systems (tin can, calcium carbonate, soda bottle, and pool ball) as *equilibrium states*. The other systems are *steady states*. Do you see a distinction between the two groups?

> We are not stuff that abides, but patterns that perpetuate themselves.—*Norbert Wiener*.

Constancy

There are three ways to maintain constancy in a system. The most obvious is to have everything fixed, so that no changes whatever are occurring. Such systems are said to be in a *static state* (sometimes called a *static equilibrium*). In terms of large-scale changes, the pool ball on the table or in a side pocket is in a static state. So are the three cones shown in Figure 15.1. No motion (no large-scale change) is occurring. However, if we examine these systems (or any other) at the molecular level, we find that motion and change are occurring. There are no known static systems at the molecular level, so I shall not discuss the static state further.

A second method of maintaining a constant system is with a *steady*

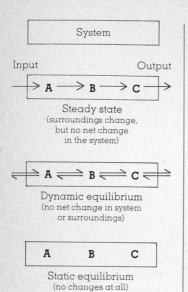

Figure 15.2
Both the steady state and
dynamic equilibrium occur
among molecules; static equilibrium never does.

state. In a steady state, change occurs at a constant rate and always in the same direction. Any new material produced in the system is immediately "used up" at the same rate, so the concentration of each substance and the properties of the system remain constant. In the light bulb, electrical energy is being converted to heat and light energy at a constant rate. Electrical energy enters the bulb at a constant rate, and heat and light energy leave at constant rates. Therefore, the temperature, brightness, and other properties of the bulb remain constant (once the steady state has been achieved). Raw materials and parts enter the assembly line of Figure 15.1 at constant rates, and finished units leave at a constant rate. The concentrations of unfinished units at various intermediate stages in the factory remain constant; the overall properties of the assembly line do not change.

A steady state must have a continuous supply of materials and energy from outside itself. It must also deliver an exactly balancing amount of materials and energy to its surroundings, so that its own content of materials and energy remains constant. If either the supply or the output of the steady state is altered, the steady state itself is altered.

A very large number of systems in nature are either steady states or very close to steady states. Most living systems are close to steady states most of the time. Because birth and death (or cell division, at least) are characteristic of all life, it should be clear that no individual living system maintains a steady state indefinitely. Eventually, some necessary supply or some essential output stops, and the system changes. A steady state, unchanging itself, always causes changes in its surroundings.

A third way of achieving constancy is called *dynamic equilibrium.* In a dynamic equilibrium, every change (forward reaction) is offset by an opposite change (reverse reaction) proceeding at the same rate. If a system in dynamic equilibrium gains any energy or material from its surroundings, it immediately offsets that gain by the loss of an identical amount of energy or material (in the same form). Any molecular change within the system is offset by the reverse change occurring elsewhere in the system. There may be small fluctuations from complete constancy. But over a period of time, all the properties of the system remain unchanged although there is constant and rapid activity at the molecular level. An equilibrium system neither undergoes nor causes any *net* change anywhere. (See Figure 15.2.)

Steady States and Reaction Rates

The concepts of rate-determining step and of steady state complement one another nicely. In a steady state, the rates of all forward reactions must be the same. In many mechanisms, the overall rate is set by the rate-determining step. Normally the rate-determining step is one of the

EXERCISE 15.1
Is a country with a constant
population at static equilibrium,
at dynamic equilibrium, or in a
steady state?

first steps in the mechanism, and it is a slow step. The concentrations of the reactants in that slow step have relatively high values. There are many more collisions between the reactant molecules than there are reactions. In other words, most of the collisions lead only to rebound, not reaction, because of strict orientation requirements or because of the lack of sufficient activation energy.

The products of the rate-determining step become the reactants for the next step. But they do not build up to a high concentration because the following reaction is "faster"; that is, it can proceed at the same rate as the rate-determining step but with lower concentrations of reactants. The rate-determining step is the step with the smallest rate constant, k. This is because a lower fraction of the collisions lead to reaction, as a result of a more demanding orientation requirement (low A value) and/or a higher activation energy (ΔE_a). (See Figure 15.3.)

If the products of a system are continually removed, back reactions are impossible and the system tends toward a steady state. One of the functions of enzymes is to set up and maintain steady-state reactions, so that there are almost constant concentrations of the intermediates between the reactants in the food and the body's waste products. These intermediates are the physiologically necessary molecules. The advantages of such an arrangement to the living system are obvious.

Steady states in chemistry are difficult to treat with precision. Usually the intermediates are present at such small concentrations that they cannot be isolated—often, not even identified. The description of the system is simplified only by the fact that all the rates are equal. The measurement or calculation of the various activation energies and of the structures of the intermediates is normally impossible with present methods. Intensive research is underway in all these areas and the promised rewards for new discoveries are great. We will not be able to go much further at the level of this book.

But we might digress a moment to mention some steady states found in societies. The principles are remarkably similar to those applying at molecular levels. Industrial assembly lines are much more efficient if they operate on a continuous, steady-state flow system. Normally one wishes every step to proceed at the same rate, and to have that rate fast enough to supply the consumer market. Hence rate-determining steps are identified, and parallel assembly lines are put in operation for slow steps, whereas single assembly lines can handle the faster steps.

Office practice is handled similarly. Every executive is faced with the problem of maximizing the output rate while minimizing expenses for both raw materials and employees. This is one of the main reasons that employees with a range of skills are so useful. They can shift from one job to another as the rate-determining steps vary.

Population levels furnish another example where steady states seem

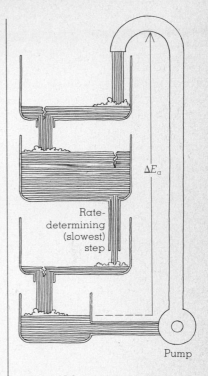

Figure 15.3
Flow analogy to a chemical steady state. All flow rates are equal. The maximum rate of flow is set by the accumulation of "potential reactants" just prior to the rate-determining step. ΔE_a is modeled by the height the water must be pumped, the A factor, by the diameter of the smallest outlet pipe.

POINT TO PONDER
More and more industries are shifting from batch-processing to continuous-flow (steady-state) systems. In batch processing a set of raw materials is processed into products, then a new batch of raw materials is started so that every step in the process is intermittent. In continuous-flow systems every step is operating at all times, using the capacities of both employees and machines more efficiently.

EXERCISE 15.2
The birth rate in freely breeding, sexually reproducing species is first order (not second order) in the population:

Rate = k [population]

What does this suggest about the number of individuals involved in the rate-determining step? Is this reasonable?

more and more desirable. The identification, discussion, and control of the rate-determining steps here may be one of the most important activities ever undertaken by human societies.

It is important to remember that, though a steady-state system does not itself change, it requires supplies from its surroundings and delivers outputs (both products and wastes) to its surroundings. Thus steady states lead to many changes in the environment. They survive on changes "elsewhere."

Dynamic Equilibrium and Random Motion

Molecules are in rapid, random motion. Their collisions with one another and with the walls of their containers lead to pressure. When the collisions are sufficiently violent, chemical reaction becomes likely. Molecular velocities are distributed over a wide range (remember the Maxwell–Boltzmann curve). Thus, a few violent collisions are likely under any normal conditions, hence a few reactions will always occur.

Escaping tendency is a particularly useful phrase to describe the likelihood of molecular reaction. No matter where an atom or molecule is, its continual random motions make it tend to escape to some other site. The pressure of a gas at constant temperature measures its escaping tendency. So does the vapor pressure of a liquid, or the concentration of a dissolved material. Thus, any increase in pressure or concentration indicates an increase in escaping tendency. We shall explore the factors that help determine escaping tendencies.

If two pure reactants are mixed in a closed container, they will initially have only one another to react with. They can only have a "forward" escaping tendency. But, as product concentrations increase in the container, collisions between product molecules become more and more likely. Some of these collisions will also be violent; they will lead to formation of the activated complex, leading back to the original reactants. Thus the reverse reaction begins to occur. The higher the concentration of products, the larger the escaping tendency in the reverse direction.

Simultaneously, the concentrations of the original reactants are decreasing, because of the forward reaction. Thus collisions between original reactants get less and less likely, and the rate of the forward reaction decreases. Lowering the concentration lowers the escaping tendency.

A time will come, with the rate of the forward reaction decreasing and the rate of the reverse reaction increasing, when the rates of forward and reverse reactions become equal. The system then comes to dynamic equilibrium. There will then be no further net change. The escaping tendencies are equal in both directions. The molecular reactions continue in their random way, but the rate of each forward step is the same as the rate of its own reverse step. The same dynamic equilibrium is reached whether one starts with pure reactants or pure products (Figure 15.4).

All molecular systems in closed containers eventually reach dynamic equilibrium. Thus all chemical equations describe two reactions; both forward and reverse reactions can and do occur.

True dynamic equilibria have no net effect on their surroundings, nor do they ever run down from frictional effects. Only molecular systems have these characteristics. Thus only molecular systems can be at full dynamic equilibrium. However, larger systems can come close to equilibrium, and the ideas developed for the molecular level can be extended to larger ones, as Figure 15.5 shows.

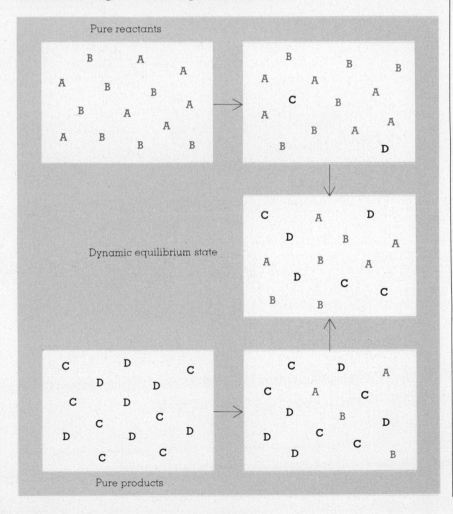

Figure 15.4
Dynamic equilibrium for the net reaction, $A + B = C + D$. The same dynamic equilibrium state is reached from either $A + B$ or $C + D$ as a starting mixture.

Figure 15.5
Some large-scale systems
that approximate dynamic
equilibria for short times.

The dynamic nature of molecular equilibria is readily demonstrated by adding a small amount of a radioactive substance to a system already at a dynamic equilibrium that includes the same substance only in a nonradioactive form. The radioactivity will distribute to all the accessible molecules. For example, add radioactive dihydrogen (HT) to the gas-phase equilibrium

$$H_2\,(g) + C_2H_4(g) = C_2H_6\,(g) \qquad (15.1)$$

Here T stands for tritium (^{3}H). The tritium will soon be present in C_2H_5T and C_2H_3T molecules, as well as in the original HT. All molecular equilibria behave similarly. All are dynamic, with both forward and reverse reactions occurring constantly.

The Law of the Perversity of Nature: Le Châtelier's Principle

In 1884, shortly after the concept of dynamic equilibria at the molecular level had been established, Henri Le Châtelier (pronounced *Luh Shot' uh lee ay*) suggested that systems in dynamic equilibrium are resistant to change. His principle can be expressed in many different ways, such as the following:

LE CHÂTELIER'S PRINCIPLE: If a stress is applied to a system in dynamic equilibrium, the equilibrium will shift in a direction that minimizes the stress.

Or, to put it another way: If you try to change a dynamic equilibrium, the system will react as if it were trying to undo your changes. Hence, we might call this principle the *law of the perversity of nature*.

Suppose that we wish to apply "stress" to a chemical system. We can do so by changing the temperature, the pressure, or the concentrations of the ingredients—in each case, we are changing the escaping tendencies of various components of the system. Both the ideal gas law ($PV = nRT$) and the rate equation (with its concentration and temperature terms) contain these stresses.

The ammonia synthesis reaction supplies some good examples of Le Châtelier's principle. The equation for the reaction is

$$N_2\,(g) + 3\,H_2\,(g) = 2\,NH_3\,(g) + 92\text{ kJ (heat)} \qquad (15.2)$$

Table 15.1 shows how applying certain stresses affects the equilibrium. For example, increasing the temperature leads to the net decomposition of NH_3—because that reaction absorbs the added heat and minimizes the rise in temperature; a heat-absorbing reaction occurs. Heat-absorbing reactions are said to be *endothermic* (from the Greek *endo*, "within," and *therm*, "heat"). Similarly, heat-releasing reactions are called *exothermic* (from the Greek *exo*, "outside").

EXERCISE 15.3
Five grams of D_2O are mixed with 6 of ordinary H_2O. What molecular formula will be most common at equilibrium? (The symbol D is used for the hydrogen isotope called *deuterium*, also expressed as ^{2}H.)

Evaporation minimizes the heating effect of the sun.

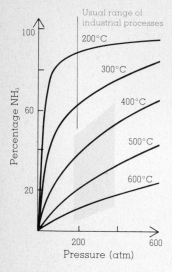

Figure 15.6
The effect of pressure and
temperature on the H_2–N_2–NH_3
gas-phase equilibrium. (From
J. T. Gallagher and F. M.
Taylor, *Ed. Chem.*, **4**, 1967, p. 30)

EXERCISE 15.4
Show that the effect of pressure
is one of the causes for the
freezing point of water being
below its triple point. See
Figure 12.6.

Table 15.1 Le Châtelier's Principle and the Synthesis of Ammonia

$$N_2\,(g) + 3H_2\,(g) = 2NH_3\,(g) + 92\ kJ\ (heat)$$

Applied stress	Equilibrium shift[a]	Reason for shift: Reaction uses up some added
Increased [NH_3]	⟵ decreases extra [NH_3]	NH_3
Increased [N_2]	⟶ increases [NH_3]	N_2
Increased [H_2]	⟶ increases [NH_3]	H_2
Increased total pressure	⟶ increases [NH_3]	Pressure
Increased temperature	⟵ decreases [NH_3]	Heat

[a]Note that each shift minimizes the effect of the applied stress.

Adding NH_3 shifts the equilibrium to the net formation of N_2 and H_2—so that the rise in NH_3 concentration is minimized and is less than it would have been if no shift had occurred. Adding N_2 or H_2 shifts the equilibrium toward the formation of NH_3, hence minimizing the increase in concentration of the N_2 or H_2. And increasing the total pressure on the system causes the net formation of NH_3—because this lowers the number of moles of gas present (4 moles of gas react to give 2 moles), and so minimizes the rise in pressure. In every case the equilibrium shifts to *minimize* the effect of the imposed change (Figure 15.6).

Let's summarize Le Châtelier's principle as applied to chemical systems in dynamic equilibrium:

1. Supplying heat leads to endothermic (heat-absorbing) reactions.
2. Raising pressure leads to reactions that diminish the volume.
3. Increasing the concentration of an ingredient leads to reactions using up that ingredient.

The reverse statements are also true:

1a. Cooling a system leads to exothermic (heat-releasing) reactions.
2a. Decreasing pressure leads to reactions that increase the volume.
3a. Lowering the concentration of an ingredient leads to reactions forming that ingredient.

Again, in general: Making a change in a system at dynamic equilibrium leads to shifts in the equilibrium that tend to offset the change.

Le Châtelier's principle is one of the more useful simple generalizations about behavior in nature. It is useful in discussing large-scale dynamic systems near equilibrium, as well as those at the molecular level. The economic law of supply and demand, for example, is remarkably parallel to Le Châtelier's principle.

Equilibrium Constants and Rate Equations

Even before Le Châtelier stated his principle, Cato Guldberg and Peter Waage had suggested (in the late 1860s) that there are some simple numerical relationships among the concentrations of species present in an equilibrium system. In fact, they had decided (using experimental evidence rather than theory) that there exists for each dynamic equilibrium a ratio of concentration terms that remains constant even though the concentrations at equilibrium vary. It is called the *equilibrium-constant expression*.

Today, we can reach the same conclusion, starting with the rate equation developed in Chapter 13. For the mechanistic step,

$$W + X \rightarrow products \tag{15.3}$$

the rate equation is

$$Rate_1 = k_1[W] \times [X] \tag{15.4}$$

Suppose that the products are Y and Z. As their concentrations (and thus, their escaping tendencies) increase, they will begin to react to form W and X (in the reverse reaction):

$$Y + Z \rightarrow W + X \tag{15.5}$$

with the rate equation

$$Rate_2 = k_2[Y] \times [Z] \tag{15.6}$$

Eventually, the system will come to dynamic equilibrium. When this happens, $rate_1 = rate_2$. Therefore (from equations 15.4 and 15.6) we know that

$$(Rate_1)_{eq} = (rate_2)_{eq} \tag{15.7}$$

and

$$k_1[W]_{eq} \times [X]_{eq} = k_2[Y]_{eq} \times [Z]_{eq} \tag{15.8}$$

I have added the subscript *eq* as a reminder that this equality holds true only for equilibrium concentrations. Now let's do some simple algebraic manipulation, "collecting" the k's on one side and the concentrations on the other side of the equation:

$$\frac{k_1}{k_2} = \frac{[Y]_{eq} \times [Z]_{eq}}{[W]_{eq} \times [X]_{eq}} \tag{15.9}$$

But k_1 and k_2 are both constants (if T is constant), so

$$\frac{k_1}{k_2} = constant = K_{eq} \tag{15.10}$$

POINT TO PONDER
Equilibrium-constant expressions (like equation 15.13) can be written by examining the chemical equation for the net reaction (equation 15.12). Rate equations cannot be written from information in the net equation. Remember, rate equations can come *only* from actual measurements of the effects of changes in concentrations on the experimentally observed rate. (Review pages 226–231.)

K_{eq} is called the *equilibrium constant*. (Lowercase k's indicate rate constants; capital K's indicate equilibrium constants.) We may now write

$$K_{eq} = \frac{[Y]_{eq} \times [Z]_{eq}}{[W]_{eq} \times [X]_{eq}} = \left\{ \frac{[Y] \times [Z]}{[W] \times [X]} \right\}_{eq} \tag{15.11}$$

So this ratio of the equilibrium concentrations of reactants and products is itself a constant (for a given temperature).

Similarly, we could derive an expression for the equilibrium constant of a more general reaction (with any number of moles of the various species involved in the reaction):

$$b\mathrm{B} + c\mathrm{C} = d\mathrm{D} + e\mathrm{E} \tag{15.12}$$

For this general case, we get the

EQUILIBRIUM-CONSTANT EXPRESSION:

$$K_{eq} = \left\{ \frac{[D]^d \times [E]^e}{[B]^b \times [C]^c} \right\}_{eq} \tag{15.13}$$

The concentration of each product molecule is placed in the numerator and raised to the power of its coefficient in the net chemical reaction. The concentration of each reactant molecule is placed in the denominator and raised to the appropriate power. (It should be obvious how to modify this general equation for net reactions that do not have exactly two reactant species and two product species.) For simplicity, the subscript *eq* is often omitted, giving a simplified equilibrium-constant expression:

$$K = \frac{[D]^d[E]^e}{[B]^b[C]^c} \equiv \frac{[D]^d \times [E]^e}{[B]^b \times [C]^c} \tag{15.14}$$

But you *must* remember that K has a particular (constant) value *only* for the concentrations of a system in dynamic equilibrium at some given temperature. Figure 15.7 gives some experimental data on the approach of the gaseous H_2–I_2–HI system to equilibrium.

For substances in solution, concentrations are expressed in moles per liter. Example: $[H^+]$ means molarity of hydrogen ions. For gases, concentrations are expressed as pressures (we shall use units of atmospheres). Example: $p(O_2)$ means pressure of dioxygen gas. In cases where the solvent (usually water) enters into the reaction, it is usually present in such great quantity that its concentration is not changed measurably by the small amount that reacts. Thus, the concentration of water remains constant and is included in the value of K. Thus $[H_2O]$ does *not* appear in the equilibrium-constant expression.

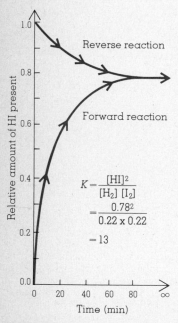

$$K = \frac{[HI]^2}{[H_2][I_2]}$$
$$= \frac{0.78^2}{0.22 \times 0.22}$$
$$= 13$$

Figure 15.7
Two approaches to equilibrium for gaseous H_2, I_2, and HI. The same equilibrium constant is found, regardless of initial concentrations, if T ($=445°C$) is constant. Net equation:
H_2 (g) + I_2 (g) $\rightleftharpoons$ 2HI (g) + 10 kJ.

Table 15.2   Examples of Equilibrium-Constant Expressions

Equation	Equilibrium-constant expression	Comment
$HgCl_2$ (aq) = Hg^{2+} (aq) + $2Cl^-$ (aq)	$K_{eq} = \dfrac{[Hg^{2+}][Cl^-]^2}{[HgCl_2]}$	
$PbCl_2$ (c) = Pb^{2+} (aq) + $2Cl^-$ (aq)	$K_{eq} = [Pb^{2+}][Cl^-]^2$	$[PbCl_2]$ (c) is a constant and is included in K_{eq}
CN^- (aq) + H_2O = HCN (aq) + OH^- (aq)	$K_{eq} = \dfrac{[HCN][OH^-]}{[CN^-]}$	Concentration of H_2O is a constant included in K_{eq}
Na (c) + H_2O (ℓ) = Na^+ (aq) + OH^- (aq) + $\frac{1}{2}H_2$ (g)	$K_{eq} = [Na^+][OH^-][p(H_2)]^{1/2}$	[Na (c)] and $[H_2O]$ are constants included in K_{eq}

Similarly, the concentration of a crystal is unchanged by reaction. The number of moles of the crystal changes, but its volume changes also—hence the concentration is unchanged. So concentrations of crystalline substances are also included in the value of K.

Concentrations of water (and other solvents) and of crystals do not appear in equilibrium-constant expressions. Their escaping tendencies do not change during reaction. Only concentrations that can be varied appear there. Table 15.2 gives some examples.

Let me remind you once more that equations like equation 15.14 were first put forward by Guldberg and Waage on the basis of experimental evidence and with no theoretical link to rates of reaction. But it is always pleasing when experimental data from various sources (here from equilibrium and from rate studies) prove to be tied together by a simple theoretical pattern. The concept of dynamic equilibrium provides that tie between rate data and equilibrium data.

Changing Concentrations and Shifting Equilibria

The equilibrium-constant expression applies only to equilibrium concentrations, but there are many possible sets of these. What equation 15.14 says is that the *ratio* in the expression—not the individual concentrations—must be constant. And as with Le Châtelier's principle, so with the equilibrium-constant expression—one can predict shifts in equilibria when various concentrations change.

Consider again the general equation

$$bB + cC = dD + eE \qquad (15.12)$$

and its equilibrium-constant expression,

$$K = \frac{[D]^d[E]^e}{[B]^b[C]^c} \equiv \frac{[D]^d \times [E]^e}{[B]^b \times [C]^c} \qquad (15.15)$$

EXERCISE 15.5
Write an equilibrium-constant expression for a limestone stalactite ($CaCO_3$) in equilibrium with aqueous Ca^{2+} and CO_3^{2-}, two ions common in hard water.

POINT TO PONDER
Science uses all techniques possible to simplify its ideas and to reduce to a minimum the need to memorize isolated, or apparently unrelated, data. Sometimes mathematical relationships are found among experimental measurements. Sometimes they are derived from theory and compared with experimental results. Sometimes a theory developed for one set of observations can be extended to describe others. The ultimate test of any theory is agreement with experiment. The hope is that theories will also suggest new experiments.

EXERCISE 15.6
Which would have the bigger
effect on the concentration of
NO_2 in a gas-phase equilibrium
with NO and O_2—doubling the
concentration of NO or doubling
the concentration of O_2?

Any change, at constant temperature, that tends to increase the denominator must also increase the numerator, and vice versa. To be specific, if some B is added to the equilibrium, this will tend to increase the denominator term [B] and decrease the ratio. But actually the equilibrium will shift to maintain the same K ratio. This can be done if B reacts with C to form more D and E. The reaction will lower the concentration of C and raise the concentrations of D and E until a new set of concentrations is reached that gives the original K ratio again.

In the same way, adding either D or E would shift the equilibrium toward B and C in such a way as to maintain the ratio value at K. In other words, the escaping tendencies all change, but K does not.

Note that the predicted direction of the shifts in concentrations are exactly those you would deduce from the application of Le Châtelier's principle. But now quantitative (that is, numerical) predictions are possible, especially if the value of K is known.

Let's consider a very simple system, the ionization of water. The chemical equation is

$$H_2O\ (l) = H^+\ (aq) + OH^-\ (aq) \tag{15.16}$$

and the equilibrium-constant expression is

$$K_w = [H^+][OH^-] \tag{15.17}$$

Note that the concentration of H_2O, 55.5 moles/liter, is so large that it is unchanged by the reaction. So its constant value, 55.5, is included in K_w. The *ionization constant of water* is used so often it is given a special subscript, as in K_w. The value at laboratory temperatures is

$$K_w = 10^{-14} = [H^+][OH^-] \tag{15.18}$$

EXERCISE 15.7
What is the concentration of H^+
in a water solution whose
$[OH^-] = 10^{-9}$ mole/liter?

Equation 15.18 contains two concentrations that can vary and one numerical constant, 10^{-14}. If either concentration is known, the other can be calculated. If $[H^+] = 1$ mole/liter, then $[OH^-] = 10^{-14}$ mole/liter. In pure water, each ionization produces one H^+ and one OH^- so their two concentrations must be equal—and each must be 10^{-7} mole/liter, because $10^{-7} \times 10^{-7} = 10^{-14} = K_w$, the ionization constant for water. Table 15.3 gives some possible equilibrium concentrations for H^+ and OH^- in water. As $[H^+]$ increases, $[OH^-]$ decreases, and vice versa.

**Table 15.3
Some Equilibrium States for H_2O**
$K_w = 10^{-14} = [H^+][OH^-]$

[H+]	[OH−]
$10^0 = 1$	10^{-14}
10^{-2}	10^{-12}
10^{-4}	10^{-10}
10^{-6}	10^{-8}
10^{-8}	10^{-6}
10^{-10}	10^{-4}
10^{-12}	10^{-2}
10^{-14}	$10^0 = 1$

Predicting Equilibrium Shifts From Rate Effects

The direction of shift for a dynamic equilibrium can be predicted by using either Le Châtelier's principle or the equilibrium-constant expression. The equilibrium-constant expression can give a quantitative prediction of effects. Equilibrium shifts can also be predicted directly from rate effects.

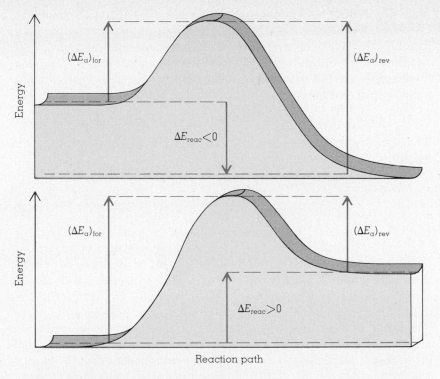

Figure 15.8
Energy changes during
reaction. Note that
$\Delta E_{reac} = (\Delta E_a)_{for} - (\Delta E_a)_{rev}$.

$(\Delta E_a)_{for}$ $(\Delta E_a)_{rev}$

$\Delta E_{reac} < 0$

Energy

$(\Delta E_a)_{for}$ $(\Delta E_a)_{rev}$

$\Delta E_{reac} > 0$

Energy

Reaction path

You will recall that increasing the temperature always increases the rate of every mechanistic step. The size of the increase is related to the size of the activation energy. (See equation 13.17.) The larger the activation energy, the faster the rate increases with rise in temperature (Figure 15.8). The activation energies of the forward and reverse reactions differ. The difference is the overall energy of reaction. Because the rate in the direction with the larger activation energy increases more with increase in temperature, the result of a rise in temperature must always be a shift favoring the direction of the endothermic net reaction. Thus, the reaction diagrammed in the upper half of the figure will shift to the right, and that in the lower half of the figure will shift to the left, when the temperature increases. The reverse shift occurs if the temperature decreases.

Raising the temperature raises the escaping tendency of all molecules, but not equally. Escaping tendencies of processes with higher activation energies increase faster with temperature than do those of processes with lower activation energies.

At dynamic equilibrium, the rate of each forward reaction exactly equals the rate of its own reverse reaction. If we set up the rate equations for each step and multiply them all together, we would find that (at equilibrium) the product of all the forward rates equals the product

POINT TO PONDER
The equations and statements of science often seem merely to summarize what you already know. You may feel like a student who, reading Shakespeare for the first time, finds only a mass of familiar quotations. But scientific equations and ideas, like passages from Shakespeare, cover many circumstances other than the ones you are familiar with. It is in their ability to shed light on new experiences that much of their power rests.

EXERCISE 15.8
Would you expect the dissociation of gaseous water into its elements to increase or decrease at higher temperatures? How about higher pressures?

of all the reverse rates. This equality equation involves the concentration of each species in the net equation raised to an exponent equal to its coefficient in the net equation. We can write the product of the rate equations at equilibrium directly from the net equation.

Consider the net equation that accounts for the slow loss of bleaching power in chlorine water (in a swimming pool, for example):

$$2\,Cl_2\,(aq) + 2\,H_2O\,(\ell) = 4\,Cl^-\,(aq) + 4\,H^+\,(aq) + O_2\,(g) \qquad (15.19)$$

The equality of the forward and reverse rates at equilibrium can be written

$$\text{Forward rates} = \text{reverse rates}$$

$$k_{\text{for}}[Cl_2]^2 = k_{\text{rev}}[Cl^-]^4[H^+]^4 p(O_2) \qquad (15.20)$$

Note that concentrations (moles per liter) are used for dissolved Cl_2, Cl^-, and H^+, while pressure, $p(O_2)$, is used for gaseous oxygen.

Changing the concentration of any reactant or product changes the rates with which it collides and reacts, as shown in equation 15.20. The changing rates of molecular collisions will cause the equilibrium to shift right or left, depending on which concentrations increase or decrease. The extent of the shift is dependent both on the change in concentration and on the exponent of the concentrations that change. A change in $[H^+]$ or $[Cl^-]$ in this system will have a much bigger effect, because of the large exponent (4), than the same magnitude of change in either $[Cl_2]$ or $p(O_2)$. Changing $p(O_2)$ will have the least effect of all; $p(O_2)$ has the smallest exponent (1). This is one reason that adjustment of acidity, $[H^+]$, in a swimming pool is even more important than the exact value of $[Cl_2]$.

It should please but not surprise you that all three methods of predicting the direction of equilibrium shifts agree. If the methods have validity, they must agree. Sometimes you will use Le Châtelier's principle, sometimes the equilibrium-constant expression, and sometimes

EXERCISE 15.9
Which direction will the gas-phase equilibrium, $H_2 + I_2 = 2\,HI + 10$ kJ, shift if the temperature is increased? If the pressure is increased? If more H_2 is added? The other variables (P, V, or T) are held constant in each case.

EXERCISE 15.10
Which of the three methods of correlating shifts in equilibria is most quantitative? Which is easiest to use?

Figure 15.9
ΔE_{reac} is independent of the presence or absence of a catalyst; only $(\Delta E_a)_{\text{for}}$ and $(\Delta E_a)_{\text{rev}}$ change. They change an equal amount: $\Delta E_{\text{reac}} = (\Delta E_a)_{\text{for}} - (\Delta E_a)_{\text{rev}}$.

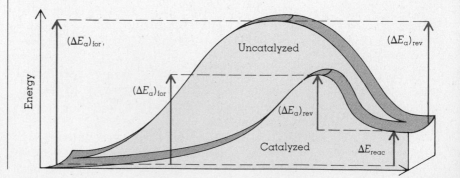

kinetic arguments. Which method you use will depend on the data at hand, the complexity of the system, and the balance needed between qualitative and quantitative predictions. The escaping tendency of any substance is increased by

1. an increase in concentration;
2. an increase in pressure (actually a concentration effect);
3. an increase in temperature—but not all escaping tendencies change at the same rate, so equilibria can and do shift with change in temperature.

Catalysis and the Equilibrium State

The presence or absence of a catalyst does not affect equilibrium concentrations. Remembering this will keep you from a common error. Knowing the reasons will be even more helpful. Figure 15.9 should make this clear.

A catalyst provides a new mechanism for a reaction—nothing more. The position of an equilibrium is determined by the differences between the reactants and the products. These differences are unchanged by the catalyst, because the catalyst affects only the path between the reactants and the products. (See Figure 15.9.) A catalyst supplies equally faster rates for both the forward and the reverse reactions. There is no shift in the position of the equilibrium. A catalyst does not change the escaping tendencies of reactants or of products. Examining Figure 15.10 may help you to remember this. The figure shows that a catalyst that did shift an equilibrium could lead to a perpetual-motion machine. That would make the present $100 billion contribution of catalysts to the economy of the United States seem like peanuts. But experiment shows the machine does not work.

Phases and Their Changes

Among the simplest equilibria are those between the various phases—crystal, liquid, and gas—of a pure substance (Figure 15.11). Especially simple, in terms of equilibrium constants, are equilibria involving the gas phase—either crystal–gas or liquid–gas transitions.

A rise in temperature increases the escaping tendency of a liquid faster than it increases the escaping tendency of the corresponding gas. Similarly, the escaping tendency of a crystal increases faster than that of the corresponding liquid (or gas).

Because heats of vaporization (liquid to gas) and of sublimation (crystal to gas) are always positive ("heat-absorbing" processes), the

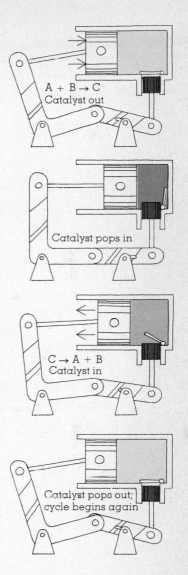

Figure 15.10
Perpetual motion results if a catalyst were to shift the equilibrium for A (g) + B (g) = C (g) to the left. Introducing and removing the catalyst would drive the piston back and forth and give work, but would use no fuel! Too bad it doesn't happen.

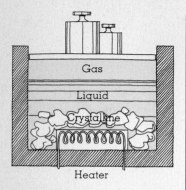

Figure 15.11
Schematic diagram of a setup
for the study of the effects of
varying temperature and pressure on the phase equilibria of
a pure substance.

vapor pressures of crystals and liquids always increase with rise in temperature, consistent with Le Châtelier's principle.

In the same way, heat is required to expel a gas from a liquid. Thus
the solubilities of gases in liquids decrease with rise in temperature.

This is one reason carbonated beverages are chilled before being
served. Both the equilibrium pressure and the rate of escape are lowered.
This allows the "tangy" taste of the carbon dioxide to remain in the
beverage longer. Warm beer, as served in English pubs, has a "smoother"
taste. (Shall we say that different palates like different tastes, and avoid
an argument over which is better?)

You should have no trouble understanding and predicting the effects
of pressure changes on phase equilibria. Increasing the pressure on an
equilibrium will always shift the equilibrium toward the lesser volume.
I defer to a later chapter the effect on phase changes of varying the concentrations by adding other materials.

Don't Hold Your Breath

It is uncomfortable to hold your breath. It may surprise you that the
problem is not so much running out of oxygen as it is building up too
high a concentration of carbon dioxide. The same thing is true of suffocation in a closed space. The atmosphere becomes toxic because of the
rising carbon dioxide concentration, not because of the falling oxygen.
In fact, the lethal level is reached when the carbon dioxide rises to about
0.05 atm. At that point the oxygen level has only fallen to about 0.15
atm, equivalent to the oxygen pressure on a mountain about 3 km
(10,000 ft) high. Plenty of oxygen remains, but the carbon dioxide is
lethal.

Let's investigate the carbon dioxide vaporization in your lungs. The
usual pressure of carbon dioxide in the atmosphere is 0.0003 atm, but
the pressure of CO_2 in equilibrium with the blood is 0.06 atm, and that
in exhaled air is about 0.05 atm. Under these conditions there will be
net flow of carbon dioxide from the blood, through the lung tissue, and
into the inhaled air, to be exhaled with the next breath.

But, if the inhaled air already contains 0.05 atm of CO_2, the rate of
reverse transfer from air to blood will almost equal the forward rate
from blood to air, and the net rate of exhalation of CO_2 will approach
zero. The CO_2 system is nearly at equilibrium. Unconsciousness and
death will follow as the concentration of CO_2 in the blood starts to rise.
Actually a concentration of even 0.03 atm of CO_2 in the atmosphere
leads to serious toxicity, even in the absence of exertion.

Breathing, like many physiological effects, maintains steady states,
not just equilibria. But equilibrium data help us to understand some of
the limits of the steady state.

Competition and the Equilibrium State

Chemical reactions occur because the random collisions between molecules break old bonds and form new ones. The equilibrium state, especially at low temperatures, favors those molecules that are most strongly bonded. Equilibrium also favors the widest possible variety of molecules, consistent with the random nature of collisions, especially at high temperatures. Thus there is continual competition between the tendency to form strong bonds and the tendency to have random distributions of atoms and energy. We shall return to this competition again and again.

For the present we shall view the competition in terms of the particles that are passed around the system. The simplest electrically neutral atom is that of hydrogen. Even more simple are hydrogen ions, H^+. Their small size and small mass make them mobile, and some of the most common equilibria involve transfer of hydrogen ions (protons) between various molecules. We shall look at this when we examine *acid–base equilibria,* and find that protons are one of the most common acids.

More complicated groups are also interchanged. The visible evidence is often the formation or disappearance of a phase—say bubbles of gas, or particles of solid (a precipitate), in a solution. Thus, solubility gives rise to common and important equilibrium states.

Electrons (e^-) are also simple particles. As you may recall, competition for electrons is called reduction and oxidation (or *redox* for short).

The next few chapters apply the ideas of equilibrium and rate to some of these competitive systems. You should not be surprised to find that many reactions in living systems involve transfers of protons and/or electrons because their rate of transfer can be fast and easily controlled.

Summary

Any molecular system left alone in a closed container will eventually come to dynamic equilibrium. In a dynamic equilibrium the random collisions of the molecules bring about a state in which every forward reaction is exactly matched in rate by its own reverse reaction. There is no net change inside the system, nor is there any net exchange with its surroundings. The escaping tendency of each species is the same throughout the system.

Changes in temperature, pressure, or concentrations of reactants and/or products can shift an equilibrium state. The direction of these shifts can be correlated and predicted using Le Châtelier's principle, the equilibrium-constant expression, escaping tendencies, or rate arguments. Quantitative correlations are often possible if the equilibrium

POINT TO PONDER
Living systems require a continuing net flow of material and energy. Thus many physiological states must avoid approaching an equilibrium condition. The zero net flow characteristic of dynamic equilibrium would lead to the death of the organism.

EXERCISE 15.11
Show that closed-cycle breathing of air by an astronaut would leave 0.15 atm of O_2 when the CO_2 level had risen to 0.05 atm. The original O_2 pressure is 0.20 atm.

HINTS TO EXERCISES
15.1 Steady state. 15.2 Only the female is involved in the rate-determining step, gestation. 15.3 H—O—D. 15.4 Applying pressure to ice tends to decrease the volume, increasing the tendency to melt. 15.5 $K_{eq} = [Ca^{2+}] \times CO_3{}^{2-}]$. 15.6 Doubling [NO]. 15.7 10^{-5} mole/liter. 15.8 Rise in temperature encourages formation of water. 15.9 Raising temperature forms H_2 and I_2; raising pressure has little effect; increasing H_2 forms HI. 15.10 Equilibrium constant is most quantitative; Le Châtelier's principle is usually easiest. 15.11 To form 1 mole of CO_2 requires 1 mole of O_2.

constants for the system are known.

Steady states also show no net change in the system, but they always require supplies from their surroundings and must always deliver products and/or wastes to the surroundings. They are much more difficult than equilibria to describe in detail, but they are very common in nature, especially in living systems. It is important to distinguish clearly whether a system undergoing no net change is a steady state or a dynamic equilibrium before predictions about it are made.

Problems

15.1 Identify or define: dynamic equilibrium, endothermic reaction, equilibrium-constant expression, equilibrium state, escaping tendency, ionization constant of water (K_w), Le Châtelier's principle, static state, steady state, stress on a system.

15.2 Which of the following constant states are equilibria and which steady states?

a. supplies on the shelves of a supermarket;
b. population of a suburban area in terms of commuting;
c. automobile assembly line;
d. depositing and lending of money by a bank;
e. atmospheric water cycle;
f. precipitation and solution of a coral reef;
g. concentration of glucose in your blood.

15.3 Show that reasoning from (a) Le Châtelier's principle, (b) rate effect, and (c) the equilibrium constant gives consistent results in predicting the effect on $p(H_2)$ of the listed changes in the equilibrium system H_2 (g) + C_2H_4 (g) = C_2H_6 (g) + 137 kJ.

a. There is an increase in $p(C_2H_4)$.
b. System is compressed at constant temperature.
c. System is heated at constant volume.
d. A catalyst is added.

15.4 Suppose H_2, I_2, and HI are placed in a sealed flask and allowed to come to equilibrium as in Figure 15.7; the temperature is then raised 10°C.

a. What happens to the pressure?
b. What happens to $p(H_2)$?
c. What happens to the equilibrium constant?

15.5 Household 3% aqueous hydrogen peroxide, H_2O_2, is quite stable. But, as its use at home became more common, carelessness led to contaminants in the bottles, and explosions of the screw-capped bottles occurred. Modern bottles have a gas vent built into them. Outline the equilibria that commonly occur and are safe in a tightly screw-capped bottle. What other equilibrium is catalyzed by the impurities to make the tight bottle unsafe?

15.6 The escaping tendency of a liquid increases faster with an increase in temperature than does that of the gas with which it is in equilibrium. Thus, vapor pressures of liquids increase as the temperature increases. Show that this is consistent with the effect on a vaporization equilibrium predicted by Le Châtelier's principle.

*15.7 Water, whether in a drop or in the ocean, is very nearly at equilibrium with gaseous water over it. Why is the humidity on the deck of a sailboat generally lower than that in a tropical jungle at the same temperature?

15.8 It has been found that putting molten iron that contains dissolved carbon and oxide into a vacuum removes the carbon and oxygen as carbon monoxide. Show that use of Le Châtelier's principle would predict this. This process can produce steels containing less than 0.03% carbon, difficult to do in any other way.

*15.9 The equilibrium constant is 10^{45} for the reaction CO (g) + $\frac{1}{2}O_2$ (g) = CO_2 (g). $K = p(CO_2)/p(CO) \times p(O_2)^{1/2} = 10^{45}$. In light of this value of K, comment on the fact that $p(CO)$ in polluted air may be 3×10^{-6} atm.

15.10 For the reaction O_2 (g) $= 2$ O (g), $K = 10^{-82}$. The pressure of both O_2 (g) and O (g) at 120 km above the earth is about 10^{-10} mole/liter. How do you correlate the two numbers?

15.11 Norbert Wiener in his remarkable book *The Human Use of Human Beings* (Avon Books, 1967) writes that "the individuality of the body is that of a flame rather than that of a stone, of a form rather than a bit of substance." Do you agree?

15.12 Hemophilia (the inability of blood to clot), achondroplastic dwarfism, cleft palate, and tuberous sclerosis (with a frequency of about 1 in 100,000) are the most common spontaneous mutations in humans. They are inheritable but tend to die out because their serious effects impair the ability of carriers to reproduce. So their frequency remains about constant. Is this a dynamic equilibrium or a steady state?

*15.13 The flour mite *Acarus siro* cannot survive if the temperature is outside the range 0–31°C, or if the humidity falls below 60%. Suggest chemical interpretations of these limits.

15.14 Concentrations of chemicals in a human cell often do not change much with time. Does this mean the cell is in dynamic equilibrium?

Figure 16.1
Some common acids and
antacids.

16 SOUR VERSUS BITTER; ACIDS AND BASES

Almost everyone has been sick to the stomach and found that the contents of the stomach taste sour. And most have learned through advertising to associate the sourness with "excess gastric acidity." We use the common expressions "sour apples" and "sour grapes" to describe both the fruits and things that have gone wrong. We talk about being sour on life and making acid comments in an almost interchangeable way. All these expressions stem from the fact that one of the earliest classes of chemicals to be recognized was the acids. Acids taste sour—often unpleasantly so, as the taste of sour milk, which contains lactic acid, demonstrates. In fact our word *acid* comes from the Latin *acidus,* "sour." Acids have also long been associated with corrosive power. A recent addition has been "acid rain." (Recall Figure 4.7.)

The word *base* has been given other uses in popular language. But in chemistry it describes the class of substances that neutralize the effects of acids. The word comes from the Latin *basis,* "foundation," and reflects the fact that early chemists used bases as raw materials for the synthesis of other compounds. Chemical bases taste bitter. The common term for base in advertising is *antacid.* When bases are added to acids the resulting compounds, called *salts,* usually taste neither sour nor bitter—most taste salty.

In this chapter we shall explore acids and bases as typical examples of the competitive nature of chemical reactions. We shall also use them to demonstrate a common happening—the existence of several theories to describe a single set of observations. Which theory you actually use becomes partially a matter of personal preference. Using personal judgment to select the most useful theory for interpreting a particular set of observations is common in science.

Way Back When

The earliest known acids and bases were what we would call *oxides,* and it was known that they could, when mixed, neutralize one another's properties. It was also found that acids and bases injected under the skin sting, and that they have characteristic tastes.

The generalization was reached that oxides of metals give basic solutions if they dissolve in water. Oxides of nonmetals give acidic solu-

When we view the reaction mixture as a macroscopic system the molecular turmoil is invisible, and what we see instead is a seemingly purposeful drive to equilibrium, rapid at first, then slower, until finally equilibrium is achieved and the chemical reaction is at an end. At the molecular level, however, the reaction is not at an end, the interactions, the transitions from state to state, and the violent disruptions being just as frequent and as confused as ever.—*B. Widom,* Science, *148, 1965, p. 1555.*

POINT TO PONDER
Many acids taste sour; bases, bitter. But taste should be used most cautiously. There are much surer and safer methods for identifying chemicals.

Table 16.1
Some Properties of Aqueous
Solutions of Acids and Bases

Property[a]	Acid	Base
Taste	Sour	Bitter
Feel	Stings	Slippery
Effect on litmus	Red	Blue
+ Zn $\longrightarrow$	H_2	—
+ Al $\longrightarrow$	H_2	H_2
Electrical conductivity	Conducts	Conducts

[a]Properties above the dashed line are neutralized (disappear) when the acid and base are mixed in suitable ratios, but electrical conductivity often does not disappear.

EXERCISE 16.1
Clean metallic aluminum reacts rapidly with H_2O. Why is the rate of reaction of an aluminum pot with water very slow in the absence of acid or base?

tions if they dissolve in water. And some substances, such as HCl, contain no oxygen but give acidic solutions. These *aqueous solutions of acids and bases* were subsequently found to be conductors of electricity. Interestingly, the property of conductivity is not always neutralized when the acids and bases neutralize their other mutual properties of taste, feel, effect on litmus, and reactivity with metals.

Table 16.1 summarizes some of the properties of acids and bases as they might have been tabulated in the 1870s, after many acids and bases had been discovered. But no one had come up with a satisfactory theory to account for what acids and bases have in common or how they neutralize each other.

Arrhenius Again

We have met Svante Arrhenius before in this book. He interpreted temperature effects on rates of reaction in terms of an activation energy. And he had a terrible time convincing his professors that ions exist in aqueous solutions. Ions were a major subject of his research thesis, which he had a great deal of trouble getting accepted. Later, on the basis of that thesis, he was awarded a Nobel Prize. (So have courage. Remember that, even though students still lose most of their arguments with teachers, information accumulates, knowledge grows, and "answers" change.)

We now interpret Arrhenius' ideas by saying that there are two kinds of electrical conduction: that by electrons, and that by charged molecules called *ions*. There are few systems other than metals, ionized gases, and certain solid semiconductors that conduct electrons. Conduction of electricity in nonmetallic systems and in aqueous solutions takes place by *ions*—whose existence was first suggested by Arrhenius.

Arrhenius believed that it is the presence of hydrogen ions (H^+) that characterizes acids, and the presence of hydroxide ions (OH^-) that characterizes bases. Neutralization involves the reaction of hydrogen ions (H^+) and hydroxide ions (OH^-) to give water—tasteless, without "feel," without effect on litmus, and unable to generate hydrogen with most metals. The residual electrical conductivity and taste, when found in water, stem from negative ions from acid solutions and positive ions from basic solutions. (For examples, see Table 16.2.)

The key reaction of neutralization in Arrhenius' theory is the formation of water from some source of hydrogen ions (a proton acid) and some source of hydroxide ions (a hydroxide base):

$$H^+ (aq) + OH^- (aq) = H_2O (\ell) \tag{16.1}$$

A brief note on net equations seems appropriate here. Each of the

Table 16.2   Some Acid–Base Reactions[a]

1. Aqueous sulfuric acid plus sodium hydroxide (caustic soda):

$$H^+ (aq) + SO_4^{2-} (aq) + Na^+ (aq) + OH^- (aq) = H_2O (l) + Na^+ (aq) + SO_4^{2-} (aq)$$

Net equation: $H^+ (aq) + OH^- (aq) = H_2O (l)$

(This is the most common net reaction for acids and bases in the laboratory.)

2. Aqueous hydrochloric acid plus magnesium hydroxide (milk of magnesia):

$$2H^+ (aq) + Cl^- (aq) + Mg(OH)_2 (c) = 2H_2O (l) + Mg^{2+} (aq) + Cl^- (aq)$$

Net equation: $2H^+ (aq) + Mg(OH)_2 (c) = 2H_2O (l) + Mg^{2+} (aq)$

(This is the reaction by which milk of magnesia lowers "excess gastric acidity.")

3. Vinegar, CH_3COOH, plus potassium hydroxide:

$$CH_3COOH (aq) + K^+ (aq) + OH^- (aq) = H_2O (l) + CH_3COO^- (aq) + K^+ (aq)$$

Net equation: $CH_3COOH (aq) + OH^- (aq) = H_2O (l) + CH_3COO^- (aq)$

---

[a]Note that soluble ionized substances are written as the separated ions that are present.

net equations in Table 16.2 is written following the outline presented in Chapter 10:

1. A list is made of the substances *initially present* and the products *actually formed*.
2. The formulas of the reacting substances are then *balanced* to conserve atoms and charge.
3. All formulas for molecules that do not change in the reaction are *deleted* (canceled) to give the net equation.

The net equation summarizes only the changes—not necessarily all the things that must be present for the reaction to occur. After all, a container is as essential as the nonreacting ions are, but neither is included in the net equation because they *do not change* during the reaction. Nonreacting ions or molecules are sometimes called *spectator ions or molecules;* they do not appear in the net equation.

Then Came Brönsted and Lewis

Before many years had passed, it became clear that the Arrhenius theory had some serious limitations. For example, many ions other than hydroxide ions can react with protons. In fact, most negative ions have a detectable tendency to add protons. Phosphoric acid (H_3PO_4) provides a good example. Phosphoric acid ionizes in water to give mainly H^+ and $H_2PO_4^-$ ions (rather than H^+ and PO_4^{3-} ions). The reason is that the second and third protons are held strongly and do not ionize to a high

EXERCISE 16.2
When aqueous $Ba(OH)_2$ and H_2SO_4, both good conductors of electricity, are mixed in a 1 mole to 1 mole ratio, the conductivity drops to a very low value and a white precipitate (solid particles) forms. Write a net equation for the change. Identify the precipitate and give the reason for the loss of conductivity.

EXERCISE 16.3
Calculate the approximate ratio of the positive charge density on a proton to that on a sodium ion. Note that the charges are the same, but the volumes of the ions differ.

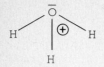

H_3O^+

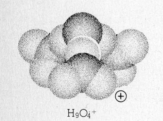

$H_9O_4^+$

Figure 16.2
Hydronium ion models. Further
hydration also occurs.

EXERCISE 16.4
What do the symbols →, =,
and ⇌ mean in this book? (See
pages 163, 268, Figure 15.7,
and Exercise 15.9.)

degree in water. This should strike you as reasonable in terms of the increasingly high negative charge from $H_2PO_4^-$ to HPO_4^{2-} to PO_4^{3-}. It should be, and is, progressively more difficult for a proton to escape from the negative ion as the negative charge gets larger.

J. N. Brönsted suggested an extension of the Arrhenius theory in which he defined bases as species that compete for protons. He also pointed out that it is ridiculous to think of protons, H^+, as actually existing in water. The electrical field around a proton would be so tremendous that it could not possibly exist as a separate ion. (Recall that the radius of a proton is 10^{-5} times smaller than that of other atoms.) Brönsted therefore suggested that protons in water be considered as attached to a water molecule, written H_3O^+ and called a *hydronium ion*.

A very reasonable electron structure can be written for hydronium ions with the "extra" proton bonded to one of the normally nonbonding electron pairs of the oxygen (Figure 16.2). We now believe that Brönsted was not only right, but conservative in his suggestion. It appears that the most common proton reaction with water produces the species $H_9O_4^+$, containing a proton and four water molecules.

I have already pointed out that water is a polar molecule and interacts strongly (hydrates) with both positive and negative ions. All ions hydrate in water. In some cases, the ion-to-water bonds are so strong that definite hydrated ions can be identified and given formulas. In many cases, the hydration layer is continually changing and a formula is hard to define.

Protons are, no doubt, more firmly hydrated than any other ion, consistent with their intense electric field. They certainly affect more than one water molecule in the vicinity. But the exact number is difficult to measure. The water molecules compete actively with one another to bond to the protons, so the protons are continually shifting from one oxygen to another. Protons in water, for example, have an average lifetime for attachment to any one oxygen atom of 10^{-3} seconds.

We shall normally represent ions by using formulas like H^+ (aq), Al^{3+} (aq), Cl^- (aq), and SO_4^{2-} (aq). Such formulas recognize hydration as an important fact, but they do not attempt to give an exact formula to the hydrated species. The case of water and the proton is so common and the existence of isolated H^+ so unlikely that we shall sometimes write H_3O^+ (aq) and talk of hydronium ions to emphasize hydration. But this is not to say that the actual formula of the reactant remains static as H_3O^+.

A modern Brönsted might write the fundamental Arrhenius equation, 16.1, as

$$H_3O^+ \text{ (aq)} + OH^- \text{ (aq)} \rightleftharpoons H_2O \text{ } (\ell) + H_2O \text{ } (\ell) \qquad (16.2)$$

acid	base	acid	base
proton donor	proton acceptor	proton donor	proton acceptor

The net equation in example 3 in Table 16.2 might be written as

$$CH_3COOH\ (aq) + OH^-\ (aq) \rightleftharpoons H_2O\ (\ell) + CH_3COO^-\ (aq) \qquad (16.3)$$

| acid | base | acid | base |
| proton donor | proton acceptor | proton donor | proton acceptor |

The reaction between acid and bicarbonate ion would be

$$H_3O^+\ (aq) + HCO_3^-\ (aq) \rightleftharpoons H_2CO_3\ (aq) + H_2O\ (\ell) \qquad (16.4)$$

| acid | base | acid | base |

Brönsted's theory emphasizes acid–base reactions as competitions for protons by two different bases, one of which is often a water molecule. In the competition for protons, an acid is a *proton donor,* a base is a *proton acceptor.* In the Brönsted theory, the presence of protons is essential, but many bases can compete to hold the protons. Conversely, if protons are being exchanged, or competed for, the Brönsted theory can describe the competition. The double arrows in equations emphasize this competition between reactants and products.

But there are plenty of substances that have properties characteristic of acids, such as the properties listed in Table 16.1, that do not contain protons. And there are plenty of nonaqueous systems that show acid–base properties though they do not contain water.

It was G. N. Lewis (whom you heard of before, in connection with electron bonds) who extended acid–base theory to include these systems. His answer to the question, "What characterizes acids and bases?" was based on his ideas of electron-pair bonds. He defined an acid as a substance able to *accept* a pair of electrons, and a base as a substance able to *donate* a pair of electrons. An acid and a base react to form a covalent bond made up of a pair of electrons.

Acid–base reactions are viewed as competitions for pairs of electrons in Lewis's theory, as competitions for protons in Brönsted's theory, and as reactions that form water from protons and hydroxide ions in Arrhenius' theory. Table 16.3 summarizes the three approaches for one common reaction. Sometimes, as here, all three theories are applicable, but sometimes two, or only one, can be used. It should not surprise you

POINT TO PONDER
The usual technique in deciding which of several competing theories to apply to a given situation is called Occam's razor. You cut out all the theories that are more complicated than you need for the data in hand, and settle for the simplest possible "explanation." I have done this in earlier chapters. For example, the neutron–proton theory of atomic nuclei, the Gillespie–Nyholm theory of molecular shape, and the ideal gas law are all known to be inexact and to lack complete generality. But each is easy to use and suffices for most common situations.

Table 16.3   A comparison of the Arrhenius, Brönsted, and Lewis Theories

Arrhenius:	H^+ hydrogen ion	+	OH^- hydroxide ion	$\rightleftharpoons$	H_2O water		
Brönsted:	H_3O^+ proton donor	+	OH^- proton acceptor	$\rightleftharpoons$	H_2O proton donor	+	H_2O proton acceptor
Lewis:	H_3O^+ electron pair acceptor	+	$:OH^-$ electron pair donor	$\rightleftharpoons$	$H{-}OH$ electron pair acceptor	+	$H_2O:$ electron pair donor

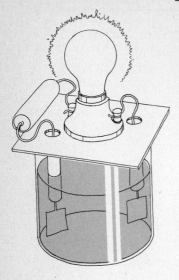

Figure 16.3
Testing electrical conductivity.
The bulb lights if there are ions
in the liquid.

EXERCISE 16.5
A solution of 0.01 mole/liter of
ascorbic acid (vitamin C) in
water is a poor conductor of
electricity. What can you
deduce from this?

to learn that there are other acid–base theories useful in even more "exotic" systems.

The problem of multiple theories for similar phenomena is common in science (and in other fields as well). There are few theories that are claimed to be "right" for everything. Most are known to be inadequate in some cases; remember and acknowledge the inadequacies when using a theory.

Much of the time we shall find Arrhenius' theory quite adequate for acid–base reactions. The rest of the time we shall use either the Brönsted theory of competition for protons, or the Lewis theory of competition for electron pairs.

Some Acids Are Strong, Some Are Weak

Even before Arrhenius, it was apparent that acids can differ in strength when present in equal concentrations. A water solution of SO_3, for example, attacks CuO much more vigorously than does a solution of CO_2 at a similar concentration. The SO_3 solution has all the acidic properties listed in Table 16.1 to a higher degree than does the CO_2 solution. Similarly, as I have indicated in Table 16.2, sulfuric acid is ionized into H^+ and SO_4^{2-}, but acetic acid, CH_3COOH, is mainly un-ionized CH_3COOH. Acetic acid is a weaker (less ionized) acid than sulfuric acid.

The presence or absence of ions and the extent of ionization are easy to measure by determining the electrical conductivity of the system (Figure 16.3). Extensive measurements show that most proton acids are weak acids—that is, they are weakly ionized. But most hydroxides of metals are strong bases. They are highly ionized, though many are not very soluble.

Many hydroxides do not dissolve in water because the interionic attractions in the crystal are high. The hydroxides that do dissolve all have ions of $1+$ or $2+$ charge, and they give strongly basic solutions. Ions with high charge create very strongly bonded crystals that are not very soluble. Recall the insolubility of jewels like zircon (Zr^{4+}, SiO_4^{4-}), and even of pearl (Ca^{2+}, CO_3^{2-}).

Table 16.4 lists the common strong proton acids and the strong hydroxide bases. It is easy to rationalize the tendency of proton acids and nonmetallic oxides to be weak acids, the tendency of hydroxides and oxides of metals to be strong bases, and the variation in acid strength from compound to compound in terms of competition for either protons or for electron pairs. The rationalization is easier if we stick with one acid, the proton, and consider it competing for electrons from various bases. Can you guess why the acids and bases listed in Table 16.4 are strong, whereas most others are not? Let's discuss it.

Charge Density and Acid Strength

Protons have a highly positive charge density. Substances that have a highly negative charge density will be good electron-pair donors and will attract a proton strongly. If such a substance (a base) has a higher negative charge density than water, it will tend to remove a proton from any H_3O^+ ions in water and form a weak (un-ionized) acid.

$$X^- + H_3O^+ = HX + H_2O \qquad (16.5)$$

Most negative ions do have charge densities higher than water, so most negative ions form weak acids. Substances with less-negative charge densities will be poorer electron donors and will attract protons less strongly.

The easiest way to estimate a relative *charge density* (or charge/volume) around an atom is from its atomic radius and its oxidation number:

$$\text{Relative charge density} = \frac{\text{oxidation number}}{\text{atomic radius}^3} \qquad (16.6)$$

I have defined atomic radii in Chapter 5 (also see the chart inside the back cover of this book). The cube of the atomic radius is proportional to the atomic volume. In order to calculate a charge density, we must have some measure of charge. We shall use the oxidation number.

The oxidation number is the apparent number of electrons lost by an atom in being converted from the element to the compound.

For a monatomic ion, the charge on the ion is the oxidation number of the ion: Ca^{2+}, Mn^{2+}, Fe^{2+}, and Zn^{2+} all have an oxidation number of $2+$, whereas F^-, Cl^-, Br^-, and I^- all have an oxidation number of $1-$. Elements in the first set have lost two electrons per atom, elements in the second have gained one electron per atom.

The oxidation number for each atom in a polyatomic species is calculated by applying the set of rules in Table 16.5. They must be applied in the order given to avoid error.

It should be clear that the result, the oxidation number of an atom, is only a crude measure of its electrical charge. Oxidation numbers are based on an essentially ionic model of bonding. They assume that the pair of electrons in a covalent bond actually reside on the more electronegative atom. Note that the signs of the oxidation numbers are consistent with the relative electronegativities. The magnitudes of oxidation numbers are consistent with position in the periodic table—that is, with group number. Oxidation numbers are very useful in correlating properties and in naming compounds.

Figure 16.4 shows how the concept of charge density can be used to correlate acid strengths of proton acids. Anything that increases the

**Table 16.4
Common Strong Proton Acids
and Hydroxide Bases**

Acids

Without oxygen

Hydriodic	H^+, I^-
Hydrobromic	H^+, Br^-
Hydrochloric	H^+, Cl^-

With oxygen

Perchloric	H^+, ClO_4^-
Sulfuric	H^+, HSO_4^-
Nitric	H^+, NO_3^-

Bases

All Group I hydroxides
(Li^+, Na^+, K^+, Rb^+, Cs^+)OH^-

All Group II hydroxides except
$Be(OH)_2$ (but none of these is
highly soluble)
(Mg^{2+}, Ca^{2+}, Sr^{2+}, Ba^{2+})OH^-

Structure	Acid strength
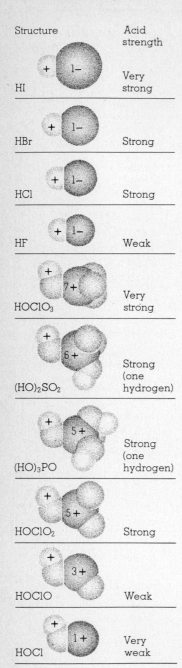 HI	Very strong
HBr	Strong
HCl	Strong
HF	Weak
HOClO$_3$	Very strong
(HO)$_2$SO$_2$	Strong (one hydrogen)
(HO)$_3$PO	Strong (one hydrogen)
HOClO$_2$	Strong
HOClO	Weak
HOCl	Very weak

Table 16.5 Calculating Oxidation Numbers

To calculate the oxidation numbers of atoms in compounds, follow these rules, in the order given. Remember that the oxidation number of an element in its elemental state is zero.

1. The oxidation number of any charged species is its charge.
2. The Group IA, IIA, and IIIA elements are $1+$, $2+$, and $3+$. (They exist as ions of these charges in their compounds.)
3. Hydrogen is $1+$. (Hydrogen is electropositive, except in its compounds with metals.)
4. Fluorine is $1-$. (Fluorine is the most electronegative element.)
5. Oxygen is $2-$. (Oxygen is the second most electronegative element.)
6. In their binary compounds, halogens (Group VIIA) are $1-$ and chalcogens (Group VIA) are $2-$.
7. Oxidation numbers of other elements are found by subtracting the sum of the known oxidation numbers from the charge on the substance.

Examples:
H_2O_2 $(1+, 1-)$ CCl_4 $(4+, 1-)$ $HClO_4$ $(1+, 7+, 2-)$
KO_2 $(1+, \frac{1}{2}-)$ Na_2SO_4 $(1+, 6+, 2-)$ $HBrO_3$ $(1+, 5+, 2-)$
F_2O $(1-, 2+)$ $K_2Cr_2O_7$ $(1+, 6+, 2-)$ HOI $(1+, 2-, 1+)$
NH_3 $(3-, 1+)$ Na_2PtCl_6 $(1+, 4+, 1-)$ MnO_4^- $(7+, 2-)$
CH_4 $(4-, 1+)$ Pb_3O_4 $(8/3+, 2-)$ SnS_3^{2-} $(4+, 2-)$
NaH $(1+, 1-)$

Note that, in general, for any element, the following statements apply:

1. The maximum positive oxidation state equals the element's group number in the periodic table. For example, $Cr = 6+$, $Mn = 7+$.
2. The maximum negative oxidation state equals 8 minus the group number. For example, $Cl = 1-$, $S = 2-$, $N = 3-$.

negative charge density of a base increases its tendency to hold a proton or to donate a pair of electrons, and thus increases the base strength. Protons are held more tightly. Anything that decreases the negative charge density of a base increases the acidity of the proton. Atomic size and oxidation number are the variables used to estimate changes in charge density. Small size and large charge (or oxidation number) give large charge densities.

Proton, Proton, Who's Got the Proton?

Proton shifts are particularly easy to follow using dyes that change color when they gain or lose protons. Litmus is such a dye, but there are many others in which the shift in color is over a narrower range of [H$^+$] and is easier to observe.

The dye bromthymol blue (like litmus) is one color in acid solutions and another color in basic solutions—yellow in acid, blue in base. In terms of molecular change, the molecule with the proton produces a yellow solution. When it loses the proton, the molecule produces a blue solution. When there is an equal concentration of each of the two species, yellow and blue, the solution is green—an intermediate color of half yellow and half blue. This is the color observed when bromthymol blue (HBb) is put into pure water, where $[H^+] = [OH^-] = 10^{-7}$ mole/liter (a neutral solution, neither acidic nor basic since $[H^+] = [OH^-]$).

In the equilibrium equations that follow, Bb^- indicates the blue ion, and HBb indicates the yellow molecule.

$$\text{Net equation:} \quad \text{HBb (aq)} = \text{H}^+ \text{ (aq)} + \text{Bb}^- \text{ (aq)} \qquad (16.7)$$
$$\qquad\qquad\quad \text{yellow} \qquad\qquad\qquad\qquad \text{blue}$$

$$\text{Equilibrium-constant expression:} \quad K_{\text{HBb}} = \frac{[\text{H}^+][\text{Bb}^-]}{[\text{HBb}]} \qquad (16.8)$$

(Note that each concentration—products in the numerator, reactants in the denominator—appears raised to a power equal to its coefficient in the net equation.) When $[Bb^-] = [HBb]$ there are equal concentrations of blue and yellow molecules so the solution is green. This occurs when $[H^+] = 10^{-7}$ mole/liter. Substituting this information into equation 16.8 gives

$$K_{\text{HBb}} = \frac{10^{-7}[\text{HBb}]}{[\text{HBb}]} = 10^{-7} \qquad (16.9)$$

So, at equilibrium in water solution, bromthymol blue follows the relationship

$$K_{\text{HBb}} = 10^{-7} = \frac{[\text{H}^+][\text{Bb}^-]}{[\text{HBb}]} \qquad (16.10)$$

Phenolphthalein (HPp), another common proton-sensitive dye, changes color at $[H^+] = 10^{-9}$ mole/liter, from colorless (more acid) to pink (more basic). So, using the same argument as used for equations 16.8 through 16.10, we conclude that

$$K_{\text{HPp}} = \frac{[\text{H}^+][\text{Pp}^-]}{[\text{HPp}]} = \frac{10^{-9}[\text{Pp}^-]}{[\text{HPp}]} = 10^{-9} \qquad (16.11)$$

The smaller value of K (10^{-9} compared to 10^{-7}) indicates HPp is a weaker acid (less readily ionized) than HBb.

Many, many dyes that undergo similar color changes are known. Two more are methyl red (with $K_{\text{HMr}} = 10^{-5}$) and methyl orange (with $K_{\text{HMo}} = 10^{-4}$). Both these dyes undergo their color change in acid solutions ($[H^+]$ is greater than $[OH^-]$), whereas phenolphthalein changes color in a basic solution ($[OH^-]$ is greater than $[H^+]$).

The dyes vary in their acid strengths. Methyl orange is the strongest acid ($K_{\text{HMo}} = 10^{-4}$), with methyl red ($K_{\text{HMr}} = 10^{-5}$) weaker, bromthymol blue ($K_{\text{Bb}} = 10^{-7}$) still weaker, and phenolphthalein ($K_{\text{HPp}} = 10^{-9}$) the weakest of the four. Conversely, Pp^- is the most effective competitor for H^+ ions (strongest base), Bb^- less so, Mo^- still less, and Mr^- is the least effective of the four in holding H^+ ions.

Figure 16.4 (opposite) Variation in acid–base strength. Protons on atoms of high negative charge density give weak acids (HF, HOCl). Small negative (high positive) charge density gives strong acids (HI, HClO$_4$). With hydroxy acids, increasing the oxidation state or decreasing the size of the "central" atom increases the acid strength by "pulling electrons away from the proton." These changes decrease the negative charge density on the oxygens and make it harder to donate a pair of electrons.

Table 16.6
Colors of Methyl Violet at
Various Values of [H⁺]

[H⁺] (moles/liter)	Color
10	Brown
1	Yellow
10^{-1}	Light blue
10^{-2}	Dark blue
10^{-3}	Violet
$<10^{-3}$	Violet

One more dye completes our list—methyl violet. Methyl violet goes through several color changes as shown in Table 16.6. Methyl violet can gain or lose more than one proton, and it changes color with each loss or gain. It is a *wide-range indicator* and is very useful for that reason. By adding a few drops of methyl violet in an aqueous solution we can identify whether the [H⁺] (in moles per liter) is nearer 10, 1, 10^{-1}, 10^{-2}, 10^{-3} or less, just from the color.

A similar effect over a greater range of [H⁺] can be obtained by mixing indicators. Suppose we mix the four indicators first discussed: methyl orange, methyl red, bromthymol blue, and phenolphthalein. Table 16.7 gives the results over a range of [H⁺] from 10^{-3} to 10^{-9} moles/liter. (Note that there is no color change, and the indicators are ineffective, outside the range.) Such a mixture is sometimes called a *universal indicator,* because of its wide range of applicability. The indicator changes color each time H⁺ changes by a factor of 10. For example, a light yellow indicates [H⁺] = 10^{-6} mole/liter, and green indicates [H⁺] = 10^{-7} mole/liter.

When all four ions (Pp⁻, Bb⁻, Mr⁻, and Mo⁻) are present they compete directly for H⁺ ions. An orange color shows that HPp, HBb, HMr, and Mo⁻ are the principal substances present, consistent with the observation that Mo⁻ is a poorer competitor for H⁺ ions (weaker base) than Mr⁻, Bb⁻, or Pp⁻.

EXERCISE 16.6
Methyl violet in 0.1 mole/liter of aqueous HCl is light blue. What can you deduce from this?

Are Most Dyes Indicators?

It would not be desirable, of course, if the dyes in your clothing changed color whenever you spilled foods of differing acidity on them. Soft drinks and fruit juices are acidic compared to water, but soaps and detergents

Table 16.7   Colors of Four Indicators and Their Mixture at Varying [H⁺]

	Indicator color at varying [H⁺][a]								
	10^{-2}	10^{-3}	10^{-4}	10^{-5}	10^{-6}	10^{-7}	10^{-8}	10^{-9}	10^{-10}
Methyl orange	←	HMo Red	Orange	Yellow Mo⁻	→				
Methyl red	←		HMr Pink	Salmon	Yellow Mo⁻	→			
Bromthymol blue	←				HBp Yellow	Green	Blue Bb⁻	→	
Phenolphthalein	←						HPp Colorless	Pink Pp⁻	→
Mixture (universal indicator)	Red	Red	Orange	Deep yellow	Light yellow	Green	Indigo	Violet	Violet
				Principal Substance Present					
	HMo	HMo	HMo, Mo⁻	Mo⁻	Mo⁻	Mo⁻	Mo⁻	Mo⁻	Mo⁻
	HMr	HMr	HMr	HMr, Mr⁻	Mr⁻	Mr⁻	Mr⁻	Mr⁻	Mr⁻
	HBb	HBb	HBb	HBb	HBb	HBb, Bb⁻	Bb⁻	Bb⁻	Bb⁻
	HPp	HPp	HPp	HPp	HPp	HPp	HPp	Pp⁻	Pp⁻

[a][H⁺] is in moles per liter.

are basic compared to water. And the difference is large. Fruit juices have a hydrogen ion concentration about 1 million times greater than that of detergents. It would be highly undesirable if our colored fabrics turned color with small changes in acidity.

But most dyes do change color if the acidity or basicity changes beyond the range normally found in foods and household chemicals. Strong acids and bases not only make holes, as we shall see—they usually first change the color of the dye. Treat such color changes as indications of trouble and quickly neutralize the offending chemical. Usually you can prevent hole formation if you do the neutralization promptly. But what to use?

Water is usually best because it is apt to be handy and is a quick diluent, but baking soda ($NaHCO_3$) is also good. Baking soda is chemically Na^+ and HCO_3^- ions, which separate and hydrate in water. The sodium ions are inert to most chemicals and belong to the class of *spectator ions*—essential to the proceedings but not actively involved in the changing scene.

Bicarbonate ions (HCO_3^-), however, can act either as acid or as base—great for neutralizing a spot caused by an unknown reactant. Here's how:

With acid: HCO_3^- (aq) $+ H^+$ (aq) $= H_2CO_3$ (aq)

$$= H_2O \ (\ell) + CO_2 \ (g) \qquad (16.12)$$

With base: HCO_3^- (aq) $+ OH^-$ (aq)

$$= CO_3^{2-} \ (aq) + H_2O \ (\ell) \quad (16.13)$$

So bicarbonate ions (HCO_3^-) can neutralize both OH^- and H^+.

Both an Acid and a Base?

In equation 16.12 the bicarbonate ion acts as a base. In equation 16.13 the bicarbonate ion acts as an acid. Water, you will recall, also can act as both acid and base. But it is much weaker, both as an acid and as a base, than bicarbonate ion is. Water's negative charge density is too low for it to be a strong base. The charge density of the OH^- ion formed if water acts as an acid is too high for water to be a strong acid. But HCO_3^- and CO_3^{2-} both have charge densities intermediate between those of H_2O and OH^-, consistent with the acid and base properties of HCO_3^-.

Substances that can act either as acids or as bases are said to be *amphoteric* (from the Greek *ampho*, "both"). Most acids and bases (including all the common ones) contain both protons and nonbonding electron pairs, so most are amphoteric toward protons as Lewis acids. Examine Figure 16.4 to see that this is so.

Table 16.8
Relation of pH to [H⁺]

[H⁺] (moles/liter)	pH
10^{-1}	1
10^{-3}	3
10^{-5}	5
10^{-7}	7
10^{-9}	9
10^{-11}	11
10^{-13}	13
2×10^{-13} ($10^{0.3} \times 10^{-13}$)	12.7

Let's Concentrate (Briefly) on Concentration

I have so far referred to concentration in moles per liter (moles/liter), often using exponents when the concentrations are small. But scientists are rather lazy and always look for an easy way to do a difficult task. A simple solution to writing concentration as moles per liter is to abbreviate it as M (formerly known as molarity). We shall use this abbreviation frequently from now on: concentration = M = moles per liter.

The second simplification is in the number itself. Concentrations of hydrogen ions are frequently referred to in *pH units*. Table 16.8 shows the correspondence between pH units and moles per liter. (Mathematically speaking, pH is the negative logarithm of the hydrogen ion concentration expressed in moles per liter. The "p" stands for German *potenz*, "exponent," of [H⁺]. The pH of a solution is the negative exponent of the [H⁺].)

The p symbol is so useful that symbols such as pOH and pK are used increasingly. Thus if the equilibrium constant $K = 10^{-5}$, pK = 5. If $K = 10^8$, then pK = −8. If pH = 6, then pOH = 8, since [H⁺] × [OH⁻] = 10^{-14}. And so it goes.

Determining Acid–Base Strength

Few acids and bases are as cooperative as the indicators in "telling" when they gain or lose protons. All the common acids and bases are colorless. But the dyes give a very easy way of determining acid–base equilibrium constants. Let's try it on vinegar, CH_3COOH, a common constituent of salad dressings. Again we appeal to the laziness (efficiency) of chemists and use the official abbreviation for CH_3CO, which is Ac (for acetyl). We therefore write AcOH, or HOAc, for acetic acid and OAc⁻ for the acetate ion, $CH_3CO_2^-$.

Vinegar tastes sour, which is one reason it is used in salad dressing. Let's measure the acidity of a vinegar salad dressing. The concentration of HOAc in vinegar, a water solution, is about 0.8 M. Adding a few drops of our universal indicator (Table 16.7) gives a red color. This means the H⁺ concentration is at least 10^{-3} M, and could be greater. Adding a few drops of methyl violet solution to another sample of vinegar gives a violet color. So [H⁺] $\cong 10^{-3}$ M. A more exact measurement gives [H⁺] =0.004 M. Note that OAc⁻ is a poorer competitor for H⁺ than any of the dyes in the universal indicator, but is a better competitor than any of the molecules in methyl violet.

The net equation in vinegar is

$$\text{HOAc (aq)} = \text{H}^+ \text{(aq)} + \text{OAc}^- \text{(aq)} \qquad (16.14)$$

Only about 1 molecule in 100 of the acetic acid is ionized. The rest remain un-ionized.

EXERCISE 16.7
Diagram the Lewis electron structure of OAc⁻.

Acetic acid is a weak acid, less than 1% ionized. This has important implications for the salad dressing. Suppose an ingredient (a base) is present that uses up some of the hydrogen ions. Recall Le Châtelier's principle. More of the acetic acid will ionize to maintain the hydrogen ion concentration at a roughly constant value. More important to the diner, this maintains the taste of the dressing.

It is easy to calculate an equilibrium constant for acetic acid. We use the net equation to get the equilibrium-constant expression.

$$\text{Net equation:} \quad \text{HOAc (aq)} = \text{H}^+ \text{ (aq)} + \text{OAc}^- \text{ (aq)} \quad (16.15)$$

Note that $[\text{H}^+] = [\text{OAc}^-]$.

$$\text{Equilibrium-constant expression:} \quad K = \frac{[\text{H}^+][\text{OAc}^-]}{\text{HOAc}} \quad (16.16)$$

Substituting, we get

$$K = 0.004 \times \frac{0.004}{0.8} = 2 \times 10^{-5} \quad (16.17)$$

Using methods similar to those discussed for vinegar, we can measure the strengths of many acids and bases. Table 16.9 lists the tendencies of

EXERCISE 16.8
Calculate [H+] in 0.01 M HOAc (in H₂O).

Table 16.9 Tendencies for Substances to Donate and Accept Protons (all species in aqueous solution)

HCl, HBr, HI, HNO_3, H_2SO_4, and $HClO_4$ are the common very strong acids, $K > 1$.

Typical equations: $\text{HX (aq)} \rightleftharpoons \text{H}^+ \text{ (aq)} + \text{X}^- \text{ (aq)}$ $K = \frac{[\text{H}^+][\text{X}^-]}{[\text{HX}]}$ $pK = \log(1/K) = -\log K$

Name	Formula	Equilibrium equation Acid (proton donor)		Base (proton acceptor)	K_{eq} (25°C)	pK
Oxalic acid	$(COOH)_2$	$(COOH)_2$	$\rightleftharpoons \text{H}^+ +$	$HOOCCO_2^-$	5.9×10^{-2}	1.23
Monohydrogen sulfate	HSO_4^-	SO_3OH^-	$\rightleftharpoons \text{H}^+ +$	SO_4^{2-}	1.6×10^{-2}	1.80
Phosphoric acid	H_3PO_4	$PO(OH)_3$	$\rightleftharpoons \text{H}^+ +$	$PO_2(OH)_2^-$	7.1×10^{-3}	2.15
Hydrofluoric acid	HF	HF	$\rightleftharpoons \text{H}^+ +$	F^-	6.8×10^{-4}	3.17
Methyl orange	HMO	HMO	$\rightleftharpoons \text{H}^+ +$	MO^-	4×10^{-4}	3.4
Most RCOOH acids		RCOOH	$\rightleftharpoons \text{H}^+ +$	RCO_2^-	$\sim 10^{-4}$	~ 4
Acetic acid	CH_3COOH (HOAc)	HOAc	$\rightleftharpoons \text{H}^+ +$	OAc^-	1.8×10^{-5}	4.75
Aluminum ion	$Al(H_2O)_6^{3+}$	$Al(H_2O)_6^{3+}$	$\rightleftharpoons \text{H}^+ +$	$Al(H_2O)_5OH^{2+}$	7.9×10^{-6}	5.10
Methyl red	HMr	HMr	$\rightleftharpoons \text{H}^+ +$	Mr^-	8×10^{-6}	5.1
Carbonic acid	H_2CO_3	$CO(OH)_2$	$\rightleftharpoons \text{H}^+ +$	CO_2OH^-	4.3×10^{-7}	6.35
Bromthymol blue	HBb	HBb	$\rightleftharpoons \text{H}^+ +$	Bb^-	1×10^{-7}	7.0
Dihydrogen phosphate	H_2PO_4	$PO_2(OH)_2^-$	$\rightleftharpoons \text{H}^+ +$	PO_3OH^{2-}	6.3×10^{-8}	7.20
Phenolphthalein	HPp	HPp	$\rightleftharpoons \text{H}^+ +$	Pp^-	1×10^{-9}	9.0
Ammonium ion	NH_4^+	NH_4^+	$\rightleftharpoons \text{H}^+ +$	NH_3	5.7×10^{-10}	9.24
Phenol	C_6H_5OH	C_6H_5OH	$\rightleftharpoons \text{H}^+ +$	$C_6H_5O^-$	1.3×10^{-10}	9.89
Monohydrogen carbonate (bicarbonate)	HCO_3^-	CO_2OH^-	$\rightleftharpoons \text{H}^+ +$	CO_3^{2-}	4.8×10^{-11}	10.32
Saccharin	$C_7H_4SO_3NH$	$C_7H_4SO_3NH$	$\rightleftharpoons \text{H}^+ +$	$C_7H_5SO_3N$	2.1×10^{-12}	11.68
Monohydrogen phosphate	HPO_4^{2-}	PO_3OH^{2-}	$\rightleftharpoons \text{H}^+ +$	PO_4^{3-}	4.4×10^{-13}	12.36
Hydroxide ion	OH^-	OH^-	$\rightleftharpoons \text{H}^+ +$	O^{2-}	$<10^{-36}$	>36

EXERCISE 16.9
Soap can be used to neutralize
acids. Write a possible equation
for the net reaction.

some molecules to donate and accept protons, and gives their equilibria and equilibrium constants. The molecules vary widely, but all compete for protons. The stronger acids (large values of K, small pK) are at the top of the table; stronger bases are at the bottom. Note that

1. Each acid–base equilibrium involves competition for hydrogen ions (or electron pairs) between water (the hydrating agent) and another base.
2. The trend in K is consistent in a general way with the trend in negative charge density on the base.
3. There is a wide variation in strengths of acids and bases.
4. Stronger bases compete more successfully for protons. (That is, they accept protons, or donate electron pairs, more freely.)

If the reactions in Table 16.9 are combined in pairs (that is, if two sets of reactants are mixed in one solution), the reaction higher in the table will go as written and the lower one will reverse. Thus a table of 19 acid strengths allows the prediction of direction for any of the almost 200 possible reactions between the reactants. Quite a powerful table! The K for the mix of any two reactions will be the *ratio* of the two Ks listed:

$$K_{mix} = \frac{K_{upper}}{K_{lower}} \tag{16.18}$$

Likewise,

$$pK_{mix} = pK_{upper} - pK_{lower} \tag{16.19}$$

Table 16.10 lists the equilibrium-constant values for some simple proton acids. Note the consistent trends with size and oxidation number, and note that most (HCl, HBr, HI excepted) are very weak acids, consistent with the highly negative charge densities (oxidation states) of the central element.

Acid Rain Is Not a Gain

For millions of years, rainwater has been acid, because of dissolved carbon dioxide. Normal groundwater has been even more acid because decaying matter generates more carbon dioxide. Both the falling rain and the groundwater have important geological effects. The effects are especially great on carbonate rocks such as limestone ($CaCO_3$) and dolomite ($CaCO_3 \cdot MgCO_3$), two of the most common rocks on the earth's surface. These are the main equilibria:

$$CO_2 \text{ (g)} + H_2O \text{ } (\ell) \rightleftharpoons H_2CO_3 \text{ (aq)}$$
$$\rightleftharpoons H^+ \text{ (aq)} + HCO_3^- \text{ (aq)}$$
$$\rightleftharpoons 2\,H^+ \text{ (aq)} + CO_3^{2-} \text{ (aq)} \tag{16.20}$$

Table 16.10 Equilibrium Constants ($K = [H^+][X^-]/[HX]$) for Some Simple Proton Acids Arranged in Periodic Table Form[a]

CH_4 10^{-58}	NH_3 10^{-39}	H_2O 10^{-16}	HF 10^{-3}	
SiH_4 10^{-35}	PH_3 10^{-27}	H_2S 10^{-7}	HCl 10^2	
GeH_4 10^{-25}	AsH_3 10^{-19}	H_2Se 10^{-4}	HBr 10^3	Stronger acids (larger size anion formed)
SiH_4 10^{-20}	SbH_3 10^{-15}	H_2Te 10^{-3}	HI 10^5	

Stronger acids (less-negative oxidation number)

NOTE: Most of the values are from P. Powell and P. L. Timms, *The Chemistry of the Non-metals*, McGraw-Hill, New York, 1974; values for hydrogen halides are from R. T. Myers, *J. Chem. Ed.*, **53**, 1976, p. 17.

[a]The acidity (K) is a function of the negative charge density. The larger the anion and the smaller the oxidation number, the smaller the negative charge density and the stronger the acid. HI is the strongest, and CH_4 the weakest, acid.

$$CaCO_3 \text{ (c)} + H^+ \text{ (aq)} \rightleftharpoons Ca^{2+} \text{ (aq)} + HCO_3^- \text{ (aq)} \qquad (16.21)$$

Equation 16.20 accounts for the normal acidity of rain, whose pH is about 6. Groundwater, with a higher concentration of dissolved CO_2, has a pH of about 4. Equation 16.21 accounts for the formation of limestone caves as the acid groundwater penetrates the limestone rock. It also accounts for the formation of stalactites and stalagmites. They form (in alphabetical order) as the carbon dioxide escapes into the growing cave and the $CaCO_3$ precipitates. (See Figure 16.5.) The colors in caves are due to precipitation of other ions, especially iron and manganese, which have dissolved in the groundwater. They precipitate as the CO_2 and the water evaporate.

Much river and spring water has run over enough rocks that it has

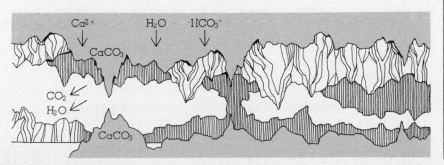

Figure 16.5
Calcium carbonate ($CaCO_3$) precipitates on stalactites, stalagmites, and other formations as carbon dioxide (CO_2) escapes from solution:
Ca^{2+} (aq) + $2HCO_3^-$ (aq) = CO_2 (g) + H_2O (*l*) + $CaCO_3$ (c).

EXERCISE 16.10
Lime (another name for lime-stone), *quicklime* (a name for calcium oxide), and *slaked lime* (a name for calcium hydroxide) are names often confused. Which substance would be safe to take internally? Which would react most vigorously with water? What does *slaked* mean? (Have you ever slaked your thirst?)

become almost saturated with Ca^{2+}, Mg^{2+}, and Fe(II or III). It is these ions that cause water to be "hard," to precipitate soap, and to leave crusty precipitates in cooking vessels and steam boilers. The removal of these ions is an important task in hundreds of communities and industrial plants, not to mention in home water-softening units. But we'll talk about that later.

So, even normal rain dissolves a great deal of rock, including construction material. But there is little we can do about it, except to choose our construction materials with care. When this has not been done—as is unfortunately the case with many of the fine medieval buildings of Europe—statues lose their features, and buildings flake away and eventually fall apart.

During the last 70 years, rain has become more acidic by the introduction of stronger acids into the atmosphere—especially SO_2 and SO_3 from the burning of sulfur-containing coal and oil. Both the rate of attack and the equilibrium states for the reaction of most stones and metals with water depend on $[H^+]$. The SO_2 and SO_3 have the additional curse of converting $CaCO_3$ and $MgCO_3$ in the rock to the sulfates, $CaSO_4$ and $MgSO_4$. These salts crystallize and further fracture the rock. So the deterioration of buildings and other structures has accelerated (as you saw in Figure 4.7). There appear to be only two solutions: change construction materials (and find new surface-protecting films), or diminish the rate of adding acids to the atmosphere. Both will no doubt occur, with incentive supplied by data such as in Figure 16.6. Many of the ancient buildings in the "undeveloped" world are in fairly good condition because there is little or no local industry (which would use fossil fuel).

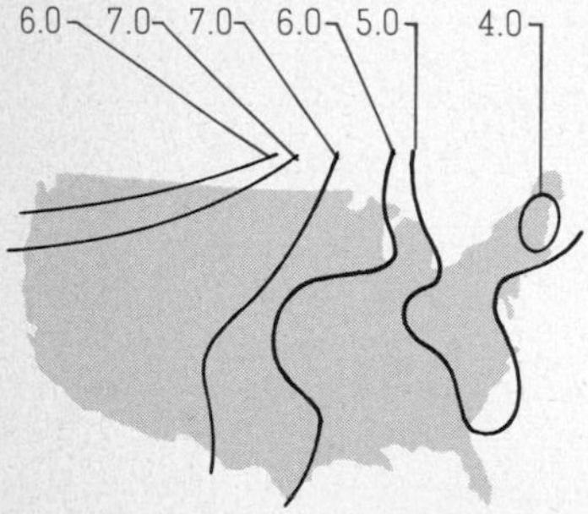

Figure 16.6
The pH of rain during June 1966. The total annual moles of acid in rain has more than doubled over much of the United States during the years since 1966. (National Center for Atmospheric Research)

Pass the Bicarb Please

Living systems are also susceptible to decomposition in high acid or base concentrations. The standard method of defleshing bones is to boil them with the concentrated sodium or potassium hydroxide. Your stomach uses a solution of hydrochloric and other acids (pH 1 to 3) to decompose protein and carbohydrate. The smaller, digested molecules can penetrate the intestinal wall to the circulating bloodstream. You could test the effects of both acid and base by placing a drop of concentrated base and one of acid on your skin. But perhaps the anticipated effect will convince you that this is one set of experiments that has been done often enough by others that you will accept their results. The well-known acid and base holes in lab coats are further evidence for acid–base effects on biological products.

But this raises an interesting set of questions about physiology. How does the stomach keep from digesting itself? What kind of a cell can synthesize almost 0.1 M hydrochloric acid in humans? What maintains the acidity different, but constant, in various parts of the body? We don't know the full answer to any of these questions, but we can get some insights on the basis of even simple acid–base theory.

Eating stimulates the stomach to collect hydrochloric acid from some neighboring cells. No one seems to know how it does this. We do know that excessive eating overstimulates the stomach and leads to "excess acidity" and "gastric distress." At such times, a dose of sodium bicarbonate may do the trick of neutralizing the acid, causing you to burp, but allowing you to keep the food where you put it instead of seeing it all again.

Excessive dosing with bicarbonate can, of course, cause its own problems—primarily excess $[Na^+]$. This is one of the main reasons for the large demand for packaged antacids—mostly $CaCO_3$, $Mg(OH)_2$ (milk of magnesia), and $Al(OH)_3$. These are all bases that are not very soluble in water, so they neutralize the acid without building up a high concentration of OH^- (Figure 16.7). Milk of magnesia also acts as a laxative. (We'll discuss that mechanism in Chapter 18.)

Part of the "logy" feeling after a big meal is due to an overfull stomach. But part is due to the *alkaline wave*. The pH of the blood rises, causing a general shift in physiological tone. Why does it rise? Synthesizing an acid normally involves synthesizing an equal quantity of base, the reverse of the neutralization process. In the cells around the stomach a likely reaction is

Na^+ (aq) + Cl^- (aq) + HCO_3^- (aq)
    = H^+,Cl^- (aq) [to stomach] + Na^+,CO_3^{2-} (aq) [to blood]   (16.22)

Adding carbonate ion to the blood makes it more basic, just as adding hydrogen ion to the stomach makes it more acidic. Eventually, transfers through the intestinal walls lead to general return to an average blood pH of 7.3 to 7.4, nearly neutral. However, both urine and feces can vary considerably in pH, from about 4.6 to 8.4

Buffers

Television advertising has unquestionably enriched the vocabulary of its viewers. (What it has done to their knowledge and wisdom is another question.) Thirty years ago, students thought that buffers were something you used to polish cars. Now almost all know that buffers can be used to control stomach acidity. Acid–base buffers are useful for many other things as well.

You have already seen that HCO_3^- and other amphoteric substances can serve both as acids and as bases. They can both donate and accept protons. And you have seen that you can make up a solution containing

Figure 16.7

$$\left.\begin{array}{l} HCO_3^- \\ CaCO_3 \\ Mg(OH)_2 \\ Al(OH)_3 \end{array}\right\} + H^+ = \left\{\begin{array}{l} H_2O + CO_2 \\ Ca^{2+} + 2\,HCO_3^- \\ Mg^{2+} + H_2O \\ Al(OH)_2^+ + H_2O \end{array}\right.$$

Figure 16.7
How antacids work in water.
(All substances are aqueous.)

EXERCISE 16.11
Would sodium hydroxide be a good antacid? NO! Why not?

both the acid and base forms of any acid–base pair. The mixture of acetic acid (a proton donor) and acetate ion (a proton acceptor) was an example. Such systems are *acid–base buffers.*

Buffer solutions, because they contain both proton donors and proton acceptors, can react with either added base or added acid, to minimize the effect of the addition in the best Le Châtelier fashion. In fact, buffer systems can be specifically constructed to resist change in acidity. They can be very effective at it. Let's see why and how.

We begin with the general acid–base equation:

$$HX \rightleftharpoons H^+ + X^- \qquad K = \frac{[H^+][X^-]}{[HX]} \qquad (16.23)$$

Or,

$$[H^+] = K\frac{[HX]}{[X^-]} \qquad (16.24)$$

If we add 0.1 mole of a strong acid to 1 liter of water, the $[H^+]$ will change from 10^{-7} M in water to 10^{-1} M after addition of the acid, a factor of 10^6. What happens if the solution is buffered?

Suppose 1 liter of solution is made up for maximum buffering effect. It will contain equal concentrations of proton donor and acceptor. The concentration of the acid, $[HX]$, will equal that of the base, $[X^-]$. Say, 1 M each. Of course, $[H^+] = K$. Suppose you added 0.1 mole of strong acid to this system. The added protons will react with X^- to give HX. If $[HX]$ were originally 1 M and 0.1 mole of protons were added, $[HX]$ would become 1.1 M. At the same time $[X^-]$, by reaction with the protons, would drop to 0.9 M. (Both calculations assume negligible change in volume.) The ratio $[HX]/[X^-]$ changes from 1:1 to 1.1:0.9, or 1.2. This means that $[H^+]$ has changed from K to 1.2 K, a factor of 20% instead of the factor of 10^6 found in the unbuffered water. Buffers *are* effective. Whether commercial buffers are equally effective in human stomachs is another question.

Titrations

Titrations are much used in chemistry to determine amounts and concentrations of unknown solutions by adding measured amounts of solutions of known concentrations (Figure 16.8). (You worked through a typical problem in Chapter 10.) During a titration there must be a sharp change in some observable property (pH, color, formation of a precipitate) at a mole ratio that can be related to the substance whose amount or concentration is to be found. A common mole ratio of acid to base (say H^+ to OH^-) is 1 to 1, and the "end point" is where the pH changes most rapidly with added base.

EXERCISE 16.12
One of the common buffers in living systems is a mixture of $H_2PO_4^-$ and HPO_4^{2-}. Write equations for its net reactions as a buffer.

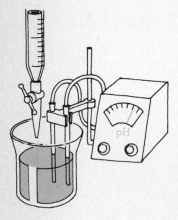

Figure 16.8
A titration, in which a pH meter is used to determine the changing $[H^+]$ values.

We can very simply calculate an end point—say, for the HOAc–OH⁻ system—and choose a suitable indicator to tell us when the pH is there. Remember that the concentration of acetic acid molecules never drops to zero in solution. There will always be some in equilibrium with the acetate ions and the protons. But, since the accuracy of most titrations is about 1 part in 1000, let's calculate the pH when the ratio of [HOAc] to [OAc⁻] is 1 to 1000.

$$HOAc\ (aq) = H^+\ (aq) + OAc^-\ (aq) \qquad (16.25)$$

$$K = \frac{[H^+][OAc^-]}{[HOAc]} = 1.8 \times 10^{-5} \qquad (16.26)$$

$$[H^+] = \frac{1.8 \times 10^{-5}\,[HOAc]}{[OAc^-]} = 1.8 \times 10^{-5} \times (1/1000) \qquad (16.27)$$

$$[H^+] = 1.8 \times 10^{-8} \qquad (16.28)$$

$$pH = 7.7\ (\text{or about 8}) \qquad (16.29)$$

Suppose this titration required 50 cm³ of base. Another drop of base would add 0.05 cm³ and would reduce the [HOAc] to [OAc⁻] ratio to 1:10,000, giving a pH of 9. The pH does change rapidly near the end point.

Reference to Table 16.7 shows that our universal indicator would be indigo at pH = 8. We could just as well use phenolphthalein alone as indicator, because it will change at pH = 9 when that one more drop of base is added. The rate of change of pH with added base is so rapid here that indicator choice is not critical.

Summary

Proton transfer and competition for protons are very common chemical reactions. This is so partly because of the small mass and high mobility of protons, and partly because of the importance of water as a solvent and reactant. Water is always in equilibrium with H⁺ and OH⁻, with an equilibrium constant, $K_w = 10^{-14} = [H^+][OH^-]$.

Arrhenius acid–base theory concentrates on the reaction of protons with hydroxide ions to give water. Brönsted theory interprets acid–base reactions in terms of proton acceptors and donors—that is, competition between bases for protons. Lewis theory concentrates on electron-pair acceptors and donors—that is, competition between acids for non-bonding pairs of electrons on bases.

Indicator dyes furnish a convenient means of measuring the strengths of acids and bases, because the indicators turn color as they gain or lose protons. Titrations allow the determination of unknown concentrations, and buffers allow the control of pH in the face of added acid or base. All these methods are tied together by straightforward calculations using equilibrium-constant expressions.

HINTS TO EXERCISES
16.1 Forms aluminum oxide, which adheres tightly to the aluminum surface. 16.2 Formation of $BaSO_4$ (c) and H_2O (l) removes ions. 16.3 Ratio of their radii cubed $= 10^{15}$. 16.4 →, direction of reaction; =, an equation; ⇌, a rapidly reversible equilibrium. 16.5 Is a weak acid. 16.6 1 M HCl is completely ionized.

16.7 $(H-)_3C-C\overset{O}{\underset{O}{\big\backslash}}^{\ominus}$ or

$(H-)_3C-C\overset{O}{\underset{O}{\big\backslash}}^{\ominus}$

16.8 4×10^{-4} M. 16.9 Soap must contain a base that can form a weak acid. 16.10 $CaCO_3$; CaO; for *slaked*, see a dictionary. 16.11 Too strong a base. 16.12 HPO_4^{2-} (aq) + H⁺ (aq) ⇌ $H_2PO_4^-$ (aq); reverse with OH⁻.

Problems

16.1 Identify or define: acid–base (Arrhenius, Brönsted, Lewis), acid rain, amphoteric, buffers, charge density, concentration (M), electrical conductivity, hydration, hydronium ions, indicators, ionization constant, net equation, neutralization, oxidation number, pH, pK, salt, spectator molecules, strong (and weak) acids and bases, titration.

*16.2 Review equation 10.5. Some friends, about to move to Southeast Asia, have asked you about washing vegetables with a solution containing permanganate ions, MnO_4^-. They believe they heard that a mixture of permanganate ions and vinegar was better for getting rid of germs and viruses. What is your advice?

16.3 When would you use vinegar to try to return a dye to its original color after a spilled chemical had caused spots? When would you use household ammonia? Write equations.

16.4 Why not add methyl violet to a solution of the four indicators in Table 16.7 to give a universal indicator covering the range of [H$^+$] from 10 to 10^{-9} moles/liter?

16.5 Emulsions are suspensions of small drops of one liquid in another, often stabilized by an emulsifying agent like soap. Adding acid may "break" such an emulsion, giving two layers. Why does added acid do this?

16.6 Self-rising flour contains $NaHCO_3$ and $CaHPO_4$. The shelf life is greatly increased by coating the $CaHPO_4$ with K$^+$, Na$^+$, Mg^{2+}, or Al^{3+} phosphates. What does this coat prevent?

*16.7 Most baking powders (which slowly generate CO_2 when wet) now contain $NaHCO_3$, $CaHPO_4 \cdot H_2O$, and $NaAl(SO_4)_2 \cdot 12 H_2O$. What is the function of each chemical?

16.8 Lab coats are often made of cotton, a long-chain carbohydrate joined by ether linkages C=O=C). Suggest how an acid might speed the hydrolysis of such a link. How about a base? Both acids and bases "chew up" most fabrics.

16.9 Nylon is a long-chain polymer with

$$\begin{array}{c} \text{H} \\ | \\ -\text{N}-\text{C}- \\ \| \\ \text{O} \end{array}$$

linkages. It is readily disintegrated by aqueous acid. Suggest which atom the proton attacks. Your stomach digests proteins similarly.

*16.10 Consider the 2+ ions of the series Mn, Fe, Co, Ni, Cu, Zn. The trend is for a decrease both in the formation constants of their complexes with other substances, and in their catalytic effects in a wide variety of circumstances. Does this trend to decrease correlate with trends in charge density for these ions? Any interpretation?

16.11 Open-hearth furnaces are lined with high-melting bricks. Which would be preferable, bricks made of Al_2O_3 or bricks made of $CaCO_3$?

16.12 A friend brings you a glass teakettle that has become coated with some insoluble material from the water boiled in it. How do you suggest it be cleaned? Would there be any change in your recommendation if it were an aluminum kettle? (Hint: what slightly soluble chemical is found in many natural waters?)

16.13 Plaster of paris is $CaSO_4 \cdot 2 H_2O$. Plastered walls are often $CaCO_3$. Which would you say is less altered by heating—that is, which would be more fire-resistant?

16.14 The isoelectric point of hemoglobin is pH 6.8. Is hemoglobin in blood positive, negative, or neutral in electric charge? The pH of blood is 7.4. (The isoelectric point is the pH at which a substance has a net electrical charge of zero.)

*16.15 The main buffer in blood is hemoglobin, but the carbonic acid system is also involved. Calculate the $[HCO_3^-]$ to $[H_2CO_3]$ ratio in blood (pH 7.4). Is this ratio able to buffer more easily against added acid or added base? ($[HCO_3^-]/[H_2CO_3] \cong 10$)

16.16 Sodium benzoate is much used as a food additive to minimize microbial spoilage. It is not effective at a pH higher than 4. Why is this? Esters like $HOC_6H_5CO_2C_2H_5$ are effective at higher pH values. Does this fact fit your theory? (Hint: pK for $C_6H_5COOH = 4$.)

16.17 Write the Lewis electron structure of formic acid, HCOOH. This is what stings when an ant bites you. Much of the sting is due to H$^+$. Calculate the [H$^+$] if the ant injects 1 μmole (micromole) into a volume of 1 μl (microliter) and $K = 2 \times 10^{-4}$. (See Appendix A if you don't know what μ means.) ($[H^+] \cong 0.01\ M$ in an ant bite)

16.18 Some brands of buffered aspirin advertise that the level of aspirin in the bloodstream increases slightly more rapidly after ingestion of their products than after the ingestion of plain aspirin. Yet a 1971 study by a panel of experts for the National Academy of Sciences–National Research Council concluded that there is no evidence that buffering increases the speed of onset of the pain-relieving effects of aspirin. Are these claims contradictory?

16.19 Saccharin is listed as an acid in Table 16.9. Yet saccharin tastes sweet, not sour. How do you account for the lack of a sour taste?

16.20 If a few sprigs of the water plant *Elodea* are put into two sealed flasks of tap water containing phenolphthalein and only one flask is illuminated, the contents of the illuminated flask turns red (first on the top of the leaves, then, slowly, everywhere). The other flask's contents are unchanged. Interpret.

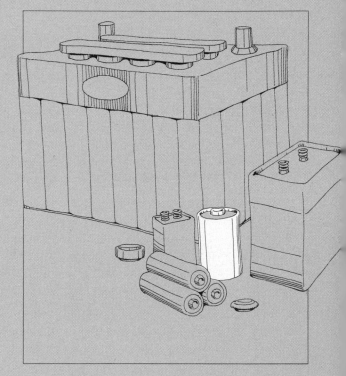

Figure 17.1
Some electric cells.

17 THE SECRETS OF CELLS - ELECTRIC CELLS, THAT IS

I'll reveal my degree of antiquity (if you haven't already guessed) by saying that I don't know what youngsters today do with all those transistor "batteries." We used to saw or knock similar "dry cells" apart and examine the nice gushy interiors. Never figured out what made them work, but there was a black rod you could write with, a black sour mush you could get on your clothes, and an enclosing chunk of metal you could bend until it broke. How could electricity come out of such a simple collection of materials? Wonderful!

I do know there is additional incentive to open electric cells today, because some contain mercury. That's more fun to play with than almost anything else. And it would never occur to a youngster that it might be dangerous. Well, it can be. Not so much to him personally as to any silver and gold, or any lead plumbing, that may be around. Mercury dissolves in and weakens these metals, but is quite insoluble in human tissue. Only long exposures to metallic mercury, such as a dentist might get, cause problems to humans. Where that mercury goes in the long run and whether it is there converted to toxic compounds is another question. But I digress from electric cells.

You already know that many reactions can be split into half-reactions, one of which gives up electrons and one of which uses up electrons. We can view such systems (called *redox,* or reduction–oxidation systems) as competitions for electrons that lead to electron transfer. The problem in designing an electric cell is to separate the competing reactions so that you can get the transferring electrons to flow through an external circuit. In Chapter 16 we discussed acids and bases as competing for protons or for electron pairs to share. The acid–base competition for electron pairs and the redox competition for electrons should sound similar to you. They are. There are many close resemblances between redox and acid–base systems. Let's explore some of them.

Cells and Batteries

Our ability to use electricity has come a long way since 1791, when Luigi Galvani caused a frog leg to twitch by touching it with two different

Atoms move in the void and catching each other up jostle together, and some recoil in any direction that may chance, and others become entangled with one another in various degrees according to the symmetry of their shapes and sizes and positions and orders, and they remain together and thus the coming into being of composite things is effected.—*Simplicius, De Caelo* (ca. 400 BC).

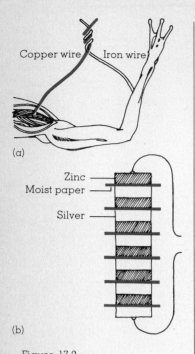

Copper wire   Iron wire

(a)

Zinc ——
Moist paper ——

Silver ——

(b)

Figure 17.2
(a) Galvani's cell, and
(b) Volta's battery.

(but connected) metals, and since 1800 when Alessandro Volta piled silver and zinc disks and acid-soaked blotting paper together to produce the first man-made battery. We would now say that Galvani used a *cell* (containing a single pair of *electrodes*) and Volta a *battery* (that is, several sets of electrodes connected in series). (See Figure 17.2.) Many of the objects you buy as batteries are really cells. Most cells deliver about 1 or 2 volts. Larger voltages are obtained when cells are connected in series in your transistor radio and other "battery-powered" devices to produce a battery. Your car has a battery of three to six cells in a single case.

If you care to learn to make the preceding distinction, fine. But the most important thing to remember is that a set of electric cells connected in series produces a voltage that is the sum of their individual voltages. The most common dry cell produces 1.5 *volts,* so a transistor radio using four of these in series works on 6 volts. Each cell in a car battery generates 2 volts. A 6-volt battery consists of three cells in series, and a 12-volt battery consists of six cells in series. You can usually check this by noting the number of connectors on the top of the battery case, or the number of caps you have to remove when checking the water level in the battery—unless your new battery is sealed and has no caps! The voltage is a measure of the force pushing electrons through the circuit.

A second electrical unit you have heard about is the *ampere.* Electrical current (that is, the rate at which electrons flow through a wire) is measured in amperes. A current of 1 ampere means that 6.24×10^{18} electrons are flowing through any given section in each second. A more commonly used explanation is that 1 ampere of current flowing for 96,500 seconds (that's 26.8 hours) transports 1 mole of electrons past any given cross-section of the conductor. One mole of electrons is called one *faraday* of electricity.

Voltages measure electrical push or pressure; amperes measure charge flow. The product of the two represents electrical *power,* most commonly expressed in *watts.* Thus:

$$\text{Volts} \times \text{amperes} = \text{watts} \qquad (17.1)$$
$$\text{Watts} \times \text{seconds} = \text{joules} \qquad (17.2)$$

So a 550-watt flatiron plugged into the 110-volt house circuit will draw a current of 5 amperes. It provides 550 joules (henceforth abbreviated J) of heat each second.

Volts, amperes, faradays, and watts will supply the quantitative measurements for this discussion of electric cells. We will be concerned primarily with the ways that chemicals can be used to generate voltage and cause current to flow. I shall emphasize the competition between chemicals for electrons.

EXERCISE 17.1
How many hours would be required for 1 faraday of electricity to pass through a section of a wire if the current is 2 amperes?

An electric cell is a device that connects two chemicals differing in their tendency to gain and lose electrons. The electrons will flow from one chemical to the other when a complete electrically conducting path is provided. They will stop flowing when the circuit is broken, as you very well know.

On Building an Electric Cell

You have already read that Charles Hall built a set of cells using his mother's canning jars to generate the voltage and current needed to electrolyze his first samples of aluminum (Figure 17.3). He actually used two kinds of electric cells. The fruit-jar cells that generated the voltage and current are called *galvanic cells.* The frying pan in which the aluminum formed was an *electrolytic cell.* Galvanic cells generate electric voltage and current. Electrolytic cells use electric voltage and current.

The relation between galvanic cells and electrolytic cells may be demonstrated by disconnecting a commercial Hall electrolytic cell from its source of power and placing a 6-volt light bulb across the cell terminals. The bulb lights! The Hall cell can "run backward," producing electric voltage and current. Thus, an electric cell when generating power is a galvanic cell; but the same cell when using up power is an electrolytic cell. Some cells—for example, the lead storage battery in your car—work equally well either way. They are *rechargeable cells.* Most dry cells work well only as galvanic cells; they are not rechargeable. But chargeable and nonchargeable cells differ primarily in the properties of their chemicals, not in their construction. So let's consider construction first, then take up the question of chargeability.

EXERCISE 17.2
Of what did Hall probably make the "other" electrode? (The iron frying pan was the cathode.)

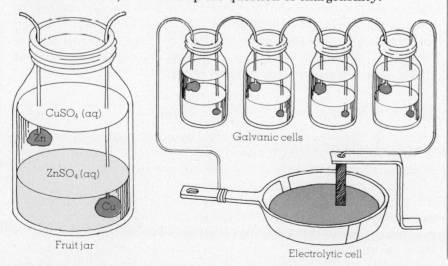

CuSO$_4$ (aq)

Zn

ZnSO$_4$ (aq)

Cu

Fruit jar

Galvanic cells

Electrolytic cell

Figure 17.3
Fruit jars (galvanic cells) and frying pan (electrolytic cell) in the Hall setup.

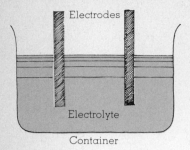

Figure 17.4
First steps in constructing an
electric cell.

Every cell must have a container, two electrodes, and an electrolyte (a conducting solution between the two electrodes). Figures 17.4–17.6 will carry us through the construction stages. The electrodes must be electrical conductors because they establish the path for electron flow from the solution to the outside circuit. In some cells they are chemically inert (such as the carbon rod of a dry cell, or platinum wires). In others they are reactive ingredients (such as the zinc casing of a dry cell). The electrolyte is an ionic solution, usually in water. The ions may be chemically inert, or they may be reactants.

When the ions are reactants, it is often important that they be confined to the region of the electrode at which they react. Therefore a separator (Figure 17.5) is usually introduced between the two electrodes. It allows an ionic current to pass but prevents general mixing of the reagents. This separator may be a *porous barrier* (usually thin plastic or wood in a lead storage battery, or paper in a dry cell). Or it may be a glass tube filled with an ionic solution such as potassium chloride (called a *salt bridge*) if the two electrodes are in separate containers.

Finally, around each electrode there must be a set of reactants that can compete for electrons. Each electrode undergoes a half-reaction. One electrode (where oxidation occurs) produces electrons and sends them to the external circuit. The electrons pass through the external circuit and finally reach the other electrode where they are used up in a reduction process. Oxidation occurs at the anode. (Note that both *oxidation* and *anode* begin with vowels.) Reduction occurs at the cathode. (Both *reduction* and *cathode* begin with consonants. Easy to remember.)

As electrons flow in the external circuit, negative ions *(anions)* and positive ions *(cations)* from the electrolyte will flow and complete the internal circuit (Figure 17.6). The word *ion* comes from the Greek word for "wanderer," consistent with the behavior here. The selection of oxidizing and reducing reactants determines the voltage and current that the cell can deliver. Let's see how and why.

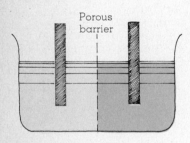

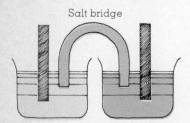

Figure 17.5
Addition of a separator, a device (either porous barrier or salt bridge) to keep the oxidizing and reducing agents from mixing.

The Lead Storage Battery (Charge It, Please)

I hope you are not bored that a car supplies so many excellent examples of chemical systems, for its battery is another one. It requires more attention than some of the other systems, but given that attention, it can last a long time.

A main use of the car battery is starting the car. This puts a heavy chemical drain on the battery, so that it would quickly become exhausted—except that it can be recharged while the car is running. It takes about 10 miles of operation to recharge the battery for each time it starts the car. If it does not get this amount of charge, it will go "dead" and the serviceman gets a call.

Batteries are usually rated in *ampere-hours*. This is the number of hours the battery can steadily deliver a current of 1 ampere. Though it takes less than 1 ampere-hour to start a car (on the average), you can see that constant starting and stopping in town driving can be hard on the battery.

Figure 17.7 diagrams a lead storage battery. All the features developed in Figures 17.4 through 17.6 are present. Each electrode is a sheet *(plate)* of a lead–antimony alloy. One is coated with spongy lead, the other with lead dioxide (PbO_2). The two oxidation states of lead are zero for the lead and $4+$ in the PbO_2. If the two plates are connected by an external wire, electrons will tend to flow from the lead to the PbO_2—consistent with their difference in oxidation state. The plates will tend to "share the wealth of electrons." If the flow continues, each plate tends to end up with the lead in the $2+$ state, the most common oxidation state for lead. In other words, the $4+$ lead in PbO_2 will compete successfully for the electrons in the metallic lead, and current will flow until either the PbO_2 or Pb is used up, depending on which is the limiting reagent.

But the current cannot flow for long unless other things also happen. If the current did flow on and on, the PbO_2 electrode would become more and more negative. At the same time the Pb electrode would become just as positive. Each of these changes would diminish the tendency of negative electrons to flow. In fact, charge effects inside a cell are so strong that almost no current will flow if the electrodes are connected only in the external circuit. The electrolyte completes the circuit internally, and the flow of ions in the internal circuit prevents charge buildup on the two electrodes.

The fact that sulfuric acid is chosen as the electrolyte also allows formation of $PbSO_4$ from the Pb(II) formed at each electrode. The $SO_4{}^{2-}$ ions neutralize the charge of the Pb(II) formed at each electrode. The H^+ ions neutralize the O^{2-} ions released by the PbO_2. No charge buildup occurs. The $PbSO_4$ is almost insoluble in the electrolyte, so it adheres to the electrodes—ready to be changed back to Pb and PbO_2, respectively, during recharging.

Here are the equations for the half-reactions:

$$Pb\ (c) + HSO_4{}^-\ (aq) = PbSO_4\ (c) + H^+\ (aq) + 2\,e^- \quad (17.3)$$

$$2\,e^- + PbO_2\ (c)$$
$$+\ HSO_4{}^-\ (aq) + 3\,H^+\ (aq) = PbSO_4\ (c) + 2\,H_2O\ (\ell) \quad (17.4)$$

$$Pb\ (c) + PbO_2\ (c)$$
$$+\ 2\,HSO_4{}^-\ (aq) + 2\,H^+\ (aq) = 2\,PbSO_4\ (c) + 2\,H_2O\ (\ell) \quad (17.5)$$

The net reaction of equation 17.5 is the sum of the two half-reactions.

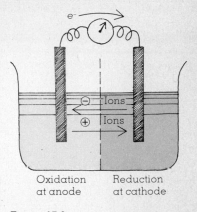

Oxidation   Reduction
at anode   at cathode

Figure 17.6
Total cell in operation after redox reactants have been added. Note directions of flow of electron and ion currents.

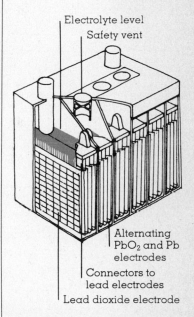

Electrolyte level
Safety vent

Alternating PbO_2 and Pb electrodes

Connectors to lead electrodes

Lead dioxide electrode

Figure 17.7
A lead storage battery.

EXERCISE 17.3
A dead battery can usually be recharged in 10 hours. What current would be used in recharging a 200 ampere-hour battery?

EXERCISE 17.4
The electrodes on a car battery
are marked + and −. To which
electrode do electrons flow in
the external circuit during
discharge according to these
markings?

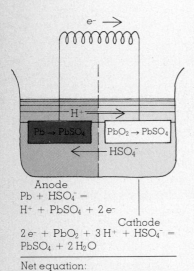

Anode
Pb + HSO$_4^-$ =
H$^+$ + PbSO$_4$ + 2 e$^-$

Cathode
2 e$^-$ + PbO$_2$ + 3 H$^+$ + HSO$_4^-$ =
PbSO$_4$ + 2 H$_2$O

Net equation:
Pb (c) + PbO$_2$ (c) + 2 H$^+$ (aq)
+ 2 HSO$_4^-$ (aq) =
2 PbSO$_4$ (c) + 2H$_2$O (ℓ)

Figure 17.8
Reactions during the discharge
of a lead storage cell.

EXERCISE 17.5
Would the green section of the
battery-test float (meaning
"charged" if the liquid surface
is at this level) be near the top
or bottom of the float?

These reactions are readily reversible in the battery, so that they go
to the right on discharge and to the left on charge. The lead storage
battery can be either a galvanic cell (on discharge) or an electrolytic cell
(on charge). Figure 17.8 diagrams the changes on discharge, when the
half-reaction of equation 17.3 occurs at the anode, and the half-reaction
of equation 17.4 occurs at the cathode.

The net effect in the electrolyte during discharge is that sulfuric
acid is used up and water is formed. The reverse occurs during charge.

A solution of sulfuric acid in water is denser than pure water, so the
degree of charge can be measured by determining the density of the
electrolyte solution. You have probably seen the serviceman do this
with a glass tube that is topped by a rubber suction bulb and contains
a little float with green, yellow, and red sections. (Many modern testers
measure the conductivity of the electrolyte. Why does this test work?)

Should your battery go dead, it is best to resist the temptation to get
a "fast charge." At these high rates of current flow, the ions have difficulty getting through the solution. This raises the electrical resistance
and the voltage, and may lead to electrolysis of water in the cells. The
consequent evolution of gas can dislodge Pb and PbSO$_4$ from the plates.
The dislodged Pb and PbSO$_4$ form a self-shorting sludge in the bottom
of the battery, ultimately causing battery failure.

Batteries have been well designed for a long time. Thirty-five years
ago I drove a car for 400 miles (including half a dozen starts for the
motor) using only the battery because the generator had broken. One
of the more interesting coincidences of my life was that the battery
finally went dead just as we crossed the crest of the international bridge
between Mexico and the United States at Laredo, Texas. We coasted
into U.S. customs glad to be there and near a new generator.

If you have been so careless as to start your car (or at least try to
start your car) with the headlights on, you have observed something
else about batteries. Their voltage drops when they deliver current; the
headlights dim. This is similar to the effect of a fast charge, but in
reverse. The starter requires a high current, so the ions must move
rapidly in the electrolyte. But if they cannot move fast enough to maintain the initial ionic concentration around the electrode, the concentration drops. This drop in concentration diminishes the voltage. We
shall explore now the relationship between concentration and voltage.

Electrode Potentials

Any solid put into any liquid builds up a charge difference between the
solid and the liquid. If the atoms of the solid do not hold electrons very
tightly, some of these atoms will enter the liquid as positive ions, leaving
negative electrons on the solid. Conversely, the solid will become posi-

tively charged if its electrons attract positive ions from the liquid.

I mentioned in the discussion of surfaces in Chapter 11 that cotton and glass both become negatively charged in contact with water. Different solids acquire charges of different magnitudes when they are immersed in the same liquid. If the solids are then connected with an external wire, current will flow. A cell has been constructed. This is what Galvani did. Most such cells are of little interest. The voltages and currents they develop are too small.

Much more interesting cells, and ones much easier to understand, are formed if a metal (M) is immersed in an aqueous solution of its own ions. The following equilibrium is quickly established:

$$\text{M (c)} \rightleftharpoons \text{M}^{n+} \text{(aq)} + n\,e^- \text{ (in the metal)} \qquad (17.6)$$

The presence in the solution of a high concentration of ions of the same metal will tend to keep the concentration of electrons in the metal low. The smaller the concentration of ions of the metal in the solution, the larger the negative charge that will accumulate on the electrode before dynamic equilibrium is established. If we construct a cell such as that in Figure 17.9—consisting of two copper electrodes, one immersed in a solution high in copper ions, the other in a solution low in copper ions—the charge on the two copper electrodes will be different, and current will flow through the external circuit as shown. The more concentrated solution of copper ions (larger $[\text{Cu}^{2+}]$) wins the competition for electrons. Such cells, with electrodes differing only in concentration, are called *concentration cells.*

The voltage generated is known as the electromotive force (EMF) of the cell. It is the difference between the electromotive forces on the two electrodes. The cell will "run," generating voltage and current, until the two solutions come to the same concentration of Cu^{2+}. Then the voltage (EMF) becomes zero. If two different electrodes, each immersed in its own ions, are connected, the voltages are usually larger than in concentration cells, because the difference in tendency to gain electrons is usually larger.

As with other equilibria, so here—the voltages depend on

1. the nature of the reactants;
2. the concentrations of the reactants;
3. the temperature.

Voltages are highly reproducible if conditions are reproduced.

Sad to relate, there is no known way of determining the EMF of a single electrode. Only differences can be measured. There is no absolute zero of EMF. So we use a defined zero and measure all cell EMFs with respect to it. Clearly such a standard should be cheap, easy to set up, highly reproducible, and rugged.

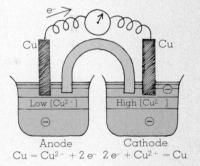

Figure 17.9
A concentration cell. Some negative ions (perhaps $\text{SO}_4{}^{2-}$) must also be present, of course.

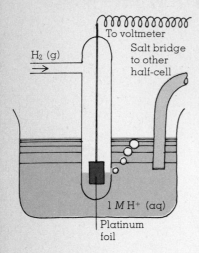

Figure 17.10
The standard hydrogen electrode. The EMF $= 0 = \mathscr{E}^0$ (by definition).

The universally accepted standard is hydrogen gas at 1 atm pressure in equilibrium with 1 M aqueous hydrogen ion. It is called the *standard hydrogen electrode*. For our purpose, we shall assume that 1 M strong acids (like HCl) contain 1 M aqueous hydrogen ion. The EMF of the standard hydrogen electrode is defined as zero. (See Figure 17.10.) The observed EMF of any cell with one electrode a standard hydrogen electrode is defined as the half-cell EMF of the other electrode. Many such half-cell EMFs have been measured.

Table 17.1 lists some half-cell EMFs when all aqueous concentrations are 1 M, all pressures are 1 atm, and the temperature is 25°C. The standard voltages are called $\mathscr{E}^0$ (read as "ee"-zero) values, and they are arranged in order of decreasing tendency to occur. Thus, under these conditions fluorine (F_2), of the substances listed, is the "best competitor" for gaining electrons, Cl_2 is next best, and so on down the line. On the other hand, all these reactions are reversible. The half-equation at the top of the list has the greatest tendency to go as written, and the

Table 17.1 Standard Reduction Potentials, $\mathscr{E}^0$, for Half-cell Reactions in Aqueous Solution at 25°C

Number	Half-equation[a] Oxidizing agents	Reducing agents	$\mathscr{E}^0$ (volts)[b]
1	$\frac{1}{2} F_2$ (g)	$+ e^- = F^-$	2.87
2	$\frac{1}{2} Cl_2$ (g)	$+ e^- = Cl^-$	1.358
3	$H^+ + \frac{1}{4} O_2$ (g)	$+ e^- = \frac{1}{2} H_2O$	1.229
4	$\frac{1}{2} Br_2$ (ℓ)	$+ e^- = Br^-$	1.087
5	Hg^{2+}	$+ e^- = \frac{1}{2} Hg_2^{2+}$	0.920
6	Ag^+	$+ e^- = Ag$ (c)	0.800
7	$\frac{1}{2} Hg_2^{2+}$	$+ e^- = Hg$ (ℓ)	0.799
8	Fe^{3+}	$+ e^- = Fe^{2+}$	0.747
9	$\frac{1}{2} I_2$ (c)	$+ e^- = I^-$	0.535
10	Cu^+	$+ e^- = Cu$ (c)	0.522
11	$\frac{1}{2} Cu^{2+}$	$+ e^- = \frac{1}{2} Cu$ (c)	0.340
12	Cu^{2+}	$+ e^- = Cu^+$	0.158
13	$\frac{1}{2} Sn^{4+}$	$+ e^- = \frac{1}{2} Sn^{2+}$	0.139
14	H^+	$+ e^- = \frac{1}{2} H_2$ (g)	0
15	$\frac{1}{2} Sn^{2+}$	$+ e^- = \frac{1}{2} Sn$ (c)	-0.136
16	$\frac{1}{2} Cd^{2+}$	$+ e^- = \frac{1}{2} Cd$ (c)	-0.403
17	$\frac{1}{2} Fe^{2+}$	$+ e^- = \frac{1}{2} Fe$ (c)	-0.440
18	$\frac{1}{2} Zn^{2+}$	$+ e^- = \frac{1}{2}$ (Zn (c)	-0.763
19	$\frac{1}{2} Mg^{2+}$	$+ e^- = \frac{1}{2} Mg$ (c)	-2.37
20	Na^+	$+ e^- = Na$ (c)	-2.87

Stronger oxidizing agents ↑ (left) Stronger reducing agents ↓ (right)

[a]Half-equations written with one-electron changes to make their addition easy.
[b]Voltage is independent of amount of reaction, so the $\mathscr{E}^0$ values are unchanged if you clear the equations of fractions.

one at the bottom of the list has the biggest tendency to go "backwards." Sodium, (Na^+) is the poorest competitor for electrons that is listed. As said, F_2 is the best competitor. Of the group, Na metal gives up electrons the most easily; F^-, the least easily.

There are 20 half-cell potentials listed in Table 17.1. They can be combined in any set of pairs (almost 200 combinations are possible), just as the acid–base equilibria in the previous chapter (Table 16.9) could be combined. When they are combined, the full cell contains two oxidizing agents (the substances listed on the left side of the equations in Table 17.1) and two reducing agents (the substances listed on the right side of the equations in Table 17.1). They will compete for electrons. The net reaction that yields the highest EMF will be between the most powerful oxidizing agent (highest in the lefthand column) and the most powerful reducing agent (lowest in the righthand column).

The lower half-equation will, of course, run "backwards" from the way it is listed in Table 17.1. This reverses the sign of its EMF. The full-cell EMF is the sum of the two half-cell EMFs, the sign of the lower half-equation being reversed. The half-equations in Table 17.1 thus allow you to predict the directions of a total of almost 200 cell reactions, the number of different possible pairs. Positive cell voltages mean that the net reaction occurs as written. Negative voltages indicate that the reverse net reaction occurs.

What voltage would be generated by a cell having electrodes made of iron and tin, each immersed in a $1\,M$ aqueous solution of its $2+$ ions? This cell is similar to a rusting tin can. Most tin cans are iron with a thin plating of tin.

Substances present are Fe (c), Fe^{2+} (aq), Sn (c), Sn^{2+} (aq), H_2O, and inert negative ions. Of these, Sn^{2+} is the best oxidizing agent (highest on left in Table 17.1). The best reducing agent present (lowest on right in Table 17.1) is Fe. Therefore, the equation for the cathode half-reaction is equation 15 from Table 17.1, and that for the anode half-reaction is the reverse of equation 17 from Table 17.1 (with the corresponding change in the sign of ($\mathscr{E}^0$):

$$e^- + \tfrac{1}{2}Sn^{2+} = \tfrac{1}{2}Sn \qquad \mathscr{E}^0 = -0.136 \qquad (15)$$

$$\tfrac{1}{2}Fe = \tfrac{1}{2}Fe^{2+} + e^- \qquad \mathscr{E}^0 = 0.440 \qquad (17)$$

Net: $\tfrac{1}{2}Fe + \tfrac{1}{2}Sn^{2+} = \tfrac{1}{2}Sn + \tfrac{1}{2}Fe^{2+} \qquad \mathscr{E}^0 = 0.304$ volt   (17.7)

or,

Fe (c) + Sn^{2+} (aq) = Sn (c) + Fe^{2+} (aq) $\mathscr{E}^0 = 0.304$ volt   (17.8)

Note that the voltage of the net reaction is independent of the number of moles that react; that is, the voltage remains the same when we

EXERCISE 17.6
Would a hydrogen electrode at 0.5 atm pressure stand higher or lower in Table 17.1 than the standard H_2 electrode, other conditions being the same?

EXERCISE 17.7
Combining a Zn–Zn^{2+} electrode
(EMF = -0.76) with a
Cu–Cu^{2+} electrode (EMF = 0.34)
gives a full-cell voltage of 1.10
volts. What is the full-cell
voltage when a Zn electrode is
combined with an Ag electrode,
both in their standard states?

double the coefficients of the net reaction. Also note that both the direction of the reaction and the sign of the voltage must be reversed when taking the equation for the anode half-reaction from Table 17.1.

The net voltage in this case is positive (0.304 volt), so the reaction does occur. Note that the iron, rather than the tin, dissolves (corrodes). This accounts for the rusting that occurs around pinholes in the tin plate on iron (steel) cans. Table 17.2 provides some more examples.

You must remember that the EMF values in Table 17.1 are those

Table 17.2   Use of $\mathscr{E}^0$ Values from Table 17.1

Example 1.   What will happen if Sn (c), Sn^{2+} (aq), and Sn^{4+} (aq) are mixed? The most likely reaction is that Sn (c) and Sn^{4+} 'will "share the wealth" of electrons:

$$\text{Sn (c)} + \text{Sn}^{4+} \text{(aq)} = 2\,\text{Sn}^{2+} \text{(aq)}$$

To find out whether this reaction will occur, we obtain appropriate half-equations from Table 17.1 (Equation numbers given here are from Table 17.1; values of $\mathscr{E}^0$ are given in volts. Note that the sign of $\mathscr{E}^0$ in the second half-equation of each set is reversed.)

$e^- + \frac{1}{2}\text{Sn}^{4+} = \frac{1}{2}\text{Sn}^{2+}$	$\mathscr{E}^0 = 0.139$	(13)
$\frac{1}{2}\text{Sn} = \frac{1}{2}\text{Sn}^{2+} + e^-$	$\mathscr{E}^0 = 0.136$	(15)

Net: $\frac{1}{2}\text{Sn} + \frac{1}{2}\text{Sn}^{4+} = \text{Sn}^{2+}$ $\mathscr{E}^0 = 0.275$

or,

$$\text{Sn (c)} + \text{Sn}^{4+} \text{(aq)} = 2\,\text{Sn}^{2+} \text{(aq)} \qquad \mathscr{E}^0 = 0.275$$

Because $\mathscr{E}^0$ is positive for the net equation (0.275 volt), we conclude that this reaction occurs as written. Sn and Sn^{4+} share the wealth of electrons to give Sn^{2+}.

Example 2.   What will happen if Cu (c), Cu$^+$ (aq), and Cu^{2+} (aq) are mixed? Let's try a "share the wealth" reaction again:

$$\text{Cu (c)} + \text{Cu}^{2+} \text{(aq)} = 2\,\text{Cu}^+ \text{(aq)}$$

From Table 17.1 we obtain the appropriate half-equations:

$e^- + \text{Cu}^{2+} = \text{Cu}^+$	$\mathscr{E}^0 = 0.158$	(12)
$\text{Cu} = \text{Cu}^+ + e^-$	$\mathscr{E}^0 = -0.522$	(10)

Net: $\text{Cu} + \text{Cu}^{2+} = 2\,\text{Cu}^+$ $\mathscr{E}^0 = -0.364$

Because $\mathscr{E}^0$ is negative for the net equation (-0.364 volt), we conclude that the reaction occurs in the reverse direction. Aqueous Cu$^+$ reacts with itself to give Cu (c) and Cu^{2+} (aq). You do not find Cu$^+$ in aqueous solutions. Some elements (such as Sn and Pb) have stable intermediate oxidation states in water. Others (such as Cu) do not.

Example 3.   Would the presence of Cd (c) prevent the oxidation of Fe (c) to Fe^{2+} (aq)? To rephrase the question, can we expect the following reaction to occur?

$$\text{Cd (c)} + \text{Fe}^{2+} \text{(aq)} = \text{Fe (c)} + \text{Cd}^{2+} \text{(aq)}$$

We find appropriate half-equations in Table 17.1:

$e^- + \frac{1}{2}\text{Fe}^{2+} = \frac{1}{2}\text{Fe}$	$\mathscr{E}^0 = -0.440$	(17)
$\frac{1}{2}\text{Cd} = \frac{1}{2}\text{Cd}^{2+} + e^-$	$\mathscr{E}^0 = 0.403$	(16)

Net: $\frac{1}{2}\text{Cd} + \frac{1}{2}\text{Fe}^{2+} = \frac{1}{2}\text{Fe} + \frac{1}{2}\text{Cd}^{2+}$ $\mathscr{E}^0 = -0.037$

or,

$$\text{Cd (c)} + \text{Fe}^{2+} \text{(aq)} = \text{Fe (c)} + \text{Cd}^{2+} \text{(aq)} \qquad \mathscr{E}^0 = -0.037$$

Because $\mathscr{E}^0$ is negative (-0.037 volt), we conclude that the reaction will occur in the reverse direction. Cd (c) will not prevent the oxidation of Fe (c) to Fe^{2+} (aq).

obtained when the materials are in the defined standard state, with—

1. crystalline materials pure;
2. all concentrations in solution $1\,M$;
3. all gases at 1 atm;
4. a temperature of 25°C.

We shall now consider what happens to the EMF if these conditions change. You should be able to guess the answers already in terms of changes in escaping tendencies, or in terms of the equilibrium-constant expression.

Every cell eventually "goes dead" if current is drawn long enough. Its voltage becomes zero when the chemicals in the cell come to equilibrium. This fits completely with what you have learned about equilibria in general. As reaction proceeds, the concentrations of the reactants drop while those of the products increase. In cells, the products compete more and more successfully for the electrons, so the tendency to continue the original flow diminishes until, at equilibrium, the tendency for net electron flow is zero. Then the cell EMF (voltage) is also zero, and no further net change occurs. High concentrations of reactants (and/or low concentrations of products) lead to high voltages. Low concentrations of reactants (and/or high concentrations of products) lead to low voltages.

The Dry Cell

The most common dry cell uses the reaction between metallic zinc and manganese dioxide, with a paste of ammonium chloride (NH_4Cl) as electrolyte. We may write as a possible equation

$$Zn\,(c) + 2\,MnO_2\,(c) + 2\,NH_4^+\,(aq)$$
$$= Zn(NH_3)_2{}^{2+}\,(aq) + 2\,MnOOH\,(c) \quad (17.9)$$

This cell provides about 1.5 volts. The reactions in the cell are known to be more complicated than that shown in equation 17.9, but the equation summarizes the major chemical changes. The zinc is the anode. A graphite cathode is used (Figure 17.11) because MnO_2 is not a good conductor. The MnO_2 is mixed with additional graphite to conduct electrons from the cathode and then with an aqueous paste of NH_4Cl, which serves as the electrolyte. The positive ions (cations) move toward the cathode, the negative ions (anions) toward the anode, as they would in any cell.

Zinc ion (Zn^{2+}) is a relatively strong Lewis acid. At the same time, chloride ions (Cl^-) and ammonia (NH_3) are weak Lewis bases, with ammonia the stronger of the two. There is plenty of Cl^- around the

EXERCISE 17.8
Can a galvanic-cell voltage ever become negative during operation?

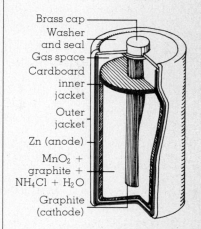

Brass cap
Washer and seal
Gas space
Cardboard inner jacket
Outer jacket
Zn (anode)
MnO_2 + graphite + NH_4Cl + H_2O
Graphite (cathode)

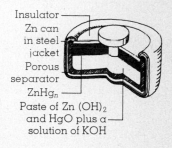

Insulator
Zn can in steel jacket
Porous separator
$ZnHg_n$
Paste of $Zn(OH)_2$ and HgO plus a solution of KOH

Figure 17.11
Schematic diagram of two kinds of dry cells. Carefully dissect one yourself and compare. The cell is not really "dry." It is not liquid, but the presence of water is essential to allow the ions to move and conduct the electric current.

EXERCISE 17.9
What is a possible net half-
reaction at the anode of the
Zn–MnO_2 dry cell?

anode where the Zn^{2+} ions form to give the reaction

$$Zn^{2+}(aq) + Cl^-(aq) = ZnCl^+(aq) \qquad (17.10)$$

This reaction lowers the concentration of Zn^{2+} ions and increases the effective potential of the Zn–Zn electrode. It also slows the migration of the $Zn(II)$, but the zinc ions are still positive and continue to migrate away from the anode.

At the same time, the MnO_2 is reacting to give MnOOH. We can write the reaction

$$e^- + MnO_2(c) + NH_4^+(aq) = MnOOH(c) + NH_3(aq) \quad (17.11)$$

Ammonia is a stronger Lewis base than is Cl^-, so the following reaction will occur when the two meet:

$$ZnCl^+(aq) + NH_3(aq) = ZnNH_3^{2+}(aq) + Cl^-(aq) \quad (17.12)$$

Actually, a single $Zn(II)$ can hold four ammonia molecules, giving $Zn(NH_3)_4^{2+}$.

This last reaction, shown in equation 17.12, is important to the operation of the cell. You may have noted that a flashlight left on for a long time will go out, but that it will often recover its ability to light if turned off for a while. One of the reasons is that the concentration of NH_3 has built up around the carbon cathode as seen in equation 17.11. With NH_3 rather than NH_4^+ around the carbon cathode, an electrically neutral layer forms, decreasing conductivity. Given time, the NH_3 diffuses through the paste and undergoes the reaction of equation 17.12. This gives additional ions and rejuvenates the ability of the cell to deliver current.

There are still other known reactions in the dry cell, but the equations given illustrate how well most of them work together to produce a practical, lightweight source of "portable electricity."

There is a tendency for Zn–MnO_2 dry cells to spring leaks as the zinc can is used up unevenly and holes form in it. Some years ago a "leak-proof" dry cell was put on the market. It has a zinc rod anode in its center, and a conducting shell of graphite inside a container as a peripheral cathode. You might try to diagram its construction.

"Center Terminal Positive"

One of the most common practical problems with cells is how to connect them to other wiring. It is to help you in these decisions that cell makers put labels such as "center terminal positive" on the cell. To connect things in series, you then join positive to negative and join negative to positive (Figure 17.12). This follows from the fact that the signs on the outside of a cell indicate the direction of electron flow in the

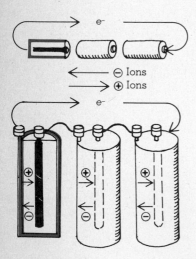

Figure 17.12
Connecting cells. External and internal currents.

external circuit. Electrons will flow from the negative $(-)$ terminal to the positive $(+)$ terminal of the cell if a conductor is placed between the two.

If you look at the full-cell diagram, you'll see that the assignment of charge to an electrode is ambiguous (Figure 17.13). One end of each electrode appears to be positive, whereas the other end appears to be negative. This is, of course, inevitable because the negative charge must flow completely around the circuit, partly as electrons, partly as anions. I bring up this point, not to confuse you, but to dissuade you from saying confidently that the cathode of a cell has a certain charge and the anode has an opposite one.

Stick with the idea that, within the cell, oxidation occurs at the anode, which provides electrons to the external circuit. The electrons pass through the external circuit to the cathode. Reduction occurs at the cathode, using up the electrons. Negative ions move away from the cathode and toward the anode, whereas positive ions move away from the anode and toward the cathode. As mentioned earlier, negative ions are called *anions;* they move toward the anode. Positive ions are called *cations;* they move toward the cathode.

Good luck when you talk to physicists, who treat current as positive and moving in a direction opposite to the electrons. History (in this case Benjamin Franklin) makes powerful and permanent impressions, but the movement of positive charges in a wire is usually almost zero. Ben guessed wrong!

How Long Will It Last?

A customer usually wants to know the lifetime of the device he is about to buy. Usually—even with good quality control—there can be no certain prediction, so minimum guarantees are issued. With electric cells, one can be a bit more exact because they normally have a single limiting reagent.

In the dry cell, it does not pay to make zinc the limiting reagent if it is also to be used for canning the cell. Thus MnO_2 is usually limiting. Suppose you wish to design a cell that will deliver 1 ampere continuously for 2 hours. How much MnO_2 must you put in the cell?

Do you recall that one faraday of electricity is 1 mole of electrons and gives 96,500 ampere-seconds of current?

Well, 2 hours is $2 \times 60 \times 60$ (seconds) = 7200 seconds. We require 1 ampere for this period, so that is 7200 ampere-seconds. You see in equation 17.11 that MnO_2 requires 1 mole of electrons per mole of MnO_2 that reacts. One mole would give 96,500 ampere-seconds/mole. To get 7200 ampere-seconds requires 7200/96,500 = 0.0746 mole of MnO_2. One mole of MnO_2 weighs 54.9 g + 32.0 g = 86.9 g/mole. So, 0.0746 mole

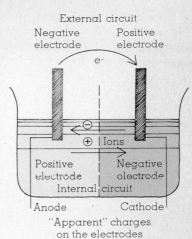

External circuit
Negative electrode   Positive electrode

e-

⊖

⊕ Ions

Positive electrode   Negative electrode
Internal circuit

Anode   Cathode

"Apparent" charges on the electrodes

Figure 17.13
Which sign is correct for which electrode? This must be decided by a "rule"—that is, by a sign convention. The cathode of an electric cell is marked positive on the *outside* of the cell.

EXERCISE 17.10
Why not make the zinc container the limiting reagent in a dry cell?

EXERCISE 17.11
How many moles of zinc will be oxidized during 2 ampere-hours? (Hint: Is there any relation to the amount of MnO_2 reacting?)

$\times$ 86.9 g/mole = 6.48 g of MnO_2. Therefore 6 g MnO_2 (to one significant figure) is needed as a minimum. Making up and testing a trial set of about 100 cells would tell you whether the MnO_2 is completely used up in the net reaction as we have assumed.

It is worth noting that the life of a dry cell could be doubled for the same amount of MnO_2 if a way could be found to reduce it to Mn^{2+}, another stable oxidation state of manganese. However, MnO_2 is cheap, and twice as much of the other reactants would be needed, so there would be little gain.

How Much Will It Cost?

I have concentrated on galvanic cells, but we can do similar calculations for electrolytic cells. How many kilowatt-hours of electricity must be used to form 1 metric ton (10^3 kg) of aluminum metal in a Hall cell? (The cell operates at about 4.5 volts.)

The unit *kilowatt-hour* is obtained by multiplying volts $\times$ amperes $\times$ hours and dividing by 10^3. (This follows from the fact that volts $\times$ amperes = watts.) Now 1 mole of electrons is 96,500 ampere-seconds, or 26.8 ampere-hours. We'll use one significant figure in our calculations; for example, we'll say that 1 mole of electrons requires 30 ampere-hours. Three moles of electrons are needed to reduce 1 mole of aluminum ions to the metal:

$$3\,e^- + Al^{3+} \text{ (in cryolite)} = Al\,(\ell) \tag{17.13}$$

To one significant figure, a mole of aluminum weighs 30 g.

$$10^3 \text{ kg} = 10^6 \text{ g} = \frac{10^6 \text{ g}}{30 \text{ g/mole}} = 3 \times 10^4 \text{ moles of Al.}$$

So, we want to form 3×10^4 moles of aluminum metal, and we'll need three times this many (or 9×10^4) moles of electrons. Each mole of electrons is 30 ampere-hours, for a total of $30 \times 9 \times 10^4 = 3 \times 10^6$ ampere-hours of electricity.

To convert to kilowatt-hours, we need to multiply by the voltage (call it 4 volts) and divide by 10^3:

$$\frac{4 \text{ (volts)} \times 3 \times 10^6 \text{ (ampere-hours)}}{10^3 \text{ (volts} \times \text{amperes/kilowatt)}} = 10^4 \text{ kilowatt-hours}$$

EXERCISE 17.12
Suppose a method were found to operate Hall cells at 3 volts. What saving in power would occur?

So we conclude that we will need about 10,000 kilowatt-hours of electricity to produce each 1000 kg of aluminum metal. A kilowatt-hour of power costs about 2 cents at the cheapest available 1976 rate, so the power alone to make aluminum costs about 20 cents/kilogram. This was about 25% of the total cost of producing aluminum in 1976.

Fuel Cells and Solar Cells

Electric cells have already made great contributions to societies, but one of their greatest contributions lies in the future. Their greatest future role should be to help provide more efficient generation and distribution of energy. Two methods seem possible: direct conversion of fossil fuels to electricity, and direct conversion of sunlight to electricity. Both can be done now, but only if cost is of little object—as, for example, in the space programs.

Burning a fuel is a very inefficient method of utilizing its energy. It can be shown conclusively that the *thermal efficiency* of such a process, in terms of converting an amount of heat (q) to work (w), is

$$\text{Efficiency} = \frac{w}{q} = \frac{T_2 - T_1}{T_2} \qquad (17.14)$$

where T_2 is the temperature of the flame (actually of the boiler in a typical steam electric plant), and T_1 is the temperature of the exhaust gases. This is a "heartbreak" equation, as simple arithmetic will show. Suppose the boiler operates at 1000 K (a high temperature for boiler operation), and the exhaust is at 400 K (a low temperature for the exhaust). The efficiency is then

$$\text{Efficiency} = w/q = (T_2 - T_1)/T_2 = (1000 - 400)/1000 = 0.60 \quad (17.15)$$

By the time friction, heat losses, and other inefficiencies are considered, this figure drops to about 0.40 in the best possible plants. So $1.0 - 0.4$, or 0.6, almost two-thirds, of the heat generated cannot be used to generate useful energy. You probably find this shocking. Perhaps you are highly skeptical about the calculations. Well, I wouldn't wish to have you never skeptical, but you will have great difficulty getting any support for your skepticism here.

Electric cells are another story. Theory and practice both show that complete conversion of chemical to electrical energy *is* possible in an electric cell. As a matter of fact, there are systems that give more work to the system than there is energy in the fuel. They do this by extracting some energy from the surroundings. Perhaps if you were only slightly skeptical of the poor yield from heat engines, you have become even more skeptical about the superb (better than 100%) yield from some electric cells! Well, rest comfortably—the inevitable occurrence of friction, heat losses, and similar difficulties gets the practical limit down to 0.8—or, in a very favorable case, 0.9—efficiency. But note that this is twice what a heat engine gives.

If it were possible to convert fuel energy directly to electricity, only half as much fuel would be needed—and the waste would drop corre-

POINT TO PONDER
The effective temperature in space is only a few Kelvins, say 5 K. This suggests that heat engines on spaceships or on the moon might approach 100% efficiency. Unfortunately, the *rate* at which heat can be lost at low temperatures in a near-vacuum (as is found in space) is very low, so heat engines are not much more attractive in space (or on the moon) than they are on earth. So the astronauts use electric cells.

EXERCISE 17.13
Would pure O_2 or air give a
higher voltage in a fuel cell?

spondingly. A most tempting goal. And that is why so much work has been done on fuel cells.

The cheapest oxidizing agent is air. The cheapest reducing agent is coal. What is needed is a cell that operates with an oxygen cathode and a coal anode. Such cells have been made, but their rate of energy production is far too low to be useful. What might be done? Did you guess "catalyze"? As in other cases, when the reaction is known to "go," the search is avidly on for the proper catalytic conditions. High temperature is also used to increase the rate—as you would guess.

Fuel cells accompany the astronauts. The cells use the reaction between pure hydrogen and pure oxygen. (The water produced is drunk by the astronauts. That's making every molecule count twice!) It is interesting to note that this is the same reaction that was used in the first fuel cell almost 100 years ago:

$$2\,H_2\,(g) + O_2\,(g) = 2\,H_2O\,(\ell) \qquad \mathscr{E}^0 \cong 1.2\,\text{volts} \qquad (17.16)$$

Other fuels have been made to work on a large scale, but they are uneconomical and difficult to use as *massive* power sources. Some are hydrazine, N_2H_4 (expensive, poisonous, hard to handle); ammonia, NH_3 (smelly, poisonous, requires refrigeration or high-pressure storage, explosive when mixed with air); methanol, CH_3OH (poisonous but easy to handle); gaseous hydrocarbons, CH_4 mainly (require high-pressure storage or refrigeration, explosive when mixed with air); coke or coal (require high temperature to give appreciable rates.)

The oxidizing agent is always oxygen, preferably as air. Practical oxygen electrodes are available but expensive. A major problem is that each dioxygen molecule must pick up four electrons. Only processes in which one or two electrons are interchanged are fast. Thus the initial steps must provide oxygen atoms that can undergo a one and/or two electron gain. These processes involve the adsorption of dioxygen on a surface with at least partial separation into two atoms. It is difficult to find surfaces that both adsorb the oxygen strongly enough to stretch or break the bond and release the product molecules readily. The bond strengths must be neither very strong nor very weak. This greatly restricts the choice of catalysts.

An additional difficulty is that oxygen electrodes work best in a strongly basic electrolyte, whereas many of the reducing electrodes (especially the carbon ones) work best in an acid electrolyte. But cells are far simpler to construct if they use a single electrolyte. So the search goes on for catalysts that will make fuel cells economical. The rewards both to the inventor and to society will be great.

Solar cells are developing rapidly. One of the rate-determining steps here involves finding a cheap receptor to cover the large areas necessary to collect enough solar energy for large-scale uses. Sunlight averages about 4 J/cm² per minute, and that is not much. It would take surfaces of many square miles to supply as much energy as that from an average power plant. On the other hand, most houses in the United States receive enough solar energy on their roofs during the average day to supply all household needs. A really cheap source for homes would solve a lot of problems. For example, fewer distribution lines would be needed.

Solar cells made from silicon, a plentiful element in nature, seem the best current bet. When very pure (eight-nines; see p. 188) silicon is cut into thin wafers and then "doped" with a small amount of boron on one side and a small amount of phosphorus on the other, a solar cell is formed. The phosphorus side has extra electrons (P has five valence electrons, compared to Si with four valence electrons). The boron side is electron-deficient because B has only three valence electrons. Exposing the phosphorus side to light drives electrons across the internal junction and leads to a potential of about 0.4 volt. Connecting many cells in series gives large voltages. Connecting them in parallel can give large currents. There are many interesting problems in developing these networks to give reasonable amounts of power.

Solar energy already keeps photosynthesis going, a mainstay of all life processes. Intensive research continues to explore and improve that not-very-efficient process to see if we can use it for electric power production. For example, biochemical fuel cells using enzyme catalysts have been built.

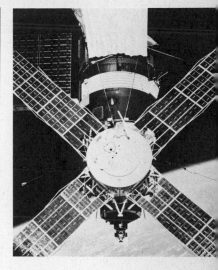

Solar panels on orbiting spacecraft

Coupled Reactions

Living systems cannot possibly be heat engines. Their efficiencies would be much too low. Nor are they simply electric batteries. But they do operate on redox systems. A main overall redox system for animals, as you know, uses oxygen to oxidize carbohydrates to CO_2 and H_2O. The reaction is often written

$$C_6H_{12}O_6 \text{ (c)} + O_2 \text{ (g)} = 6\,CO_2(g) + 6\,H_2O\,(l) + 2800 \text{ kJ} \quad (17.17)$$

But this is far from the true situation. A better representation is

$$C_6H_{12}O_6 + 6\,O_2 + 4\,H^+ + 38\,\textcircled{P}O^- + 38\,ADPH$$
$$= 6\,CO_2 + 44\,H_2O + 4\,ATP + \text{energy} \quad (17.18)$$

Don't worry about the ADPH and ATP symbols. Just admire the wealth of things that must happen. The full description is, of course, much more complicated.

Equation 17.17 merely shows the generation of heat from the overall redox process. Actually, many other processes must be fueled: growth, metabolism, structure maintenance, reproduction, locomotion. These

EXERCISE 17.14
Calculate the efficiency of a human acting as a heat engine if the average external temperature is 25°C.

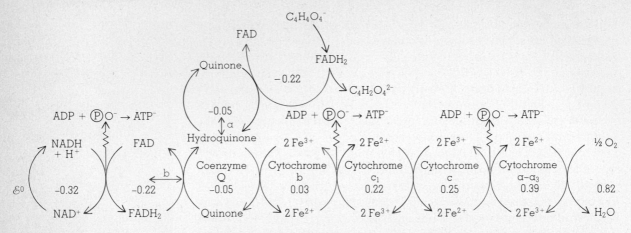

Figure 17.14
Some redox sequences common
in many living systems. Oxygen
is used to form ATP⁻, a common
molecule in which energy is
stored for future use. The
colored arrow in each cycle
indicates the oxidation step.
The numbers indicate $\mathcal{E}^0$ values.

and other functions require energy, which can be stored in the bio-chemicals ATP and ATP⁻ produced by equation 17.18. The ATP and ATP⁻ then help fuel the necessary *coupled reactions*. Humans appear to use about 40% of their energy for these processes, with the other 60% generated as heat. This is a better efficiency than most man-made power plants.

Redox in living systems acts through a series of coupled reactions. In photosynthetic plants, the input energy comes from sunlight. In most other forms of living systems, it comes from the oxidation of carbo-hydrates by oxygen. Instead of setting up little electric cells with anodes and cathodes, living systems seem to carry out their redox reactions so that a high fraction of the energy loss in each step is captured immedi-ately in another molecule.

Many of these processes are cyclical. You may have heard of the Krebs cycle, a principal mechanism for generating and storing energy in living cells. Its net reaction is

$$CH_3(CO)_2OH + 2\,H_2O + 4\,NAD^+ + FAD + GDPH + \text{\textcircled{P}}O^-$$
$$= 3\,CO_2 + 3\,H^+ + 4\,NADH + FADH_2 + GTP + energy \quad (17.19)$$

I shall not discuss the meaning of the symbols, nor the functions of the molecules for which each symbol stands. Suffice it to say that many coupled reactions are involved. About 10 enzymes are employed in the separate steps.

Figure 17.14 summarizes another set of coupled reactions that occur in almost all living systems. Ten coupled reactions are shown for which the net equations are

$$C_4H_4O_4{}^{2-} + \tfrac{1}{2}O_2 + 2\,\text{\textcircled{P}}OH^- + 2\,ADPH$$
$$= C_4H_2O_4{}^{2-} + 3\,H_2O + 2\,ATP^- + energy \quad (17.20)$$

$$H^+ + NADH + \tfrac{1}{2}O_2 + 3\,\text{\textcircled{P}}OH^- + 3\,ADPH$$
$$= NAD^+ + 4\,H_2O + 3\,ATP^- + energy \quad (17.21)$$

Again, do not concern yourself with the details of the symbolism. The point is that, complicated as the net equations are, each step is relatively simple, has an $\mathscr{E}^0$ associated with it, and occurs between known molecules. Because each step is simple, the overall rate can be fast. The system is flexible enough to meet various demands on it. The overall reaction, though very complex in the changes it accomplishes, occurs reproducibly and controllably. The organism can survive.

Conduction of nerve impulses may remind you of the current that flows in the external circuit of a battery. In the battery circuit, the current is electrons flowing along the wire. In the nerve it is ions flowing *perpendicular* to the nerve. In fact the ions merely penetrate the cell wall of the nerve, then stop and are pushed back outside so they can relay the next signal. Aqueous systems (including nerves) cannot conduct electrons, and ions move through water much too slowly to carry nerve signals at the rate of about 1 m/millisecond found in nature.

The actual mechanism involves a layer of positive ions (Na^+) on the outside of the nerve. When the nerve receives an impulse from a sensor or another nerve, the cell wall suddenly becomes permeable to the ions. They enter the cell. This permeability is transmitted down the nerve wall so there is a wave of discharge of the ionic layer moving about 1 m/millisecond. The ions are quickly pumped back out again, and the cell is made ready to transmit the next signal should it come a few milliseconds later. So the ions move only the thickness of the cell wall, about 10 nm. But the signal moves the length of the nerve. (See Figure 17.15.) Muscle cells go into contraction when Ca^{2+} ions enter in a similar way. In a very real sense, living cells are also electric cells.

Galvanic Protection

You have read about the use of metal plating or external pieces of metal to protect iron from corrosion (Chapter 14). The zinc coating on galvanized iron and the external pieces of magnesium or aluminum are better reducing agents than iron. When there is corrosive attack by an oxidizing agent, the zinc or magnesium gives up electrons more easily than the iron so the iron remains uncorroded. The other metals are known as *sacrificial metals*. Should the iron begin to corrode, the other metals tend to reduce the oxidized Fe^{2+} back to iron. In other words, iron is a better competitor for electrons than zinc, aluminum, or magnesium.

Tin plate on iron is another story. You have seen from Table 17.1 that tin is a better competitor for electrons than iron. Tin holds electrons more tightly than iron does. The role of tin on a tin can is the same as that of paint. It covers the iron and prevents the oxidizing agent from reaching the iron. It is very important for the tin plate to be free of pinholes, through which water might get in contact with the iron. When

EXERCISE 17.15
Suppose nerve signals traveled at 1 m/s, still pretty fast. What difference would that make to you?

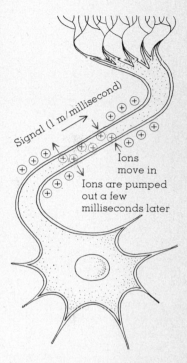

Figure 17.15
A nerve cell. Ions moving perpendicular to the cell membrane (through it) pass the nerve signal the length of the nerve.

EXERCISE 17.16
Aluminum metal is more powerful as a reducing agent than is zinc, but it is less powerful than magnesium. For the reaction $3e^- + Al^{3+} = Al$, what can you say about $\mathscr{E}^0$?

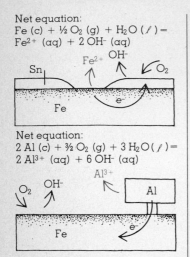

Net equation:
$Fe\,(c) + \frac{1}{2}\,O_2\,(g) + H_2O\,(\ell) = Fe^{2+}\,(aq) + 2\,OH^-\,(aq)$

Net equation:
$2\,Al\,(c) + \frac{3}{2}\,O_2\,(g) + 3\,H_2O\,(\ell) = 2\,Al^{3+}\,(aq) + 6\,OH^-\,(aq)$

Figure 17.16
Tin and aluminum in relation to the corrosion of iron.

HINTS TO EXERCISES
17.1 13.4 hours. 17.2 A carbon rod. 17.3 20 amperes. 17.4 Electrons flow from − to +. 17.5 Bottom. 17.6 Higher. 17.7 1.56 volts. 17.8 No, stops at $\mathscr{E} = 0$. 17.9 Zn forms Zn^{2+} or $Zn(NH_3)_4{}^{2+}$. 17.10 Cell would disintegrate. 17.11 0.0373 mole of zinc. 17.12 A one-third saving. 17.13 Lower. 17.14 Efficiency is 3.8%; humans are better than this. 17.15 About 3 seconds required for reaction to a nerve stimulus.

this happens, rapid corrosion can occur. The iron loses electrons and goes into solution at the pinhole. The electrons then move to the tin, where they can reduce oxygen to hydroxide ion as shown in Table 14.4, step 2. Similarly, cadmium plate cannot protect iron if there are pinholes in the plate, as shown in Table 17.2, example 3.

Note that it is not necessary for oxygen and iron to come into contact for corrosion to occur. Water around the iron can provide for ionic movement. This sets up an electric cell in which the processes of oxidation and reduction can occur at widely separated points. For example, external magnesium and aluminum are not good protectors of iron unless the surroundings are reasonably good conductors. They protect in ocean water, but not in distilled water. They protect in moist soils, but not in dry soils. (See Figure 17.16.)

Summary

Oxidation is the apparent loss of electrons. Reduction is the apparent gain. Redox processes can often be made to occur in separate places, and the electrons can be passed through an external conductor to form an electric cell. The cell must also contain an electrolyte through which ions can move between the anode (site of oxidation) and the cathode (site of reduction).

The direction of current flow is determined by the relative strengths of the chemicals present in competing for electrons. Those that are strong competitors for gaining electrons are good oxidizing agents. Oxygen is a good example. Those that give up electrons easily are good reducing agents. The tendencies to give up electrons in a defined standard state can be tabulated as $\mathscr{E}^0$ values.

Living systems use redox systems extensively. They often use coupled-reaction systems, in which part of the energy is released as heat but part is used to synthesize molecules essential for life or to maintain the physiological processes.

Problems

17.1 Identify or define: ampere, anode, battery (lead storage), cathode, cell (dry, electric, electrolytic, fuel, galvanic, nerve, rechargeable, solar), coupled reactions, electrolyte, faraday (96,500 amperes × seconds = 1 mole of electrons), galvanic protection, hydrogen electrode, standard voltage ($\mathscr{E}^0$), thermal efficiency, volt, watt.

17.2 A set of flashlight cells used intermittently will light the flashlight over a somewhat longer time than if used continuously. Why?

*17.3 Most dry cells eventually leak a sticky liquid. What is it? Used dry cells can serve as a source of several hard-to-obtain chemicals for schools in some of the emerging nations. List possible chemicals and how you would obtain them in reasonably pure form with no special equipment.

17.4 The nickel–cadmium (NiCad) cells use Cd and NiO_2 electrodes in aqueous KOH. Write possible equations for the two electrode reactions.

Should this cell be rechargeable? This is one of the common "alkaline" cells.

*17.5 Street hawkers sometimes sell a black liquid as a rejuvenator for transistor dry cells. They inject a few drops through a nail hole in the bottom of a "dead" cell and, sure enough, the cell lights a bulb (though it could not before the treatment). Suggest what chemicals are being sold. Would you recommend selling this fluid in a store you owned?

*17.6 You have purchased a "dry-charge" lead storage battery. It is charged but contains no liquid electrolyte. When you come to use it, the bottle of sulfuric acid has been lost, but you do have equally concentrated nitric, hydrochloric, and phosphoric acid. You need the battery badly. Look up the properties of these acids and their reactions with lead and lead compounds. Arrange the substitutes in order of most to least satisfactory, writing a possible net reaction for each one during discharge if put into the battery.

17.7 Lead storage batteries are often rated in ampere-hours. What is the minimum amount of Pb, PbO_2, and H_2SO_4 in a 200 ampere-hour battery? [$\sim$3000 g plus case and connectors]

17.8 The voltage of a lead storage battery decreases slowly as it is used. The voltage of the Edison battery hardly decreases at all. The Edison battery uses the reaction

$$Fe \, (c) + Ni_2O_3 \, (c) + 3 \, H_2O$$
$$= 2 \, Ni(OH)_2 \, (c) + Fe(OH)_2 \, (c)$$

The electrolyte is usually NaOH. Why is this voltage more constant than that of the lead battery?

17.9 It has been suggested that electric cars have no exhaust, hence cause no pollution. Comment in detail on the electric car as a solution to pollution.

17.10 Copper can be purified by oxidizing an electrode of impure copper and reducing the resulting ions onto a cathode of pure copper in an electrolytic cell. Sketch a diagram of the cell. What happens to the gold and silver impurities in the copper? To the zinc? In many places the reclaimed gold and silver pay for the process.

*17.11 The oxide content of molten steel in a steel-making furnace can be measured by determining the EMF of an oxygen electrode immersed in the molten steel. The half-equation is $\frac{1}{2}O_2 \, (g) + 2e^- = O^{2-}$. Would the voltage of the reaction become more, or less, positive as the oxide concentration decreased?

17.12 A CH_4–O_2 fuel cell operates with concentrated aqueous KOH as electrolyte. It produces CO_3^{2-} and OH.

   a. Write an equation for a possible net reaction.
   b. Write an equation for each half-reaction.
   c. Identify the anode and cathode reactions.
   d. In what mole ratio should CH_4 and O_2 enter the cell? [1 mole CH_4/2 moles O_2]

17.13 You are working for a jeweler and are told that a teaspoon is to be silver-plated to a thickness of 10^{-5} cm using a current of 0.1 ampere. How long should the plating operation take? (Hint: About what is the area of a spoon?) [$\sim$30 seconds]

*17.14 Electrolysis of an aqueous suspension of very finely divided $Fe(OH)_3$ precipitates the hydroxide at the cathode. Why so? Addition of a small amount of Na_3PO_4 precipitates the suspension. Any similarity to electrolysis?

17.15 In what common kind of conducting system do only the negative charges move?

17.16 Many car bodies now get their initial paint coat by total immersion in an aqueous suspension of electrically charged paint drops that are electroplated onto the car. What advantages has this method over spray painting? Should the paint drops be positive or negative in charge?

17.17 The tin coat on steel for cans was at one time applied by running the steel sheet through molten tin ("hot-dipping" process). Almost all such plating is now done by running the steel, as a cathode, through an electroplating bath of alkaline sodium stannate, SnO_3^{2-}. Cite some possible advantages.

*17.18 It does not pay to mix copper and galvanized iron plumbing, unless a nonconducting coupling is placed between the two systems. Why is a mix acceptable if the coupling is used but not otherwise?

Figure 18.1
Success (?) in the search for a
universal solvent.

18 LIKE DISSOLVES LIKE: SOLUBILITY

The alchemists searched for the philosophers' stone to transmute other metals into gold, and they searched for the universal solvent. Whether they worried about finding the completely insoluble container to hold the universal solvent does not seem to have been recorded (Figure 18.1).

What we have discovered is that substances do dissolve in one another given an opportunity. Sugar dissolves in coffee, milk dissolves in tea, salt dissolves in food juices, candy dissolves in your saliva. Chapter 16 discusses the dissolving of limestone in groundwater, and of buildings in acid rain. Solubility does not have to be large to be observable.

Table 6.1 lists some of the gases dissolved in the atmosphere. Table 6.2 indicates that the ocean is a solvent for many substances and lists some of these substances. As a matter of fact, many of the substances known to be dissolved in the ocean and the atmosphere are not listed in the tables because they are present in such small concentrations. I have also pointed out that both cotton and glass dissolve enough in water to give a negative charge to the solid. Apparently even though the whole substance does not dissolve, some of it does.

In fact, we now know that everything is somewhat soluble in everything else. But for a high degree of solubility to occur, the substances involved must follow a rule: Like dissolves like. Whether the rule is helpful to you depends on whether you can tell which things are "like."

The degree of "likeness" can be correlated with energy effects as substances dissolve. The overall energy change is given the sumbol ΔH. (Originally, ΔH stood for heat content, now called *enthalpy;* we shall usually just refer to ΔH.) If energy is required, ΔH is said to be positive; if energy is generated by the change, ΔH is negative. Like substances (say both polar) tend to form bonds to one another, generating energy on solution—so ΔH is negative. Unlike substances require more energy to disperse than they generate on bonding so ΔH is positive. Figures 18.2 and 18.3 illustrate these effects.

Water, Water Everywhere

The closest thing we have to a universal solvent is probably water. This is partly due to the fact that water is by far the most common liquid on

Science is a great many things, ... but in the end they all return to this: science is the acceptance of what works and the rejection of what does not. That needs more courage than we might think. It needs more courage than we have ever found when we have faced our worldly problems.
—*Jacob Bronowski.*

Figure 18.2
(a) Ionic NaCl and (b) non-polar I_2 molecules dissolving in water. In each case:

1. The bonds in the crystal break (ΔH is positive).
2. The water molecules separate to create holes (ΔH is positive).
3. The water and solute form bonds (ΔH is negative).

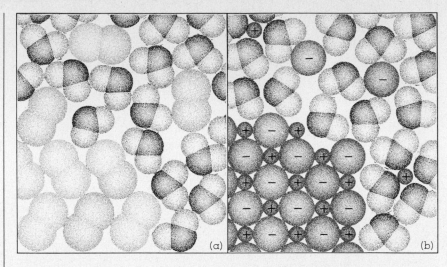

(a)   (b)

EXERCISE 18.1
Draw a diagram of the bonds you would expect calcium ions to make in H_2O.

the surface of the earth. But it is also due to the electrical properties of the water molecule. Water is polar, so it dissolves many polar materials well. If a polar substance is liquid, it is quite possible that water will dissolve it in any proportion. Such substances are said to dissolve without limit, or to be completely *miscible* (a substitute spelling for the less-used *mixable*).

Water is also sufficiently polar to dissolve many ionic substances readily (Figure 18.2a), and most of them measurably. The bonds formed between water and the ions are often almost as strong as the bonds between the ions in the crystal. There is no great change in energy in going from the two pure substances to the solution. Nor does the insertion of the ions in the water, with consequent breaking of water–water bonds, hinder the solution process greatly. The ionic substances that are most insoluble in water are those having the most highly charged ions (which give the most tightly bonded crystals, as we saw when we looked at gemstones).

Even nonpolar substances are appreciably soluble in water (Figure 18.2b). This is how fish get oxygen, bacteria get nitrogen, and humans get rid of oil slicks on the ocean. It is true that these solubilities tend to be low, because it is difficult for the nonpolar molecules to force their way into water.

The problem is that the dissolving molecules must shove the water molecules apart (make a hole in the water for themselves) and then form such bonds as they can with the water. The bonds between nonpolar substances and the small water molecules are much weaker than the bonds that water forms with itself. The net process of inserting a nonpolar molecule into water requires a considerable expenditure of energy. This minimizes the chance the process will occur. The solubil-

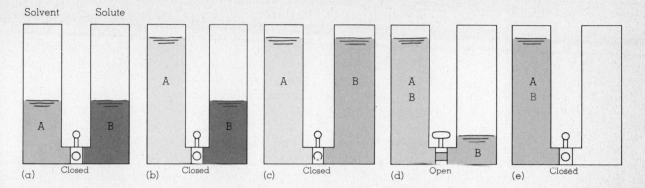

Solvent   Solute

A   B   (a) Closed
A   B   (b) Closed
A   B   (c) Closed
A
B   B   (d) Open
A
B   (e) Closed

Figure 18.3
A theoretical interpretation
of the formation of a solution.
From (a) to (b): Solvent A
expands to final volume, a
bond-breaking process (ΔH is
positive). From (b) to (c): Solute
B expands to final volume, a
bond-breaking process (ΔH is
positive). From (c) to (d) to (e):
Expanded A and B mix to give
the solution, a bond-forming
process (ΔH is negative).
$\Delta H = \Delta H_{cde} + (\Delta H_{ab} + \Delta H_{bc})$.
The more negative ΔH is, the
higher the solubility.

ities of nonpolar substances in water are lower than those of polar or ionic substances, which can and do form stronger bonds with the water molecules.

As in acid–base, redox, and other reactions, so here in solubility reactions: we find a competitive situation. You must consider both the reactants and the products in correlating solubilities.

Figure 18.3 schematically presents a sequence that describes the formation of a solution. That system is favored in which the bonds are the strongest. That reaction is favored for which the overall energy change, ΔH, is most negative.

In some cases ΔH is zero. This is true whenever the average of the intermolecular bond energies in solute and solvent is the same as those in the solution. Two substances that dissolve with no loss or gain in energy ($\Delta H = 0$) are said to form an *ideal solution*. They will usually be soluble in each other without limit. Gases mix readily and form ideal solutions (assuming they do not react). The various liquid hydrocarbons dissolve in one another to form ideal solutions without limit. So do many salts when melted together.

But water forms no ideal solutions, except in very dilute solutions (though some water solutions are not far from ideal and do not involve big energy changes). The properties of water are too unusual to be matched by the same type of bonding in any other substance. Water forms many solutions, but we will not find it easy to develop a simple model, or arithmetic, for what goes on. There is no ideal solution rule for water. But don't give up hope. There is a great deal that can be correlated.

The Solubility of Ionic Crystals in Water: The Solubility-Product Constant

One of the most common types of reaction, both in the laboratory and in nature, is the solution of an ionic substance in water. You may have noted in Chapter 6 that many ores are sulfides and oxides, and in Chapter 7 that hydrothermal deposition appears to have been common. This is consistent with the very low solubilities of most metallic ores in water. Let's consider the ore silver chloride, AgCl. It can be deposited hydrothermally.

Now, geothermal processes seldom occur at 25°C but, unless specifically listed otherwise, we shall do equilibrium calculations with equilibrium constants tabulated for 25°C. What we lose in truth to nature, we will gain in ease of calculation. So let's consider the equilibrium between AgCl and its aqueous solutions.

If pure AgCl is in contact with water for enough time, it will come to the following equilibrium:

$$AgCl\,(c) = Ag^+\,(aq) + Cl^-\,(aq)$$
$$K_{AgCl} = [Ag^+][Cl^-] = 1.7 \times 10^{-10} \quad (18.1)$$

Now a value of 1.7×10^{-10} is a small value for an equilibrium constant, so AgCl is quite insoluble in water—at least at 25°C. Solubility relationships for slightly soluble substances are often presented in terms of equilibrium constants (K). They are called *solubility-product constants*.

EXERCISE 18.2
Write the expression for the solubility-product constant for AgI in H$_2$O.

We may calculate the solubility of pure AgCl in water from the net equation and the solubility-product constant (equation 18.1) if we note that dissolving AgCl in pure water must give equal numbers of Ag^+ and Cl^- ions. Therefore,

$$[Ag^+] = [Cl^-] \quad (18.2)$$

so

$$K_{AgCl} = 1.7 \times 10^{-10} = [Ag^+][Cl^-] = [Ag^+][Ag^+] = [Ag^+]^2 \quad (18.3)$$

and

$$[Ag^+] = \sqrt{1.7 \times 10^{-10}} = 1.3 \times 10^{-5}\,M = [Cl^-] \quad (18.4)$$

The solubility of AgCl in water is $1.3 \times 10^{-5}\,M$ (assuming that only Ag^+ and Cl^- ions form). This solubility value, though small, is significant. Furthermore, it increases with temperature, so it is easy to believe that AgCl can be transported hydrothermally.

What will happen if extra Cl^- is added to the equilibrium system of equation 18.1? Le Châtelier's principle tells us that some solid AgCl will be formed, because the equilibrium will be shifted "toward the left." In fact, because K_{AgCl} is so small, most of the AgCl will precipitate from the solution.

Suppose that the hydrothermal stream of dissolved AgCl passes over a deposit of sodium chloride, which is readily soluble in water. The value of $[Cl^-]$ will increase greatly—let's say it reaches 0.1 M. What will the equilibrium concentration of Ag^+ become? The value of the equilibrium constant does not change, so

$$K_{AgCl} = 1.7 \times 10^{-10} = [Ag^+][Cl^-] = [Ag^+] \times 0.1 \qquad (18.5)$$

$$[Ag^+] = \frac{1.7 \times 10^{-10}}{0.1} = 2 \times 10^{-9} \, M \qquad (18.6)$$

The value of $[Ag^+]$ has changed from $1.3 \times 10^{-5} \, M$ to $2 \times 10^{-9} \, M$ (which is $0.0002 \times 10^{-5} \, M$). More than 99% of the Ag^+ must have precipitated as AgCl (c). This is how some silver ores form. Many other ores form similarly.

Silver bromide is more insoluble than AgCl, and AgI is still more insoluble.

AgBr (c) = Ag^+ (aq) + Br^- (aq)
$$K_{AgBr} = [Ag^+][Br^-] = 5.0 \times 10^{-13} \quad (18.7)$$

AgI (c) = Ag^+ (aq) + I^- (aq)
$$K_{AgI} = [Ag^+][I^-] = 8.5 \times 10^{-17} \quad (18.8)$$

Let's ask which ionic concentration would limit the value of $[Ag^+]$ in the ocean: $[Cl^-]$, $[Br^-]$, or $[I^-]$? The ocean concentration values are $[Cl^-] = 0.5 \, M$, $[Br^-] = 8 \times 10^{-4} \, M$, and $[I^-] = 4 \times 10^{-7} \, M$. Substituting these values in the corresponding equilibrium-constant expressions, we can compute the maximum concentration of Ag^+ that could be in equilibrium with each negative ion.

AgCl: $$[Ag^+]_{max} = \frac{K_{AgCl}}{[Cl^-]} = \frac{1.7 \times 10^{-10}}{0.5} = 3 \times 10^{-10} \, M \qquad (18.9)$$

AgBr: $$[Ag^+]_{max} = \frac{K_{AgBr}}{[Br^-]} = \frac{5.0 \times 10^{-13}}{8 \times 10^{-4}} - 6 \times 10^{-10} \, M \qquad (18.10)$$

AgI: $$[Ag^+]_{max} = \frac{K_{AgI}}{[I^-]} = \frac{8.5 \times 10^{-17}}{4 \times 10^{-7}} = 2 \times 10^{-10} \, M \qquad (18.11)$$

If these were the only equilibria in seawater (and, of course, they are not), we might conclude that $[I^-]$ would act as a limit on the maximum possible value of $[Ag^+]$. That is, if $[Ag^+]$ slightly exceeds $2 \times 10^{-10} \, M$, solid AgI will begin to form and $[Ag^+]$ will decrease. Of course, most ocean water is cooler than 25°C, and the values of $[Ag^+]_{max}$ we have computed are very nearly equal. Therefore, we would need more accurate data before making a definite conclusion about the limiting ion concentration in real seawater. However, the observed concentration of silver in seawater is in fact $2 \times 10^{-10} \, M$. So it looks as though all four ions are close to values consistent with the solubility-product constants.

EXERCISE 18.3
Suppose 1 liter of water per year passes through a rock containing AgCl (c), then later picks up further chloride ions and redeposits the AgCl (c). How many years would be required to transport 1 mole of AgCl? Does the process seem feasible in terms of the age of the earth?

POINT TO PONDER
In many chemistry books and
articles, the word *un-ionized* is
written without the hyphen. So,
if you read elsewhere that a certain compound is "unionized,"
you shouldn't conclude that it
has joined the Teamsters! Of
course, an un-ionized molecule
does have a certain union
about it.

Limited and Unlimited Solubility and the Solvent

Solubility without limit—that is, complete miscibility—can occur only between substances that are in the same phase: gas, liquid, or crystal. I have already said that all gases are soluble (in one another) without limit if they do not react. I have also pointed out (in the discussion of Dalton's law of partial pressures) that the ideal gas law provides good descriptions of the properties of gaseous solutions.

The cases of limited solubility of ions can be handled well by using the equilibrium-constant expression, as long as the concentrations are lower than about 0.1 M. This technique covers, for example, ionic crystals in water or the substitutions in the silicate minerals. At higher concentrations, the ions interact so strongly with one another that the solubility-product constant, K, seems to vary.

The extent of the solubility of un-ionized materials cannot be correlated with an equilibrium constant such as the solubility-product constant, because the molecules do not usually dissociate when they dissolve. Actual data on solubility must be tabulated, together with the variation in solubility with temperature. The extent of solubility depends on the degree to which the polarities of the solvent and solute match. Random motion always tends to encourage mixing, so that the principal variable is the energy required to accomplish the solution process. You have already seen that the more exothermic the process, the more likely it is to occur. Formation of bonds in the solution that are as strong as those in the pure solute and solvent ensures a high degree of solubility.

Le Châtelier's principle applies to solubility equilibria as much as to any other equilibrium. Le Châtelier's principle tells you that exothermic processes are favored at low temperatures, and endothermic processes at higher ones. A rise in temperature favors those processes that absorb heat—that is, endothermic ones. Most slightly soluble substances are slightly soluble "because" the process of solution is endothermic. So most slightly soluble substances become more soluble as the temperature is raised.

The dissolving of a gas is almost always an exothermic process, so gas solubilities decrease with rise in temperature. In fact, the solubility of a gas is usually zero in a liquid at its boiling point. This gives a method of purifying liquids from dissolved gases: boil the liquid. The common gases can all be expelled from water in this way.

Similarly, the solubility of a gas rises as its pressure increases. You could predict such an increase in solubility of a gas with increase in gas pressure using either Le Châtelier's principle, or escaping tendencies, or kinetic arguments. Couldn't you?

All these approaches treat the solvent as an inert material whose

EXERCISE 18.4
You are on a camping trip
and find a spring whose water
contains traces of hydrogen
sulfide. It stinks! How could you
make the water more pleasant
to drink?

main function is to keep the solute molecules separated. I shall use most of the rest of this chapter to discuss changes in the *solvent* when substances are dissolved in it.

The distinction between solvent and solute is really quite arbitrary. (This is especially obvious in cases of unlimited solubility.) The term *solvent* is always applied to a substance that does *not* change phase when the solution forms. (For example, if a gas and a liquid form a liquid solution, the original liquid would be called the solvent.) Where neither substance undergoes a phase change during solution, I shall apply the term *solvent* to the substance that is present in the greatest concentration. In the examples here, we shall concentrate on solutions where the concentration of the solute is less than 1 M.

Let's Start With the Pure Solvent

The solution of a solute in a solvent involves more than the intermingling of the two substances, even if they form an ideal solution. The formation of an ideal solution does not result in any changes in the average intermolecular forces, but it does change many of the properties of both the solvent and the solute. If only liquids and crystals are involved, both become more diluted (less concentrated) than when they were pure. The escaping tendencies of both liquid and crystalline substances decrease when they dissolve in one another. If the system is one of limited solubility, it will finally form a saturated solution in which the escaping tendency of the dissolved solute is the same as that of the pure solute. Equilibrium between the solution and the pure solute is then reached.

The escaping tendency of the solvent in the solution is always less than that of the pure solvent, because its concentration is lower. It is the changes in the escaping tendency of the solvent—its tendency to "go elsewhere"—that we shall now investigate. The properties of osmosis, vapor pressure, boiling temperature, and melting temperature (among others) of the solvent are all related to escaping tendency. Each of these involves the solvent moving from one phase to another. Let's see what happens to these tendencies when a solute is present.

Vapor Pressures of Solvents and Concentrations of Solutes

Perhaps the simplest measure of escaping tendency is pressure. The higher the pressure, the higher the tendency of the substance to go elsewhere in its random movements. So, the higher its escaping tendency.

What happens to the vapor pressure of a solvent when something dissolves in it? Let's assume this something is itself so nonvolatile that

POINT TO PONDER
Sometimes a very tiny concentration of solute can produce detectable effects in the solution. When the Kodak Research Laboratories developed a new kind of glass for lenses in the 1930s, an undesirable dark color, caused by impurities (mostly metal oxides) dissolved in the glass as it was being made, persisted. Even when platinum containers were used in the manufacturing process, the glass had a distinct yellow color. After many years of research, colorless glass was obtained by using gold crucibles. To get this colorless glass, the impurities had to be reduced to less than 1 part per billion—a purity of eight-nines.

POINT TO PONDER
The escaping tendency from any phase or compound always increases with increasing temperature and/or increasing concentration. But every reaction consists of a forward and a reverse possibility. Shifts in equilibria depend on which escaping tendency, forward or reverse, increases more with the change in temperature or concentration.

we need not be concerned with its presence in the gas. Thus the total pressure over the solution will be due to solvent molecules.

Adding such a solute to a solvent reduces the concentration of the solvent, reduces the escaping tendency of the solvent, and must therefore reduce the vapor pressure of the solvent. We get the same result using a kinetic argument. Adding solute molecules must dilute the solvent and decrease its rate of vaporization. The vapor pressure must decrease until the rate of condensation again becomes equal to the rate of vaporization.

An identical conclusion will be reached using Le Châtelier's principle. Adding a solute dilutes the solvent. That reaction will occur that will tend to increase the concentration of the solvent. Its gas molecules will tend to condense into the solution, thus lowering the vapor pressure.

The extent of lowering of the vapor pressure in an ideal solution is given by a simple equation:

$$\text{RAOULT'S LAW:} \quad p_A = X_A p_A^0 \qquad (18.12)$$

where p_A^0 is the vapor pressure of the pure solvent (A), X_A is the mole fraction of the solvent in the solution, and p_A is the vapor pressure of the solvent over the solution. *Mole fraction of the solvent* means the fraction of the molecules that are solvent molecules.

Raoult's law should strike you as reasonable. It says that if only 95% of the molecules in the solution are able to vaporize, the vapor pressure will be 95% of that which would be found if all could vaporize. The escaping tendency has been lowered to 95% of that of the pure solvent, so the vapor pressure falls to 95% of that of the pure solvent. Or, in kinetic terms, the rate of vaporization is 95% of that of the pure solvent, so the condensation rate must decrease a like amount, resulting in the same decrease in vapor pressure. (See Figure 18.4.)

It is this effect of lowering the vapor pressure of water when it forms solutions that makes candy sticky in humid weather. Gaseous water hits the surface of the candy and forms a solution. The tendency to vaporize is low because the solution is concentrated. Thus, there is further net condensation of water, and a layer of sticky solution forms on the candy. Manufacturers who sell food products, including candy, in tropical countries must use special water-impervious packing to keep the condensation of water from ruining the food in transit.

A similar effect accounts for the rust spots on chrome plate near the ocean. Salt spray lands on the chrome. On some days the spray may dry and just leave an innocuous salty layer. But on humid days there will be condensation of water to form a salt solution. This provides an electrolyte in any pinholes in the chrome plate, and rusting occurs. In a dry climate, or in the absence of salt, the rusting would be very slow. Car

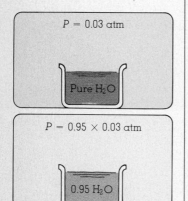

Figure 18.4
Raoult's law. The pressure of gaseous solvent in equilibrium with liquid solvent equals the mole fraction of the liquid solvent times the pressure of the pure solvent.

bodies stay unrusted and last a long time under one set of conditions, but only a few years under the other.

Boiling Temperatures and Concentrations

The boiling temperature is the temperature at which the vapor pressure of a liquid just equals the pressure of the surrounding atmosphere. On top of a mountain the atmospheric pressure is lower, so the boiling point is lower. Hence the difficulty in cooking potatoes by boiling on camping trips in the mountains.

The presence of a solute lowers the vapor pressure of the solvent. If the solvent is to boil, its vapor pressure must equal atmospheric pressure. The boiling point of a solution of a nonvolatile solute must be *higher* than that of the pure solvent. The rise in boiling point, ΔT_{bp}, is directly proportional to the mole fraction, X_B, of the solute (B). The more solute present, the larger the increase in the boiling point.

$$\Delta T_{bp} = K_{bp} \times X_B \qquad (18.13)$$

The value of K_{bp} is characteristic of the solvent, and independent of the nature of the solute in ideal solutions. For historical reasons it is not common to use mole fractions in this calculation. Special concentration units are used. We are more interested in water as a solvent so I will write the equation in terms of moles per liter, M. In water,

$$\Delta T_{bp}\,(^\circ C) = 0.51 \times M \qquad (18.14)$$

This value of the constant works quite well in water solutions that are not more concentrated than 1 M. Notice that the effect is not large, only half a degree Celsius per mole of solute in a liter of water.

If you wish to raise the boiling temperature of a liquid, place the liquid in a boiler having an adjustable pressure valve. This is much easier than adding a solute. The boiling temperature may then be set to any desired value by adjusting the pressure valve to maintain the vapor pressure corresponding to that value of temperature. Kitchen pressure cookers, for example, have their valves set for about 1.5 atm. At this pressure the boiling temperature of water is about 112°C or 385 K (Table 18.1).

Melting Points and Concentrations

Though boiling points can be adjusted easily by varying the pressure, melting points are only slightly affected by pressure. But melting points are sensitive to variation in solute concentration. (This is one effect that lowers the melting point of ice below its triple point. See Figure 12.6.)

EXERCISE 18.5
Why is the time required to bake potatoes not very sensitive to altitude, whereas the time to boil them is? Would it make any difference if you punch holes in the skins before baking the potatoes?

EXERCISE 18.6
Some sugar is added to water. The boiling point rises from 100.0°C to 100.1°C. What is the concentration of sugar in the solution?

Table 18.1
Vapor Pressure of H_2O
Versus Temperature

P (atm)		T (K)
0.01		280
0.03		298
0.05		306
0.1		319
0.2		333
0.4		349
0.6		360
0.8		367
0.9	Standard	370
1.0	boiling	373
1.1	point	376
1.2		378
1.4		383
1.6		387
1.8		391
2.0		394

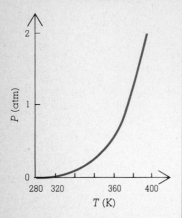

P (atm) vs T (K)

2 · · · 1 · · · 0

280 320 · 360 · 400

T (K)

Vapor pressure of water plotted versus Kelvin temperatures. Data from Table 18.1.

The solute, assuming it is soluble only in the liquid and not in the crystalline phase, will lower the escaping tendency of the liquid solvent. This makes it more difficult for the solvent to freeze, a tendency that can be offset by lowering the temperature. Lowering the temperature also lowers the escaping tendency of the crystal. The temperature at which the escaping tendency of the solvent in solution again becomes equal to the escaping tendency of the pure crystalline solvent will be lower than the normal melting point. Again there is a simple linear relation. For water it is

$$\Delta T_{\mathrm{mp}}\,(^\circ C) = -1.86 \times M \qquad (18.15)$$

When using equations 18.14 and 18.15, remember that M refers to the *total* concentration of *all* solute molecules actually present. For example, solutions of NaCl contain Na^+ and Cl^-, solutions of $CaCl_2$ contain Ca^{2+} and $2\,Cl^-$. Thus, there are 2 moles of ions per mole of NaCl and 3 moles of ions per mole of $CaCl_2$ that will affect the boiling and melting points of their aqueous solutions.

Melting-point lowering is an important factor in some large-scale applications. If you make ice cream at home, you add salt to ice to lower the temperature sufficiently to freeze your ice cream. Similarly, manufacturers often circulate a cold solution of calcium chloride around the containers of water to be frozen in ice-making plants or ice-skating rinks. The calcium chloride keeps the circulating water from freezing while it, in turn, freezes the ice by absorbing the heat given out in the process. This heat is called the *heat of fusion*. The heat of fusion of water is 6 kJ/mole.

Both sodium chloride and calcium chloride are scattered on ice-covered roads to melt the ice in winter. Only recently has anyone worried about the effects of the salty runoff water on neighboring fields and streams. But then, it is only recently that really large-scale de-icing programs have been carried out. A more obvious effect is the increased rusting of car bodies.

Osmosis

Osmosis comes from a Greek stem meaning "push." In popular usage the word means "acquisition with little effort," as in "They learn by osmosis." In science it refers to the process in which a solvent moves (pushes) through a *semipermeable barrier* (or membrane) into a solution. The barrier (or membrane) allows solvent molecules to penetrate but does not pass solute molecules.

Cell membranes in living systems are excellent osmotic membranes for water as a solvent. They are not merely osmotic membranes, because they actively transport certain substances in one direction and just as

EXERCISE 18.7
Cooks get vegetables such as carrots to be crisp by immersing them in water. Why is this effective? Would salt water be preferable to pure water? Hot water to cold water?

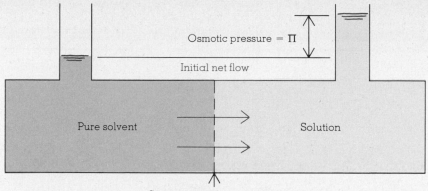

Figure 18.5
An osmotic-pressure apparatus.

Osmotic pressure = Π

Initial net flow

Pure solvent

Solution

Semipermeable membrane

actively transport others in the reverse direction. But water seems to flow readily through cell walls in both directions, solely in terms of its escaping tendencies on the two sides of the cell. In other words, water diffuses in and out all the time, but most of the other chemical species present do not. At equilibrium, which is most of the time, the two rates of flow of water are equal and the cell is in osmotic equilibrium as far as water is concerned. A solution that is in osmotic equilibrium with a cell is said to be *isotonic* with that cell. For example, blood plasma is isotonic with blood cells.

Figure 18.5 diagrams a simple osmotic apparatus. The osmotic pressure is the pressure that must be applied across the barrier (with pure solvent on one side and a solution on the other) to give equal rates of solvent flow in the two directions. In the apparatus shown, this pressure will build up by net flow of water from pure solvent into solution. When equilibrium is reached, net flow will cease, and the system will be at osmotic equilibrium. Remember that the barrier is selected so solute molecules *cannot* penetrate it.

The mathematical description of osmotic pressure (Π) is as simple as the other three equations relating vapor pressure, boiling temperature, and freezing temperature to concentration:

$$\text{Osmotic pressure} = \Pi \ (\text{atm}) = R \times T \times M \qquad (18.16)$$

where the constant, R, is 0.082. At 25°C, this equation becomes

$$\Pi \ (\text{atm}) = 0.082 \times 298 \times M = 24 \times M \qquad (18.17)$$

As in the earlier equations, M represents the total concentration of solutes in moles per liter.

Pressure is easy to measure accurately, even when it is small. Osmotic equilibria are not hard to attain for many systems. And look at the size of the constant in equation 18.17—it is 24 atm/(moles/liter). A 1 M solution gives an osmotic pressure of 24 atm! This means that a

EXERCISE 18.8
A sample of 4 g of a new polymer is dissolved in 1 liter of water. The osmotic pressure is 2.4×10^{-2} atm. What is the molecular weight of the polymer?

EXERCISE 18.9
It has been suggested that most present students would not be alive if hard candy in the old-fashioned corner store had not had such a high osmotic pressure against water. The argument goes that flies, which carry many deadly germs, walked on the candy, yet the candy was still sterile by the time your ancestors ate it. Complete the argument.

small concentration of solute gives a more readily observable effect in an osmotic cell than on any of the other three properties related to escaping tendency.

So osmotic pressure is not only vital to living systems. It is also important to the research scientist interested in determining the molecular weight of large molecules such as are found in living systems.

The occurrence of osmosis in living systems ensures an equal "concentration" of water in most living cells. A cell placed in an isotonic solution loses and gains water at equal rates. If that cell is placed in distilled water, the cell will swell. If it is placed in a solution less concentrated in water (more concentrated in solutes than the isotonic one), or on a solid that is soluble in water, the cell will shrink. This raises some interesting questions about how steelhead, salmon, and other fish that move from fresh water to the sea and back again adjust their osmotic balances during the transition periods.

Many molecules move through body membranes by selective paths requiring expenditure of energy. This is known as *active transport*. But many other molecules, like H_2O, move by osmosis. The laxative action of Mg^{2+} comes about because the intestinal walls will not let this ion pass. So Mg^{2+} increases the osmotic pressure across the membrane and water flows into the intestine making the contents more fluid.

An Unlikely Trio: Milk, Genes, Osmosis

Some 75% of the world's adults are made ill by drinking milk. They get diarrhea, gas, and general abdominal pain. Theirs is probably a genetic problem, as suggested by the data in Table 18.2. This effect, called *lactose intolerance,* has been medically recognized only since 1963. The gene that directs the synthesis of the enzyme lactase is inactive in lactose-intolerant individuals. Lactase in the intestine catalyzes the splitting of lactose (milk sugar) into glucose and galactose (two simpler sugars). The simple sugars can penetrate the intestinal wall, but lactose cannot. So unsplit lactose increases the osmotic pressure across the wall, and water flows into the intestine, causing diarrhea. The lactose is digested by the intestinal bacteria to form carbon dioxide, causing gas and pain.

Lactose intolerance once caused serious problems for some recipients of the overseas food program of the United States—when dried milk was used to enrich local diets. The problem is solved either by putting the enzyme lactase in the dried milk or by removing the lactose.

Pure Water by Osmosis

An interesting application of osmosis, currently commercial but still expensive, is in the purification of water. *Reverse osmosis* can be used

EXERCISE 18.10
Few babies are lactose-intolerant. Also, African and Asian herders are seldom intolerant as adults. How do you interpret these data?

**Table 18.2
Lactose Intolerance of Adults of Various Ethnic Groups[a]**

Country	Group	Percentage lactose intolerant
Australia	Aborigines	85
	Europeans	4
United States	Indians	100
	Negro	95
	Caucasian	12
Other	Thai	98
	Chinese	93
	Bantu	89
	Finnish	18

NOTE: Data from McCracken (1971) and Simons (1970).

[a]Sample size varied from 20 to 250.

either to desalinate seawater or to remove some of the ions remaining in water that has been through a sewage treatment plant and is to be reintroduced into the water system. If there is no external pressure applied to an osmotic cell, water flows from where it is more concentrated (pure water) to where it is less concentrated (the solution). However, if pressure exceeding the osmotic pressure is applied to the solution side, there will be net flow of water from solution to pure water. The barrier will not pass ions and they will be "filtered out" from the solution. Pure water will result. (See Figure 18.6.)

The need for high-quality water grows rapidly. Some predictions suggest that all the streams in the United States will be fully utilized by the end of the century if current trends continue. Reverse osmosis is one method of quickening the natural water cycle and making the liquid pure enough for reuse. Very large sums of money are being spent in research on suitable barriers. The current problems, and probably the only serious long-term ones, are that the barriers are not highly permeable to water, do not last long, and are expensive to replace. The invention of a cheap, rugged, high-rate-of-flow barrier for a reverse-osmosis plant would help solve man's need for one of the most critical of all substances.

EXERCISE 18.11
What does a reverse-osmosis plant produce in addition to purified water?

Solubility and Living Systems

Almost all forms of life are based on aqueous-solution chemistry. How do living systems assemble and maintain themselves using water as the circulating fluid and yet not dissolve every time it rains or a wave washes over them? This situation is, perhaps, even more remarkable if you con-

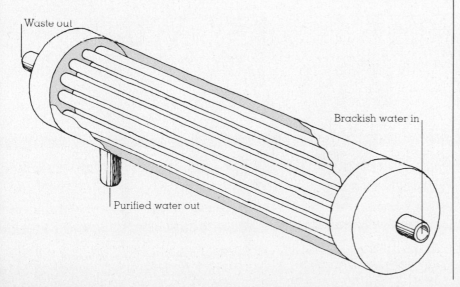

Waste out

Brackish water in

Purified water out

Figure 18.6
Reverse osmosis. When pressure that exceeds the osmotic pressure is applied, H_2O is forced to flow from the solution to the pure H_2O. Commercial plant and detail of osmotic unit.

sider that early life forms arose in the sea. Clearly they solved the problem of controlled solubility several billion years ago.

We have already discussed carbonate equilibria. You know that seashells are calcium carbonate. You have probably seen spiral shells with spikes (Figure 18.7) and noted that the inhabitant must be able both to precipitate fresh seashell and to dissolve old shell when the spiral growth brings one of those spikes into the living area. This is done by pH control. Calcium carbonate dissolves at low pH (acid solutions) and precipitates at high pH.

Humans must control precipitation and crystal solution reactions in growing bone and teeth. When the rate of precipitation gets too great, or occurs in the wrong places, we complain of calcium deposits or "stone" formation; both are painful. But we are delighted that our bones can "figure out" when they are broken and grow back together again. These involve controling complex phosphate solubility equilibria. Let's leave it at that.

But consider the problem of lipid (formerly called *fat*) transport. You have seen that cell membranes consist of long lipid chains with polar groups at the ends. The stability of cells attests to their low solubilities in body fluids. How did the lipids get to the cell wall, then? You may also have noted that the natural fats and oils occur only in a narrow range of hydrocarbon chain lengths—about 16 to 18 carbons (see Figure 11.8). Well, one reason for this limit is that longer chains are not soluble enough to be transported in physiological fluids—and shorter chains, although more soluble, attack cell membranes and dissolve them.

You may also have noted that natural lipids contain an even number of carbon atoms. This is because they are synthesized—two atoms at a time—from acetate ion, CH_3COO^-. The formate ion, $HCOO^-$, would

EXERCISE 18.12
Should a shellfish increase or decrease the local pH in order to grow additional shell?

Figure 18.7
Shelled animals must be able to precipitate and to dissolve $CaCO_3$.

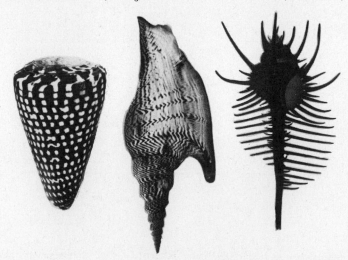

require twice as many steps and has no methyl group (much used in physiologically important molecules). Acid ions with more than two carbons are too unreactive to form chains readily. If you look at the other soluble two-carbon molecules, you find that alcohol is less reactive, aldehyde is too reactive, and both are volatile so would tend to escape from the system. So our lipid chemistry, and much of our other chemistry as well, is based on the acetate ion and its behavior in water. The fact that the acetates of all the common cations are soluble also helps, by minimizing the change of precipitating acetate in the "wrong" places.

Conversion of small molecules to large ones in living systems is not restricted to precipitation of rigid or ionic substances. A common reaction is *polymerization* (and its reverse, depolymerization). The main structural materials of plants are made up of polymerized glucose, a sugar of high water-solubility. The main structural materials of animals (other than their skeletons) are proteins, made up of α-amino acids, which are also water-soluble. Digestion converts these *polymers* to small molecules of water-soluble glucose or α-amino acids. These are readily transported by aqueous body fluids, penetrate the cell walls, and are then polymerized into the long-chain insoluble polymers. I shall discuss polymers at length in a later chapter. Their insolubility is due to their great length and great surface areas—which allow tangling, cross-linking, and other intermolecular forces (hydrogen bonds, polar bonds—see Chapter 9). These effects stabilize the polymer phase and give a low solubility.

Nucleation

I have alluded to rate effects several times in this chapter, but mainly as they determine equilibrium conditions. Remember that equilibria may be favorable for changes, yet the changes may not occur. In fact, there are many living systems that survive because some equilibria that "should" be attained are not. For example, some fish caught in the deep water off Labrador will freeze solidly if touched with a piece of ice. Their body fluids are cooled well below their freezing points. But freezing does not occur in the natural habitat.

The freezing of a liquid converts a random collection of closely packed molecules into an ordered crystal. If the crystal is complicated, and if the liquid is viscous (difficult to move through), it will be difficult for the ordered pattern to occur merely through the bumping together of randomly oriented molecules. It is quite easy to cool most liquids well below their freezing points to produce *undercooled* (sometimes called *supercooled*) liquids. If the crystal pattern is introduced (as by touching the fish with ice), nuclei with the correct pattern begin to form, and crystal growth occurs from those points. This is called *nucleation*.

EXERCISE 18.13
If you put the sharpened end of
a pencil into a soft drink, there
is rapid evolution of gas. The
tip of a ball-point pen does not
usually give this effect. Why
the difference?

Nucleation provides the "orientation factor" required for a reordering process to proceed at an appreciable rate.

In the same way, many liquids (especially if they are very pure) can be heated above their equilibrium boiling temperatures. There can be a high activation energy for the formation of a hole in a liquid (we call it a bubble). But if the liquid does not wet its container well, or if some porous material is present with gas bubbles on its surface, these bubbles can serve as nucleation centers and equilibrium boiling occurs.

It is also quite easy to prepare solutions whose concentrations are greater than the equilibrium, or saturation, value. They are said to be

Figure 18.8
Nucleation in a supersaturated solution. Nucleation occurred at two separate times, hence the two different-sized clumps of growing crystals. (Courtesy Mrs. Fraser P. Price)

supersaturated. Adding a tiny amount of the missing phase, say a grain of the appropriate crystal (or one with a similar crystal pattern) leads to rapid crystallization (Figure 18.8). Again we find the effect of orientation on rate. The rapid evolution of gas from a poured carbonated beverage (such as a soft drink or beer) is a good example of nucleation (by air bubbles) of a solution supersaturated with a gas. You know it doesn't pay to shake a bottle of carbonated beverage just before opening it.

The one type of phase transition that never seems to be held up by rate effects is the melting of a crystal. In spite of repeated attempts, it has been impossible to heat a crystal above its melting point. The fact that the transition is from an ordered to a disordered state, with very little change in the separation of the molecules, accounts satisfactorily for the lack of both orientation and activation-energy during melting.

It is quite likely that supersaturated solutions are common in living systems. Note that net precipitation can occur only if a solution is supersaturated. A *saturated* solution would only maintain a dynamic equilibrium. An *unsaturated* one would dissolve material into itself. In fact there is evidence that many living systems contain molecules that encourage the existence of supersaturated states by increasing the viscosity of the fluids or by coating the particles so they cannot collide. These effects would be very useful during the transport state. Once the particle got to the place where it was to precipitate or react, the "patterns" would be presented to it, and supersaturation would disappear.

Summary

Whenever two substances are placed together, solutions form because of the random motion of the molecules. If there is little change in the average intermolecular bond strengths between the two pure substances and the solution, they will be quite soluble in one another. If the energy change is zero, the substances are said to form an ideal solution. Gaseous solutions are nearly ideal unless reaction occurs. The concentrations of ions in a solution in equilibrium with crystals are related by the equilibrium constant.

Trends in solubility with temperature and with pressure can be predicted by using Le Châtelier's principle or kinetic arguments, or by considering the effects on escaping tendencies.

The vapor pressure, boiling temperature, freezing temperature, and osmotic pressure of the solvent are directly proportional to the concentrations of the solutes. These effects have important applications in living systems, in industry, and in the research laboratory.

Rate effects involving phase changes in solutions and pure substances can often be interpreted in terms of nucleation, which is dependent on orientation and activation-energy effects.

HINTS TO EXERCISES
18.1 Would attract the oxygen. 18.2 $K_{sp} = [Ag^+] \times [I^-]$. 18.3 8×10^6 years. 18.4 Boil the water. 18.5 Temperature of boiling water depends on pressure, that of ashes doesn't. 18.6 Concentration of sugar 0.2 M. 18.7 Pure water enters the cell faster than it leaves; with salt the process could reverse; hot water destroys cell membranes. 18.8 4×10^{-3} mole. 18.9 Bacteria would be dehydrated. 18.10 Drinking milk stimulates the body to produce necessary enzyme. 18.11 A rather concentrated solution of impurities. 18.12 Increase pH. 18.13 Trapped air bubbles serve as nuclei.

Problems

18.1 Identify or define: active transport, boiling-point rise, heat of fusion, ideal solution, isotonic, lactose intolerance, melting-point lowering, miscible, nucleation, osmosis, polymer, Raoult's law, reverse osmosis, saturated, semipermeable membrane, solubility-product constant, solute, solvent, vapor pressure.

18.2 Is homogenized milk homogeneous? Give evidence.

18.3 Whiskey that is 100 proof was originally defined as having a concentration of water such that gunpowder drenched in the whiskey would, when ignited, catch fire as the whiskey burned away. A lower proof whiskey would not let the gunpowder burn. Show why in terms of molecular behavior.

*18.4 Clouds may sometimes be seeded to initiate rain or snow by sprinkling dry ice (solid carbon dioxide) or crystalline silver iodide in them. Suggest mechanisms for these effects. Under what conditions would you recommend each method?

*18.5 The presence of enzymatically hydrolyzed cornstarch (which gives syrup and glucose) is used to inhibit the crystallization of sucrose in jams, jellies, candies, ice cream, and so on. Why does inhibition occur?

18.6 It is rather easy to cool liquids below their freezing points without causing crystals to form. But no one has ever succeeded in heating a crystal above its melting point. Why the difference?

18.7 Lettuce put into a refrigerator freezes unless wrapped or placed in a "crisper." Why?

18.8 "It isn't the heat, it's the humidity" is a common phrase to describe discomfort. The relative humidity in Antarctica often exceeds the relative humidity in the tropics, yet people living in Antarctica almost never complain about humidity there. Why?

*18.9 The equilibrium constant for dissolving CO_2 in H_2O is 0.03 (mole/liter)/atm. Soft drinks have a bottled pressure of 2 atm of CO_2. Calcu-

late the concentration of CO_2 in the soft drink. [$\sim 0.06 \, M$]

18.10 Freshly boiled water tastes flat. Its taste can be restored by pouring it from one container to another a few times. Why the change?

18.11 Maple sugar is made by boiling the sap (a very dilute solution) of the sugar-maple tree (*Acer sacrum*) until the boiling point has risen by 4°C. Estimate the moles per liter and weight of the sugar ($C_{12}H_{22}O_{11}$) in the final solution. [~ 3000 g/liter]

18.12 At 4000 m altitude the pressure is about 0.6 atm and the boiling temperature of water about 85°C. Why is the pressure less than atmospheric and the boiling temperature of water less than 100°C?

18.13 a. Why does it take longer to boil an egg in the mountains?
b. Why is a stream purified when it flows over rocks?
c. Why is it often harder to open a refrigerator door that has just closed than one that hasn't been opened for a while?
d. Mushrooms are often washed in salt water, but the same treatment would wilt lettuce. Why the difference?
e. Will frozen laundry dry?
f. What happens to a pond when the surface water cools off at night?
g. Why do farmers put large tubs of water in their fruit-storage bins during a cold winter?
h. Why does freezing fermented cider produce applejack, a liquid of higher alcohol content than the cider?
i. Why do the raisins in raisin toast burn your fingers more than the toast does?

18.14 Your younger brother comes in with large patches of oil from a street "oiling" on his leg and shoe. How do you treat him (to get the oil off, that is)?

*18.15 Dehydration is by far the oldest means of preserving foods. Food is spoiled primarily by bacterial and/or enzyme action. Why does drying stop spoilage? Much dried food smells of "sulfur." This is because it has been the practice, since antiquity, to sulfite food to stop non-enzymatic browning. Do you think it is really sulfur that you smell?

*18.16 Freezing is a more and more popular method of preserving foods. It is also being actively explored as a means of preserving living species—sperm for example, but milticellular species are also under investigation. Either low or high rates of freezing damage the cells so they cannot be revived. The problem in rapid freezing is the formation of internal ice, which destroys the internal organization of the cell. The damage due to slow freezing is quite different. It tends to decrease cell volume lethally. Why the decrease?

18.17 A major change in commercial canning of food occurred in 1861 when canned food was heated in a calcium chloride solution rather than in water. The time required to sterilize the contents dropped from 4 hours to 25 minutes. Why did the $CaCl_2$ work?

18.18 Phosphate effects on the environment have led to replacement of phosphates by sodium carbonate and silicate as additives to increase the the cleaning power of detergents. The additives often cause a white scale deposit in washing machines and even on the clothes. What are the deposits?

18.19 Large quantities of hydrochloric acid are often introduced into "worn-out" oil wells to activate them. Suggest a possible interpretation of the effects.

*18.20 Sodium dissolves in mercury to form a solution, called an *amalgam,* that generates hydrogen only slowly when placed in water. Suggest why the rate is low.

18.21 Study of the rate at which water bound (hydrated) to a cation interchanges with the bulk water shows for the ions listed the following trends in rates:

$$Ca^{2+} > Cu^{2+} > Zn^{2+} > Mn^{2+}$$
$$> Fe^{2+} > Co^{2+} > Mg^{2+} > Ni^{2+}$$

Can you explain why?

18.22 The melting point of glycerol is 20°C, yet very few people have ever seen crystalline glycerol. Why? Some years ago a glycerol plant in Canada had all its glycerol freeze in the winter. The stored glycerol was melted but froze again every cold day until summer. Why? There was no trouble the next winter. Why?

*18.23 Silicate-glass fibers are as strong as any large-scale fiber produced, but they become weak if scratched, especially if the air surrounding them is moist. Immersed in plastic, they retain their great strength. Suggest reasons for these effects in terms of SiO_2–CaO–Na_2O composition of glass.

*18.24 Glass made of SiO_2–CaO–Na_2O is soluble in water to the extent of about 30 ppm. Treating the surface with gaseous sulfur produces a surface insoluble in neutral or acid solutions after it is washed once. Such containers are good for bottled medicines. The sulfur is said to "combine with the alkali." Suggest what happens and why the surface is insoluble. This process saves purchasers of medicine about $85 million per year.

18.25 Estimate the temperature inside a pressure cooker that is set to 1.3 atm.

18.26 Garlic and onion odors in breath are not removable by tooth brushing or mouth washing since they are due to molecules like allyl disulfide, $(CH_2{=}CHCH_2)_2S_2$, and allyl propyl sulfide, $CH_2{=}CHCH_2SC_3H_7$, which are absorbed from food in the intestines, transported to the lungs by the blood, and then exhaled. What is there about the structure of these molecules that lets them transport as they do in the body?

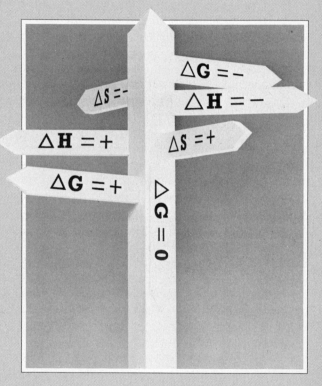

Figure 19.1
Which way?

19 WHICH WAY WILL THEY GO? THE TENDENCY TO REACT

You should be convinced by now that molecules move in rapid random patterns—colliding, rebounding, making new bonds, breaking old ones, constantly moving toward dynamic equilibrium. Net reaction ceases at equilibrium, but all the random processes continue.

All pure substances dissolve in one another to some extent. They also react with one another to some extent. They compete for electrons, for protons, and for other atoms. All these effects are due to the randomness of molecular collisions and their wide range of energies. You have learned to describe the equilibrium state with equilibrium-constant expressions. And you can estimate, by using Le Châtelier's principle, kinetic theory, or escaping-tendency arguments, how an equilibrium will shift with changing temperatures, pressures, and concentrations.

You know that sizes and bond angles can be assigned to individual atoms and are repeated from molecule to molecule. Similarly, the bond energies and polarities between two atoms are quite constant from one molecule to another.

Are there similar regularities that allow the tabulations of stabilities of individual molecules, and their tendencies to react? Such tables would enable you to calculate the position of the equilibrium state for any set of molecules, perhaps even under any conditions. What a great simplification of the data that would allow!

Think how much would be gained if one could tell ahead of time that a given combination "ought to react" (or cannot react appreciably) to give some desired product. How much easier it would then be to avoid the costly errors involved in "choosing" to do something impossible. Such tables exist. They tabulate the *free energies of formation of individual compounds*. It is easy to use free-energy tables. Only multiplication, addition, and subtraction are used. The tables have greatly simplified the design of chemical processes, and have markedly reduced the number of "impossible" things that get tried. They have also enhanced our understanding of living systems. I shall discuss their development and use in this chapter.

There are quite a few arithmetical calculations used in this chapter,

Order and simplification are the first steps toward the mastery of a subject—the actual enemy is the unknown.—*Thomas Mann.*

some involving logarithms. It is not nearly so important that you follow every detailed calculation as that you see the general relationships and find they "make sense."

Compounds From Elements

The simplest comparison state for reactivities might be separated neutrons, protons, and electrons—because all the atoms and molecules can be made from them. But these particles are hardly standard reactants in the chemistry laboratory, and we don't really know enough about their equilibrium states to use them as a standard.

But we *can* visualize all compounds as synthesized from the elements. We can even directly measure the equilibrium constants for this process in many cases. If a compound has a great tendency to form from the elements, the equilibrium constant is large. The larger the equilibrium constant, the more stable the compound. Let's explore the possible use of equilibrium constants as measurements of chemical stabilities.

The equilibrium synthesis of NH_3 from the elements is easy to measure. It shows that a mixture of N_2 and H_2 (each at 1 atm pressure at 298 K) would be in equilibrium with 800 atm pressure of NH_3. The equilibrium clearly favors forming ammonia.

Thus, ammonia is much more stable than its constituent elements at 298 K. It would be possible to assign it a relative stability of 800 atm, compared to hydrogen and nitrogen each at 1 atm pressure. One possible equilibrium state would have equal numbers of molecules of N_2 and H_2 and 800 times that many molecules of NH_3.

If we choose some other temperature or another standard pressure for hydrogen and/or nitrogen, the stability pressure for ammonia will change, consistent with the equilibrium constant. If we are to compare relative stabilities of compounds with respect to the elements, we must define a *standard state for the elements*. We shall use the same state we used in discussing standard cell EMFs, or $\mathscr{E}^0$ values: 298 K for temperature, 1 atm pressure for the gaseous elements, and the pure liquid or crystalline state at 1 atm pressure for the other elements. If we wish to talk about an element (or another substance) in solution, we shall use an ideal 1 M solution as the standard state.

On this basis, the relative stability of NH_3 with respect to the elements from which it can be made is 800 atm. That is a rather high equilibrium pressure, and very little H_2 and N_2 remain at equilibrium (Figure 19.2). Yet NH_3 is far from being one of the most stable molecules.

There are some difficulties with using the equilibrium pressure as the measure of stability. For one thing, we seldom describe concentrations of crystalline or liquid substances in terms of pressures, so values

800 = NH$_3$
1 = H$_2$
4 = N$_2$

Figure 19.2
Schematic representation of
a possible equilibrium state.
The molecular ratio is
1 N$_2$/1 H$_2$/800 NH$_3$.

of pressure for nongaseous substances would be difficult to measure or to visualize. For another thing, many reactions go almost to completion (with equilibrium pressures in thousands or millions of atmospheres), and other reactions occur hardly at all (with equilibrium pressures in minute fractions of an atmosphere). A standard way to deal with such a wide-ranging scale of values is to compress the range of possible numbers by using some relation based on logarithms (see Appendix A).

A measure more useful than equilibrium pressure emerged from the work of J. Willard Gibbs early in this century. Gibbs defined a quantity called the *free energy* of a substance as a measure of the substance's tendency to react. We now call this quantity the *Gibbs free energy*, abbreviated as G and expressed in units of energy such as kilojoules. There is no absolute zero for this quantity, so Gibbs arbitrarily defined the value of G as zero for any pure element in the standard state. We will mainly refer to G^0, using the superscript zero as a reminder that the equations have been simplified by assuming that all substances are measured and all reactions occur in the standard state.

The quantity that is useful as a replacement for equilibrium pressure is the *difference* in G^0 between the products and the reactants of a given reaction. Gibbs wrote this difference as ΔG^0, and defined it to give

$$\Delta G^0 = -2.303 \, RT \, (\log K) \tag{19.1}$$

where T is the absolute temperature, K is the equilibrium constant for the reaction, and R is a constant (actually the gas constant of $PV = nRT$). In the standard state, $T = 298$ K. If we use units for R that are appropriate to obtain an answer expressed in kilojoules, we can rewrite equation 19.1 in this form:

$$\Delta G^0 \, (\text{kJ}) = -5.708 \log K \tag{19.2}$$

If K is very large (indicating that the equilibrium strongly favors the products), ΔG^0 has a large negative value. (This follows from equation 19.2 because the log of a number greater than 1 is positive.) If K is less than 1 (indicating that the equilibrium favors the reactants), ΔG^0 has a positive value.

For every K (representing the equilibrium constant for the formation of a compound from its elements) we can calculate a corresponding value of ΔG^0 with a subscript f standing for *formation*. This ΔG_f^0 (Gibbs free energy formation of the substance) is a measure of the substance's tendency to react. Let's try a few simple calculations to show how ΔG^0 values might be obtained.

For the formation of NH_3 from its elements (N_2 and H_2):

EXERCISE 19.1
What is the value of ΔG^0 for a reaction with an equilibrium constant K of 10?

$$N_2 \, (g) + 3 \, H_2 \, (g) = 2 \, NH_3 \, (g) \qquad K = \frac{[NH_3]^2}{[N_2][H_2]^3} \tag{19.3}$$

We have seen that the equilibrium pressure of N_2 can be 1 atm, of H_2, 1 atm, and of NH_3, 800 atm. So

$$K = \frac{800^2}{1^2 \times 1^3} = 800^2 = 6.4 \times 10^5 \tag{19.4}$$

Using equation 19.2, we find that

$$\Delta G_f^0 \, [\text{of } NH_3 \, (g)] = -5.708 \, (\log 6.4 \times 10^5) \tag{19.5}$$

$$= -5.708 \times 5.9 \tag{19.6}$$

$$= -34 \, \text{kJ}/(2 \text{ moles } NH_3) \tag{19.7}$$

Note that this value (-34 kJ) is for 2 moles of NH_3. You will find a more accurate value, -16.5 (kJ/mole NH_3), listed in Table 19.1. It means that the Gibbs free energy (in the standard state) of NH_3 is 16.5 kJ/mole less than that of the N_2 and H_2 from which it can be made.

When ΔG_f^0 is negative, the compound is more stable than its constituent elements (in the standard state). The value $\Delta G_f^0 \, (NH_3) = -16.5$ kJ/mole means that if we mix equal numbers of molecules of NH_3, N_2, and H_2 in the standard state (each gas at 1 atm and 298 K), we expect a net reaction that forms NH_3 and uses up N_2 and H_2 (Figure 19.3). Because NH_3 is more stable than the elements, ΔG_f^0 of NH_3 is negative. Experimental studies confirm that this reaction occurs, though the rate is slow in the absence of a catalyst.

This example emphasizes the fact that changes in standard Gibbs free energy (ΔG^0) indicate only the direction of likely reaction, not the rate of reaction. A negative value of ΔG_f^0 in Table 19.1 indicates that the substance will tend to form from its elements in the standard state,

Figure 19.3
At 1 atm pressure and 298 K, N_2, H_2, and NH_3 will react to give net NH_3; $\Delta G^0 = -16.5$ kJ/mole NH_3. So NH_3 tends to form. At equilibrium, $\Delta G = 0$; no net reaction occurs.

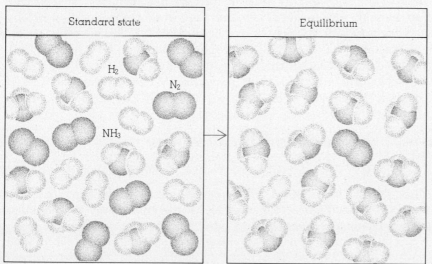

Standard state	Equilibrium

H_2

N_2

NH_3

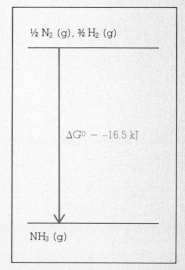

½ N_2 (g), ³⁄₂ H_2 (g)

$\Delta G^0 = -16.5$ kJ

NH_3 (g)

Table 19.1
Relative Stabilities of Some
Substances in the Standard State
at 298 K (ordered from least
stable to most stable)

Substance	ΔG_f^0 (kJ/mole)	
C_2H_2 (g)	209.2	Less stable than elements
O_3 (g)	163.2	
C_6H_6 (g)	129.7	
HCN (g)	120.1	
NO (g)	86.6	
C_2H_4 (g)	68.1	
NO_2 (g)	51.7	
C_8H_{18} (g)	13.1	
HI (g)	1.3	
All elements	0	
NH_3 (g)	−16.5	More stable than elements
C_3H_8 (g)	−23.5	
C_2H_6 (g)	−32.9	
CH_4 (g)	−50.8	
HNO_3 (g)	−74.8	
HCl (g)	−94.9	
H_2O_2 (ℓ)	−120.4	
CO (g)	−137.2	
H_2O (g)	−228.6	
H_2O (ℓ)	−237.2	
SO_2 (g)	−300.2	
SO_3 (g)	−370.5	
CO_2 (g)	−394.4	

but even a large negative value does not imply that the rate of reaction will necessarily be fast.

The important point to remember from this section is that ΔG^0 measures the *difference* in free energy (tendency to react) between products and reactants in the standard state. A negative value of ΔG^0 means the reaction proceeds from "left to right."

More Examples

In Chapter 15 you saw some data for the equilibrium between hydrogen iodide (HI) and its constituent elements (H_2 and I_2). Let's calculate the Gibbs free energy of formation for hydrogen iodide. At 298 K, experiment shows that $K = 0.352$. From this value of K, and equation 19.2 we get $\Delta G^0 = 2.60$ kJ/(2 moles HI), for the reaction:

$$H_2 \text{ (g)} + I_2 \text{ (g)} = 2\,HI \text{ (g)} \qquad \Delta G_f^0 = 2.60 \text{ kJ} \qquad (19.8)$$

Note that this value has been obtained for 2 moles of HI forming from the elements. Dividing by 2, we find that ΔG_f^0 of HI (g) = 1.30 kJ/mole HI, which agrees with the value shown in Table 19.1.

Because the value of ΔG_f^0 is positive, we conclude that HI will not form from H_2 and I_2 if the three gases are mixed in equal amounts in the standard state. Instead, the net reaction will involve dissociation of HI to form more H_2 and I_2 (Figure 19.4).

Let's look at another reaction. This time $K = 10^2$.

$$NH_3 \text{ (g)} + HI \text{ (g)} = NH_4I \text{ (c)} \qquad K = 10^2 \qquad (19.9)$$

We cannot calculate ΔG_f^0 of NH_4I (c) directly from this value of K because the reactants in equation 19.9 are not elements. However, we can calculate a value of ΔG^0 for the reaction shown in equation 19.9 by using equation 19.2. It is $\Delta G^0 = -11$ kJ/mole NH_4I. This value of ΔG^0 means that NH_4I (c) is 11 kJ/mole more stable than NH_3 (g) + HI (g). We have already found that the values of ΔG_f^0 for NH_3 (g) and HI (g) are −16.5 and +1.30 kJ/mole, respectively. That is, the reactants of equation 19.9 are a total of −16.5 + 1.30 = −15.2 kJ/mole more stable than their respective elements. Because NH_4I (c) is 11 kJ/mole *still more stable* than these gases, we conclude that

$$\begin{aligned}\Delta G_f^0 \,[\text{of } NH_4I \text{ (c)}] &= \Delta G_f^0 \,[\text{of } NH_3 \text{ (g)}] + \Delta G_f^0 \,[\text{of HI (g)}] + (-11)\\ &= -16.5 + 1.30 - 11 = -26 \text{ kJ/mole } NH_4I \quad (19.10)\end{aligned}$$

That is, NH_4I (c) is 26 kJ/mole more stable than its elements in the standard state. Figure 19.5 summarizes the free-energy relationships of the various substances.

As this example shows, we can compute values of ΔG_f^0 for tables such

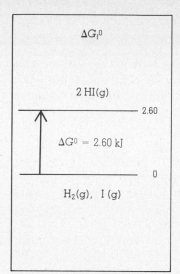

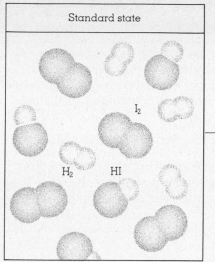

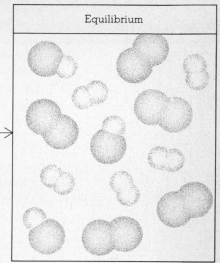

as Table 19.1, even in cases where we cannot measure K directly for a proposed reaction. The Gibbs free energy of formation may be calculated from the value of K for the direct reaction of the elements to form the compound. However, it may also be calculated from any reaction that forms the compound, if the free energies of formation of the other reactants and products are known.

Table 19.1 tabulates a small fraction of the thousands of known values of free energies of formation. These values measure the relative stabilities of each substance. From these values, you (and I do mean *you*) can calculate tendencies to occur (and equilibrium constants) for any reaction involving only these substances. That is a large number of reactions! Table 19.1 summarizes a great amount of chemical information.

Substances near the top of Table 19.1 can react spontaneously to form the elements and the substances lower in the table. Those at the bottom do not react spontaneously to form the elements or the substances at the top of the table.

Note that Table 19.1 lists relative stabilities of substances in a way similar to that in which Table 16.9 lists acid strengths and that in which Table 17.1 lists $\mathscr{E}^0$ values. Each table summarizes the equilibrium results of the random competitions found at the molecular level: competition for atoms of the elements in Table 19.1, for protons in Table 16.9, and for electrons in Table 17.1.

The relationship between Table 16.9 of acid–base strengths (in terms of equilibrium constants, K) and Table 19.1 is given by equation 19.2, $\Delta G^0 = -5.708 \log K$. That between Table 17.1 of redox potentials

Figure 19.4
Standard state reaction: H_2, I_2, and HI (all at 1 atm pressure) will react to give net H_2 and I_2 at 298 K.

Figure 19.5
Some ΔG^0 values allowing the calculation for NH_4I (c) of ΔG_f^0 = − 26 kJ/mole. Circled values are obtained by adding other values found from direct experimentation.

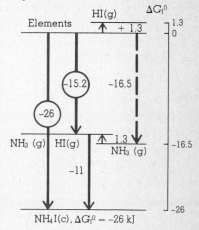

EXERCISE 19.2
Where in Table 19.1 do you
find the substances used as
fuels? The substances produced
in combustion?

(in terms of $\mathscr{E}^0$) is given by the equation

$$\Delta G^0 \text{ (kJ/mole of electrons exchanged)} = -96.4\,\mathscr{E}^0 \quad (19.11)$$

Thus, ΔG^0 can tie together all relative stabilities. And ΔG_f^0 for any substance gives the relative chemical stability of that substance with respect to all others; ΔG_f^0 is a most useful number in understanding, predicting, and controlling all chemical changes.

Negative Values of ΔG^0 Mean Go!

It is very important to remember when using Table 19.1 that a negative value of ΔG^0 means that the spontaneous reaction will proceed to the right (from reactants to products) if you mix all reactants and all products in the standard state. The initial pressure of each gas is 1 atm; any crystalline or liquid substance (whether reactant or product) is initially present in its pure state; and the temperature is 298 K. A positive value of ΔG^0 means that the spontaneous reaction will proceed to the left (from products to reactants) in the standard state.

Testing the Direction of the Net Reaction

Let's try some quick calculations using Table 19.1. Remember that a negative value of ΔG^0 means that in the standard state the net reaction will occur spontaneously as written (although the rate may be slow or fast).

Example I. Hydrogen cyanide, HCN, is a much-used intermediate in synthesizing carbon–nitrogen compounds. Can HCN (g) be synthesized directly from the elements in the standard state?

$$\tfrac{1}{2}H_2 \text{ (g)} + C \text{ (c)} + \tfrac{1}{2}N_2 \text{ (g)} = HCN \text{ (g)} \quad (19.12)$$

From Table 19.1, we find that

$$\Delta G_f^0 \text{ [of HCN (g)]} = 120.1 \text{ kJ/mole} \quad (19.13)$$

The positive value indicates that the synthesis is not possible. The reaction is not spontaneous. In the standard state, HCN would spontaneously undergo decomposition into its constituent elements. You can arrive at the same conclusion more quickly merely by noting that HCN (g) is above the elements in Table 19.1.

Example II. Perhaps it would be possible to synthesize HCN (g) from CO (g) and NH_3 (g). We can write a net equation for such a reaction:

$$NH_3 \text{ (g)} + CO \text{ (g)} = HCN \text{ (g)} + H_2O \text{ (g)} \quad (19.14)$$

Will this reaction occur in the standard state? To find out, we sum the

+120.1 ——— HCN

$\Delta G^0 = 120.1$ kJ

0 ———
Elements
H_2, C, N_2
This reaction is
not spontaneous.

values of ΔG_f^0 (from Table 19.1) for the reactants and then subtract this sum from the sum of the ΔG_f^0 values for the products:

EXERCISE 19.3
Try C_6H_6 (g) and N_2 (g) as possible reactants to give HCN (g). Would you recommend this process as worth trying for a direct (one-step) synthesis?

$$NH_3 \text{ (g)} + CO \text{ (g)} = HCN \text{ (g)} + H_2O \text{ (g)}$$

ΔG_f^0 (kJ/mole)	-16.5	-137.2	120.1	-228.6

Sum for the
  reactants -16.5 -137.2 $= -153.7$

                                               $\downarrow$ increase

Sum for the
  products 120.1 $-228.6 = -108.5$

Net change (ΔG^0) $+45.2$ kJ/mole

The value of ΔG^0 ($+45$ kJ) is positive. Therefore, the reaction will not proceed to form HCN in the standard state. If liquid water forms as a product instead of gaseous water, ΔG^0 will be less by 8.6 kJ. It will be 36.6 kJ/mole, slightly smaller, but still positive. We cannot make this synthesis "go" at 298 K. (I call your attention to the tabular form in which I carried out the calculation. If you follow this form of using the chemical equation to organize the thermodynamic quantities you will make few mistakes in chemistry or in arithmetic.)

Example III. Ethylene (C_2H_4) does not occur in large deposits in nature, but it is in heavy demand for the manufacture of polyethylene and other compounds. Can ethylene be produced from the following reaction?

$$C_2H_6 \text{ (g)} = C_2H_4 \text{ (g)} + H_2 \text{ (g)} \qquad (19.15$$
ΔG_f^0 (kJ/mole) -32.9 68.1 0

The hydrogen gas is an element ($\Delta G_f^0 = 0$), so ΔG^0 for this reaction can be calculated simply by subtracting ΔG_f^0 for C_2H_6 (g) from ΔG_f^0 for C_2H_4 (g):

$$\Delta G^0 = 68.1 - (-32.9) = 101 \text{ kJ/mole} \qquad (19.16)$$

Again ΔG^0 is positive, so we conclude that C_2H_4 does not form in the standard state by reaction 19.15. You can reach the same conclusion at a glance by noting that C_2H_4 (g) is above C_2H_6 (g) in Table 19.1.

However, this at-a-glance method can be used only when the reaction involves no more than one reactant and one product that are not elements, and when the molecular ratio of these substances is 1 to 1. The next example illustrates the care that must be taken in drawing quick conclusions from the table.

Example IV. Benzene (C_6H_6) is used in synthesizing a large number of carbon compounds. Can C_6H_6 form from C_2H_4? The net reaction would be

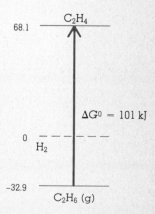

EXERCISE 19.4
Pick out two compounds in
Table 19.1 whose direct
syntheses from the elements
seem possible.

$$3\,C_2H_4\,(g) = C_6H_6\,(g) + 3\,H_2\,(g) \qquad (19.17)$$
$$\Delta G_f^0\ (\text{kJ}) \quad 3 \times 68.1 \qquad 129.7 \qquad 3 \times 0$$

In this reaction 3 moles of C_2H_4 are needed for each mole of C_6H_6 that forms. The values in Table 19.1 are expressed in kilojoules per mole, so we will have to multiply ΔG_f^0 of C_2H_4 (g) by 3 before using it in our calculation. Because ΔG_f^0 for $H_2 = 0$, it does not appear in the calculation.

$$\Delta G^0 = \Delta G_f^0\ [\text{of } C_6H_6\ (g)] - 3 \times \Delta G_f^0\ [\text{of } C_2H_4\ (g)]$$
$$= 129.7 - (3 \times 68.1) = 129.7 - 204.3$$
$$= -74.6 \text{ kJ/mole } C_6H_6 \qquad (19.18)$$

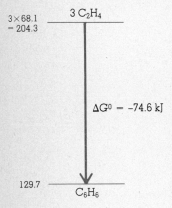

Because ΔG^0 is negative, we conclude this synthesis is possible. However, many bonds must be broken and formed in the reaction, so a complicated mechanism is likely. The reaction will probably be very slow at 298 K. Note that C_6H_6 (g) is above C_2H_4 (g) in Table 19.1—but it would be below the value for 3 C_2H_4 (g) if that value were listed in the table.

Escaping Tendencies and ΔG_f^0; Constant m,P,T Systems

I have a strong personal preference for the phrase *escaping tendency* to describe the tendency of a substance to react. It is much more descriptive than are the more technical terms introduced here, the *Gibbs free energy* or the *change in Gibbs free energy*. Escaping tendency also has an exact mathematical definition. But we shall not need it. Instead just note (and remember) that, as the Gibbs standard free energy of formation becomes more negative, the escaping tendency decreases; a positive ΔG_f^0 means high escaping tendency to the elements. Similarly a negative value for the change in the Gibbs standard free energy, ΔG^0, means the escaping tendency is decreasing when the reaction proceeds in that direction. So this will be the direction of net, or spontaneous, reaction in the standard state. In a closed system at constant temperature and pressure, every net reaction tends to lower the escaping tendency of the system—that is, lower the Gibbs free energy.

As I have derived and used the Gibbs free energy, the equations apply only to systems of constant temperature (T) and constant pressure (P) that are not gaining or losing atoms and so have constant mass (m). I shall call such systems *constant* m,P,T *systems*. Note that a constant temperature is specifically included in equation 19.1. Equations are known that allow calculations to be made when temperature, pressure, and masses of material change, but they will not concern us.

If you go on in science, you will become more and more impressed with the contribution of Gibbs and the usefulness of the free energy in correlating and predicting the direction of net reactions in *closed systems* (those that conserve atoms) at constant temperature and pressure. Note that constant m,P,T conditions are common. They are even common in living systems over short time spans, though all actual living systems are *open systems*—that is, they interchange atoms with their surroundings.

Why Does ΔG^0 "Work"?

I hope you are suitably impressed by now that it is possible to assign to every substance a single number, the Gibbs free energy of formation, to represent its reactivity with *any other* substance in the standard state. Such information has amply repaid the millions of dollars spent in obtaining it. And note, though we have applied the data only to 298 K and 1 atm, there are rather easy methods to extend the calculations to temperatures ranging from a few K to 5000 K (or even higher) and to pressures of thousands of atmospheres.

Why does the free energy "work"? You may well respond "because ΔG_f^0 values can be calculated from experimental equilibrium constants by a simple logarithmic conversion." You are correct. But then I ask what factors determine the relative concentrations at equilibrium?

One answer is that the equilibrium state (and the free-energy change) are determined by the tendency to form strong bonds while simultaneously achieving randomness. This is the approach I shall take.

If the atoms in a system always formed the molecules with the strongest bonds, those molecules would use up all the reacting atoms and only strongly bonded substances would form. On the other hand, if the atoms were only to distribute randomly among all the possible molecules, then every molecule would be present at a concentration independent of its bond strength. Both these extremes seem unreasonable, and you already know that equilibrium states achieve neither of them. You also know that equilibria can be shifted, which would not be the case if either extreme held total sway.

Each idea is correct. Every system tends to

1. form strong bonds;
2. form as many kinds of molecules as possible.

The shifts in equilibrium state depend on which tendency is fostered by the change imposed. The equilibrium state is the one for which the tendency toward strong bonds and the tendency toward randomness just balance at the set of conditions prevailing. (See Figure 19.6.)

The equation that describes this balance for a closed system at constant pressure and temperature in the standard state is

$$\Delta G^0 = \Delta H^0 - T\,\Delta S^0 \qquad (19.19)$$

where ΔH^0 is a measure of the change in bond strength, T is the Kelvin temperature (we shall stick with 298 K), and ΔS^0 is the change in

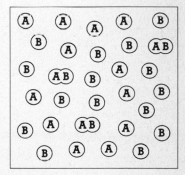

Figure 19.6
Strong bonds favor AB, randomness favors A + B. All forms are present at equilibrium, with strong bonds predominating at low temperatures and randomness predominating at high temperatures.

Table 19.2
Values of ΔH_f^0 and S^0 for Compounds Listed in Table 19.1

Substance[a]	ΔH_f^0 (kJ/mole)	S^0 (J/mole $\times$ K)
C_2H_2 (g)	226.7	200.9
O_3 (g)	142.7	238.8
C_6H_6 (g)	80.7	269.2
HCN (g)	130.5	201.7
NO (g)	90.3	210.7
C_2H_4 (g)	52.3	219.4
NO_2 (g)	33.5	239.8
C_8H_{18} (g)	−224.1	425.2
HI (g)	25.9	206.3
All elements	0	[b]
NH_3 (g)	−46.1	192.3
C_3H_8 (g)	−103.8	269.9
C_2H_6 (g)	−84.7	229.5
CH_4 (g)	−74.8	186.2
HNO_3 (g)	−135.1	266.3
HCl (g)	−91.9	186.8
H_2O_2 (l)	−187.8	110
CO (g)	−110.5	197.5
H_2O (g)	−241.8	188.7
H_2O (l)	−285.8	70.1
SO_2 (g)	−296.8	248.1
SO_3 (g)	−395.2	256.7
CO_2 (g)	−393.5	213.7
C (diamond	1.9	2.4
C (graphite)	0	5.7
I_2 (c)	0	116.7
S_8 (c)	0	255.0
H_2 (g)	0	130.6
N_2 (g)	0	191.5
F_2 (g)	0	202.7
O_2 (g)	0	205.3
Cl_2 (g)	0	223.0
I_2 (g)	62.3	260.6
S_8 (g)	222.8	432.1

[a]The ordering of substances in this table (above the dashed line) is the same as that in Table 19.1. Note that values of ΔH_f^0 and S^0 do not fall in exactly the same order as do values of ΔG_f^0, but that the trend for ΔH_f^0 is similar. Values for various forms of some elements are listed below the dashed line. Note the trends in values of S^0 with state changes, increasing molecular complexity, and higher molecular weights.
[b]Given below the dashed line.

entropy (a measure of randomness) during the reaction.

Some values for ΔH_f^0 and S^0 are listed in Table 19.2. The symbol ΔH_f^0 stands for the heat absorbed when the substance forms from the elements. You learned in Chapter 9 that this is the same as the difference between the bond strengths in the elements and the bond strengths in the substance. The symbol S^0 stands for the entropy of the substance. Both refer to the standard state. Let's see if we can have them "make sense" before we apply equation 19.19. Otherwise you will merely be putting numbers into equations with no comprehension or feel for what is going on.

Enthalpy and the Drive for Strong Bonds

The symbol ΔH^0 (read as "delta H super-zero") means "the change in the standard enthalpy." *Enthalpy* (from a Greek word meaning "heat") is the name represented by the symbol H. In chemistry, changes in enthalpy are due primarily to the breaking and making of bonds; ΔH_f^0 is the difference between the heat required to break the bonds in the elements and the heat released when the bonds form in the compound. Let's look at a few examples, comparing average bond strengths (Table 9.1) with enthalpies of formation (Table 19.2). Note that both tables are expressed in kilojoules per mole. Also note that ΔH_f^0 (like ΔG_f^0) is defined as zero for every element (in its most stable phase or form).

Example V. Consider the formation of NH_3 from N_2 and H_2:

$$\tfrac{1}{2}N_2\,(g) + \tfrac{3}{2}H_2\,(g) = NH_3\,(g) \qquad (19.20)$$

Let's rewrite the equation to show the bonds involved and list the bond strengths from Table 9.1.

$$\tfrac{1}{2}N\equiv N + \tfrac{3}{2}H{-}H = N{\overset{\displaystyle H}{\underset{\displaystyle H}{-}}}H$$

Bond strengths (kJ/mole)	945	435	389
Heat absorbed to break bonds	$\tfrac{1}{2}\times 945$	$\tfrac{3}{2}\times 435$	
	$\parallel$	$\parallel$	
	473 +	652	$= +1125$
Heat released as bonds form		$-(3\times 389)$	$= -1167$
Net change (kJ/mole)			$-\ 42$

The negative value, −42 kJ, indicates a net *release* of heat. In Table 19.2, we find that ΔH_f^0 of NH_3 (g) is listed with a value of −46.1 kJ/mole.

The values in Table 9.1 are averages for similar bonds in a variety of different compounds; most values for particular compounds differ by less than 5 kJ per bond from the values given in the table. In this example, the difference between heat release predicted from average bond strengths and that predicted from ΔH_f^0 is only 4 kJ from the three bonds in NH_3. The agreement is considered excellent confirmation that ΔH^0 primarily represents changes in bond strengths (Figure 19.7).

Example VI. Now let's consider the reaction

$$C_2H_6 \text{ (g)} = C_2H_4 \text{ (g)} + H_2 \text{ (g)} \qquad (19.21)$$

Rewriting the equation to show the bonds involved and listing the bond strengths from Table 9.1 gives

½ N_2 (g), ³⁄₂ H_2 (g)

$\Delta H^0 = -46.1$ kJ

NH_3 (g)

Figure 19.7
ΔH_f^0 of NH_3 with respect to the elements.

$$
\begin{array}{c}
H \quad\quad\quad H \quad H \quad\quad\quad H \\
\mid \quad\quad\quad\quad \mid \quad\, \mid \quad\quad\quad\quad \mid \\
H-C-C-H = \quad C=C \quad + H-H \\
\mid \quad\quad\quad\quad \mid \quad\, \mid \quad\quad\quad\quad \mid \\
H \quad\quad\quad H \quad H \quad\quad\quad H
\end{array}
$$

Bond strengths (kJ/mole)	$[6\,C\!-\!H + C\!-\!C]$	$[4\,C\!-\!H + C\!=\!C] + H\!-\!H$
Heat absorbed to break bonds	$(6 \times 414) + 347$	$= \quad 2831$
Heat released as bonds form		$-[(4 \times 414) + 611 + 435] = -2702$
Net change (kJ/mole)		$+ \ 129$

Using ΔH_f^0 values from Table 19.2, we subtract the total ΔH_f^0 of reactants from the total ΔH_f^0 of products to calculate ΔH^0 for the reaction (the same procedure we used earlier to compute ΔG^0 from ΔG_f^0 values):

$$\Delta H^0 = (52.3 \text{ kJ} + 0) - (-84.7 \text{ kJ}) = +137 \text{ kJ/mole } C_2H_4 \quad (19.22)$$

Again, the value obtained from average bond strengths (129 kJ/mole) and that obtained from ΔH_f^0 values (137 kJ/mole) agree to within ± 5 kJ per bond change, further confirming that ΔH^0 represents the difference between the heat required to break reactant bonds and the heat released as product bonds form (Figure 19.8).

Example VII. Because the enthalpy of reaction and the difference between the heat to break and form bonds are the same, ΔH_f^0 can be estimated from bond-strength tables, even if the enthalpies of formation are not known for all the reactants and products. For example, Tables 19.1 and 19.2 contain no data for HF, but we can use bond strengths to estimate ΔH_f^0 of HF. Let's consider a reaction in which HF is the only compound for which we do not know ΔH_f^0.

$$HF \text{ (g)} + \tfrac{1}{2}Cl_2 \text{ (g)} = HCl \text{ (g)} + \tfrac{1}{2}F_2 \text{ (g)} \qquad (19.23)$$

We can use bond strengths to estimate ΔH^0 for this reaction. The reaction involves breaking 1 mole of H—F and ½ mole of Cl—Cl bonds, and the formation of 1 mole of H—Cl and ½ mole of F—F bonds. Therefore, from Table 9.1 (in kilojoules):

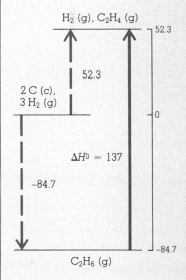

H_2 (g), C_2H_4 (g)   52.3

52.3

2 C (c), 3 H_2 (g)   0

$\Delta H^0 = 137$

-84.7

C_2H_6 (g) -84.7

Figure 19.8
Relative ΔH_f^0 values for C_2H_6 and $C_2H_4 + H_2$ (kJ/mole).

$$\text{Heat absorbed} \; = 565 + (\tfrac{1}{2} \times 240) = 685$$
$$\text{Heat released} \; = -[430 + (\tfrac{1}{2} \times 160)] = -510$$
$$\text{Estimated } \Delta H^0 = 685 - 510 = 175 \text{ kJ/mole HF}$$

with the positive sign indicating net absorption of heat. Now we know that ΔH_f^0 for gaseous Cl_2 and F_2 is zero and that for HCl it is -91.9 kJ/mole (from Table 19.2). And we know that ΔH^0 for equation 19.23 equals the total ΔH_f^0 of reactants subtracted from the total ΔH_f^0 of products:

$$(-91.9 + 0) - \{\Delta H_f^0 \,[\text{of HF (g)}] + 0\} = 175$$

so

$$\Delta H_f^0 \,[\text{of HF (g)}] = -91.9 - 175 = -267 \text{ kJ/mole HF} \quad (19.24)$$

(See Figure 19.9.) The experimental value for ΔH_f^0 of HF (g) is -269 kJ/mole. This differs by about 2 kJ/mole from our estimate—quite good!

The bond-strength method is less accurate than the ΔH_f^0 method of determining the enthalpy change in a reaction—because the bond-strength values in the tables are averages for many different compounds. However, bond strengths can be used to estimate enthalpy changes when no other data are available. Such estimates are certainly better than a total lack of information.

There are no simple generalizations that allow the prediction of the strengths of particular bonds from formulas or theories. We must use tables that are summaries of experimental measurements. However, it is generally true that a single bond between a pair of unlike atoms is stronger than the average of the single-bond strengths between like pairs of the same atoms. For example, compare the strength of the H—Cl bond with the average of H—H and Cl—Cl strengths in Table 9.1. Make similar comparisons for other binary hydrogen compounds.

Entropy and the Drive to Spread Out

You have seen that the random collisions of molecules tend to cause substances to dissolve in one another and to cause atoms to exchange among molecules. These random collisions lead to pressure in gases and liquids, and to variation in molecular velocities. They also lead to the interdiffusion of substances and to their tendencies to spread out into larger volumes and into more combinations.

Entropy (S) is the quantity we use to measure the randomness in a system. The name *entropy* comes from Greek stems meaning "a change within." The value of ΔS measures the universal drive of substances to spread out or to become more and more random.

You may have noted that absolute values for the entropies are given

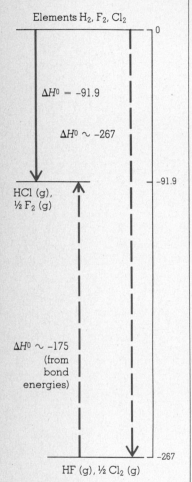

Elements H_2, F_2, Cl_2

$\Delta H^0 = -91.9$

$\Delta H^0 \sim -267$

HCl (g), $\tfrac{1}{2} F_2$ (g)

$\Delta H^0 \sim -175$ (from bond energies)

HF (g), $\tfrac{1}{2} Cl_2$ (g)

0

-91.9

-267

Figure 19.9
Relative enthalpy of HCl–$\tfrac{1}{2}F_2$ and HF–$\tfrac{1}{2}Cl_2$ (kJ/mole) compared to one another and to the chemical elements.

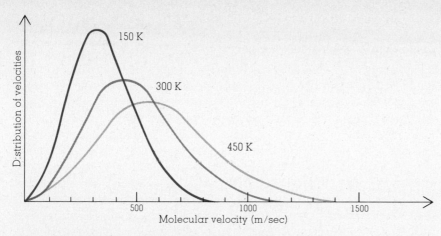

Figure 19.10
Maxwell–Boltzmann distribution
of velocities.

in Table 19.2. But only relative values (referred to the elements in their standard states as arbitrary zeros) are given for ΔH_f^0 and ΔG_f^0. This should remind you of the discussions of pressure and temperature measurements in Chapter 12. When possible, scientists prefer absolute scales. Why are absolute values possible for entropy but not for enthalpy or free energy?

For an answer to this question, we shall rely on your knowledge of crystals and of the Maxwell–Boltzmann distribution summarized in Figure 19.10. Note that the distribution of molecular velocities becomes narrower and narrower as the temperature falls. Many experiments show that, at zero Kelvin, all molecules of a given substance would have the *same* energy. Near zero Kelvin there is almost no randomness in the distribution of the molecular energy.

Similarly, all substances crystallize as the temperature drops toward zero K, and all crystals become more and more perfect. Every atom settles more and more into the same kind of position as other corresponding atoms in the crystal. This, too, is confirmed by experiments carried out at low temperatures.

You will recall that it is impossible to attain a temperature of 0 K. But even at attainable low temperatures, the atomic order in pure crystals is so high (and the randomness so low) that we can see the entropy approaching a zero value. There is no "spreading out" at absolute zero. In a pure crystal, all molecular energies are identical and all molecular positions are identical at 0 K. This fact is expressed as the

THIRD LAW OF THERMODYNAMICS: The entropy of any pure crystalline substance is zero at absolute zero (0 K).

The increase in disorder (or randomness, or spreading out) is then measured as the substance is heated, and the entropy is expressed as a function of temperature. Because disorder always increases as temper-

POINT TO PONDER
The third law identifies 0 K as
one condition of zero entropy.
But other conditions are conceivable. For example, if all the
molecules in a gas had the same
energy and all were equally
positioned from one another,
their entropy would be zero.
The Maxwell–Boltzmann curve
shows that the actual distribution of energies and the positions
in a gas are random—consistent
with the high entropies found in
gases.

EXERCISE 19.5
Is the entropy of a crystalline
substance larger at 100 K or
at 300 K?

ature rises, the entropy of every substance is always positive, and it
always increases with a rise in temperature.

Note in Table 19.2 that the most complicated substances have the
highest entropies. Gases have higher entropies than do liquids, and
liquids higher than crystals. Polyatomic gaseous molecules have larger
entropies than diatomics, which in turn are higher than monatomics.
All these values are consistent with what you have already learned
about the random nature of molecular distributions and molecular
energies. The more ways the atoms and energies can be distributed in a
system (the more spread out the energies and the atoms), the larger the
entropy becomes. Figure 19.11 gives examples.

Change in entropy measures change in atomic disorder or randomness. Conversion of crystal to liquid (or liquid to gas) is accompanied
by an increase in entropy. Decomposition of a molecule into fragments
causes an increase in entropy. Distributing increasing amounts of energy
among atoms causes an increase in entropy.

Let's check these ideas using some of the data in Table 19.2. In each
case look at the chemical equation first and guess whether ΔS^0 will be
positive or negative before you look at the numbers. You should be able
to guess the sign of ΔS^0 in most cases. The units for S^0 in Table 19.2 and
here are joules/(mole $\times$ K). Thus $T \Delta S^0$ is in units of joules per mole,
not kilojoules per mole.

Figure 19.11
The entropies at 298 K of 1
mole of atoms (both crystalline
and gaseous) plotted versus
family in the periodic table.
Note that (1) gases have higher
entropies than crystals; (2) soft
crystals (graphite) have higher
entropies than hard crystals
(diamond); and (3) the trends
in both rows and families are
generally regular.

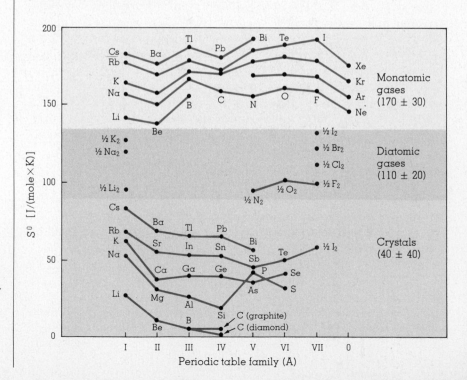

Example VIII. Guess the sign of ΔS^0 for the reaction

$$N_2 \text{ (g)} + 3\,H_2 \text{ (g)} = 2\,NH_3 \text{ (g)} \qquad (19.25)$$

(Don't look ahead. Think about it and make a reasoned guess before reading on. The practice will be helpful.) The value of ΔS^0 for the reaction can be computed in a fashion similar to that used for ΔH_f^0 computations. Using values of S^0 from Table 19.2, we find the total S^0 of reactants and the total S^0 of products, then determine ΔS^0, the change in entropy (Figure 19.12):

$$N_2 \text{ (g)} + 3\,H_2 \text{ (g)} = 2\,NH_3 \text{ (g)}$$

Entropy of reactants
[J/(mole × K)]
$$191.5 + (3 \times 130.6) = 583.3$$
$\downarrow$ decrease

Entropy of products
$$2 \times 112.3 = 384.6$$

Net change: $\Delta S^0 = 384.6 - 583.3 = -198.7 \text{ J/(2 moles NH}_3 \times \text{K)}$
$$= -99.4 \text{ J/(mole NH}_3 \times \text{K)}$$

Did you guess correctly that a decrease in the number of moles of gas (4 moles reactants → 2 moles product) would decrease the randomness of the system, and so decrease the entropy, S^0? If not (and even if so), keep trying. In rather short order you can become quite expert in guessing entropy changes. You can even be somewhat quantitative in your guesses by noting that the formation of 1 mole of gas often involves ΔS^0 of about 100 to 150 J/(mole × K).

This ability to make predictions about ΔS^0 can be valuable, as we discuss more fully in the next chapter. To anticipate that discussion slightly—a very large number of ecological problems have arisen and are resisting solutions because little attention has been paid to entropy changes.

For each of the following examples, try to guess the sign of ΔS^0 for the reaction before reading the computation of its value.

Example IX. What is ΔS^0 for the condensation of gaseous water?

$$H_2O \text{ (g)} = H_2O \text{ }(l) \qquad (19.26)$$

From the table, S^0 of H_2O (g) is 188.7 J/(mole × K) and S^0 of H_2O (l) is 70.1 J/(mole × K); S^0 decreases from 188.7 to 70.1. Therefore,

$$\Delta S^0 = 70.1 - 188.7 = -118.6 \text{ J/(mole} \times \text{K)}$$

Note that a decrease of entropy results as the gas condenses to liquid.

Example X. What is ΔS^0 for the reaction

$$C_2H_6 \text{ (g)} = C_2H_4 \text{ (g)} + H_2 \text{ (g)} \qquad (19.27)$$

Looking up S^0 values in the table, we find that

$$\Delta S^0 = (219.4 + 130.5) - 229.5 = 120.4 \text{ J/(mole} \times \text{K)}$$

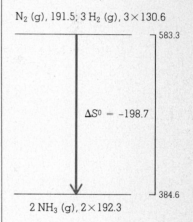

N_2 (g), 191.5; 3 H_2 (g), 3×130.6

583.3

$\Delta S^0 = -198.7$

384.6

2 NH_3 (g), 2×192.3

Figure 19.12
Entropy changes for N_2–H_2 to NH_3 [J/(mole × K)].

The increase in entropy is consistent with the change from 1 mole of gaseous reactant to 2 moles of gaseous products.

Example XI. What is ΔS^0 for the reaction

$$NH_3 \text{ (g)} + HI \text{ (g)} = NH_4I \text{ (c)} \tag{19.28}$$

Table 19.2 does not include a value of S^0 for NH_4I (c). However, we certainly expect ΔS^0 to be negative, because 2 moles of gas are condensed to the crystalline state in the reaction. In fact, because "disappearance" of each mole of gas usually decreases the entropy by about 100 to 150 J/(mole $\times$ K) (see the preceding three examples), we can estimate that ΔS^0 is probably about -200 to -300 J/(mole $\times$ K).

EXERCISE 19.6
Guess the sign of ΔS^0 for the reaction

$$C_6H_{12}O_6 \text{ (c)} + 6\,O_2 \text{ (g)}$$
$$= 6\,CO_2 \text{ (g)} + 6\,H_2O(\ell)$$

Strong Bonds and/or Disorder (Security Versus Freedom)

We are now in a position to analyze why the Gibbs free energy of formation data are so useful for closed systems at constant T and P. It is because ΔG^0 summarizes in one term the net result of the drive to maximize bond strengths and the simultaneous drive to maximize the randomness of the system. Any combinations of ΔH^0 and ΔS^0 that give a negative value to ΔG^0 will drive the net reaction to the right. Any combinations that produce a positive value of ΔG^0 will drive the reaction to the left. Table 19.3 summarizes the possible combinations. Remember that T is always positive, but ΔH^0 and ΔS^0 can be either positive or negative.

An analogy to the human drives for security (strong bonds) and freedom (spreading out) may help here. Every society achieves some balance between these forces. No society has achieved both. It is doubtful if any society could. Most human laws are efforts to define the proper balance, and there are always individuals who break the laws because they disagree. Regardless of the rigidity or laxity of laws, there are

Table 19.3   Direction of Reaction from $\Delta H^0 - T\,\Delta S^0 = \Delta G^0$, Using Only the Signs of ΔH^0 and ΔS^0

ΔH^0	ΔS^0	$(-T\,\Delta S^0)$	ΔG^0	Conclusion: reaction proceeds to	
$-$	$+$	$-$	$-$	right	always
$-$	$-$	$+$	$-$	right	if ΔH^0 term outweighs $T\,\Delta S^0$ term
			$+$	left	if $T\,\Delta S^0$ term outweighs ΔH^0 term
$+$	$+$	$-$	$-$	right	if $T\,\Delta S^0$ term outweighs ΔH^0 term
			$+$	left	if ΔH^0 term outweighs $T\,\Delta S^0$ term
$+$	$-$	$+$	$+$	left	always

always some who seek more freedom. Regardless of the degree of freedom, there are always some who seek more security. Every individual has a preferred balance. The direction of change in a society is a partial reflection of these desires.

All Possible Reactions Occur (But Not Equally)

If one or more of the products of a reaction are initially absent, the random molecular collisions will ensure that the reaction will "go" until some product is produced. It may well be that only a tiny amount of product is formed before equilibrium is reached, so that the concentration at equilibrium is very small. But some concentration must be present to satisfy the equilibrium constant. Once equilibrium is reached, both forward and reverse reactions continue at equal rates; the system is in dynamic equilibrium.

Similarly, if one reactant is absent the reaction will proceed in the reverse direction until some of the reactant has been produced and the concentrations reach values that will satisfy the equilibrium-constant expression.

All experimental observations are consistent with these predictions. Equilibrium requires the presence of some of each of the reactants and products.

Thus, all possible reactions occur. By *possible,* I mean any reaction for which a chemical equation can be written. And by *equation,* I mean a statement that specifically satisfies the conservation laws for charge, mass and energy, and (in most cases) atoms. If you have taken some physics courses, you know that other conservation laws exist (for example, conservation of momentum, of angular momentum, of electron spin), but we will not worry about these other conservation laws in this book.

The fact that all possible reactions can and do occur is consistent with the random motions of molecules. They will assume some most unusual configurations, given enough time. Such reactions are unlikely, but not impossible. And a typical sample contains so many molecules that there may be quite a few molecules of even very rare configurations. The more unstable (less tightly bound) molecules will have fleeting existences and probably will not be detectable, but you can be quite sure that they do form.

At equilibrium there is no net tendency for reaction, so

$$\Delta G = 0 \qquad (19.29)$$

then,

$$\Delta G = 0 = \Delta H - T\,\Delta S \qquad (19.30)$$

and

$$\Delta H = T\,\Delta S \qquad (19.31)$$

EXERCISE 19.7
In which direction would the net reaction go (in the standard state) for the following gaseous system?

$$Cl_2 + I_2 = 2\,ICl$$

EXERCISE 19.8
Which is the more likely molecule, Kr or Kr_2, in the equilibrium $2\,Kr\,(g) = Kr_2\,(g)$ at 298 K? Use ΔH and ΔS ideas in your answer. Any change if $T = 100$ K?

EXERCISE 19.9
Many liquids have an entropy
of vaporization (ΔS^0) of about
85 J/(mole × K). This generalization is called *Trouton's rule*.)
Estimate the enthalpy of vaporization of a liquid boiling at
400 K and one boiling at 300 K.
Is the trend of ΔH with boiling
point "reasonable"?

Thus the tendency to form strong bonds (represented by ΔH) exactly
equals the tendency to spread out (represented by $T \Delta S$) and no further
net change occurs.

Let me close on a note of hard reality. Chemists recognize the fleeting
existence of many species. But professors give students little credit for
listing such species in writing chemical equations for usual reactions.
So stick with stable structures that satisfy the Lewis rules. For example,
no doubt there are some molecules of H_2Cl_2 in any sample of hydrogen
chloride, and some molecules of O_4 in any sample of·oxygen. But they
exist at a level far less than 1 part per 1000, so you can expect to receive
less than one-thousandth of full credit for writing such formulas. Fair
enough?

Summary

The standard Gibbs free energy of formation, ΔG_f^0, is a measure of the
reactivity of a compound. All compounds can be made from their elements, and the elements are not generally interconvertible, so the free
energy of formation of elements in their standard states is defined as
zero. Tabulations of ΔG_f^0 exist for thousands of compounds at a variety
of temperatures and concentrations. More are added daily. Thus, known
or yet to be measured, there exists for every chemical under any set of
conditions a Gibbs free energy of formation that represents the reactivity of that compound with all other compounds. A most useful number!

The difference between the free energy of formation of the products
and that of the reactants indicates the direction of the net reaction. A
negative value for ΔG indicates that the reaction will go as written.
A positive value indicates that the reverse reaction is the spontaneous
one. That is, the reaction proceeds in the direction that will decrease G.

The change in free energy summarizes in one number the relative
tendencies to form strong bonds and to spread out or achieve randomness. Enthalpy measures the bond-strength effects. Entropy measures
the effects of randomness. The three terms are related by the equation

$$\Delta G = \Delta H - T \Delta S \qquad (19.32)$$

HINTS TO EXERCISES
19.1 −5.708 kJ. 19.2 Fuels at
top. 19.3 $\Delta G^0 = 590.9$ kJ/mole
C_6H_6. 19.4 Any compound for
which ΔG_f^0 is negative.
19.5 300 K. 19.6 ΔS^0 should
be +. 19.7 Toward ICl.
19.8 $\Delta S^0 = 85$ J/(mole × K);
larger T_{bp} indicates stronger
intermolecular bonds in liquid.
19.9 ΔS^0 is −, ΔH^0 is −.
Kr_2 forms only at very low
temperatures.

This equation allows you to calculate the direction of spontaneous
change for a closed system at constant pressure and temperature. These
conditions are among those most commonly met, so the equation is
much used.

The random collisions of molecules ensure that all possible reactions
will occur. Actual equilibrium states at low temperatures consist mainly
of substances held together with strong bonds, but the drive to spread
out atoms and energy ensures that some of every possible substance will
be present.

Problems

19.1 Identify or define: closed system, constant m, P, T system, drive to form strongest bonds, drive to spread out, enthalpy of formation (ΔH_f^0), entropy (S^0), free energy change $[\Delta G^0 \text{ (kJ/mole)} = -5.708 \log K = \Delta H^0 - T\Delta S^0]$, free energy of formation (ΔG_f^0), spontaneous reaction, stability, standard state $(P, T, \text{concentration})$, third law of thermodynamics $(S^0 = 0, \text{at zero Kelvin})$.

19.2 Calculate ΔG^0 for the following reactions and identify those that are spontaneous as written at 298 K:

  a. $C_6H_6 \text{ (g)} = 3\,3C_2H_2 \text{ (g)}$
  b. $CH_4(g) + C_2H_2(g) = C_2H_6(g) + C \text{ (graphite)}$
  c. $2H_2O \text{ (g)} + CH_4 \text{ (g)} = CO_2 \text{ (g)} + 4H_2 \text{ (g)}$
  d. $2HCl \text{ (g)} + \frac{1}{2}O_2 \text{ (g)} = H_2O \text{ (g)} + Cl_2 \text{ (g)}$
  e. $SO_2 \text{ (g)} + \frac{1}{2}O_2 \text{ (g)} = SO_3 \text{ (g)}$
  f. $C_3H_8 \text{ (g)} = CH_3CH{=}CH_2 \text{ (g)} + H_2 \text{ (g)}$

  [Yes: (b) -200 kJ, (d) -40 kJ, (e) -70 kJ. No: (a) 500 kJ, (c) 100 kJ, (f) $\sim$100 kJ]

19.3 Which of the following are chemical equations for *possible* reactions?

  a. $H_2 + He = 2He + H$ b. $Cl_2 + Br_2 = 2BrCl$
  c. $I_2 + BrCl = IBa + ICl$
  d. $C_8H_{18} + HCl = C_8H_{16}Cl + H_2$
  e. $Pt + Ar = PaRt$ f. $H_2O + T_2 = HOT + HT$

*19.4 For the reaction

  $$2\,BO \text{ (g)} + 3\,H_2O \text{ (g)} = B_2H_6 \text{ (g)} + \tfrac{5}{2}O_2 \text{ (g)}$$

  ΔH^0 is 700 kJ; ΔS^0 is -200 J/K at 298 K. Do these values seem reasonable? What will be the net reaction if the system is mixed in its standard state?

19.5 The value for ΔG^0 is 33 kJ/mole for the reaction

  $$C_2H_6 \text{ (g)} = 2\,C \text{ (c)} + 3\,H_2 \text{ (g)}$$

  Does this mean that no decomposition of pure C_2H_6 can occur at 298 K? Draw a stability diagram (similar to Figure 19.9) for the system. [$p(H_2) \cong 10^{-2}$ atm]

*19.6 Tobacco mosaic virus protein depolymerizes in an aqueous Ca^{2+} solution at 5°C, but repolymerizes when the solution is heated to room temperature. The process can be repeated indefinitely. Interpret at the molecular level.

(From J. C. McMichael and M. A. Lauffer, *Arch. Biochem. Biophys.*, **169**, 1975, pp. 209–216.)

19.7 How are ethylene and propylene produced from petroleum? High or low pressure? Temperature? Any other products? Write a possible equation, assuming $C_{12}H_{26}$ as an average formula for petroleum.

*19.8 Some years ago chemists at the Montana School of Mines were measuring heats of wetting of various tars on a series of gravels. Any use for such data?

19.9 Highly volatile gasoline can cause vapor lock (formation of bubbles in the fuel line) in a hot engine, or can cause carburetor icing on cold, moist days. What causes the ice to form? It is minimized by additives in the gasoline that prevent the ice adhering to the metals in the carburetor.

*19.10 A major problem in developing pure graphite (free from boron) for nuclear piles was solved by thermodynamic calculations. Scientists successfully predicted that passing Freon 12 (dichlorodifluoromethane) over the carbon would remove the boron as BF_3. List some types of data needed for this calculation.

19.11 Attempts have been made to lower coke costs by mixing methane, CH_4, with the air entering the blast furnace. Actually, this causes the furnace temperature to drop, and other methods had to be used. Show that this is calculable in terms of the reaction $C \text{ (c)} + CH_4 \text{ (g)} = C_2H_4 \text{ (g)}$. Use of thermodynamics can save lots of time and money. [$\Delta H_{reac}^0 \cong 100$ kJ]

19.12 Cell division may be written as

  $$1 \text{ cell} + \text{nutrient} = 2 \text{ cells} + \text{waste}$$

  or as

  $$\text{Nutrient} = 1 \text{ cell} + \text{waste}$$

  The first cell emerges essentially unchanged (it is a catalyst). Comment on the thermodynamics of cell production. The experimental enthalpy loss (ΔH^0) is 10 kJ/g for *E. coli*. (From S. Bayne-Jones and R. S. Rhees, *J. Bacteriology*, **17**, 1934, p. 123.) How does this figure relate to your argument?

CHEMICAL REACTIONS

Random molecular motion is a principal property of gases.

Random motion is also found in liquids, but is smaller in crystals.

Reaction occurs when the random motion fulfills three conditions.

GAS

LIQUID

CRYSTAL

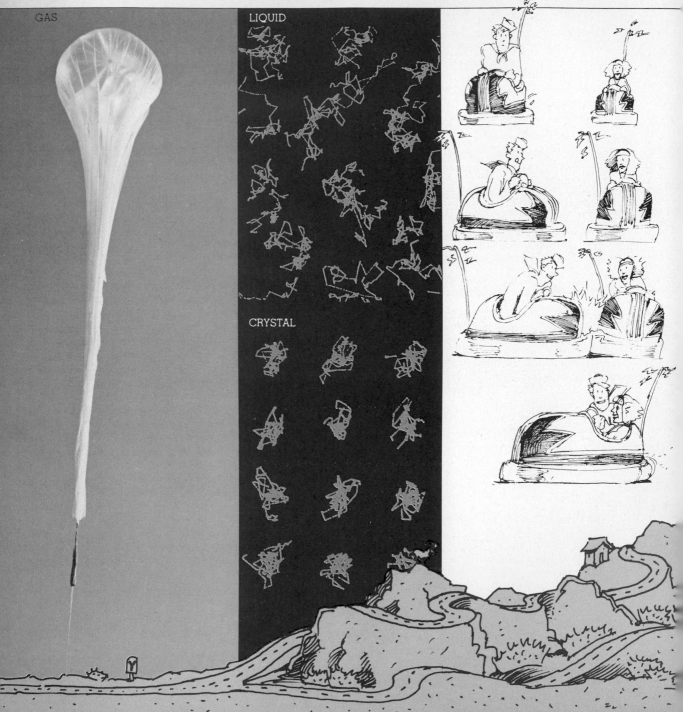

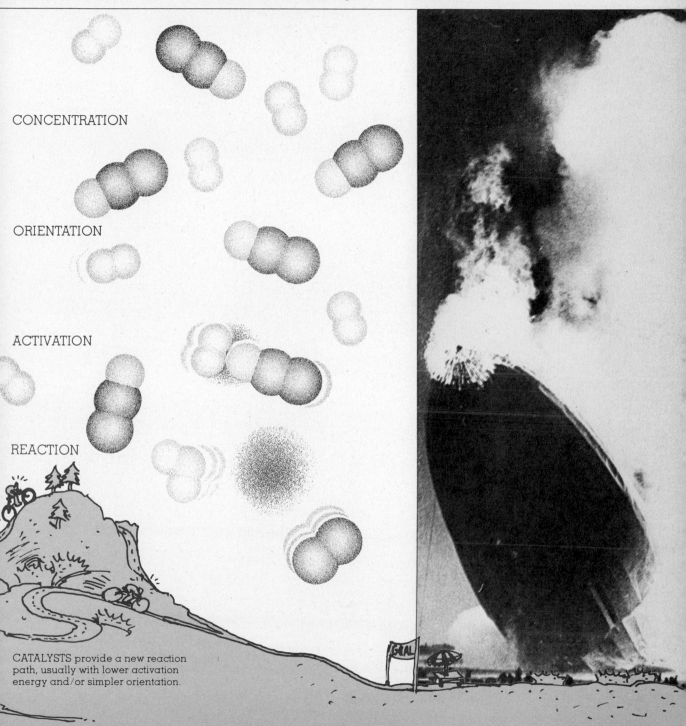

Most collisions lead to simple rebound, but... suitable concentration, orientation, and energy can lead to net change... sometimes a violent reaction. (Photo: UPI)

CONCENTRATION

ORIENTATION

ACTIVATION

REACTION

CATALYSTS provide a new reaction path, usually with lower activation energy and/or simpler orientation.

Figure 20.1
"All the king's horses and all
the king's men cannot put
Humpty together again"—but
a hen can.

20 NEAT OR MESSY?

It is raining as I write. The drops splatter as they strike the earth. The streams of water spread into the lawn. The wind sifts the leaves across the yard. Sunlight, diffused by the clouds, penetrates into every corner. Water evaporates. The line of writing slowly extends as ink leaves pen for paper. Smoke eddies up from the fire that heats the room. I breathe sitting in one place, yet I do not reinhale only the same air. What common characteristic do all these changes share?

Or consider yourself. The light leaves the paper and enters your eye. There it is converted to electrical signals that are transmitted to your brain. Your latest meal, decomposed into small molecules, spreads out through your body via the bloodstream. The harder you study, the more heat (and sweat) you lose to the room.

Or, on a larger scale, consider the presumed history of a star. Atoms coalesce into molecules, molecules into dust, dust into larger particles. Each time particles stick together they release energy, most of which radiates into the surrounding space. But as the clump gets larger, energy cannot leave so easily and the local temperature rises, often reaching millions of degrees. Atomic collisions become so violent that nuclear reactions occur. These generate more energy, which streams out from the star. Sometimes the temperature reaches such heights, and nuclear reactions occur so rapidly, that there is a very rapid further rise in temperature, resulting in a nova. Perhaps even a supernova, which explodes into a rapidly expanding cloud of tiny particles and dust.

Do you discern any general pattern that characterizes all these changes? If so, try your generalization on some other changes. Does it work?

The Tendency to Spread Out

Observation of each total change mentioned—indeed any *total* change— in terms of what is happening to mass and energy shows that either mass or energy (or both) is spreading out.

Energy tends to flow outward from regions in which it is concentrated—for example, from hot regions into cooler surroundings, or from the bright sun into dark space. We do *not* observe heat spontaneously flowing from cold to hot objects or light gathering from dark space to illumine the sun.

Atoms (mass) also tend to spread out: water evaporates, stars explode, exhaled breath diffuses, rain splatters, new combinations form

Knowledge is usually gained by the orderly loss of information, not by accumulating it; by the filtering out of noise, not by the piling up of data.
—*Kenneth Boulding, 1972.*

Figure 20.2
Who's been messing up my
room?

among the atoms. Of course, mass also comes together: Raindrops form from water vapor, stars from dust, cells from protoplasm. However, in each of these aggregating processes, energy spreads out into the surroundings.

Thus, we arrive at a generalization: In any total change, mass and/or energy spread out. Both may do so, or only one. (See Figure 20.2.) If mass aggregates, energy must spread out. A local concentration of energy results in only part of the *total* energy being in the "concentrate." Some spreads out. The process of concentrating energy is *always* offset by the greater spreading out of the energy that eluded the concentration process. This is often accompanied by a spreading out of the mass as well. This spreading out, or *entropy change,* in all processes is described by the

SECOND LAW OF THERMODYNAMICS: The entropy (S) of the universe increases in any total change.

(You may recall that the word *entropy* comes from Greek stems meaning "the change within.")

Because mass and energy are conserved in the universe, we do not use up either of them. We merely "process" them and in so doing hasten their spreading out. Our actions hasten the increase in entropy—that is, hasten the changes within the universe.

The astronomical universe is a bit large for most people to comprehend. So we define an *isolated system* as one with *no* mass or energy exchanges with its surroundings, and we restate the

SECOND LAW OF THERMODYNAMICS: The entropy (S) of an isolated system increases in any total change.

The fact that the entropy increases as an isolated system approaches equilibrium shows that the entropy will reach a *maximum* in an isolated system at equilibrium.

Some Systematics of Systems

We shall spend most of this chapter learning of the great predictive power of the second law of thermodynamics, and of the strict limitations it puts on human and other activities. But first let's review some things about mass and energy, and about *open, closed, and isolated systems*.

You have used over and over again the idea that energy is conserved—yet you know that energy is leaving the sun, so it is not conserved there. And you know that your body, among many other systems, does not conserve energy.

A correct statement about the conservation of energy is commonly

POINT TO PONDER
The law that entropy always increases—the second law of thermodynamics—holds, I think, the supreme position among the laws of Nature. If [your pet theory of the universe] is found to be contradicted by observation—well, these experimentalists do bungle things sometimes. But if your theory is found to be against the second law of thermodynamics I can give you no hope; there is nothing for it but to collapse in deepest humiliation.—*Sir Arthur Eddington,* The Nature of the Physical World, *1928 (Ann Arbor Paperback, 1958, p. 74).*

EXERCISE 20.1
Can you think of any actual small system that is essentially an isolated system?

Table 20.1   Systems and Conservation of Mass and Energy

| Conserve | | | | |
Mass	Energy	System	Direction of change	At equilibrium
Yes	Yes	Isolated	ΔS is positive	S = maximum; $\Delta S = 0$, no net flows
Yes	No	Closed	ΔG is negative	G = minimum; $\Delta G = 0$, no net flows
No	No	Open		Equilibrium not reached in an open system, though a steady state may be reached

known as the

FIRST LAW OF THERMODYNAMICS: Energy is conserved in an isolated system.

(You may recall from Chapter 4 that mass and energy are interconvertible, but this does not invalidate the law because mass can be regarded as one of the forms of energy.)

It turns out that this first law is as much part of the definition of an isolated system as it is a statement about the nature of energy. This is because the three types of systems with which we deal—open, closed, and isolated—differ in whether or not they conserve mass and/or energy (Table 20.1).

From Table 20.1 you see that entropy has a role in the description of isolated systems similar to the role of Gibbs free energy in the description of closed systems. For any net change in an *isolated* system, the change in entropy must be positive. For any net change in a *closed* system (constant m,P,T), the change in Gibbs free energy must be negative. Clearly it is important to identify the type of system under consideration before attempting to determine how it will behave. Many serious errors in planning have occurred because this distinction was not understood and taken into account.

It is perhaps interesting to note in passing that the three laws of thermodynamics—

1. conservation of energy
2. universal increase in entropy
3. entropy equals zero for a pure crystal at 0 K

all apply to isolated systems. One of the great powers of "thermo" (the common nickname for thermodynamics, the study of heat changes) is that its developers have found ways of extending the ideas in a precise

EXERCISE 20.2
Note that there is one possible combination *(no–yes)* of conservations not mentioned in Table 20.1. Do such systems exist?

and accurate manner to include closed systems. Lately the ideas have been extended to open systems, but I shall not discuss those applications.

How Many Ways?

Just as Gibbs defined free energy using a logarithmic term (equation 19.1), so Ludwig Boltzmann defined entropy similarly as

$$S = 2.303 \, k \log W \tag{20.1}$$

where k is the Boltzmann constant, and W is the number of ways the system may arrange itself. The term W includes the number of ways the atoms can distribute in the system and the number of ways the energy can distribute among the atoms. Determining W is often a very difficult task. We shall usually not undertake it.

But do note that equation 20.1 is consistent with the third law of thermodynamics. There is only one way of arranging the system at 0 K. All molecules in a pure crystal have the same energy and are in well-defined positions at 0 K. No interchanges of energy or position occur. Therefore, at 0 K, $W = 1$, $\log W = 0$, and $S = 0$.

Note also that equation 20.1 gives absolute values for S, because S really is zero at 0 K. If W can be calculated for any other condition, the calculation of S is simple using equation 20.1; W is always greater than one in real systems so S is always greater than zero.

We can correlate the tendency to spread out by calculating ΔS when some state 1 (achievable in W_1 different ways) changes to some state 2 (achievable in W_2 different ways). For a mole of substance

$$\Delta S \, [\text{J}/(\text{mole} \times \text{K})] = 18.2 \log (W_2/W_1) \tag{20.2}$$

The change in entropy (ΔS) is related to the ratio of the number of ways of arranging state 2, compared to the number of ways of arranging state 1. If the system spreads out (becomes more random), during a change, the value of W will increase (W_2 will be greater than W_1) and ΔS will be positive. This method of calculating S or ΔS requires only the determination of W values for use in Boltzmann's very simple equation 20.1.

Figure 20.3 shows the entropy of a mole of atoms of many elements at 298 K and 1 atm. Note the general trends and the entropy values—hence similar W values. Crystals have low values; liquids, higher values; gases, still higher values. Diatomic gases have lower values per mole of atoms (but higher values per mole of molecules) than do the monatomic noble gases. All the crystalline elements have similar entropies, so all must have about the same degree of order and disorder (same value of W) at 298 K. And, of course, all values of S are positive because an increase in temperature (here from 0 to 298 K) always increases randomness, or spreading out.

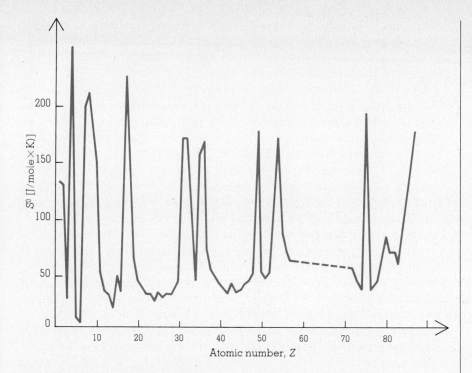

Figure 20.3
Standard entropies for the
elements (1 mole of atoms of
the element in its standard
state).

Photosynthesis and Rotting

A forest is a wonderful place to wander, and to ponder. It also is a remarkable example of chemistry in action. Thousands—more likely, millions—of different chemical reactions are occurring daily. I have already mentioned one set of reactions—photosynthesis. And I have pointed out that the process is really much more complicated than is indicated by the net reaction usually written for the process. However, that simplified net equation does describe some major chemical changes involved in photosynthesis:

$$6\,CO_2\,(g) + 6\,H_2O\,(\ell) = 6\,O_2\,(g) + C_6H_{12}O_6\,(c) \qquad (20.3)$$

At the same time, throughout the forest (even on parts of most trees that are actively photosynthesizing), rotting is occurring (Figure 20.4). The net equation for rotting is just the reverse of equation 20.3:

$$6\,O_2\,(g) + C_6H_{12}O_6\,(c) = 6\,CO_2\,(g) + 6\,H_2O\,(\ell) \qquad (20.4)$$

Equations 20.3 and 20.4 may remind you of their parts in the natural cycles outlined in Chapter 2 for water, oxygen, and carbon.

How can net growth occur in one place when only a few inches away net destruction is also occurring? What kind of a system is this, in which the same reaction can be occurring but in opposite directions, even though the external conditions are the same and the two opposed re-

Figure 20.4
A tree grows and rots.

EXERCISE 20.3
What "reactant" that we often
specifically list in equations is
omitted from equations 20.3
and 20.4?

actions are clearly not participating in a dynamic equilibrium? (Recall that no net changes can be observed in a system at dynamic equilibrium, but here we can see growth and rotting taking place as distinct changes in different parts of the system.)

Part of the problem is that we have not defined the system. Let's start to do this by drawing an imaginary boundary around the forest, on all sides and above and below. Then let's assume this boundary is impervious to molecules—say it consists of a big polyethylene sack (Figure 20.5). Will the forest still grow? Yes, at least for a long time. It can recycle its atoms, conserve atoms and mass, and act as a closed system. Can it operate as an isolated system? No. Turn off the sunlight for any appreciable length of time and the forest will die. Furthermore, prevent the heat generated in rotting from leaving, and again the forest will die. So I will describe the forest as a closed system—atoms are conserved, but energy is not.

The analysis of the system has pointed up an important fact about the energy balance. True, equation 20.3 is the opposite of equation 20.4, but I did not include the energy terms. And in the forest itself, the energy term in equation 20.3 must be in the form of sunlight, whereas the energy term in equation 20.4 is in the form of heat. It could easily be that the magnitude of these energy terms would be identical—the forest would gain as much energy from sunlight as it loses as heat. Its own

Figure 20.5
A forest system surrounded by transparent plastic: open, closed, or isolated?

energy content remains constant. But can the system "run backward"? Can it take in heat and emit sunlight? The answer is no. It is not sufficient merely to supply the energy; the energy must be provided as light.

The conversion of light to heat, which is the net reaction in a primeval forest (or in our forest in the plastic bag), is common. Your skin accomplishes it whenever you are at the beach on a sunny day. Even the beach sand can do this conversion. Yet the reverse process is never observed. No one and no system has ever been found capable of converting heat completely into visible light.

It is not possible to reverse the normal and commonly complete conversion of visible light into heat. It is possible to keep the trees growing and rotting, but the surroundings must change.

Energy Is Conserved—To Heat

A little thought will convince you that you know many processes in which energy can move from one form to another but cannot spontaneously move back again. In fact, it does not take long to realize that every energy change has this property. The direction of energy flow is not spontaneously reversible.

But wait a minute. What about the rechargeable storage battery? It can deliver electrical energy, then can be recharged by having electricity passed through it in the reverse direction. A little measurement will convince you that your first conclusion was correct. The voltage of the cell on discharge (about 1.8 volts per cell) is less than the voltage required to charge it (about 2.1 volts per cell). The reversed current can be the same as the forward one if one is careful. But the total energy is voltage times ampere (energy $= V \times A$), and the energy given out on discharge is never sufficient to recharge the cell. An additional source of energy is required. It is possible to get the cells back to their original condition, but the surroundings will have changed.

The amount of energy in the universe is unchanged by the forest, by the photosynthesis and rotting, by charging and discharging electric cells, or by any other changes we observe. But the availability of the energy continually decreases. Heat flows from high temperatures to low until the temperatures are equal. Electrons flow from high voltages to low voltages until the voltages are equal. All forms of energy tend in time to be converted into heat, and into heat at relatively low temperature. Energy spontaneously flows in a single direction. Net energy flow stops when equilibrium is reached. We never observe spontaneous reversal of the normal direction of net energy flow from more random to less random systems, and we can always describe the flow of energy in terms of entropy changes. This is what the second law does. It describes the fact that energy tends to spread out, become less concen-

EXERCISE 20.4
A forest absorbs sunlight at a rate of about 4 J/(cm² × min), or about 3000 kJ/(m² × day). About how much heat does it lose per 24-hour day per square kilometer?

POINT TO PONDER
A particularly dramatic demonstration of the second law of thermodynamics is the running of almost any motion picture backward. You quickly recognize that it is "backward." The processes you see may be made to occur, but only when an additional source of energy is available. The normal course is for the "forward" change to occur. "Forward" in a total system always means in the direction of an increase in the total entropy.

EXERCISE 20.5
About how many kilojoules do you "emit" per day? (One food calorie or 1 kilocalorie is equal to about 4 kJ.) What happens to this heat?

EXERCISE 20.6
What happens to the energy
generated when a river goes
over a waterfall?

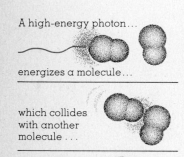

A high-energy photon...

energizes a molecule...

which collides
with another
molecule . . .

so the photon's
energy is spread
over two molecules
(and soon over
more).

trated, more random, and less available for use. Entropy increases.

The entropy increase can be treated quantitatively, and an ample number of equations exist to describe and calculate the change in entropy for many possible changes. But we shall use very few of these. A qualitative feel for entropy in terms of increasing randomness and the tendency of mass and energy to spread out will usually be more than adequate to understand change.

Our initial equations for photosynthesis (equations 20.3 and 20.4) are faulty in three respects. They do not specify the type of system, the amount of energy involved, or the kinds of energy. The atoms can be recycled through the reactions indefinitely, but the light energy goes through once and comes out as heat. Some of the solar energy is converted to heat to accomplish reaction 20.3, the rest is converted to heat in reaction 20.4. That energy cannot spontaneously cycle through this system again.

The energy comes into the system as a relatively small number of high-energy photons of visible light. It leaves as random thermal motion of many molecules. The energy of each photon becomes distributed over many molecules; W increases; ΔS is positive. The small units of molecular energy do not spontaneously recombine to give back all the high-energy photons even though the total energy is unchanged.

Heat and Temperature

If all forms of energy are converted to heat, it is reasonable to ask what happens to the heat. You have seen that heat increases the random motion of molecules. In a pure substance, with no phase change occurring, this increased randomness at the molecular level causes a rise in temperature. This is consistent with everything you have learned about heat and temperature—including Charles' law, the kinetic theory, and the Maxwell–Boltzmann distribution curves (Figure 20.6). (A brief review of Chapter 12 might be helpful for you at this point.)

The conversion factor from ΔH to ΔT is called the *heat capacity, C.* It is expressed as joules/(mole $\times$ K):

$$\Delta H = n \times C \times \Delta T \tag{20.5}$$

where n is the number of moles. The heat capacity (C) of water is 75 J/(mole $\times$ K). If 3.0×10^3 J (ΔH) of heat is supplied to 2 moles (n) of water, the temperature rises 20 K (ΔT):

$$\begin{aligned}
\Delta T &= \Delta H/(n \times C) \\
&= (3.0 \times 10^3 \text{ J})/[2 \text{ moles} \times 75 \text{ J}/(\text{mole} \times \text{K})] \\
&= 20 \text{ K, the rise in } T \tag{20.6}
\end{aligned}$$

Tables of heat-capacity data for various substances are available,

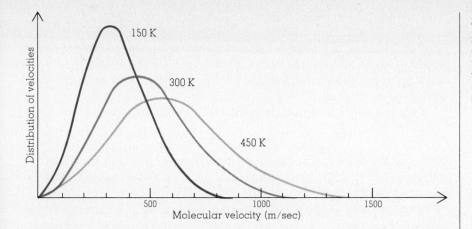

Figure 20.6
The increasing variety of
molecular velocities as temperature increases. For example,
the fraction of molecules with
velocities above 700 m/sec is
0.01 at 150 K, 0.04 at 300 K, and
0.30 at 450 K.

of course. An easy thing to remember is that the heat capacity of most metallic elements is about 25 J/(mole of atoms $\times$ K). This generalization is known as the *law of Dulong and Petit*.

Is ΔS Always Positive?

The second law is quite firm in describing *any total change* in an isolated system as resulting in an increase in the entropy. There are no known exceptions. The Boltzmann definition (equation 20.1) interprets this as an increase in the number of ways the system can arrange itself, and I have described this as an increase in randomness. But you know many systems that are becoming more ordered.

After every football play, the variable tangle of players is sorted out and an ordered lineup attained. The leaves that were blowing across the yard appear as neat piles after the arrival of the gardener (or you). Ice crystallizes from fluid water. Crystals can be cooled toward 0 K with a continuous decrease in their entropy. Yet as we have already discussed, none of these systems is isolated. Each is losing energy to the surroundings. They can be defined as closed systems. The number of football players, leaves, water molecules, or atoms in the crystal remains constant. But there is always an energy (and entropy) change in the surroundings. There is always a larger entropy increase in the surroundings, which offsets any entropy decrease in the closed systems.

These observations may be summarized in a simple relation of great use and interest:

$$\Delta S_{\text{total}} = \Delta S_{\text{system}} + \Delta S_{\text{surroundings}} > 0 \qquad (20.7)$$

According to the second law, ΔS_{total} (from here on, ΔS_{tot}) must be positive for any change. There are no known experimental exceptions to this statement. The second law seems to be a good description of our universe and its behavior.

EXERCISE 20.7
Which would provide you
with more heat on an aching
stomach—a container filled
with 1 kg of water at 50°C, or
the same container filled with
1 kg of iron at 50°C?

EXERCISE 20.8
When 1 mole of water freezes,
6 kJ of heat is released. (Recall
equation 19.30.) Calculate ΔS
for H_2O in the change from
water to ice. Comment on the
sign of ΔS. Why does a freezer
need some external source of
energy to make it operate?

EXERCISE 20.9
An isolated system is usually at constant volume. If our universe is expanding, does that effect tend to increase or to decrease the entropy of the universe?

Forward?

However, although ΔS_{tot} must be positive, either ΔS_{system} (ΔS_{sys}) or $\Delta S_{surroundings}$ (ΔS_{sur}) could be negative—but *not* both. This is, of course, what we observe. If a change causes a decrease in the entropy of some system (that is, an increase in its order), there *must* be a larger increase in the entropy of its surroundings—so that the total entropy of the universe does increase as a result of the change.

If the system is at equilibrium with its surroundings,

$$\Delta S_{sys} = -\Delta S_{sur} \qquad \text{and} \qquad \Delta S_{tot} = 0 \qquad (20.8)$$

If the system is the entire universe, this equation could apply only after the entire universe had reached equilibrium, with no further net changes occurring anywhere.

Equation 20.7 is an important statement that is beginning to be appreciated, understood, and used explicitly in decision-making groups. It has been implicit in intelligent decisions for a very long time. We may paraphrase it this way: If you create order in one place, you will simultaneously create more disorder somewhere else. Local entropy can *decrease;* total entropy can only *increase.*

Time and the Second Law

The fact that there are no known exceptions to the second law—that total entropy always increases—and the fact that time never seems to reverse have attracted the attention of many thoughtful people. A. S. Eddington, a British astronomer, linked the two: "Entropy is time's arrow."

All those science-fiction stories to the contrary, no one yet has found a way to "run time backward." It is easy to tell whether a standard motion picture is running forward or backward. In the same way no one yet has found an isolated system in which the entropy is decreasing. All systems are running down, energy is getting less available, useful forms of energy are being converted into heat—the random motion of molecules. The "heat death" of the universe was the subject of many discussions in the 1930s, possibly enhanced by the pessimism brought on by the worldwide depression. But there is little discussion now, even though most scientists are more firmly convinced than ever that the prediction is correct. However, some idea of the time scale is also now available.

The evidence is quite strong that the earth, in the absence of massive human intervention, will maintain roughly its present climate and favorable environment for life for about 5 billion more years. Universal equilibrium is at least that far off. So there is little reason to worry about changes on such a long time scale when more immediate problems demand consideration. Table 20.2 may give you some insight into

EXERCISE 20.10
Suppose that the universe changed so the second law of thermodynamics read "The entropy in an isolated system always decreases." Would the other two laws necessarily change?

the time scale, converting the world's past to a scale of 24 hours. The earth is no more than halfway through its lifetime.

It is true that man could considerably shorten this time scale of the future, especially for himself as a species, if he pays no attention to environmental problems. More than 99.99% of the species that have existed on earth have become extinct. The average life of a species appears to be about 100,000 years. It is interesting that biologists suggest that living species with high survival value are those that are successful at minimizing their rate of entropy production in their surroundings. These observations do not encourage one to predict a long life for a species that does not remain adaptable, or for one that fouls its own ecological niche or nest.

We may not be able to turn time backward nor to decrease the entropy of an isolated system, but we are quite sure that there is plenty of time to work out reasonable solutions to problems as they arise. The heat death of the earth is not an imminent possibility. I shall shortly discuss the probable effects of human intervention in the steady-state systems that currently prevail in the major chemical cycles on earth.

Cycles, Wastes, and Reversibility

The earth is essentially a closed system (Figure 20.7). True, it gains meteors and cosmic radiation, and it loses gas from the top of the atmosphere. But these processes are insignificant compared to the overall mass of the earth, even over very long periods of time. Energy, on the other hand, pours in and pours out at a great rate. Available data indicate that, although the temperature of the earth's surface fluctuates with a period of about 10^5 years, the average temperature has remained constant for a long time. (See Figure 20.8, which is based on isotopic analysis of geological deposits and ice volume.)

On the average, the solar heat striking the surface of the earth plus the heat generated inside the earth (from radioactive changes) almost exactly equal the heat lost to space by radiation from the earth's surface. The earth is in an energy steady state and has been for a long time. There is no reason to believe this steady state will change much in the next few billion years—unless man intervenes. Again, I shall put off discussion of this intervention for a few chapters.

The flow of energy sets up flows of mass and atoms. I have already mentioned the winds and tidal currents, which are caused primarily by the greater amount of energy received at the equator than at the poles. These currents of air and water are the major factors in determining climatic variation, especially local temperatures and water supplies. Coupled with available sunlight, these determine the likelihood of life and the types of life that can survive in various areas.

Table 20.2
Geological Time Scale:
5×10^9 Years Compressed
Into 24 Hours

Time	Event
−24 hours	Earth forms
−12 hours	Life appears
−1 hour	Animals well established
−0.5 hour	Precursors of man appear
−1 minute	*Homo* species appear
−1 second	*Homo sapiens* (modern man) appears
−0.1 second	Written history begins
0	Today
24 hours	At least an equal time for the future

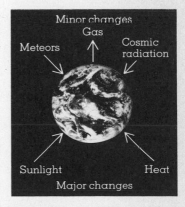

Figure 20.7
Earth is a closed system (almost).

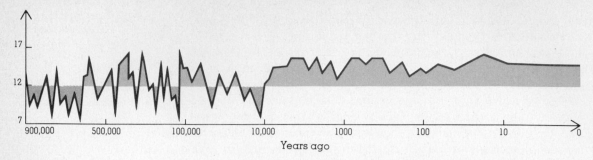

Figure 20.8
Average air temperature at
midlatitudes for the last million
years. The present average is
15°C, warmer than usual as
you can see, in fact near the
historical maxima. (Data from
Understanding Climatic Change,
National Academy of Science,
Washington, D.C., 1975)

POINT TO PONDER
A hen is an egg's way of making
another egg.—*Samuel Butler.*

EXERCISE 20.11
Coal is often referred to as
stored sunshine. Why?

The circulating air and water are also the primary means by which
the wastes inevitable from a living system are swept away from their
origin to places where they can be recycled, using solar energy, into
molecules that can then serve as food to keep the cycles going.

You have already seen that all chemical equations represent two
reactions—the forward and the reverse. Natural systems are excellent
at providing mechanisms for both reactions. Many natural reactions
proceed both ways in natural systems, as pointed out earlier with
growth and rotting.

Chemists have become adept at operating closed systems of much
smaller scale, and we now know there are no impossibilities in reversing
any chemical reaction. But to do so, an energy source must always be
available. And we know that energy cannot be reversed in its flow, nor
can it be recycled. This is one of the great lessons of the last 150 years,
and only now is it being taken into account in the making of major
decisions.

In formal terms we can summarize: Closed systems are reversible;
isolated systems are not.

Humpty Dumpty cannot be put back together by the king and his
men but, as Figure 20.1 points out, feed him to a hen and out pops
another Humpty Dumpty. But the hen needs an external source of
energy to reverse the shattering of Humpty's fall.

The hen should remind you that human inability to reverse a process
does not mean that the process cannot be reversed. In fact, more and
more scientists are using the chemical abilities of other life forms for
human ends. Penicillin is synthesized by molds, alcohol is made by
yeasts, oil spills are cleaned up by bacteria, enzymes are used to catalyze
industrial processes, and food is made by plants and animals. We are
even trying to isolate or "synthesize" bacteria that can degrade plastics.
Such a synthetic bacterium especially good at degrading oil spills was
announced in 1975.

A waste material is sometimes defined as a chemical out of place, but
most wastes can be characterized as substances with high entropy compared to the desired product. They may be convertible into materials

that can serve as raw materials for further processes. But the lowering of the entropy of the wastes requires energy and a corresponding increase in the entropy of the surroundings.

We cannot prevent the increase in the entropy of the universe. What we can do is minimize the rate of entropy increase, control the type of entropy that is created, and choose the place in which the entropy is generated and stored. Every really long-lived species is adept in these three areas, and no one even mentioned the entropy principle to them. They learned the hard way—experimentally. The nonlearners of the past are extinct.

Information and Entropy

There is a fable about a race of excellent master masons who built the most beautiful buildings ever seen. Their sense of design was faultless and their construction techniques meticulous. But the process of forming handmade blocks took a long time.

One of the more inventive builders, analyzing the rate-determining step, conceived of a machine that produced bricks every bit as beautiful and with as much variety of shape. Exquisite bits of potential architectural detail began to pile up in the storage yards. Builders came and helped themselves. Magnificent structures sprouted at a rising rate. The brickmakers, inspired by the loveliness rising around them, redoubled their output. The storage yard grew, the multiplicity of forms (all exquisite) was a joy to behold.

True, the builders had increasing difficulty finding just the right shape for their needs, but perseverance paid off for a while. So proliferation increased until a new rate-determining step was reached—the search for the proper piece. Eventually, the inevitable. The search became too difficult. Buildings were put together from what was readily available. Grotesque combinations predominated. The brickyard workers, discouraged by the results, produced standard forms—easily stored and retrieved. Moral: There are various ways to brick yourself in.

At least two points can stand explicit mention: (1) the acquisition and application of knowledge do not, of themselves, guarantee a satisfactory long-term outcome and (2) a more subtle point: the aquisition of knowledge simultaneously produces entropy in the surroundings just as any other change does.

When you learn something, you are producing an ordered array. Ideas that used to be confused are clarified. Symbols and phrases that had no meaning become useful. All these involve a lowering of the local entropy and, of course, an increase in the entropy of the surroundings. Whenever you learn something, that very act makes it impossible ever to learn something else. By increasing the order or organization in one

EXERCISE 20.12
Where does most of the entropy appear that is generated by a blast furnace producing iron? Is there any part of the system that is lowered in entropy?

Chaos in the Brickyard

EXERCISE 20.13
You hear "static" when you turn up the gain (volume) of a radio. Much static is due to random motions of electrons in the set; it is an entropy effect. Suggest one way static could be lessened.

EXERCISE 20.14
List some of the clear-cut increases in entropy associated with a supersonic airplane. List some of the decreases.

place, you have irrevocably increased even more the disorder or disorganization somewhere else. G. N. Lewis (of paired-electron and acid–base fame) had this in mind when he wrote as a summary of the second law, "Gain in entropy is loss in information."

It thus becomes important that decisions be made as to what is most worth learning and doing, and what is least worth the effort. Commonly an increase in entropy appears as heat and waste products in your immediate surroundings. The heat and molecular waste are then "taken away" and become of small concern to you, certainly compared to the greater facility acquired in using your own brain and the greater comfort in your life. The decisions become more critical when large numbers of people and very difficult problems are involved. "Away" becomes hard to locate.

How much disorder in the surroundings can be justified by the hoped-for gain in order in the system? Was the man-in-space program worth the enormous expenditure of human effort and raw materials involved? Is an SST a net gain in terms of entropy balance? Do these and other undertakings minimize the rate of increase in entropy, produce tolerable forms of entropy, and put the inevitable increase in a reasonable place? What alternative efforts should be considered? The question is not "Neat or messy?" but "Neat where, messy where, and how much messiness can we stand?"

The Quest for Purity

Fresh air, pure water, and clean soil—or rather the absence of them—are much in the news. Humans have always created wastes, and wastes have always caused problems. (Nor have any substances ever been completely pure.) But the problems seem worse now. Actually many of them are not.

Sulfur-dioxide pollution of city atmospheres was worse in the United States 50 years ago. In most of the world, drinking water is now much safer than it was 30 years ago. It is the fact that air and water have become so good that makes us aware when they begin again to be bad.

What concerns the most thoughtful persons is the number of new chemicals being synthesized, and the very large amounts of these chemicals that must be used because of the large populations. For example, there are about three times as many people alive today as when I was born—4 billion compared to over 1 billion. The world is the same size, so any use of a chemical on a per capita basis is three times as concentrated. The grouping of masses of people in cities multiplies the concentration problems many times over. So does the necessity of feeding more and more people at ever greater distances. And the second law makes clear that regardless of how chemicals are used they tend to spread out. The chemical DDT applied to kill malarial mosquitoes in homes gets into streams and fields. Polychlorinated compounds designed for plastics or electrical transformers appear in the oceans. Waste gases from steel mills spread over neighboring towns.

But science also gives guidance on attacking the problems:
1. The only way to have zero pollution is to have no pollutant produced.
2. If a chemical must be produced, it is far easier to control at its source before it has a chance to spread out.
3. If an existing chemical must be controlled, this can be done—but there is no way to reduce the pollution to zero.
4. The more dilute the pollutant, the greater the cost of reducing the concentration further. In fact one can show from the second law that the cost will be about constant each time the concentration is reduced by a factor of 10. At low concentrations the actual cost rises even faster. In paper-mill waste, it costs 30 times as much to remove 1 g of pollutant at 95% removal as at 70% removal.
5. If you stop producing a chemical, you must either find a substitute (often more expensive) or live with the problem the chemical was supposed to ease.

Conclusion? Stop the syntheses and live with the absence, or set a tolerable level and live with the pollutant.

Summary

The second law of thermodynamics is one of the most powerful generalizations available to humans. There are no known exceptions to it, and it applies to all net changes. The tendency for the entropy to increase in any process in an isolated system can be measured quantitatively using the Boltzmann definition:

$$S = 2.303\, k \log W \qquad \text{and} \qquad \Delta S = 2.303\, k \log(W_2/W_1)$$

Changes in entropy can be understood qualitatively in terms of the universal tendencies of mass and/or energy to spread out; W increases. If there is a local decrease in entropy in one system, there must be a larger increase in entropy in the surroundings.

There is little need to worry about the long-term "running down" of the earth or the universe many billions of years in the future. But there are many reasons to remember entropy considerations in making contemporary decisions. Are we minimizing the rate of entropy production? Are we producing entropy in a tolerable form, and in a tolerable place? What are the alternatives? These questions have faced humans ever since they began to formulate decisions. They now have available to them a powerful guide and help in the second law.

POINT TO PONDER
A soap manufacturer once had its product analyzed. The chemists reported all the ingredients they could identify and summed the total to 99.44%. From then on the product was sold as "99 and 44/100ths% pure."

HINTS TO EXERCISES
20.1 Any system at equilibrium.
20.2 No. 20.3 Energy.
20.4 3×10^9 kJ/(km² × day).
20.5 10^4 kJ/day. 20.6 Heat is generated. 20.7 Water has many more atoms per kilogram.
20.8 $\Delta S = 20$ J/(mole × K).
20.9 Increase. 20.10 No.
20.11 Photosynthesis provides solid carbon compounds.
20.12 Heat and gases.
20.13 Lowering the temperature.
20.14 Your choice.

Problems

20.1 Identify or define: energy flow and avail- ability, heat, heat capacity (C), irreversible change, reversible reaction, $S = 2.303\ k\ \log W$, $\Delta S = 2.303\ k\ \log(W_2/W_1)$, systems (open, closed, isolated), temperature, waste.

20.2 The "three laws of thermodynamics" are some- times states as follows: (1) You can't win. (2) You can't break even except at absolute zero. (3) You can't get to absolute zero. Translate each of the humorous (?) phrases given into a sentence or two that a person who has never heard of thermodynamics might understand.

*20.3 Two identical clock springs, one wound and one unwound, are dissolved in a mixture of hydrochloric and nitric acids (aqua regia). What happens to the stored energy in the wound spring?

20.4 Liquid and gaseous H_2O at 298 K have entropy values of 70 and 189 J/(mole $\times$ K) respectively. Calculate W_2/W_1 for a mole of H_2O (l) forming a mole of H_2O (g). Does the ratio seem reason- able in terms of your ideas about gases and liquids? [$W_g/W_\ell \cong 10^6$]

20.5 A roulette wheel in rotation, with the rolling ball running around inside it, has many pos- sible arrangements. Yet eventually it comes to rest with the ball in one socket. Its entropy has decreased. How is this possible?

20.6 Select any change with which you are familiar, and show whether or not it leads to an increase in total entropy. If so, would it be worthwhile to minimize the rate of entropy production? If so, how would you do so?

20.7 It is probable that the most deadly poisons, the most dangerous viruses, and the most effective cancer-causing substances known are synthe- tic—produced by humans. Does the possible knowledge that can be so gained justify the potential risk of these substances escaping from the labs?

20.8 The European Age of Exploration followed close on the Renaissance and seemed to justify the Renaissance feeling that there were few if any boundaries to human accomplishments. Some have claimed that the astronauts have launched a similar burst of human expansion. Support or attack this position in terms of chemical supplies or limitations.

20.9 At Alice's tea party, the Mad Hatter and the March Hare continually moved around the table to clean sets of dishes. When Alice asked what would happen when they completed the circle—they did not answer. What problem of increasing interest today might the author have had in mind? Any answer for Alice?

20.10 Groups differ in their opinions about the likeli- hood or imminence of the world "coming to an end." During the last 100 years there has been an increasing belief that this "end" may be far distant. Discuss possible relations between this shift and our rising ecological concerns.

20.11 Albert Einstein, in discussing scientific laws, once said, "The Lord is subtle, but he isn't simply mean." What bearing could *subtleness* or *meanness* have on scientific efforts to dis- cover regularities?

20.12 "In a very real sense we are shipwrecked passen- gers on a doomed planet. . . . we shall go down, but let it be in a manner to which we may look forward as worthy of our dignity." So writes Norbert Wiener in *The Human Use of Human Beings* (Avon Books, 1967). What kind of "doom" would a physical scientist like Wiener find so inevitable? When is such a doom likely to occur?

*20.13 While on a camping trip in the desert, you have drained the radiator of your car to keep it from freezing. The storage buckets are knocked over during the night and you have no addi- tional water. Estimate the distance it would be safe to drive the car in search of a source of water. (Use only one significant figure because you must "guess" much of the data needed to solve this problem.) [~1 km or 1 mile]

20.14 Crystals are certainly more orderly than the liquids from which they form. Outline how the entropy of an isolated system can increase if water in it is freezing to ice. Hint: What must happen to the water if it is to freeze?

20.15 A sample of a new element, M, was prepared in 1886 and found to have a specific heat of 0.35 (J/g × K). What approximate atomic weight was assigned to the element? The formula of the oxide was MO_2. What element do we now call it?

*20.16 You may have noted, in Table 20.1, that no system is described for which energy is conserved but mass is not. Present an argument for why no such systems are known, based on what you know about the energy distribution among molecules.

*20.17 Isaac Newton assumed gases were made of particles with identical energies all spaced equally throughout the occupied volume. Show why the entropy of such a gas should be zero at every temperature.

20.18 The standard free energy of formation (ΔG_f^0) of many carbohydrates of formula $C_6H_{12}O_6$ (c) is about -850 kJ/mole. Use data in Table 19.1 to decide whether equation 20.4 or 20.5 is the spontaneous one in the standard state. Does this fit what you know about photosynthesis and rotting?

Figure 21.1
Almost everything you see in
this picture is a macromolecular
material.

21 ONE FROM MANY: POLYMERS

Most of the substances we have looked at so far have been made up of small molecules having molecular weights of about 100 or less. But most of the substances that surround you, that you use, and on which your health and comfort depend are made up of large molecules (macromolecules), many of them polymers. Their molecular weights range into the millions and higher.

Concrete, metal, wood, glass, paint, tile, clothing, carpets, draperies, upholstery, carbohydrate, protein, paper, plastics, rubber—all are macromolecular. Polymeric substances can be chemically classified as *metals, silicates* (concrete, glass), *carbohydrates* (wood, cotton, paper, rayon, linen, food), *proteins* (wool, silk, food), *hydrocarbons* (rubber, polyethylene, polypropylene), *vinyls* (many synthetics), *polyesters* (Dacron), and *polyamides* (nylon). Other examples and other classes are manifold. And all of these, except some wood and a few foods, undergo chemical processing before use.

One reason macromolecules are so widely used is that they are structurally strong. The covalent and ionic bonds holding the mass together are much stronger than polar and van der Waals bonds. In big molecules with large surface areas, even the van der Waals forces can become large. There is little slippage between the molecules when stress is applied. Their large size and their tendency to tangle and interlink give them low solubility and high resistance to chemical attack. They often have high resilience and resist breaking because they can stretch and recover. Small wonder there are some 300 trade-named plastics on the market.

One of the most useful characteristics of macromolecules is their infinite variety, reminiscent of the interchangeable parts discussed in Chapter 7. Almost all macromolecules, both natural and synthetic, can be modified over a wide range of properties. One of the chemist's jobs is to correlate these properties with molecular structures and then to synthesize molecules that have the desired properties.

I have seriously neglected the difficulties found and the clever solutions invented in synthesizing desired compounds. My emphasis has been, and will continue to be, the interpretation of properties in terms of molecular behavior. But more chemists are engaged in synthesis than in any other single activity. Perhaps you will, in this chapter, get some slight feel for the problems they face and the methods they use.

Men who love wisdom must be acquainted with very many things indeed.—*Heraclitus*.

From Pelts to Plastics—5000 Years

The earliest macromolecular materials used by humans were silicates and wood (and also foods, which I shall consider in another chapter). Early shelters and weapons were made from things at hand, and wood and rocks were at hand. Humans learned to fashion both wood and rock for more effective use, and by the end of the Stone Age they had become quite skilled in manipulating these materials.

During the same interval, fibers and cloth were invented. Early fibers were used to lash things together, and early cloth was a substitute for fur pelts and skins. The origins of the carding and spinning of fibers into threads and the weaving of threads into cloth are lost. Wool was probably the first woven material, and cotton fiber and cloth were present in India by at least 3000 BC and in Peru by 2500 BC. For the next 4000 years there was little change in the uses made of the natural macromolecules. Metals were developed, additional fibers were processed, and special breeds of sheep and other wool-bearing animals were developed.

Large-scale developments in the manipulation of macromolecules followed the harnessing of coal and steam power in the eighteenth century and the drive to understand properties in terms of molecular behavior, which began about the same time. Most of the synthetic plastics have been developed in the last 50 years, and a busy 50 years it has been.

Few structures still use natural macromolecules that are the same as those of 200 years ago. Even the occasional "natural" wood or stone building is frequently made up of materials treated to improve or enhance the natural properties. The wood in the building may even be from a hybrid variety of tree that did not exist 100 years ago. Very few of the macromolecular materials you see around you were available for purchase when your parents were born.

EXERCISE 21.1
Does it seem reasonable that wool was made into fabrics before cotton? Why?

Paper: A Modified Natural Polymer

You as a student may be more familiar with paper than with most macromolecular substances. But have you ever looked closely at a piece? Say this one? It's grainy. If you tear off a piece (not too big a piece, please), you see fibers projecting from the edge. You may even notice that it is easier to tear in some directions than in others. If you work hard at it, you will find that the individual fibers vary in length from about half a centimeter down, and they are about one-hundredth as wide. They are matted together, almost but not quite at random angles, which is why they tear differently in different directions. And if you work even harder, you will find that if the paper is placed in water and the fibers are floated off, some nonfibrous materials will settle to

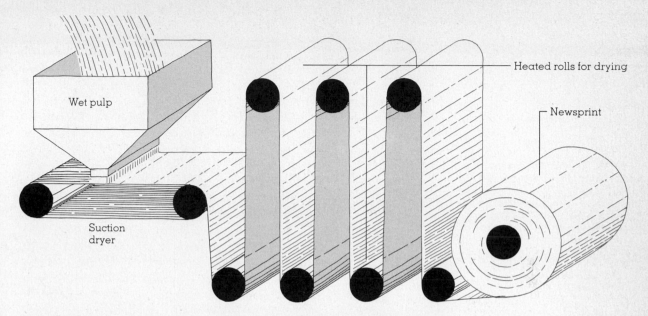

Figure 21.2
From wood pulp to newsprint.

the bottom. As a matter of fact, there are several chemicals in a piece of paper. It is a highly processed material, with its main properties based on the macromolecular properties of cellulose.

Wood, the source of most paper, contains mainly *cellulose* (a long-chain carbohydrate) and lignin. In the simplest papermaking process, logs or sticks of wood are pressed against a rotating pulpstone, which tears the wood fibers apart. A stream of water carries away the fibers and the lignin, forming a watery mixture called *pulp*. The pulp is spread into thin sheets, and the water is squeezed out or evaporated to leave dry paper—mostly cellulose. See Figure 21.2. Paper produced by this mechanical process is not very strong, not very white, and because of chemical changes in the remaining lignin, it quickly darkens and becomes brittle as it ages. Most paper produced by mechanical pulping is used for newsprint.

Better-quality papers are made by a chemical pulping process. Chips of wood are placed in big cooking vessels with chemical solutions and cooked at high temperature and pressure to separate the lignin from the fibers. The lignin is removed with the solution as a waste product, and most of the lignin is burned. There are two different processes commonly used in preparing the pulp—one uses sulfate ions and the other uses sulfite ions. The sulfite process produces paper that is almost pure cellulose; this paper can be made very white (by chemical bleaching), but it is not strong. Furthermore, the sulfite process requires a lot of chemicals and can cause serious water pollution when waste solutions are discharged from the processing plant.

EXERCISE 21.2
Assume that there are 100
paper-pulp manufacturing plants
in the United States using the
sulfate process to produce 10^6
kg of paper per day at each
mill. What is their average
hourly sulfur emission if they use
the cleanest available process?

POINT TO PONDER
Glucose is produced by most
plants, and it is a major energy
source for animals. It must be
stable to be synthesized, yet it
must be reactive to be digested.
A major factor is seen in its
structure, a puckered hexagon
with a small H atom and a larger
—OH or —CH_2OH group at
five corners. Note in Figure 21.3
that all the —OH groups are as
far apart as possible (around the
rim of the hexagon). All other
carbohydrates are less stable
because they are more crowded.
(See Problem 19.13.) At the
same time, the polar —OH
groups in the glucose rings and
the ether links, even when they
are in a long chainlike starch or
cellulose, are exposed and can be
attacked by digestive enzymes.
Even so, the process is slow, as
shown by the necessity of multiple "stomachs" in cows (plus
chewing of cud).

Figure 21.3
Cellulose—a string of glucose
units connected by other
linkages.

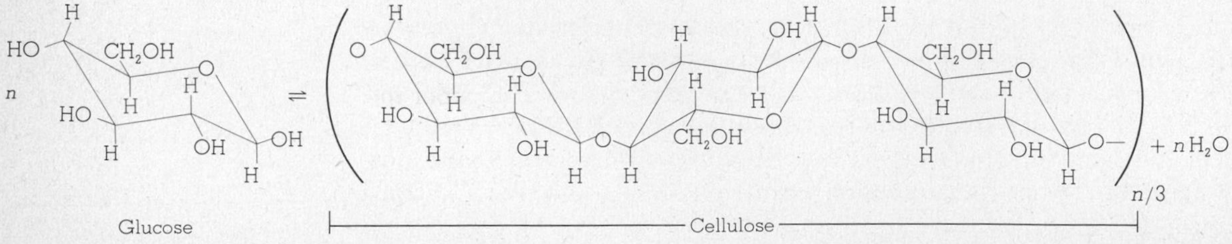

The sulfate process is more commonly used and produces exceptionally strong paper (pulp and paper produced in this way are called *kraft,*
from the Swedish word for "strong"). This process is apt to pollute the
atmosphere with odorous mercaptans (hydrocarbons containing an
—SH group). Some 20% of the paper industry's investment in recently
built plants has been for pollution-control equipment. The industry is
highly capital-intensive (about $1.50 of invested capital is needed for
each $1.00 of annual sales), so costs have risen considerably, contributing
to recent price increases for books and magazines.

The rivers and breezes near paper plants are now cleaner, but there
are still problems to solve. The mercaptans (the same odorous compounds emitted by skunks) can be smelled at a concentration of 1 part
per 100 million. Present techniques that reclaim and recycle the pulping
chemicals have lowered sulfur emissions to the air from 25 kg to 2 kg
for each 1000 kg of paper produced. The aim is to get down below 0.05 kg
per 1000 kg. Because the annual production of paper in the United
States is over 50×10^9 kg, even such "ideal" plants would dump about
2 million kg of sulfur compounds into the air each year. Further improvements seem imminent through the use of such processes as hastening
the pulping with high-pressure oxygen, combining mechanical pulping
with chemical treatments, and forming the paper from fibers fluffed
with air rather than with water.

Figure 21.3 shows the chemical structure of cellulose. It is a *polymer*
of the sugar glucose. You can see that the Greek stems (*poly,* "many,"
and *meros,* "units") are well chosen. Many glucose units join together
to make cellulose. The linkage is an oxygen bridge, like the oxygen
bridge in the anesthetic ether; hence it is called an *ether linkage.* You
are unable to cleave (digest) this ether linkage if you eat cellulose. But
herbivores (animals that thrive on cellulose) either synthesize the
appropriate enzyme themselves or have in their digestive tracts bacteria
that can cleave the ether linkage and give glucose. Cellulose is the most
abundant, readily renewable carbon compound on earth. There would
be a very great increase in available food carbohydrate if we could find
some direct way to convert wood, straw, and other cellulose matter to
glucose for human consumption.

Glucose   Cellulose

The cellulose molecules form strong hydrogen bonds to one another and so build up the fibers such as those in this paper. Because there can be as many as five hydrogen bonds per glucose unit, cellulose is stiff, does not stretch, absorbs water, and swells and disintegrates into smaller molecular units if immersed in water long enough. Thus, pure cellulosic paper does not weather well. It is attacked by water, oxidized by air, and digested by bacteria, as you can see by the rapid disintegration of facial tissues, which are almost pure cellulose. On the other hand, multiwall bags used for shipping and storing are extremely resistant to weathering. They can be immersed in water or stored outside for more than a year without serious deterioration. The differences between these two products lie in the nature of the additives used to treat the basic cellulosic fiber mat.

If pure cellulose pulp is used, the fiber mat formed when the water is drained and evaporated away is bonded together only by the polar, van der Waals, and hydrogen bonds that can form between the fibers. Water easily breaks the fibers apart. Addition of white pigments like clay or titanium dioxide fills in the pores, produces a hard white surface for writing, and gives more binding area to hold the fibers together. Resins, water-resistant compounds, and special surface treatments (including surface laminates) provide packaging and covering materials ranging from ice-cream and milk cartons through multiwall sacks to roofing paper.

The cellulose provides the fundamental structure and most of the bulk and strength of paper. The additives (about 10% of the weight of the paper) provide the specific properties needed for special jobs. The chemical problem is to combine the additives and the cellulose to achieve optimum service at minimum cost.

About one-third of all cut timber now goes into paper. As costs rise, so does interest in recycling paper. Some of this interest is based on the increasingly severe waste-disposal problem (though paper can always be burned efficiently), some of it is based on concern for disappearing forests (though all the large paper companies run constant-growth forests that are really tree farms), and some of it is based on the general desire to reduce waste.

About 20% of paper products (such as glassine, parchment, grease-proof and roofing paper, and wallboard) are not readily recyclable because of their specialized additives. Over 20% of paper products are now recycled. This percentage is smaller than that recycled in the past, primarily because of the increased labor cost of collecting the paper for recycling as compared to the cost of obtaining materials for virgin pulp. Most towns in the United States used to have door-to-door junkmen who paid householders for their waste materials, including paper. But no more. The cost of reversing the second law's tendency of things to spread out increases each year. (See Table 21.1.)

EXERCISE 21.3
Recycled paper is frequently used for the central bag in multiwall bags, but not for the outer one. Suggest a reason.

EXERCISE 21.4
Write an equation for the complete nitration of one glucose unit in a cellulose polymer (Figure 21.3). How many nitrate groups result? Would this material be a smoky or smokeless explosive? How many nitro groups per glucose unit would just give a smokeless explosive?

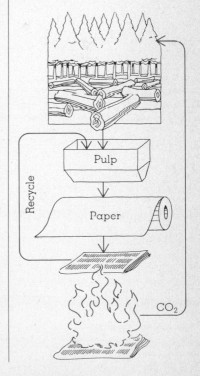

Recycle

Pulp

Paper

CO_2

Table 21.1
Decline (and partial recovery)
of Recycling in the United States:
Percentage of Production
Derived From Recycled
Materials[a]

Item	Year	Percentage
Paper	1960	26.2
	1969	17.8
	1975	19.8
Iron	1951	36.1
	1969	30.2
	1975	26.0
Aluminum	1951	7.7
	1969	4.2
	1975	23.4

NOTE: Data are from the National Association of Recycling Industries, Inc.

[a]Figures for the 1920s and 1930s would probably be higher.

The second law describes two more problems encountered in recycling paper. Some of the additives (such as printing inks) are strongly bonded and hard to remove, so they accumulate more in each cycle. The recycled paper gets grayer and grayer from the rising nonbleachable content. Furthermore, the cellulose fibers get smaller and less capable of forming strong hydrogen bonds, so the strength of the paper decreases in each cycle. In spite of these difficulties, there will be continuing and rising interest in recycling.

What we need are (1) an efficient and cheap procedure for collecting used paper and (2) initial additives that are readily removed in the recycling process. Recycling should be part of the initial design of the product. Table 21.2 gives some cost data on new pulp versus 25% recycled pulp for a plant producing 100 tons of packaging paper per day. Recycling may not save money, but it can minimize the load on resources and on waste-disposal systems. And trash, the "raw material" for recycling, is growing at 8% per year. The 1975 United States level was 17 billion cans, 38 billion bottles, 4 billion kg of plastic, 8 million television sets, and 35 billion kg of paper.

Modified Celluloses— If You Can't Beat Them, Join Them

Cotton, rayon, cellulose acetate, and *cellulose nitrate* are all produced on a large scale to make fibers and films. All these substances (even most cottons nowadays) are modified from their cellulose origin, though all retain much of the original structure, including many of the original ether linkages.

Some of the modifications are made by reacting the hydroxy groups on the glucose units with nitric acid. Water splits out and $-O-NO_2$ groups replace the OH groups. Heavily nitrated cellulose is a powerful explosive *(guncotton).* Less complete nitration produces *lacquers,* as well as the material that served as the base for early photographic film, including movie film.

Cellulose nitrate films (celluloid) are highly flammable. Life in the projection booths of early movie theaters often got more exciting than the events portrayed on the screen. The fact that the fumes are toxic (carbon monoxide and oxides of nitrogen) added to the hazards. No nitrate film is now used, and the old films are either stored with great care or are copied onto modern film base and the originals are destroyed.

Cellulose acetate was invented partly as a replacement for nitrate film. It is made by treating cellulose with acetic anhydride, $(CH_3CO)_2O$, or another acetylating agent that attacks the hydroxyl groups and leaves an $-O-CO-CH_3$ (acetate) group instead. Water is split out in the process. The degree of acetylation can vary (80–97% is the usual

Table 21.2
Comparative Costs for New Versus 25% Recycled Pulp ($/ton)

Item	New pulp	25% recycled pulp
Wood	38.00	28.50
Recycle	—	9.10
Chemicals	3.50	3.50
Processing	37.00	38.50
Storage	—	.59
Total	78.50	80.19

range), giving various properties to the cellulose acetate, used in making both fibers and films. *Cellophane* is a common term for such films.

Both films and fibers are made by extruding the hot molten material or a solution of it. Thus the fibers have the long cellulosic chains oriented in roughly parallel fashion along the fiber axis. The film similarly has the fibers somewhat oriented parallel to the extrusion direction. You can often detect this direction by seeing in which direction the film tears most readily.

Rayon is modified cellulose. It is cotton that has been dissolved in a solution of carbon disulfide (CS_2) and sodium hydroxide. The resulting viscous solution, which contains the long cellulose molecules in soluble ionic form, is extruded into a bath that reprecipitates the cellulose into long, continuous filaments. These may be used as *monofilament rayon,* or they may be chopped into short lengths (usually crimped to improve properties), spun into multifilament fibers and cords, and woven into fabrics.

The solution and casting processes allow easy modification of the separated molecules and the incorporation of other molecules as integral parts of the cast fiber. These additives can be dyes, molecules that are easy to dye (that is, that react chemically with specific types of dyes that would not react with cellulose), plasticizers (to make the fibers more flexible), stiffeners (to make the fibers less flexible), antioxidants (to increase resistance to air oxidation), flame retardants, and numerous other property-modifying and property-enhancing substances.

King (?) Cotton

Cotton, once it is deseeded and dewaxed, is almost pure cellulose. It has been used for spinning fibers and weaving cloth for at least 4000 years. But modern cotton is far different from the original fiber. The plants have been selected and hybridized to produce a range of fiber types, and the processing has made essentially new fibers compared to those of even 50 years ago. Yet the basic structure remains unchanged. Almost all the modifications involve the hydroxy groups and the presence of additives.

Shrinkage and *wrinkling* may not strike you as chemical problems, yet modern no-shrink and no-iron cottons (and other fabrics) owe these traits primarily to chemically added cross-links between adjacent fibers. Both shrinkage and wrinkling are due to slippage of the wet fibers (many of their interfiber hydrogen bonds having been destroyed by the presence of water). Cross-linkage prevents slippage. So no-shrink, no-iron fabrics result.

The chemical problem is made more complicated by the need to establish links that are tight enough to withstand laundering and

POINT TO PONDER
The cost of disposing of waste-paper is typically paid in the form of city taxes or bills for garbage collection; it is not included in the cost to the paper company of using virgin pulp. Therefore, the company may find it cheaper to use virgin pulp, whereas the society as a whole would find it cheaper to use recycled paper. Many of the current laws aimed at pollution control can be regarded as attempts to impose all the costs associated with a particular production process on the producers and consumers of that product. Would you choose less-expensive products if forced to pay a purchase price that included pollution control and other costs not formerly included in the cost of the goods?

POINT TO PONDER
Hilaire de Chardonnet (working as an assistant to Louis Pasteur) found a way to make fibers of partially nitrated cellulose. His "artificial silk" was introduced at the Paris Exposition of 1891. He called it *rayon* because it was very shiny. But, like celluloid, it was flammable. In fact, it was popularly known as "mother-in-law silk," since gift dresses in a time of open fires might diminish the number of unpleasant mothers-in-law. An English scientist, Joseph Swan (trying to find a way to make lasting filaments for electric light bulbs) later found ways to produce less flammable (but less strong) rayon.

EXERCISE 21.5
"No-iron" cotton usually requires touch-up ironing after a few launderings. Why?

POINT TO PONDER
Federal regulations require that all children's underwear in sizes 0 to 14 be made flame-retardant. The most effective compound for this treatment (2,3-dibromo propyl phosphate, or Tris) is now suspected of causing cancer. All risks must be considered.

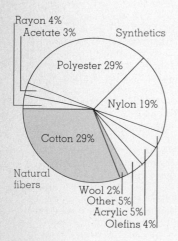

Figure 21.4
The U.S. fiber market in 1975: total 2.2 × 10¹⁰ kg.

bleaching. Yet they must be loose enough not to destroy the drape, "hand" (the technical term for the pleasant soft feeling of a good fabric), and water-permeability that make cotton desirable. The cross-links must also resist air oxidation, yellowing in ultraviolet light, and soiling. *Permanent press* goes one step further. It is accomplished by further cross-linking after the final shape of the garment, including creases and pleats, has been formed.

Consider the steps a good fiber must undergo: cropping, cleaning, carding, spinning, weaving, dyeing, cutting, sewing, pressing, wearing, laundering, bleaching, altering, and—finally—disposal. It must have good drape, hand, moisture-permeability (but water-resistance is sometimes desirable), and resistance to fire, mold, insect larvae (moths), rot, soil, abrasion, tearing, and a growing list of environmental chemicals. It must not be toxic and should not even cause allergic reactions. It also must not "pill" (the technical term for forming "fuzzballs"), and it must not build up static electricity. In fact, a good fabric must come close to being a completely insoluble, chemically inert substance. But it should be biodegradable when worn out! The net result is that fabrics is a highly competitive field, and great pressure is put on chemists to come up with improvements.

Cotton was for a long time the number-one fiber in terms of kilograms produced, but it is now almost equaled by a single synthetic (polyester) and will probably drop further in usage (Figure 21.4). Its per capita use in the United States has decreased greatly, from a high of about 15 kg in 1950 to less than 10 kg today. As with kings of nations, so here: "Uneasy lies the head that wears the crown."

Linen, from the flax plant, is a minor (specialty) cellulosic fiber. So also are jute from the hemp plant, coconut fiber, bagasse from sugarcane, and a number of others. Plant fibers in general owe most of their properties to cellulose. Most of these fibers are used with minimum chemical modification.

All (?) Wool

Wool is viewed by many as the ideal natural fiber. It has good moisture-permeability but holds up well to being wet. It wears better than most fabrics and provides great warmth per unit weight. (This is one of the main reasons wool is not used more in the centrally-heated, air-conditioned United States.)

Wool has a scaly surface, with the scales pointing toward the end of the fiber, away from the skin from which the wool grows. Thus wool shrinks badly when these fibers slip past one another, because the scales prevent reverse slippage. Once wool is well-matted by wear, however, both shrinkage and wrinkling are at a minimum. So wool, like cotton, is often much modified before it gets to the user. The proud advertising phrase "All Wool" gets stretched more each year.

Dozens of varieties of sheep have been bred to yield a wide choice of fibers. Goat, camel, llama, and vicuna hair are also used. Each is a complicated protein with roughly parallel chains of α-amino acids joined by nitrogen to carbon peptide linkages (Figure 21.5). The chains are held together mainly by hydrogen bonds, with occasional disulfide bonds giving further cross-bonding. It is the opening and closing of disulfide bonds that allows the curling and uncurling of human hair. The wave-setting lotion opens disulfide and hydrogen bonds, and the neutralizer closes new bonds in the curled position. It is also the opening and closing of hydrogen bonds that accounts for the ability of water to give a short-term curling or uncurling effect to hair.

Wool is probably on its way to becoming a specialty fiber. This shift is partly because synthetics and cotton can be more easily modified, but also because animal products are rapidly increasing in expense.

Nylon—First of the Wonder Fibers

One day in 1935 a group of chemists at the Du Pont laboratories watched chemist Wallace Carothers squeeze a heated fluid from a hypodermic syringe. The fluid solidified in the cool air, forming a clear fiber. Two years later it was on the market in toothbrushes; 4 years later, in women's stockings. And today, 40 years later, nylon is produced in several varieties with a total annual U.S. production of about 10^9 kg, an average of about 4 kg per person. This makes it the fourth most used fiber, after cotton, rayon, and polyester. (Polyester, though discovered only in 1943, passed nylon in amount produced by 1970.)

Nylon was the first fiber to result from a deliberate search (begun in 1928) into the relationships between molecular architecture and bulk properties with the intent of designing a fiber. The textile industry throughout the world has not been the same since. Most polymer discoveries are now based on a similar combination of pure research into molecular properties followed by design of the cheapest possible way to produce a salable material.

Nylon is a *polyamide* held together by the same linkages as found in proteins. It is tough, has all the appropriate "resistances" and a high tensile strength. And it is sheer! It differs from proteins in that a hydrocarbon chain of variable length intervenes between the linkages. Nylon 66, for example, is made up of two six-membered carbon intervals (Figure 21.6). It can be prepared from several raw materials—one of them oat hulls, as shown in Figure 21.7.

Nylon is a strong fiber but still has only 10 to 20% of the ultimate strength it would have if the molecules could be aligned in perfect

EXERCISE 21.6
Discuss the "warmth" of wool as influenced by its scaly surface.

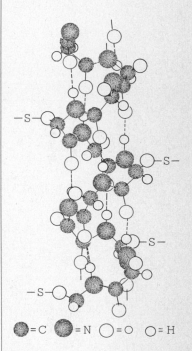

●=C ●=N ○=O ○=H

Figure 21.5
Schematic structure of wool. Hydrogen bonds are represented by dashed lines, interchain sulfide bridges by —S—

POINT TO PONDER
Here is a good place to note again that (though inventions are still made by the Edisonian approach of trying everything) guidance from a wide knowledge of science becomes more important each year. Similar scientific guidance on social and political problems also increases annually.

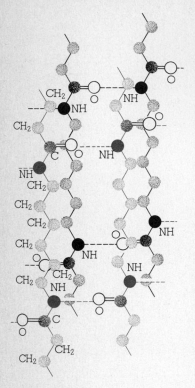

Figure 21.6
Structure of nylon 66. Hydrogen bonds are indicated by dashes. (Adapted from *Chemistry in Britain*, **11**, 1975, p. 174)

parallel order. An "ultimate strength" nylon fiber 1 cm² in cross section could support about 350,000 kg.

Other chemical modifications involved in producing polymers have been in molecular weight and chain-terminating groups. Fiber cross-section (both size and shape), relative amounts of crystalline and non-crystalline regions in the fiber, and degree of nonlinearity and crimp in the finished fibers are also varied. It should be clear that, with so many variables to work with, a wide variety of properties can be obtained with fibers of great similarity in chemical composition. (Again you should remember Chapter 7 and the value of interchangeable parts.)

And So We Get Poly-Polymers

After 1937, a new commercial polymer entered the market every second or third year, though the pace has slowed since 1960. The slower pace is due partly to the intensive research in that 20 year period, and partly to the massive capital outlay required to put a new polymer into production. New polymers are discovered frequently. Usually their properties are not sufficient improvements nor their costs sufficiently low in comparison with the established group to justify large-scale commercial development. The ideal fabric is not with us yet, but the present ones

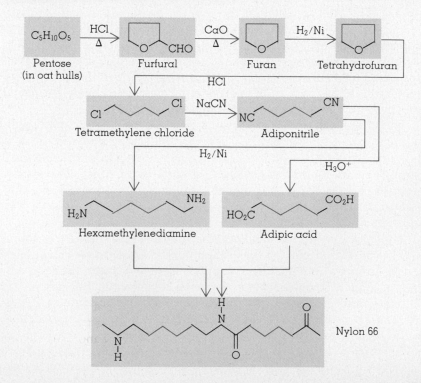

Figure 21.7
Synthesis of nylon 66.

Table 21.3   Some Polymers Produced in the United States

Polymer generic name	Year of U.S. commercialization	Monomer or polymer source	Annual U.S. production[a] (kg/person)			
			1870	1920	1970	1976
Olefin	1961	Ethylene, propylene	—	—	18	158
Vinyl[b]	1927		—	—	16	47
Polyester	1953	Dihydric alcohol, terephthalic acid	—	—	3.2	33
Rubber	1930	Butadiene, isoprene vinyls	—	—	9.1	22
Nylon	1939	Polyamide	—	—	3.2	21
Cotton	Brought by colonists	Glucose	6.5	15	9.1	14
Phenolic	1919	Phenol, formaldehyde	—	—	2.3	13.4
Acrylic	1950	>85% acrylonitrile	—	—	1.4	6.2
Rayon	1910	Glucose	—	0.1	4.1	5.5
Acetate	1924	Cellulose acetate	—	—	2.3	3.0
Wool	Brought by colonists	α-amino acid	2.5	1.9	0.5	0.4
		Total[c]	9	18	70	345

[a]Although the population of the United States makes up only 6% of the world population, U.S.
  production of the polymers listed ranges from 30 to 70% of world production.
[b]Includes vinyl chloride (1927), styrene (1930), acrylic (1950), Saran (vinylidene chloride; 1941),
  Modacrylic (35–85% acrylonitrile; 1949), methacrylate (1936), and Anidex (acrylic ester; 1969).
[c]Includes other synthetic polymers: fiberglass (SiO_2, CaO, Na_2O; 1936), Vinyon (>85% vinyl
  chloride; 1939), metallic fibers (metal, coated metal, metallic coated; 1946), and Spandex
  (urethane; 1959).

are very good indeed. Current betting is that the source of the biggest advances in the future will be modification of present fibers rather than completely new molecules.

Table 21.3 lists some current polymers: generic name, monomer or polymer source, date of commercial introduction in the United States, and quantities produced from 1870 to 1975. The vinyls are such a large family that Table 21.4 is added to outline some currently on the market.

The fastest growing polymer is *polyester,* made by copolymerization of terephthalic acid ($HOOCC_6H_4COOH$) with a dihydric alcohol, $HO—(CH_2)_n—OH$, to produce a linear polyester. For *Dacron,* the alcohol is $HOCH_2CH_2OH$, ethylene glycol.

From Monomer to Polymer

The word *polymer* means "many parts." If all the parts are the same (as are the glucose units in cellulose), a *simple polymer* results. If the parts are different (as in nylon) a *copolymer* results. One of the reasons for the tremendous variation in properties of polymers is the indefinitely large number of possible copolymers. The proteins found in living systems are an excellent example, and we shall study them in a later chapter. But all the thousands, probably millions, of natural proteins—each

EXERCISE 21.7
Nylon monofilament, after it comes from the extrusion process, can easily be stretched to about twice its extruded length. It will not contract when the tension is released nor can it be stretched appreciably further without breaking. Why?

POINT TO PONDER
Two of the main pressures to develop vinyl chloride polymers came from naval disasters. Recovery of a sunken submarine showed the vinyl electrical insulation intact, but the rubber had disintegrated. The other disaster was the crippling of the German battleship *Graf Spee* when burning rubber insulation destroyed the ship's electrical controls.

Table 21.4 Vinyl Polymers of the type n

$$\begin{matrix} A \\ \\ B \end{matrix}\!\!\diagdown\!\!C\!\!=\!\!C\!\!\diagup\!\!\begin{matrix} E \\ \\ D \end{matrix} = \left(\!=\!\!\underset{B}{\overset{A}{C}}\!-\!\underset{D}{\overset{E}{C}}\!-\!\underset{B}{\overset{A}{C}}\!-\!\underset{D}{\overset{E}{C}}\!-\!\underset{B}{\overset{A}{C}}\!-\!\underset{D}{\overset{E}{C}}\!-\! \right)_{n/3}$$

Monomer		A	B	D	E	Trade name	Used in
Common name	IUC name[a]						
Ethylene	Ethene	H	H	H	H	Polyethylene	Films, coatings, containers
Vinyl chloride	Chloroethene	H	H	H	Cl	Geon, Tygon	Lab tubing, insulation
Vinylidene chloride	1,1-Dichloroethene	H	H	Cl	Cl	Saran	Films, wrappings, tubing
Chlorotrifluoroethylene	Chlorotrifluoroethene	F	F	F	Cl	Kel-F	Oils, greases, insulation
Tetrafluoroethylene	Tetrafluoroethene	F	F	F	F	Teflon	Lab ware, seals, bearings
Propylene	Propene	H	H	H	CH_3	Nalgene, polypropylene	Lab ware, household objects
Methyl vinyl ether	Methyl ethenyl ether	H	H	H	CH_3O	Poly (vinyl methyl ether)	Adhesives
Styrene	Ethenylbenzene	H	H	H	C_6H_5	Polystyrene	Insulating foams, molded objects
Vinyl acetate	Ethenyl acetate	H	H	H	CH_3CO_2	Polyvinyl acetate	Adhesives
Methyl methacrylate	Methyl 2-methylpropenoate	H	H	CH_3	CH_3O_2C	Lucite, Plexiglas	Glass substitutes, moldings
Acrylonitrile	Propenonitrile	H	H	H	NC	Orlon	Wool substitutes

[a]Systematic International Union of Chemistry name.

with a limited number of highly specialized functions—are made by copolymerizing 20 different α-amino acids into chains of various lengths.

Polymerization itself usually occurs in one of two ways. You have seen one of them in the polymerization of glucose and of nylon. This is called *condensation polymerization* and involves the splitting out of a small molecule (H_2O, CO_2, NH_3) between two units, which then join to extend the polymer. The other method, characteristic of vinyl and most rubber polymerizations, is called *addition polymerization*. It involves the cleaving of a bond (either a cyclic bond or a double bond)

EXERCISE 21.8
Show that addition of 1 mole of H_2O per glucose unit is required to convert cellulose to glucose. (See Figure 21.3.)

and the subsequent addition of the fragment to the growing polymer. Thermodynamics helps interpret the tendency to polymerize: Splitting out small molecules gives a positive ΔS, and converting a double bond to two single bonds gives a negative ΔH. What about the rate of reaction?

Good (that means "well-behaved") monomers do not polymerize on mixing—otherwise how could you store the monomer? Thus polymerizations are processes that normally have high activation energies and strict orientation requirements that are bypassed by a catalyst (or initiator) to provide a more rapid mechanism when polymerization is desired.

Catalysts may be either acids or bases that react with the monomer and accept or donate a pair of electrons, thus activating the monomer to act in turn as an acid or base toward another monomer. The process is repeated, with the chain lengthening by one unit per time, until the desired length is reached or until all the monomer is used up.

Redox catalysts react with the monomer by donating or accepting single electrons, creating a molecule with an uneven number of electrons. These molecules are called *free radicals*. The unpaired electron is, of course, in a half-filled orbital that will interact with electrons in neighboring monomer units to form an electron-pair bond and will leave another unpaired electron at the growing end of the polymer chain.

The amount of catalyst determines the average chain length by fixing the number of polymers that are growing at any time. Many polymerization "catalysts" are not true catalysts at all. They are consumed in the process and become end groups on the growing polymer chain. They should be, and often are, called *initiators*. Remember that a true catalyst must emerge regenerated at the end of the reaction.

You can see that for a polymer to grow, there must be at least two reactive sites per monomer molecule. The monomer must be difunctional. One site adds to the growing polymer, the other serves as the point of addition for the next monomer. Such linear polymers are much like a string of beads. They are semicrystalline, and they soften on heating because the chains are not cross-linked. They are called *thermoplastic* polymers.

If there are more than two reactive centers per monomer, growth can occur in three dimensions. Nonlinear polymers form. They will be more rigid and less extensible because of the cross-linking between what would otherwise be linear chains. Three-dimensional plastics do not become readily extensible when heated. They are said to be *thermoset* polymers (Figure 21.8). They are more difficult to recycle compared to the thermoplastics.

The degree of cross-linking can often be varied after a linear polymer has formed, especially if the polymer still contains double bonds. *Vulcanization* of natural and synthetic rubbers *creates cross-links* between adjacent chains by attacking the residual double bonds with sulfur at

Figure 21.8
Thermoplastic and thermoset
polymers.

Thermoplastic; few cross-links   Thermoset; highly cross-linked

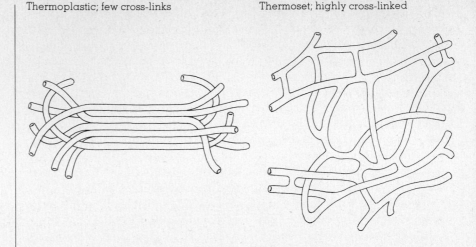

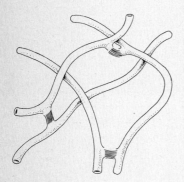

Figure 21.9
Vulcanization. Disulfide
(—S—S—) links are colored.

POINT TO PONDER
Industrial patents, on the
average, require 3 to 4 man-
years of effort in development.
Major breakthroughs may follow planned research requiring
10 times that amount of work—
or they may result from an
accident whose causes are
investigated.

elevated temperatures. Sulfur will form bridges two atoms long by
adding to the double bonds (Figure 21.9). Hard rubbers, as in combs
and battery cases, are heavily vulcanized so that most of the double
bonds are used up. Rubber bands and highly extensible rubber objects
have only a small percentage of the double bonds cross-linked. They
retain the possibility of the chains sliding past one another until they
are brought up sharply by the interchain sulfur bonds. Note the sudden
change in ease of extension if you stretch a rubber band.

Polymerization is easy to understand in terms of linking monomers
together. But it is often very difficult to find a suitable catalyst and set
of conditions. Many have been discovered by accident.

The Ziegler catalyst used to produce high-density polyethylene of
very high molecular weight resulted from cooperation between a metallic impurity in the reaction vessel and the aluminum alkyl that was
the intended catalyst. The discoverer, Karl Ziegler, got a Nobel Prize,
and, I must add in fairness, did a great deal of other first-class work.
(Do you recall the Alexander Fleming story?) Teflon was another acci-
dental discovery—it resulted from work on refrigerator coolants. But
still, "chance favors the prepared mind."

The Tendency to Tangle

Polymerization is very similar to stringing beads, or to elephants forming
their head-to-tail circus parade—at least if there are only two reactive
positions per monomer. But the bonds formed are usually tetrahedral
in angle. The randomness of the collisions and the thermal twitching of
the growing polymer both tend to give a twisted molecule rather than
a stretched-out one.

The second law of thermodynamics is again useful in describing the situation. There is only one way for the molecule to be stretched out straight, but there are many ways for it to twist. The entropy of a twisted set of molecules is higher than that of straight ones; twisting is more likely. Furthermore, there is usually little energy preference for one form over the other. The polymer can form equally strong bonds with other sections of itself as with neighboring molecules, so the ΔH of stretching is close to zero. With ΔH close to zero and ΔS positive for the twisted form, the value for $\Delta G = \Delta H - T\Delta S$ will be negative for forming the twisted from the stretched form. The twisted form will be more common.

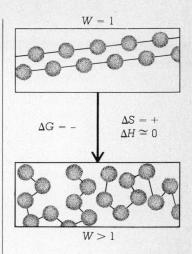

The molecules will not only twist, they will intertwine and tangle. A good analogy for a polymer is a plate of spaghetti or noodles. The extent of polymerization determines the length of the individual noodle and the degree of tangling. The temperature and the extent of three-dimensional cross-linking determine the resistance of the mass to deformation and elongation. The strength of any one noodle to compression or extension is small, but the overall strength from the many noodles present can be great.

If the polymer is dissolved in a solvent that is "like" it, the polymer will tend to twist its way throughout the surrounding solvent. The ΔH_{soln} effect is small. If the solvent is not "like" the polymer, ΔH_{soln} is positive and the polymer will tend to form a compact ball, binding to itself rather than penetrating the solvent.

Silicone oils provide an interesting example of the tendency to tangle. Their chains contain alternating oxygen and silicon with two hydrocarbon groups on each silicon. The chains tend to ball up at low temperatures, because they can form better polar bonds with themselves than with neighboring chains. As the temperature rises, the increasing entropy possible if the chains are somewhat extended becomes important. The chains extend more and more with rising temperature and tend to tangle with one another. The increase in tangling prevents the usual decrease in viscosity with rise in temperature. As a result, the viscosity (resistance to flow) stays almost constant over a wide temperature range. Therefore the silicones are excellent special-duty oils for cold-engine starts.

Designing Molecules

All polymers are macromolecular. It is easy to see why the linear polymers are less rigid than the three-dimensional ones, but why is there so much variation among the linear ones? Why is rubber so different from polyethylene when both are long-chain hydrocarbons?

The simplest measurement that characterizes polymer properties is the melting point. For those substances that cannot be crystallized, it

the chinese solution
to eating noodles....
depolymerize.

POINT TO PONDER
Stretch a rubber band and hold it to your forehead to estimate its temperature change. Take it away from your forehead, let it contract, then return it to your forehead,. What has happened to the temperature? Which process, extension or contraction, is exothermic? Would a somewhat-stretched rubber band get longer or shorter if heated? (Recall Le Châtelier's principle.) Explain your results in terms of entropy changes in the rubber.

is the *glass point* (the temperature at which they harden on cooling or soften on heating). Above these temperatures the polymers are rubbery and extensible, with stretch and recovery. Below these temperatures they will bend and deform slightly, but they will not stretch and recover. Plastics are polymers with melting points above room temperature. Rubbers are polymers with melting points below room temperature. It was natural rubber's stiffness when cold and stickiness when hot that made disasters out of efforts to use it prior to vulcanization (in 1840).

You may recall that, at the melting point, the tendency to convert the ordered solid to disordered liquid is opposed by the tendency to form stronger bonds in the solid—the stickiness of the molecules toward one another. In polymers it is the differences in the strength of the inter-chain forces that primarily cause variation in melting points, and hence cause variation in polymer properties. High intermolecular forces give high melting points and rigid plastics. Low intermolecular forces give rubbery materials. Vulcanization converts rubbers into plastics by increasing the interchain forces.

Polyethylene is a linear polymer in the crystalline state, as shown in Figure 21.10 (see also Figure 21.12). Only van der Waals forces between the CH_2 groups hold the crystal together. If compact and regularly spaced side groups are added, the melting point rises because of greater

Figure 21.10
Structure of polyethylene, $(CH_2-CH_2)_n$. (Adapted from *Chemistry in Britain*, 11, 1975, p. 172)

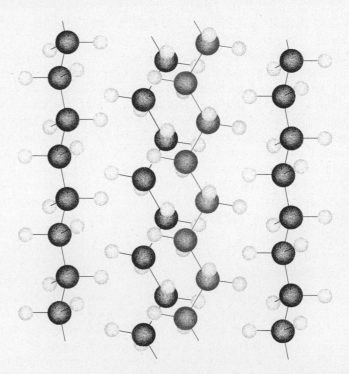

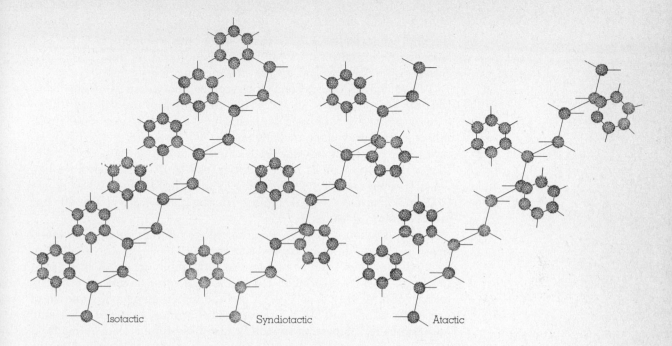

Isotactic   Syndiotactic   Atactic

surface area for interaction. If the side groups get long and flexible, packing becomes difficult, contacts decrease, and the melting point goes down. The presence of polar groups, especially hydrogen-bonding groups that are close together, raises the melting point.

Polymerized vinyls have a side chain on each monomer unit in the chain. If these side chains are all positioned on the same side of the chain, the polymer is *isotactic*. If they alternate regularly from side to side, the polymer is *syndiotactic*. If they are randomly positioned, it is *atactic*. (See Figure 21.11.) The isotactic polymer has the highest melting point; the syndiotactic polymer, an intermediate melting point; and the atactic polymer, the lowest melting point. Until the discovery of stereospecific catalysts, which are capable of producing a purely isotactic polymer, vinyl polymers were tacky, rubbery, atactic polymers of little use.

Teflon (polymerized tetrafluoroethylene) is an interesting exception to the generalization that simple linear chains have low melting points. Teflon is the basis of the nonstick surfaces in cooking utensils and of a water-repellent material used to coat fabrics. It has a uniformly negative molecular surface because of the high electronegativity of fluorine. So, we would not expect it to have strong interchain forces—and it doesn't. But it does have a very rigid chain. The fluorine atoms are just the right size to force the chain into a slowly spiraling linear structure.

Figure 21.11
Isotactic, syndiotactic, and
atactic vinyl polymers.

Flexing is difficult. (See Figure 21.12.) So the chains have a minimum of thermal randomness in the crystal. They tend to remain well packed. Even in the liquid or glassy state, the chains will be rigid. Hence the entropy of melting will be low. So the melting point is high.

I have shown illustrations of crystallized polymers to demonstrate their packing problems. In actual plastics, much of the volume will be noncrystalline—twisted and tangled. But polymers that can pack well in the crystal (have high melting points) will also tend to be stiff, non-crystalline (amorphous) materials. Furthermore, there are relatively large crystalline regions in all plastics, especially stretched ones like fibers. (See Figure 21.8.) So most plastics must have "plasticizers" added to them to increase their flexibility. It's the plasticizers you smell that give the "new car" odor.

I have spoken of copolymers that have different monomers alternated in the chain. It is also possible to make blocks of polymers from a single monomer, then to mix two sets of such blocks and graft them together. This gives chains, called *block grafts,* having alternating lengths of different properties. Or it could give chains having one set of properties at one end and a different set of properties at the other end. Detergents for dispersing mutually insoluble liquids can be made in

Figure 21.12
Structure of Teflon, $(CF_2CF_2)_n$, compared to polyethylene, $(CH_2CH_2)_n$. Note crowded fluorines in Teflon leading to stiffness in the polymer. (Adapted from *Chemistry in Britain*, 11, 1975, p. 176)

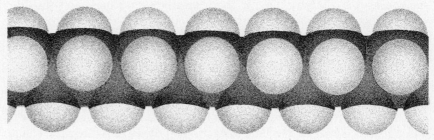

Polyethylene $(CH_2CH_2)_n$

Teflon $(CF_2CF_2)_n$

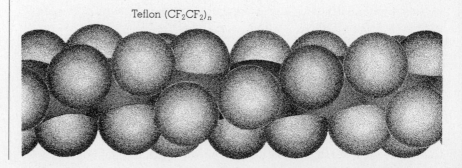

this way. One end dissolves in one liquid, the other end dissolves in the other liquid, creating a bridge just as tensides do in water systems.

The possibilities of designing various sets of useful properties into single molecules are almost endless. Polymer chemists will be busy for a long time to come.

Tires

There are few modern products that come closer to meeting their design criteria than vehicular tires. Tires are an excellent wedding of the skills of mathematicians with toruses, physicists with flexible shells full of fluid, chemists with macromolecules, and engineers with complex problems of assembly. In fact, one of the biggest problems is what to do with an annual supply of some 250 million used but still sturdy tire carcasses. Most of them, maltreated though they were, are only deficient in tread from the cornering, starting, and stopping. On the average rotation, a tire loses less than one atomic layer.

A 1927 set of four tires cost about $7 per 1000 miles, and flats and/or blowouts were almost a certainty. A modern set costs about $5 per 1000 miles, and flats or blowouts are unlikely. If you use an inflation factor for the currency, the cost has dropped to about one-third of what it was 50 years ago. The world now rolls on low-cost rubber, much of it synthetic.

Even vulcanized rubber can be tacky. So tires contain carbon black and silica, which form tight bonds to the rubber, reduce tack, and increase wear. Other additives allow smoother blending, increase resistance to air oxidation, minimize the leakage of air, and adjust road friction. A major advance (1951) was the blending of oil with the rubber. Most car tires are now made of *oil-extended rubber*. They are cheaper, and they give better friction and a softer ride. Tires also contain reinforcing cords made of rayon, nylon, polyester, glass, or even steel. See Table 21.5.

It may be that the next big innovation in tires will be a cast tire that does not require reinforcing cords. This would also minimize the recycling problem. Such tire carcasses could more readily be pyrolized to carbon black or ground to particles for incorporation in road pavement—two likely solutions to the disposal problem of the future.

Rubber: Natural and Synthetic

Natural rubber is a polymer of *isoprene,* $CH_2=C(CH_3)CH=CH_2$. But it is a special polymer of isoprene: head to tail, all *cis,* and with 1,4 addition. Figure 21.13 shows four possible isoprene polymers in their most *extended forms.* A fifth possibility is a random sequence of head–tail,

EXERCISE 21.9
Check, by rough calculation (to one significant figure), the statement that a tire loses less than one atomic layer per average rotation.

**Table 21.5
Typical Contents of
a Rubber Tire**

Ingredient	Percentage
Rubber	45
Sulfur	1
Carbon, silica	23
Oil	11
Chemical additives	9
Cord	7
Bead	4

Head to tail,
all *cis*

Head to head,
all *cis*

Head to tail,
all *trans*

Head to head,
all *trans*

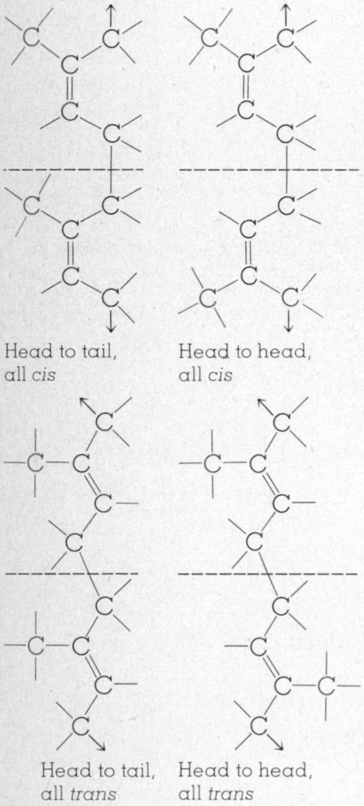

Figure 21.13
Some isoprene isomers. The
"head" CH₃ group is colored
for easy identification.

EXERCISE 21.10
Draw a head-to-tail, all-*cis*
representation of chloroprene.

cis–trans arrangements. The illustration shows the side methyl group as the tail of the monomer. *Cis* means that the two bonds continuing the chain are on the *same side* of the double bond. *Trans* means they are on *opposite sides* of the double bond. And *1,4 addition* means addition to the first and fourth carbons of the isoprene monomer. It is also possible to get 1,2 addition (like a vinyl) across only one of the end double bonds, so the other one ends up in a side group. A 1,2 addition leads to isotactic, syndiotactic, and atactic possibilities.

The head-to-tail, all-*trans* form of polymerized isoprene is also found in nature. It is called *gutta percha*. Unlike rubber, it is hard and plastic.

The C_5H_8 units in both rubber and gutta percha are flat. All five carbon atoms are in the same plane because of the 120° bond angles (remember the Nyholm–Gillespie rule in Chapter 8) and their inability to rotate about a double bond. The large zigzags of the all-*cis* form do not pack into a crystal structure as well as the smaller zigzags of the all-*trans* form. So the all-*cis* form has weaker interchain forces, lower ΔH_{fusion}, a lower melting point, and rubberlike qualities at room temperature. A major chemical triumph of the mid-1950s was the discovery of catalysts that could produce these stereorubbers—that is, polymers with a regularly repeating sequence.

The residual double bonds in rubber molecules give extra flexibility to the chains (compared, say, to polyethylene). They also allow attack by sulfur to give vulcanization. The double bonds still present after vulcanization are reactive to oxygen, especially in the form of ozone. It is this reaction that leads to the cracks and checks found in old rubber objects. In fact the rate of appearance of these cracks is one of the standard tests for the concentration of oxidizing smog.

Neoprene, the first good synthetic rubber, is polymerized chloroprene (isoprene with a chlorine in place of the side methyl group). The —Cl and —CH₃ side groups are about the same size, so the polymeric properties are similar—but chloroprene rubbers are more resistant to hydrocarbon solvents.

During World War II two copolymers of *butadiene* (CH₂=CHCH=CH₂) were developed—one with *styrene,* the other with *acrylonitrile.* The structures of these two vinyl compounds are shown in Table 21.4. The vinyls provided the side groups. The butadiene provided the residual double bonds for vulcanization. At about the same time, copolymers of isoprene and another vinyl, 2-methyl butene, CH₃CH₂(CH₃)C=CH₂, led to *butyl rubber,* especially useful in inner tubes and linings for tubeless tires because of its low permeability to air.

Since then the number of rubbery copolymers has increased. But, so far, most are vinyls plus isoprene or butadiene. Each has special properties for special uses. Among the current hopes for the future are the urethanes, based on the —NCO (isocyanate) sequence. Many difunctional isocyanates have been polymerized with a wide range of properties and uses—from foam mattresses to cast tires. The number and amount in the marketplace will grow.

Fixed Reactants

A common problem in chemistry is how to separate the products of a reaction. Polymers give an interesting solution, and the *home water softener* is an example. Hard water, containing Ca^{2+}, is passed over an ionic polymer covered with Na^+. The $2+$ ions adhere to the polymer, releasing harmless Na^+ into the solution.

Laboratory ion exchangers are often a mixture of two types, one covered with H^+; the other, with OH^-. When the positive ions are adsorbed by the polymer, they release protons; similarly, when negative ions are adsorbed, they release hydroxide ions. The protons and hydroxide ions then react to give water—very pure water, in fact better than most distilled water.

In both cases, the undesired products are still adsorbed on the solid polymer and can be either thrown away or driven off into a smaller amount of water by reversing the equilibrium. The separation is quick, easy, and not too expensive. Similar techniques are commonly used to separate chemicals similar to one another. The desired chemical is adsorbed, the undesired ones pass on, and the desired chemical can then be reclaimed. (See Figure 21.14.)

Many catalysts are attached to inert polymers over which the reactants flow. The catalysts stay in place and are easy to reclaim and renew when the need arises. Enzymes can be firmly attached to tiny, porous polymer beads and allowed to catalyze the desired reaction. They can then be separated from the products by a simple filtration.

An interesting recent innovation is for the sweet-toothed person who cannot eat sugar. A sweet molecule is tightly bonded to an indigestible polymer. The tongue tastes it as sweet, but the polymer and attached sweetener pass unchanged through the digestive system. They give sweetness to the food and no calories at all to the pleased dieter!

Summary

Most of the substances you come in contact with are macromolecular. Practically all are synthetic or at least appreciably modified from their natural states. Many are polymers made by condensation or addition of small multifunctional molecules into long strings or three-dimensional arrays. The number of possible monomers is large, as are the possible

EXERCISE 21.11
Why does Ca^{2+} displace Na^+ in an ion exchanger?

Figure 21.14
A fixed-bed reactor. A dilute solution of material is poured into the jacketed tube. The solid in the central tube adsorbs the desired material (and sometimes catalyzes a desired reaction), and the liquid is discharged. The adsorbed material can then be removed with another liquid.

Figure 21.15
Production of some macro-
molecular substances.

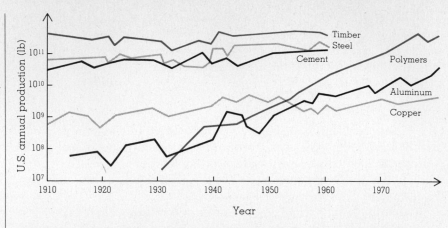

21.1 Long-haired sheep probably existed before long-
stranded cotton. 21.2 20 kg
S/hour for each plant.
21.3 Insufficient strength.
21.4 Three ONO_2 groups per
unit; $2\frac{1}{2}$ would need to be
nitrated, on the average.
21.5 Early washing removes the
least tightly bound of the
permanent press molecules.
21.6 Air (a good insulator) is
trapped. 21.7 Pulling tends to
straighten out the molecules.
21.8 Oxygen bridges break and
add water. 21.9 3×10^{-7} cm
lost per rotation, about three
molecular layers. 21.10 See
Figure 21.13. 21.11 Larger
charge, smaller size of Ca^{2+}.

combinations of chain length, degree of cross-linking, end groups, additives, and processing techniques. In a sense, the United States is a
copolymer with 50 units! *E pluribus unum.*

It is now possible to design and create macromolecules for most
applications. The biggest unsolved problem is how to prepare macromolecules that remain stable at high temperatures. Some other goals
are improved laminated plastic films, simulated wood-graining, plastic
beer and soft-drink bottles, injection molding of thermosets, plastic
coatings on metals, nonwoven fabrics, nonsewn clothing, fabrics that
can change color, truly no-care fabrics, readily recyclable polymers,
and greater flame resistance.

A further problem is to convince users that although natural fabrics
and polymers can be thrown into the weather to rot, the new polymers
cannot. Different waste-disposal methods must be developed, including
deliberately minimizing the amount of waste produced. The growing
magnitude of the problem can be seen in Figure 21.15. Remember that
everything made gets thrown "away" or lost sometime.

Problems

21.1 Identify or define: block grafts, cellulose, *cis,*
copolymer, initiator, monomer, polyester, polymer, shrinkage, thermoplastic, thermoset,
trans, vulcanization.

21.2 If you visit the dye house of a modern textile
mill, you are quite apt to see white, visually
unpatterned cloth enter a single dye bath and
emerge with a three-color pattern on it. If you
wait, you may see the dye vat changed and the
same cloth dyed with the same pattern but in
three other colors. How is this possible?

21.3 Retailers anticipate the day when you can select a colorless shirt for its design and then return the next day to find it finished in the
three-color pattern you selected from the
dozen or so available. How could such a shirt be
constructed at the factory and processed by the
retailer?

*21.4 Most substances become more fluid when
heated, but eggs solidify. Interpret the effects
at the molecular level.

21.5 What sources may be available for plastics, medicine, and liquid fuels when petroleum is exhausted?

21.6 Average annual yields from plantations of rubber trees are 0.1 kg/m^2. Treatment of the trees with a slow release of ethene, C_2H_4 (from 2-chloroethene phosphonic acid), doubles the yield. Some yields go to 0.5 kg/m^2 or higher. Could rubber latex become a source of "oil"? In 1977 rubber sold for about \$0.60/kg. [Oil costs about \$0.07/kg]

21.7 Metals and plastics are both macromolecular. Why are plastics replacing metals in so many uses?

21.8 Polymerizations are very sensitive to impurities, which must be held below 1 part in 10,000 in most cases. Why?

21.9 A linear polymer that does not have a regularity in the arrangement of units in the chains has a lower melting point than the same polymer when it is regularly arranged. Why? Early polypropylene was useless, for example, because it was irregular and tacky.

*21.10 When a formed piece of thermoplastic—say, a lucite chair—is heated, it will return to its original sheet or block form. Discuss the interpretation of this "plastic memory" at the molecular level.

*21.11 Measure the ratio of the relaxed length to the fully stretched length of a rubber band. Account for this in terms of changes at the molecular level. About what fraction of the double bonds are vulcanized: 0.8, 0.5, 0.2, 0.1, 0.05, or 0.01? [less than 0.1 vulcanized]

21.12 Why is it desirable to sprinkle clothes or to use a steam iron when ironing?

21.13 What type of treatment is used to give wet-strength to paper? Why does a simple paper mat have so little wet-strength?

21.14 Is polyethylene a thermoplastic or a thermoset? How about a polymer based on glycerin?

21.15 Thermosets are much more difficult to recycle than thermoplastics. Why?

21.16 You have a chance to invest in either a plastics general-recycling plant or a plastics pyrolysis venture. Which would you choose and why?

21.17 Polystyrene was discovered in 1839, but nothing was done with it until about 100 years later, after Nobel Prize winner Hermann Staudinger's work. Why then?

21.18 Extruded Saran serves as a shrink-film when heated. Why does it shrink?

21.19 Which seems more likely to you—that polyethylene and polypropylene are dyed by forming covalent bonds with a colored molecule, or that they are dyed by incorporation of the molecule into the cast plastic or fiber?

*21.20 Phenol,

can react with formaldehyde, $H_2C=O$, to form as many as three CH_2 bridges per phenol molecule to other phenols. The phenol reacts at positions 2, 4, and 6 in the model shown. This plastic, called Bakelite, has a wider range of properties than any other plastic. What properties would you expect if copolymerization involves 1.5 moles H_2CO per mole of C_6H_5OH? Write an equation.

21.21 Epoxies are formed mainly from bisphenol A,

and epichlorohydrin,

Write an equation for a probable copolymerization. Discuss why epoxies make such good glues.

21.22 Why is Saran Wrap so self-sticky, when polyethylene is not?

21.23 One "self-cleaning" surface treatment for fabrics is a block-graft polymer of sections of Teflon with sections of polyethylene oxide, $(CH_2—CH_2—O)_n$. Normally the Teflon is on the outside, the polyethylene oxide on the fabric. The presence of a cleaning agent is said to reverse this, turning the polymer inside out and freeing the dirt. The polymer is said to reduce initial soiling also. Are these reasonable claims?

22 SOME ESSENTIALS FOR LIFE

Some years ago I took one of the greatest trips in the world—the descent of the Grand Canyon by boat. For 10 days and 300 miles, I carried what I needed through some of the most magnificent country the world has to offer. I approximated a small isolated system. My personal belongings were limited to 40 pounds of essentials and there was no problem in keeping within the limit, even though I brought four cameras and related gear. How simple life was—or at least seemed to be.

I say life *seemed* to be simple because it was easy to forget, in that semi-isolated state, the very large support systems that had provided the "essentials" I brought. The cameras, clothes, sleeping bag and tent, the large rubber raft itself—all were products of a complex industrial society. The raft was left over from World War II, where it would have served as a pontoon for a military bridge. The food was the product of a highly mechanized agricultural system. The "essentials" I had with me were a tiny fraction of the essentials needed to produce them and to maintain me. And I haven't mentioned the helicopter that flew in to pick up the body of one of the boatmen, who drowned when a raft overturned in Upset Rapids.

Consider for a moment your own essential requirements. They will make a rather short list. But then think of that set of support systems and consider how many of your items are so essential as to justify the existence of the means used to produce them. Does your list shorten?

In this chapter, I shall discuss some chemicals truly essential to human life. All are found in nature; most can also be synthesized in the laboratory or factory. But the only necessary support systems are the natural cycles driven by solar energy. I shall concentrate on chemicals that are inputs or foods to living systems rather than on those synthesized by the living systems for their own use. The list will not be exhaustive, but without these substances human beings could not survive in a normal fashion. Studying this truly essential list may help you evaluate the larger list of "essentials" most people cherish.

$C_{105}H_{609}O_{256}N_{25}Ca_2P_2S$ and You

You may recall that Table 1.1 lists the composition of a human in terms of chemical elements. This composition approximates the formula $C_{105}H_{609}O_{256}N_{25}Ca_2P_2S$ plus some ionic and trace elements. But I now

> The [living] cell finds a way to make the reaction go at just the right moment and at the desired rate. The reaction may be improbable, but if it is thermodynamically possible, the cell will find a way to use it.
> —*Albert Szent-Györgyi.*

Figure 22.1
Some inputs, rhythms, and
outputs that affect you.

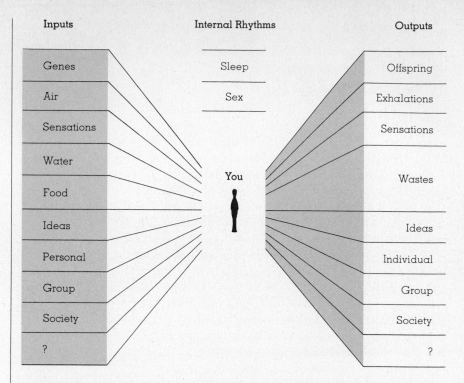

want to become much more specific and to discuss particular molecules and some of their roles. The organizational scheme in Figure 22.1 shows some of the inputs, internal rhythms, and outputs you experience. The inputs listed are ordered according to how long you could survive without them: genes the shortest time, food a longer time.

A second possible organizational scheme is shown in Figure 22.2, which gives the elements thought to be essential to life in terms of their positions in the periodic table. About 20 chemical elements are known to be essential to the direct operation of the human body as shown in Table 1.1, though others from the list may be needed by you. But all living systems in the world are interrelated, so at least in this sense all the elements marked in Figure 22.2 are essential to you.

A third possible organizational scheme includes the elements listed in Table 1.1 plus the molecules known to be essential to humans. There are fewer than 30 molecules known to be essential to humans—14 vitamins, 10 α-amino acids, and 2 fatty acids. Table 1.2 lists most of these. More may well be found, but for the moment we think the body can synthesize the other substances it needs from these molecules and suitable sources of the essential elements. All these substances, of course, are found in a well-balanced diet. In fact, that is the definition of a *well-balanced diet*—one that contains the essentials for life in sufficient

EXERCISE 22.1
How many of the inputs in
Figure 22.1 seem to you to have
a clear molecular correlation?

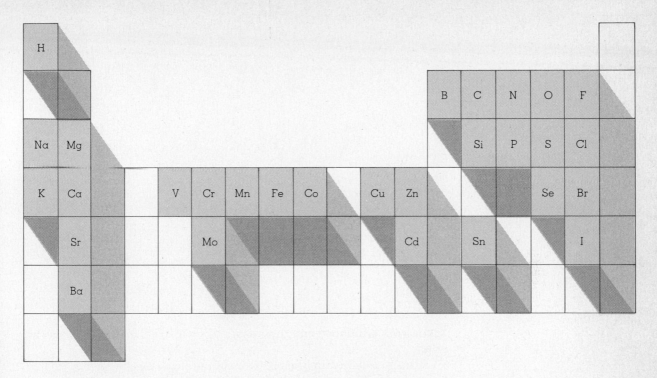

Figure 22.2
Elements essential to life.

quantity to ensure both the maintenance of life and vigorous health. (The coke–hamburger–vitamin pill combination does not qualify.)

A fourth scheme might return to the "system method" of Chapter 2. Anatomists and physiologists commonly treat humans as assemblages of interrelated systems: skeletal, muscle, circulatory, digestive, reproductive, enzymatic, nervous, endocrine, immune, and so on.

I shall use ideas from each organizational scheme throughout this chapter, but for the most part I shall follow Figure 22.1.

The Magic Moment

A most important chemical event in your life occurred at the moment of your conception—when one, and only one, of the millions of available sperm from your father entered one of the 400 eggs with which your mother was born. At the moment of conception it was determined whether you would be male or female, whether you would have one of the several thousand metabolic diseases (such as mongolism or sickle-cell anemia) that affect humans, and what internal limits would be set on your abilities in almost every endeavor undertaken throughout your life. The development of your abilities and traits to some fraction of their full potential is, of course, a later function of your genetic inheri-

Figure 22.3
The 46 human chromosomes. Males have both X and Y; females, a pair of X chromosomes. Mongoloid children have an extra number 21 chromosome. Egg and sperm each have only one chromosome from each pair, so each has half as many chromosomes (23) and genes.

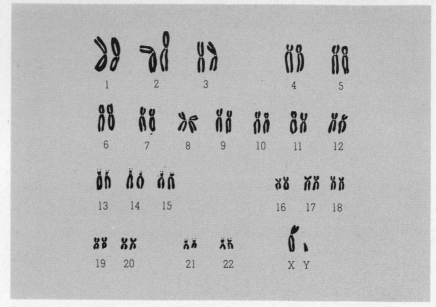

tance, your childhood environment, and your social environment as an adult.

A normal individual has 23 pairs of chromosomes in each cell (Figure 22.3). Most pairs consist of similar individual chromosomes, but one pair might consist of a small Y chromosome paired with a larger X chromosome. If this is the case, you are male. If you happened to get two X chromosomes instead of an X and a Y, you are female. Each cell in your body contains a set of copies of the 46 chromosomes that first came together when the sperm and egg joined at your conception. The only exceptions to this rule are the egg or sperm cells that your body produces. Each of these cells contains only one chromosome from each pair—that is, 23 individual chromosomes. The selection of which chromosome from each pair gets into any one sperm or egg is random. So, assuming a simple sorting process, any one individual might produce some 2^{23} or 10^7 different kinds of eggs or sperm. Actually the number is larger than this because small parts of chromosomes can be exchanged between the members of a pair before the sorting occurs.

As you can see, even after a single egg has been presented any one of millions of individuals might develop—depending on which sperm gets there first. Each of the chromosomes carries about 10^3 genes. These determine many characteristics. So it becomes easy to understand the great variety of individuals we see about us.

Identical twins are (almost) identical because they develop from the splitting of a single fertilized egg into two fertilized halves, both of which then develop into individuals. Because their prenatal positions

EXERCISE 22.2
Every parent has an average of about 10 defective genes that will be damaging to offspring if a damaging gene from the mother pairs with a similar one from the father. Suppose both husband and wife have such a lethal gene in one of their number-20 chromosomes. Show that only one out of four of their children will be affected on the average.

and access to nutrients will vary, even identical twins are not quite identical. One is almost always appreciably larger at birth, is born first, and is favored in further development by this better start. But the twin less favored by genetics may actually do better in the long run by coming closer to full genetic potential through favorable environmental effects. The evidence is strong that most individuals fail to achieve their genetic potentials (often for chemical reasons).

You may consider your chromosomes as a set of tape recordings containing the directions for most of your chemical development. Each of the 46 chromosomes is found in cells throughout the body, but only some of the genetic information is actively used in each cell. A single *chromosome* contains many individual "tapes," each consisting of long sections (paragraphs) called *operons* made up of *genes* (sentences). Each gene consists of many *codons* (words) of three *nucleic acids* (letters), from a set (alphabet) of four possible nucleic acids differing in their carbon, hydrogen, oxygen, nitrogen, and phosphorus chemical structure (strokes). (See Table 22.1.) Shifting the atoms in any one nucleic acid could make nonsense of the directions that should be imparted by a whole operon. In fact, such a shift in one molecule could kill you.

It should be clear that the system must be assembled with care, with many fail-safe mechanisms. The individual steps must be simple to give the high reproducibility actually found. If all the operon tapes in one of your cells were stretched out linearly, they would total only about 1 m in length. Yet from the original 1 m you have synthesized some 10^{14} cells, including 10^{14} m of genes. Each cell contains a full set of chromosomes, and most sets are identical with the original set. Very few errors in transcription—none fatal in the short term—have been made. The consistently correct copying of chromosomal material is one of the most remarkable chemical syntheses known. And of course, if the entropy is to be low in the genes (information content high), there must be a corresponding increase in entropy in the surroundings. A price is paid for those 1-m "tapes."

Decoding the Codons

The early 1950s were an especially exciting time for those interested in the chemistry of living systems. Analytical techniques, both chemical and instrumental, had developed to the stage where it became possible to determine the atomic structure of large molecules. A great deal had been learned about the rules of genetics and inheritance, and it had become established that the long molecules in the chromosomes were deeply implicated.

The best general account of this research is probably *The Double*

**Table 22.1
Information Flow in
Human Genetics, Showing the
Number of Individuals at
Each Level**[a]

Society ($\sim 10^3$)
↑
Person (4×10^9)
↑
System (~ 10)
↑
Organ (~ 15)
↑
Cell ($\sim 10^{14}$)
↑
Chromosome (46)
↑
Operon (?)
↑
Gene ($\sim 10^5$)
↑
Codon (64)
↑
Nucleic acid (4)
↑
C, H, O, N, P

[a]Numbers apply to normal humans.

POINT TO PONDER
Because each codon is three nucleic acids long, the code-reading mechanism simply assumes a space after each third letter. For example, you can readily convert a message like BUYANYBIGDOG into the sensible instruction BUY ANY BIG DOG if you know that each word is three letters long. But suppose that the typist acidentally puts an *A* on the front, typing the message as ABUYANYBIGDOG. You would now decode to get the useless nonsense, ABU YAN YBI GDO G. Similar nonsense would result from a dropped letter. A substituted letter might be less serious because it would change only one word, but it could still change the whole meaning of the message—for example, BUY ANY BIG HOG.

Helix by James D. Watson (Atheneum, New York, 1968). He and Francis Crick received a Nobel Prize for their contributions to the solution of gene structure, and the book is his personalized account. Its discussions of the rivalries, the interchange of ideas, and the secretiveness that developed when an idea seemed especially important is a much more useful account of science and scientists in operation than any textbook discussion of the "open, unbiased, truth-seeking, total-disclosure scientific method" so often outlined. **You should get a copy and read it!** Two to three hours should be enough to give you an excellent exposure to the human nature of creative scientists.

One of the key bits of information was the analysis by Erwin Chargaff (presented in Table 22.2). Genes were known to be long strands of these four nucleic acids—adenine (A), thymine (T), guanine (G), and cytosine (C)—polymerized together in deoxyribonucleic acid (DNA) chains. Any model of genes would have to account for the remarkably constant ratio of 1 to 1 found by analysis for both the adenine–thymine and the cytosine–guanine pairs (A–T and C–G, respectively).

Watson and Crick were helped in the solution of this problem when they cut out cardboard outlines of these four molecules and used some recent x-ray information on hydrogen-bond formation to show that the pairs fit in one and only one way (Figure 22.4). One of the reasons this worked is that all four molecules happen to be flat, so are easy to model with cutouts. Ah, the wonderful coincidences of science! And so they arrived at the double-helix model for DNA, which you have seen in texts and popular articles (Figure 22.5).

A general pattern began to fall into place. The methods of synthesis of the four nucleic acids and of their combination into the DNA chains were discovered in stepwise detail. The general means by which the linearly stored information guides the synthesis of proteins were unraveled. The roles of proteins in controlling cell metabolism, organ function, and individual development are now being outlined. Each step

EXERCISE 22.3
The circumference of the world is about 4×10^7 m. If all the genes in your cells were stretched out as a single thread, how far around the world would it go?

Table 22.2   Nucleic Acid Composition of DNAs of Various Organisms[a]

Organism	Nucleic acid			
	Adenine (A)	Thymine (T)	Guanine (G)	Cytosine (C)
Sarcina lutea	13.4	12.4	37.1	37.1
Escherichia coli	24.7	23.6	26.0	25.7
Wheat	27.3	27.1	22.7	22.8
Chicken	28.8	29.2	20.5	21.5
Cow	28.0	28.0	22.0	21.0
Sheep	29.3	28.3	21.4	21.0
Man	30.9	29.4	19.9	19.8

NOTE: Data from Erwin Chargaff.
[a]Note that A = T and G = C (within experimental uncertainty) in all cases.

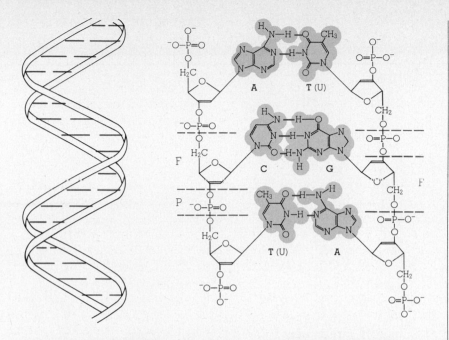

Figure 22.4
Hydrogen bonding (colored dashed lines) of nucleic acid pairs in DNA. Symbols: P, phosphate; F, furanose (a sugar); A, adenosine; C, cytosine; G, guanine; T, thymine (in DNA); U, uracil (in RNA). Uracil differs from thymine only in its lack of a methyl group, —CH₃, on its ring. (See also Figure 22.5.)

(as you might anticipate) ordinarily involves a bimolecular collision on a catalytic enzyme, with one bond formed and one broken. The activation energies and orientation requirements lie in the range of values giving rapid yet controllable rates at body temperature.

In general, each gene carries the information required to synthesize one enzyme. The information is transferred from the DNA gene to a series of RNA (ribonucleic acid) linear polymers, which are "inverse" copies of the DNA. Each cytosine of DNA becomes a guanine in RNA; each guanine, a cytosine; and each thymine, an adenine. Adenine in DNA is matched in RNA by uracil, a very similar molecule to thymine as you can see in Figure 22.4. Thus the A–T, C–G code of DNA is converted into the U–A, G–C code of RNA.

The letters U, C, A, and G constitute the four-letter code used in Table 22.3. Four letters are needed because the function of the RNA is to guide the condensation copolymerization of enzymes from the 20 α-amino acids mentioned in Table 1.2. This could be done with 20 different "letters," one for each amino acid, but an overall economy of molecular synthesis is achieved with a four-letter code and three-letter codons (words).

There is a different set of codons for each of the 20 amino acids, as shown in Table 22.3. Because there are 64 possible codons and only 20 amino acids, considerable redundancy occurs. For example, there are 4 RNA codons (GCU, GCC, GCA, GCG), for alanine. But note that most shared codons have the same first two letters. This minimizes the

POINT TO PONDER
It has been pointed out by James Bonner that the breakdown of food in the body is a little like burning wood in a wooden oven. Have you ever considered how you are able to eat and digest stomach (called *tripe*) without digesting your own stomach?

EXERCISE 22.4
What amino acid is coded by the DNA nucleotide structures shown in Figure 22.5, if they read from the T ends?

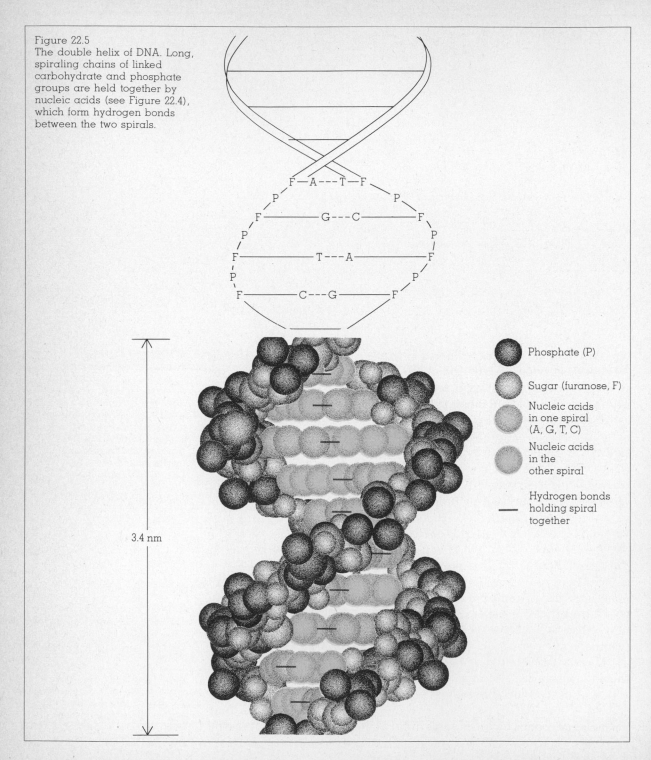

Figure 22.5
The double helix of DNA. Long,
spiraling chains of linked
carbohydrate and phosphate
groups are held together by
nucleic acids (see Figure 22.4),
which form hydrogen bonds
between the two spirals.

3.4 nm

Phosphate (P)

Sugar (furanose, F)

Nucleic acids
in one spiral
(A, G, T, C)

Nucleic acids
in the
other spiral

Hydrogen bonds
holding spiral
together

Table 22.3   Codons in RNA for Amino Acids, Based on the Four "Letters"—
U (Uracil), C (Cytosine), A (Adenine), and G (Guanine)

Amino acids		RNA codons
Symbol[a]	Name	
A	Alanine	GCU, GCC, GCA, GCG
C	Cysteine	UGU, UGC, UGA
D	Aspartic acid	GAU, GAC
E	Glutamic acid	GAA, GAG
F	Phenylalanine	UUU, UUC
G	Glycine	GGU, GGC, GGA, GGG
H	Histidine	CAU, CAC
I	Isoleucine	AUU, AUC, AUA
K	Lysine	AAA, AAG
L	Leucine	UUA, UUG, CUU, CUC, CUA, CUG
M	Methionine	AUG
N	Asparagine	AAU, AAC
P	Proline	CCU, CCC, CCA, CCG
Q	Glutamine	CAA, CAG
R	Arginine	CGU, CGC, CGA, CGG, AGA, AGG
S	Serine	UCU, UCC, UCA, UCG, AUU, AUC
T	Threonine	ACU, ACC, ACA, ACG
V	Valine	GUU, GUA, GUC, GUG
W	Tryptophan	UGG
Y	Tyrosine	UAU, UAC
*	"End-of-chain" signal	UAA, UAG

[a]Note that A, C, and G have double meanings. They are used to identify both nucleic acids and
amino acids.

chance of errors in transcription. So the sequence of amino acids in the enzymes reflects the sequence of codons in the RNA. These in turn are inversions of the three-"letter" sequences in the DNA. Thus the sequence CGA ACA in DNA gives GCU UGU in RNA and alanine–cysteine in the product amino acid. Many long sequences of codons have been determined. An example published in 1976 is the 3569 nucleic acids (1189 codons) in the RNA from the bacteriophage MS2.

Many enzymes are involved in the transcription processes. Some catalyze the condensation polymerization of the DNA, some repair breaks and errors in the chain, and some assist the transcription to RNA. Others are watchdogs for other critical steps and become active when the system starts to break down. All in all, it is a remarkably interwoven system, and it makes the checks and balances of the American Constitution seem trivial. Your very existence proves its effectiveness. Figure 22.6 indicates some of the steps in the synthesis of cytochrome c, a much-studied molecule.

A Ubiquitous Substance: Cytochrome c

Figure 22.7 shows the structure of *cytochrome c,* which can be thought of as a protein enzyme surrounding a *coenzyme* (the flat group, called a

POINT TO PONDER
Genes may be transferred from one species to another. If the gene for making rabbit hemoglobin is introduced into a frog egg, the egg synthesizes rabbit hemoglobin. If the gene specifying honeybee venom is introduced, the frog egg makes honeybee venom. Transfers of this type suggest that genetic engineering (the introduction into an individual of new genes for desired traits) may be possible, even in humans.

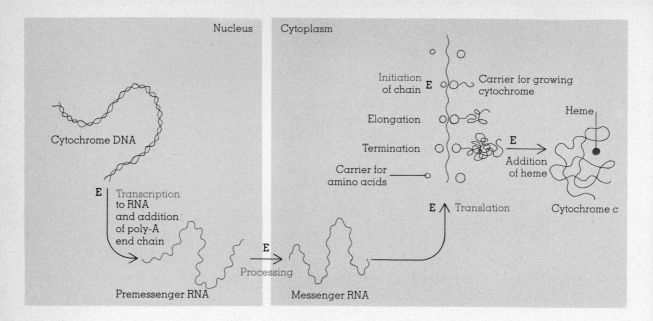

Figure 22.6
Some schematic steps in the synthesis of cytochrome c. The instructions from genetic DNA direct the polymerization of amino acids. (E means that enzymes·are needed to complete the step.)

heme, which has an iron atom in its center). The iron atom is the center of the reactive group in cytochrome c. This association of enzyme and coenzyme to enhance the reactivity of an atom or group of atoms is common in enzyme catalysis.

Cytochrome c is found in every living organism that has been studied, and it has the same essential function in each. But the detailed sequence of the amino acids varies in different organisms. Some places in the molecule—for example, positions 70 to 83—are identical in all species. These are the places in the twisted cytochrome that are near the reactive iron. They must always be the same if the reactant molecules are to be brought into just the right position to react. Other sequences in the cytochrome (say, 49 to 69) show variation—sometimes considerable variation. These lengths of the molecule must be present, and random variation is not allowed, but the overall function of the molecule is maintained as long as the substitutions are of amino acids with similar chemical character, especially similar polarity. (Recall the βMSH of Figure 7.15.)

It should be clear that an enormous amount of research has gone into assembling the information in Figure 22.7. You might be interested in the general approach, which is surprisingly straightforward. The large protein is treated with enzymes and other chemicals that split it into 20 or 30 soluble fragments. The solution is then allowed to diffuse along a piece of paper. The process is called *paper chromatography.* The paper is rotated 90°, and a different solvent is allowed to diffuse at right

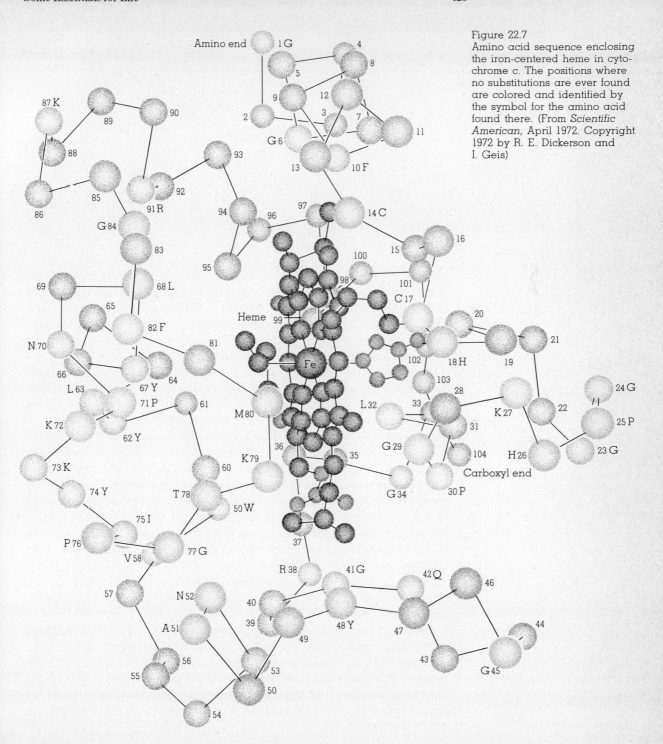

Figure 22.7
Amino acid sequence enclosing the iron-centered heme in cytochrome c. The positions where no substitutions are ever found are colored and identified by the symbol for the amino acid found there. (From *Scientific American*, April 1972. Copyright 1972 by R. E. Dickerson and I. Geis)

Figure 22.8
Paper chromatography comparison of two human hemoglobins: (left) Hb A (a common form) and (right) Hb Oβ (a form found in the Nile Delta). (From K. A. Karmel et al., *Science*, **156**, 1967, p. 397; copyright 1967 by the American Association for the Advancement of Science)

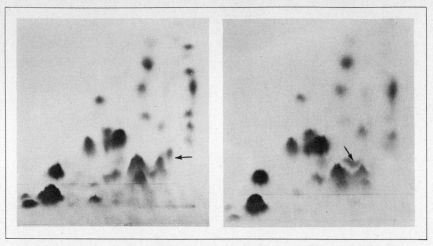

angles to the first. The result for a hemoglobin is shown in Figure 22.8. If the fragments from two samples are identical, they diffuse identically through the paper. If not, they don't. It is then easy to see where the differences lie, but still difficult to find *what* the differences *are*. Careful work involving further degradation gives data such as those outlined in Figure 22.7 and Table 22.3.

An interesting outcome of the study of these variations is general confirmation, and some detailed rectification, of the postulated evolutionary sequence of species. Figure 22.9 shows some chemical evidence on evolution, based on the number of differences in amino-acid sequence in cytochrome c in the species indicated. The assumption is that mutations (changes) occur in a random way, with some being incorporated into long-lived species and others leading to extinction. The more differences between two existing species, the longer the interval back to when they had a "common ancestor." Note that the plot of number of changes versus the time of geological occurrence of the species gives a straight line within the experimental uncertainty, consistent with the assumptions. The average frequency of tolerable (viable) mutation in cytochrome c is one every 20 million years.

Other enzymes are constituted similarly to cytochrome c. Many plants can synthesize all 20 amino acids, but humans cannot synthesize 10 of them—so these 10 are essential in human diet (as pointed out in Table 1.2). It is likely that the inability of humans to synthesize these amino acids can be attributed to the loss of some of the genes that generate the enzymes—that is, direct their synthesis as in plants. Individuals can survive the loss of a gene as long as the amino acid is abundant in the normal food supply of the species. However, individuals die following loss of a gene responsible for an enzyme producing something not in the diet.

EXERCISE 22.5
Use the data in Tables 22.2 and 22.3 to draw some conclusions about the relative frequency of proline occurrence in proteins in humans, and in *Sarcina*.

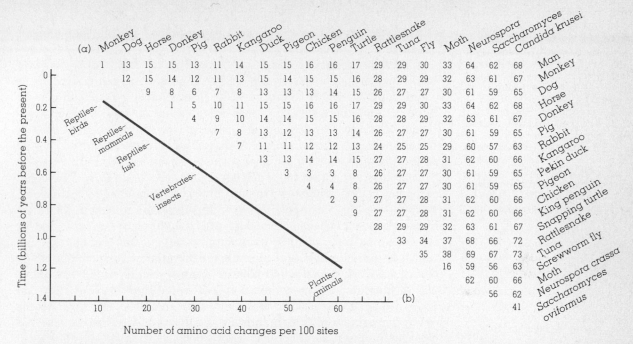

It is presumably random changes in the genes (partly due to the influence of high-energy radiation on the sperm- and egg-synthesizing organs) that lead to evolutionary changes. The change could be in a single codon, or it could involve the loss or gain of a whole gene, or even a whole chromosome. After all, humans have 46 chromosomes, but most simpler animals have fewer. An increase in gene and chromosome number permits encoding the additional information necessary for a more complicated species.

Much of modern genetic study involves organisms that are deficient in one or more genes. In fact, there is no reason in principle why geneticists cannot isolate individuals, called *mutants,* deficient in any trait—and so study the relationships between observable traits and chemical composition of the individual. But I have digressed into internal body chemistry. Let's return to inputs.

The Breath of Life

The gene–enzyme systems bring the fetus from conception to birth. Chemicals from the bloodstream of the mother are processed in orderly developmental sequences, which are the subject of the field we call *embryology*. Normal infants at birth have the organs and systems they need to meet the new environment. They also apparently have all the brain cells they are going to get. Females have all the egg cells they will

Figure 22.9
(a) Number of mutations postulated in a total of 312 gene nucleotides to account for cytochrome c variations between several species. (b) Data from (a) plotted as number of mutations versus billions of years before the present (from the fossil record) when several evolutionary differentiations of living species occurred. For example, birds and reptiles differentiated some 0.2 billion years ago and may differ by about 12 mutations on the average.

EXERCISE 22.6
Refer to Figure 7.15 and deduce one or two codon differences between the human and cow genes that produce βMSH.

POINT TO PONDER
The number of enzymes identified and characterized (as to
their functions) is growing
rapidly: 80 by 1930, 200 by 1947,
660 by 1957, 1300 by 1967, and
about 3000 by 1976. Many have
been purified and crystallized.
Some (such as cytochrome c)
have had their structures determined in considerable detail.
Enzymes have been incorporated
in detergents to digest dirt, and
they have been used industrially
to catalyze large-scale processes.

People known to be deficient
in certain enzymes may be put
on special diets; they then lead
otherwise normal lives. Some
2000 human diseases are due
to known lacks of particular
enzymes. The absence of vitally
needed enzymes can be detected
in a fetus before birth for about
200 of these diseases, and a
prediction can be made of the
severity of the affliction. In
other words, enzymes control
many rate-determining steps.

EXERCISE 22.7
Estimate the effective altitude
to which jet passenger planes
are probably pressurized.

EXERCISE 22.8
Suggest how you could decide
whether cold and heat are
sensed by the same nerve
ending.

ever have. Some other cell systems may also be complete as to number.
But almost all cells will continue to develop. For this they require
chemicals.

The first chemical action of a newborn infant is probably to adapt
to the use of gaseous dioxygen, rather than the oxygen-rich aqueous
solution provided by the mother. The change must be made in a short
interval, and thereafter the need for gaseous oxygen is maintained
throughout life. Oxygen is the chemical that most quickly leads to
trouble when the body is deprived of it. Few adults can survive loss of
oxygen for more than 5 minutes without permanent brain damage.

Furthermore, either too much oxygen *(hyperoxia)* or too little oxygen
(hypoxia) causes problems. One of the first astronauts to venture outside a spaceship became very happy and resistant to suggestions that
he return, possibly because of a euphoria brought on by too rapid and/or
too deep breathing. Meditators sometimes rely on this technique.

Mountain climbers and aviators are familiar with hypoxia, which
sets in at an altitude of about 3000 m for most people. At 2000 m, about
20% of humans have some initial loss of visual function, but they can
adapt on longer exposure. Between 3000 and 5000 m, most people
develop shortness of breath and loss of full use of the visual sense. No
one can function satisfactorily above 6000 m without an additional
supply of oxygen.

The Chemical Senses

It may have struck you as strange that, as listed in Figure 22.1, you can
go a longer time without food, or even water, than without stimulation
of the senses. People *totally deprived* of all sensory stimulation inputs
appear to have great difficulty going more than about 10 hours without
serious hallucinations and feelings of great insecurity. There is some
argument as to whether it is the deprivation or the constrained condition of the subject that causes the difficulties. In any case, the experience is more unpleasant than going without water or food the same
length of time. It has been suggested that the absence of external signals
leads to random firing of the nerves. They get tired of "waiting for something to happen." Let's look at the senses.

We normally speak of the five senses (sight, taste, smell, sound, and
touch). The first three are identified as chemical. Actually there are
known to be several more senses—those that focus on heat, cold, pain,
vibration, direction of gravity, linear and rotational acceleration, and
body position (where is my arm right now?). And there are many detec-
tors of the internal state of the body, as we shall soon see.

It is probable that all the senses involve signals originated by
mechanical deformation, light, temperature changes, or the presence

or absence of particular chemical substances. These signals are converted into electrical signals transmitted by the nerves as short-lived "spikes" of current (Figure 22.10). All spikes are the same size. The message is carried by the frequency and spacing of the spikes and by the particular nerves that are carrying current to the central nervous system (which includes the brain).

Scientists have invented and are using instruments to extend the unaided senses, but eventually the information must be presented in a form the senses can transmit. Not one of the senses is fully understood, but let's look into sight, taste, and smell to see some of what is known.

Sight and Molecular Flip-Flop

We have known for about 10 years that the eye does not act as a camera with each activated eell transmitting a signal about intensity and color to the brain. Rather, the eye detects edges, corners, and patterns—with several receptor cells connected to each nerve that transmits signals to the brain. The nature of these connections is far beyond this text (in fact, the details are still beyond the researchers), so we will stick to the primary event—the effect of the photon. After very diligent and difficult research, that effect turns out to be simple. In the light-receptor cell, there is a light-absorbing pigment. Each molecule of pigment is made of a molecule of *cis*-retinal attached to a larger protein molecule (Figure 22.11). When a pigment molecule absorbs a photon, the *cis*-retinal is converted to *trans*-retinal within about 6×10^{-9} second. This isomeric flip initiates a complicated (not fully understood) set of reactions, which may generate a spike signal in the nerve leading away from the receptor cell.

The human eye contains three kinds of color-sensitive receptor cells—each kind having a different pigment. (The pigments differ in the nature of the protein attached to the retinal.) One pigment absorbs

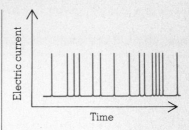

Figure 22.10
Electric current "spikes" found in conduction by nerves. See also Figure 17.15.

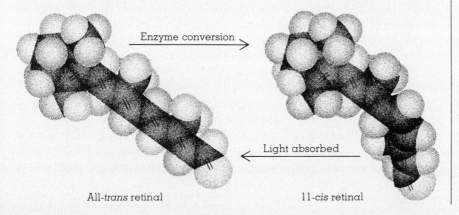

All-*trans* retinal   11-*cis* retinal

Figure 22.11
The reaction that leads to vision. Double bonds are indicated by =, as usual.

EXERCISE 22.9
Fluorescent lights go on and off
60 times each second, as you
can tell by looking at one while
waving the spread fingers of
your hand between the lamp
and your eye. Why don't you
usually see the light flicker?

Vitamin A

−2H (enzyme) ↓

Retinal

Monosodium
glutamate

mainly red light; another, blue light; and the third, yellow light. Thus, the ratios in which the three kinds of cells are triggered by incoming light provide information about the color of the light. The signals from the cells are processed in a complex nerve network within the eye to produce "summarized information" about colors that is sent on to the brain. For example, if all three kinds of cells are stimulated equally, the brain receives signals that we "see" as white color.

Finally, an enzyme-catalyzed reaction flops the *trans*-retinal back to the *cis* form. The *cis*-retinal then spontaneously re-forms a molecule of pigment ready to absorb another incoming photon. A receptor cell can produce a spike signal about 2.5×10^{-5} second after photons strike the cell's pigment, but recovery of the pigment takes about 100 times that long (a few milliseconds). Furthermore, the brain is not ready to accept a new signal from a given nerve for about one-thirtieth of a second after receipt of a signal. That is, a visual image "persists" in the brain for about one-thirtieth of a second before it can be replaced by a new one. This is why you do not usually notice the flicker between the successive still images of a motion picture, even though the screen is totally dark about half the time.

Retinal is synthesized by the body from vitamin A. But vitamin A cannot be synthesized by humans, so it is one of the essential foods. Absence of vitamin A in the diet can cause serious problems with vision. The initial effect is on vision at low illumination (dark vision), but extended deprivation can extend to the whole visual range. It is interesting to note that a "warning" is issued, by loss of the less important dark vision, before daytime vision is affected.

Salt, Sour, Bitter, and Sweet (Plus Some Others)

The four common tastes are salty, sour, bitter, and sweet, but others are recognized—metallic, soapy, and astringent, for example. The chemical basis of none of these is known, though a great deal has been done on the physiology.

Note that only one of the tastes—sweet—urges the taster to keep going. There are not many common sweet substances that are toxic. The other tastes are much more apt to warn of undesirable consequences. The fact that mother's milk contains the sweet carbohydrate lactose is probably more than coincidental. It is found only in milk.

It is interesting to note that monosodium glutamate, MSG, was formerly added to baby food to "improve the taste." However, it is quite likely that it only improved the taste for the parents and left the baby quite unimpressed. When it was found recently that glutamate may diminish sensory acuity and even cause birth defects in some animals, it was dropped from baby food. It is still widely used in homes and restaurants (especially oriental ones—"Chinese restaurant syndrome") as a flavor enhancer. Attempts to find some effect of glutamate on the taste nerves in the tongue have failed. Perhaps MSG only works through the sense of smell, as is true of most of what we "taste."

What we call flavor is not only taste. It is as much, or more, odor and texture. Almost 50% of the food additives approved by the Federal Food and Drug Administration (FDA) are flavors. Some 1100 different additives are used in the United States; 750 of them are synthetic.

A major problem for many people is a simultaneous craving for sweets and for a slim figure. In others there is a real physiological intolerance of sugars, yet a desire for food and drink to taste "good." So we have the synthetic sweeteners. Some can be used in very small concentrations because they are thousands of times as sweet as sucrose, the most common sugar. But many have come under recent criticism as possible *carcinogens* (cancer-causing agents). Actually there is little if any evidence that reducing the sugar intake by the amount required to sweeten food has any effect on obesity. But the soft-drink market is a big one, and the drive for a safe "dietetic" sweetener is attracting a lot of money.

Because there is no accepted theory of taste and no known correlation between chemical structure and taste (Figure 22.12), the search for synthetic sweeteners is mostly by trial and error. One synthetic sweetener, sucaryl, was discovered in the 1940s by a graduate student smoking in the lab when he wasn't supposed to. He found that some of his recently prepared compounds were getting from his fingers to his cigarettes, making them taste sweet. He tasted all his preps and "invented" sucaryl. A very dangerous method!

If It Smells, It's Got to Be Chemical

For years chemistry was associated with assaults on one's nose. Then, during the days when atmospheric pollution was at a minimum, the nasal assault was minimized—only to grow again after the mid-1950s. Until recently there were about as many theories of smell as there were "basic" smells. No one theory is fully accepted yet, but all designate a set of elementary smells that individually and in combination give the varied responses we detect. One experimental problem is the inaccessibility of the very short nerves that run from the sensitive area at the top of the nasal passage directly to the brain a few centimeters away.

Recently John Amoore presented a seven-odor analysis. He attributes each odor to the reaction of a receptor of a particular shape and size in the nose. The actual signal may be caused by a shift in the normal vibrational frequency of the cavity. Figure 22.13 outlines the suggested cavity sizes and the odor associated with each. Composite odors result

EXERCISE 22.10
Sugar and bread have essentially the same "caloric content." Giving up one piece of bread per week would allow about how many cups of sugar-sweetened coffee to be drunk by a dieter?

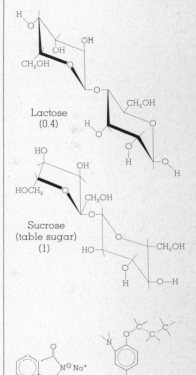

Lactose (0.4)

Sucrose (table sugar) (1)

Sodium saccharinate (saccharin) (300)

2-Amino-4-nitro propoxybenzene (4000)
The sweetest substance known

Figure 22.12
Some molecules that taste sweet. Note the lack of similar structures. Relative sweetnesses are given in parentheses.

Primary odor	Chemical example	Molecular shape		Receptor site
Camphoraceous	Hexachloroethane Camphor	Spherical		0 1 2 Size (nm)
Musky	Xylene musk Pentadecanolactone	Disk		
Floral	α-Amyl pyridine Phenylethyl methyl-ethyl-carbinol	Disk and tail		
Pepperminty	Menthol	Wedge		
Ethereal	Diethyl ether	Rod		
Pungent	HCOOH Formic acid	Simple (electrophilic)		
Putrid	H_2S, C_2H_5SH	Simple (nucleophilic)		

Figure 22.13
The receptor cavities proposed in a theory of odor based on size, shape, and polarity.

when a molecule can fit into two or more cavities.

In order to be smelled at all, the molecules must be volatile enough to enter the nose. They must also be appreciably soluble in the nasal mucus. But the nose is remarkably sensitive. It can detect ethyl mercaptan, C_2H_5SH, at a concentration of about 0.001 ppm. It may be that fewer than 10 molecules can excite the human nose. It is this sensitivity that accounts for the fact that so much of flavor is actually odor. The volatile molecules from the food enter from either the front or the rear nasal passages and are adsorbed on the receptors. Some can be so tightly adsorbed that they prevent other odors from registering for long periods. Others are adsorbed lightly and have only fleeting effects. Table 22.4 gives some of the compounds responsible for several familiar odors.

Chemical Messengers

Genes, oxygen, and sensory stimulation are all necessary inputs for life. In some cases, special information is associated with the function—especially with the genes, but also with some of the tastes and smells, as we have already suggested. Most of the messages involving sight are learned—reading, for example. But let's look at unlearned messages—messages that are carried by chemicals much smaller than genes and are interpreted without prior learning or conscious effort of the individual.

EXERCISE 22.11
Use the Amoore theory to guess the smell of ethyl benzoate, $C_6H_5CO_2C_2H_5$

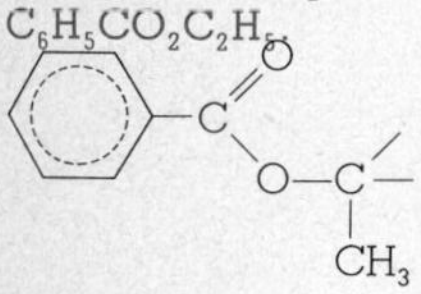

planar

Table 22.4   Some Naturally Occurring Odorous Compounds
(all belonging to the class called *esters*)

Ester[a]	Characteristic odor	Occurs in
(structure)	Blackcurrant	Orange
(structure)	Apple	Strawberry
(structure)	Apple	Honey
(structure)	Apple	Pineapple
(structure)	Pineapple	Pineapple
(structure)	Fruity	Strawberry
(structure)	Banana	Wood oil
(structure)	Wintergreen	Wintergreen
(structure)	Wintergreen	Currant

[a]The peripheral hydrogen atoms are not indicated.

Rapid messages requiring prompt response are sent by nerves at a speed of about 1 m per millisecond. Otherwise, for example, burns would become serious before contact could be broken. Even these fast messages involve chemical messengers, but the chemicals need move only very short distances—say, 10–50 nm—so the overall rate of transmission is rapid. Slow messages (especially messages requiring a long-acting response, or action at a distance from the sender) are commonly sent by molecules, not by nerves.

EXERCISE 22.12
A distance of 100 m can be run in about 10 seconds. How far does the runner get during each cycle of nerve impulse (foot to brain and back again)? Does the 10 second/100 m distance seem reasonable?

It is convenient to divide chemical messengers into four groups:

1. those synthesized by regulatory genes (enzymes are a good example);
2. repressors (enzyme products or molecules specifically designed to minimize some reaction);
3. hormones (stimulants released internally in response to external or internal signals);
4. pheromones (substances released to elicit societal response). Two groups of pheromones exist: (a) those that elicit an immediate response, like sex attractants and trail and alarm indicators, and (b) those that initiate a delayed response. Pheromones of the latter group emitted by a male mouse will cause a female mouse to abort if she smells them during the first 4 days of pregnancy—but only if the male mouse is a stranger.

The first three classes are not usually chemical inputs to the individual but are synthesized internally. However, we should look at them briefly because of their common "messenger" roles. More and more consideration is being given to their possible use as inputs to control undesirable behavior—for example, illnesses.

The Merry-Go-Round of Feedback

Your genes direct the synthesis of thousands of enzymes. The enzymes, each a chemical messenger that says "synthesize a certain substance," catalyze the conversion of food into the structural and operational molecules that keep you going. In the absence of appropriate foods the enzymes cannot act; deficiencies result. But how do the enzymes know, food being plentiful, when they have produced enough of the needed products?

The most common answer is the action of *repressors,* which inhibit one of the early steps in the enzyme synthesis. The product of the enzyme-catalyzed reaction is the easiest repressor to understand. It often adheres to a site on the enzyme, distorting the original reactive site. Then the enzyme can no longer act as a catalyst. When the concentration of product reaches an adequate level, its adsorption equilibrium with the enzyme is such that almost all the enzyme molecules are repressed. Sometimes the product even reacts with the gene producing that enzyme and represses further enzyme synthesis. Systems in which a product carries a signal back to some of the early steps either to slow them down or speed them up are called *feedback systems.* Your physiology contains thousands of feedback systems.

The messenger role of enzymes and of repressors is well illustrated by the synthesis of two of the essential amino acids—threonine and isoleucine. The common starting material is aspartic acid, of which you

Aspartic acid

(enzymes, missing in humans)

Threonine

(enzymes, missing in humans)

Isoleucine

POINT TO PONDER
An interesting example of a feedback-repressor system is involved in your inability to tickle yourself.

make plenty. But you do not have the enzymes that say, "Convert aspartic acid to threonine." So you cannot perform this synthesis, and threonine (essential for your protein synthesis) becomes an essential in your food.

Threonine, in turn, is converted by plants into isoleucine by a multiple-enzyme sequence of reactions. You do not have the necessary enzymes for this synthesis either, so isoleucine is also an essential in food. In plants that can perform the synthesis, the isoleucine inhibits the first enzyme in the sequence (the one that reacts with threonine) and represses its own synthesis. Isoleucine also represses the gene synthesizing the enzyme—a fail-safe system!

"Come Hither" and Other Messages

Figure 22.14 shows the structure of some similar molecules. The natural pheromone is a powerful sex-attractant produced by the red-banded leaf roller. The female exudes the volatile chemical, which is borne by the wind. When the male detects its presence (a main use of the large antennae some male insects possess), he flies upwind along the concentration gradient until he finds the female. Similar "perfumes" *(pheromones)* are used by many species. Some of them are sex attractants, some of them are alarms or territorial markers, and some of them are trail markers. Ants use pheromones to lay out and follow those narrow paths you see them use. Note in Figure 22.14 the loss of sensitivity if small changes occur in molecular structure.

Each species has its own pheromones. If similar species use the same molecules, they mix them in different, but fixed, proportions. So pheromones are highly specific to species. They are also detected in minute quantities. Some can bring about the appropriate response when present at a concentration of 1 molecule/cm^3 and when only 1 molecule/second is received. Pheromone receivers are among the most sensitive chemical analysts in the world. But their range of abilities is a bit narrow. It has been suggested that there are human pheromones also; they might be the basis of detecting good and bad "vibes," for example.

Figure 22.15 outlines the endocrine glands and some target organs. These glands synthesize and secrete the hormonal messengers regulating much of your body chemistry and many of your physiological rhythms. Endocrinology is the study of what these glands do. It is not at all clear how the glands themselves are regulated, though feedback systems and external light sources are both known to be important. The pineal gland is sensitive to light, and the hypothalamus–pituitary is involved in many feedback loops.

Figure 22.16 shows some of the output of the pituitary in initiating responses in other endocrine glands. Not long ago, the thyroid releasing

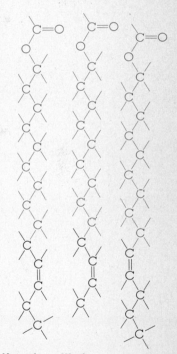

Natural   Weak
pheromone pheromone Inactive

Figure 22.14
A pheromone produced by the red-banded leaf roller *(Argyrotaenia velutinana),* and some similar molecules.

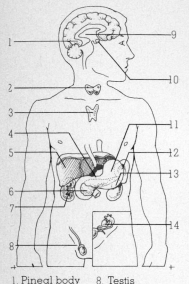

1. Pineal body   8. Testis
2. Thyroid and   9. Hypothalamus
   Parathyroids   10. Pituitary
3. Thymus   11. Stomach
4. Duodenum   12. (spleen)
5. (liver)   13. Adrenals
6. Pancreas   14. Ovary
7. (kidney cortex)

Figure 22.15 (above)
Part of the human endocrine
system. Some typical target
organs are given in
parentheses.

Figure 22.16 (above right)
Some message lines in your
endocrine system, and the
structure of thyroid releasing
factor, TRF.

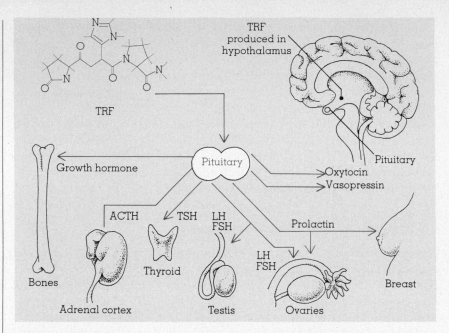

factor (TRF) was isolated, identified, and synthesized. The isolation started with 10^6 kg of brains from 5×10^6 sheep to produce 1 mg of TRF product. Now that TRF can be synthesized as needed, research into its effects is much easier. Human TRF, incidentally, is chemically identical to that of sheep.

Some interrelationships between light and hormonal action are suggested in Table 22.5. Normally the white-throated sparrow is stimulated to migrate north or south by changes in the length of the day, which probably affect the pineal gland through light received on top of the head. Eyes are not needed; blinded sparrows respond to the light changes. But if the head of a sighted sparrow is covered or tattooed black, the light is not received and preparation for migration does not occur.

Table 22.5   Effects on the White-Throated Sparrow (Zonotrichia albicollis) When Corticosterone Dosage Is Followed by Prolactin Dosage

	Interval between doses		
	4 hours	8 hours	12 hours
Fat deposition	Yes	No	Yes
Orientation	South	None	North
Reproductive system	No effect	No effect	Activated
Bird "thinks" season is	Fall	Summer (or winter)	Spring

NOTE: Data from Albert H. Meier, *American Scientist*, **61**, 184 (1973).

The same effects may be stimulated chemically by administering first the adrenal hormone corticosterone and then the pituitary hormone prolactin, with from 4 to 12 hours between dosages, as shown in Table 22.5. Essentially the full range of behavioral responses necessary for migration is initiated by the hormones, regardless of actual time of year or light fluctuations.

There is growing evidence—for example, from sleep patterns and experiments with humans who stay in total blackness (caves) for a long time—that humans (probably by way of the hormones) may also be affected by light intensity and its changes. *Circadian* (almost 24-hour) *rhythms* and other rhythms triggered by light are common in plants and animals.

The endocrine system is closely coupled with the nervous system. Specific molecules carry the slow messages, and nerves carry the fast messages. It is likely that the nervous system developed after the endocrine system and in close linkage with it. Plants, for example, have chemical messengers, but no nerves. But rudimentary nervous systems exist in many of the simplest "animals."

One of the simplest chemical messengers in plants is ethylene, $H_2C{=}CH_2$. Many plants can produce 0.5 to 5 cm^3 of ethylene per hour for each gram of their own weight, and its presence, absence, or change in concentration drastically alters plant physiological processes. For example, the addition to the trunks of rubber trees in Malaysia of a chemical releasing ethylene has increased the crop by over 20%, adding millions of dollars to the income of the country.

A host of other plant hormones are also known—examples are auxins, gibberellins, and traumatic acid (used in healing wounds). You may well have used some of these in easing the shock of transplanting or in rooting some cuttings. Much shipped food is now ripened in transit, for example, by treatment with ethylene.

Input Equals Output

The chemical messengers signal and operate efficiently, but they do not violate the conservation laws. Each atom that is needed must be supplied in the air, water, or food. In the steady state, each atom becomes waste and must be eliminated. The energy generated must also be eliminated. For example, you require about 1 kg of food (dry weight), 1 kg of water, and 10 kg of air each day. From these you generate more than 2 kg of liquid and solid waste plus about 8×10^6 J of energy.

Thus, in a year you generate 700 kg of waste material (not counting the air) and 3×10^9 J of waste heat—quite a load on your environment. And, of course, these are only the essential wastes. They do not include the 10 to 20 kg of other wastes (trash) generated daily per person in the

EXERCISE 22.13
Some people are convinced that if they love their plants and talk to them the plants do better. Can you suggest a possible molecular interpretation? Is your idea testable?

POINT TO PONDER
Humans have always been threatened by the effects of their own wastes, especially as people have become more concentrated in large cities. It is interesting to see more and more cities exploring the ancient technique of using human sewage as agricultural fertilizer. But nowadays it will be rendered sterile before such use.

EXERCISE 22.14
Justify the figure given of a daily air-intake requirement of about 10 kg.

Figure 22.17
The essentials for life?

HINTS TO EXERCISES
22.1 Certainly the upper five.
22.2 Possible pairs: good–good,
good–defective, defective–good,
defective–defective. 22.3 Would
encircle the world 2×10^6
times. 22.4 UCA = serine;
UGA = cysteine. 22.5 Proline is
coded by CCU, CCC, CCA,
CCG; *Sarcina* has more C, so it
may have more proline.
22.6 K to R need only involve
AAA to AGA, or AAG to AGG;
S to E perhaps UCA to GAA,
or UCG to GAG. 22.7 1000 to
1500 m. 22.8 Use tiny heated
and cooled probe to touch skin.
22.9 Eye retains image for
about one-thirtieth of a
second. 22.10 Guess weight
ratio; 1 slice = 5–10 helpings of
sugar, so 5–10 cups of coffee.
22.11 Tail–disk molecule, so
odor is fruity. 22.12 Almost 3 cm
per nerve impulse cycle.
22.13 Some people may exude
growth stimulators. 22.14 Count
breaths, each about 500 cm³,
air about 1 g/liter.

United States. I shall discuss each of these factors in terms of some of their roles, and as possible limits to growth, in the next chapters.

Living systems have highly reliable techniques for controlling both inputs and outputs. Remarkable numbers of dangerous materials are detected as having bad tastes or bad odors. The human stomach can, and often does, signal the injection of too much food, or of a dangerous food, by ejection of most of it! Other sensors let you know when more water or air or food is needed, and the excretory systems have similar signaling systems. The existence of 4 billion humans—far more individuals than any other large species has ever maintained—is overwhelming evidence of the effectiveness of these controls.

Summary

The essential inputs and outputs for life are very few, if by life we mean survival of the individual and the species. In humans, the genes from the parents set out the major characteristics of the offspring. Genes carry most of the information for converting the one initial cell into the 10^{14} cells that make up the adult. The DNA messages transcribe to RNAs, which generate enzymes. These enzymes catalyze the conversion of air, water, and food into the molecules necessary for survival, including the enzymes and the genes themselves. They set up and maintain the steady state and its feedbacks for metabolic synthesis and destruction—and for energy generation and loss, which have so successfully ensured the development of humans. The senses provide stimulation, contact with the environment, warnings of danger, and guidance toward rewards.

Human life has come to mean much more than the survival of individuals and the species. There are many inputs and outputs of ideas, interpersonal interactions, social interactions—some say even astrologics, vibrations, or numerous other phenomena. The search for interactions, mechanisms, and projections accelerates. The rewards in terms of greater health, longer life, and fuller life are attested to by the continually growing population.

Acquisition of knowledge reveals threats as well as benefits. There is ample evidence that change does not always mean progress. I will discuss some of the possibilities in the next chapters. But it is doubtful, as Figure 22.17 suggests, that all the "essentials" of modern life are really essential.

Problems

22.1 How far back can you trace your genetic ancestry? How many ancestors did you have in that generation? Julius Caesar traveled a good deal. What is the chance that Caesar is one of your genetic ancestors? How many descendants could he be reasonably expected to have today? [Most of the people alive today are related to Caesar unless he was sterile.]

22.2 Estimate the total volume if one set of genes from each individual of the present world human population were pooled. [$\sim 10^{-2}$ cm^3]

22.3 According to C. D. Turner and A. Bagnara in their *General Endocrinology* (Saunders, Philadelphia, 1972), a single mutation could cause phenylalanine to replace isoleucine, and a single mutation could also cause lysine to replace arginine. Do you agree with these two statements? If so, what mutation would be required? (Hint: Use Table 22.3.)

22.4 Enzymes requiring Na$^+$ cannot be activated by K$^+$, nor can enzymes requiring K$^+$ be activated by Na$^+$. However, NH$_4^+$ will sometimes activate the enzymes "requiring" K$^+$. Why the differences? [Radius of NH$_4^+$ is 0.143 nm.]

22.5 The number of hours required for fruit fly eggs to hatch varies with the temperature of the egg:

T (°C)	13.5	15.5	18	26	30	32
t (hours)	80	60	40	20	15	18

Plot time versus temperature and account for the shape of the curve. [Data from Andrewartha and Birch, 1954.]

22.6 It has been suggested that banning all food additives would lower the world's food supply by 10%. Do you think the statement is true? Why or why not?

*22.7 Some male moths can detect the arrival of pheromones (sent out by a female) at a concentration rate of 1 molecule/cm^3 per second. The female may be as much as 1 km away. Assume a breeze of 10 km/hour and uniform dispersal of the pheromone downwind from the female in a half-cone ($V = \pi r^2 h/4$) that is 100 m across at 1 km distance. How many molecules must the female emit per hour? [$\cong 10^{14}$ molecules/hour.]

22.8 Spun soybean protein is low in methionine. Considerable effort has been expended to produce a polymerized methionine additive, because the monomer's taste is objectionable. What evidence is there that the methionine would not taste bad in a polymer?

22.9 Plastics usually contain added small molecules, called *plasticizers,* that increase flexibility. Suggest a relationship between these molecules, the odor of new plastics (the "new car smell"), and the brittleness that comes with age in a plastic.

*22.10 Given two hemoglobin molecules, A and B, with A holding three molecules of O$_2$ and B holding no O$_2$ molecules (though each can hold a maximum of four), the probability of an O$_2$ molecule sticking to A is 70 times that of it sticking to B. Any advantage in this to the oxygen-transport system? It's like the Christian idea of "To him who hath shall be given."

22.11 About 2×10^6 kg/year of colorings and some 1100 different flavoring additives (750 of them synthetic) are used in the U.S. food industry. What are some advantages and disadvantages of this wealth of additives?

22.12 Birth defects (of which about 80% are thought to be of genetic origin) are a leading cause of infant mortality and decrease life expectancy by about 35 years on the average. Heart disease, cancer, and strokes decrease life expectancy by 8, 4, and 2 years, respectively, on the average. How would you divide the national budget for medical research among these four diseases?

22.13 Phenylketonuria is an enzyme-deficiency disease in infants that causes serious permanent brain damage leading to an average lifetime care cost of $400,000 per case. Providing general prenatal screening (called *amniocentesis*) costs about $20,000 per case detected. Tay-Sachs disease is caused by another such deficiency. The corresponding costs here are $40,000 for maintenance (due to the short life expectancy—about 5 years) and $5000 for screening. A cost ratio of about 10 to 1 for maintenance versus detection is typical. What role should scientists take in presenting these alternatives to the public?

Figure 23.1
"It says 100% additives."

23 FOOD? MEDICINE? DRUG?

Little of what we take into our bodies today is as readily classified chemically as it would have been 50 years ago, so I title this chapter with question marks. Common usage identifies foods as materials required by almost everyone to maintain a reasonably healthy life. Medicines are exceptionally pure chemicals designed to restore health. Drugs are chemicals used to alter mood. Pollutants might be classed as negative drugs or antidrugs—they diminish health, and they certainly alter moods!

Two hundred years ago, the air was often referred to as "miasmal" and leading to ague—fever and chills. Windows were tightly locked at night to exclude the air. The water was often unsafe to drink, and waterborne plagues were common. Food poisoning and dietarily deficient foods were rife. Only 60 years ago, 20 million people (550,000 in the United States) were killed in an airborne flu epidemic. The further you go back in history, the worse the reports get. But humans fought back. They did more than lock the windows at night—they learned to bathe, bury garbage, put lime on sewage, cook foods thoroughly, and establish quarantines.

Much of the recent extension of life can be traced to chemical treatment of the water we drink, the food we eat, and the wastes we generate. It is not particularly surprising that we have fallen into the trap of "If a little bit is good, more is better." So some "foods" are a combination of food, medicine, and drug (sometimes intentionally, occasionally unintentionally).

The Sea of Air Revisited

I have discussed some of the problems relating to air pollution in Chapters 12 and 14 and have given an average composition of "pure" air in Table 6.1 We are much more experienced in adding undesirable chemicals to air than in removing them. And about the only attempts to treat the air to make it more desirable than it is when pure involve the pheromones, including perfumes. I doubt that I need discuss further their effects or desirability. It might pay to say that perfumes probably arose to counterbalance the effects of the early human idea that washing was unhealthful. You may remember that perfumes were made as early as 3600 BC.

Potable, n. Suitable for drinking. Water is said to be potable; indeed, some declare it our natural beverage, although even they find it palatable only when suffering from the recurrent disorder known as thirst, for which it is a medicine. Upon nothing has so great and diligent ingenuity been brought to bear in all ages and in all countries, except the most uncivilized, as upon the invention of substitutes for water. To hold that this general aversion to that liquid has no basis in the preservative instinct of the race is to be unscientific—and without science we are as the snakes and toads. —Ambrose Bierce, The Devil's Dictionary, 1906.

POINT TO PONDER
Sprays to "freshen the air" usually dull the sense of smell rather than remove annoying chemicals.

Table 23.1   Some Contaminants of the Atmosphere Introduced by
Human Activities in the United States

Substance	Amount introduced (10^9 kg/yr)	Toxic level (ppm)	National air quality standards (ppm)	
			Average	Maximum
CO_2	3000	5×10^4	—	—
CO	91	40	9	35
CH_x	29	High	0.24	—
SO_x	30	0.25	0.030	0.14
NO_x	19	5	0.05	—
Dust	26	—	0.075	0.26

Table 23.1 lists estimates of massive contamination of the atmosphere by humans. These numbers are very large. They give 2–3 kg/day per person of contaminants in the United States. Yet the atmosphere is large—it weighs about 10^{19} kg and contains 3×10^{20} moles of gas. If all the contaminants were uniformly dispersed, we would probably find few detectable effects from them, except for dust and CO_2. But the contaminants tend to be concentrated over cities, where the people are. And there we find plenty of effects. (The solid-waste problem in the United States is far greater, about 20 kg/day per person—10 times the atmospheric load.)

The worst atmospheric contamination faced by any large group of citizens in the United States probably occurs in certain industrial atmospheres and parking garages. Rapidly improving safety standards are making such unsafe atmospheres more and more rare.

The next most concentrated pollution is probably in the "smoke-filled room." Smoking is now forbidden in many public places. And air conditioning can keep the contamination low. Many smokers no longer make the mistake of confusing close contact with friendship. The effects of smoking on the smoker are becoming more apparent (Figure 23.2 and Table 23.2), and secondary effects on nonsmoking companions have

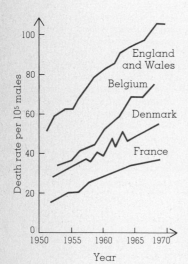

Figure 23.2
Deaths from lung cancer in some Western European countries. (Adapted from J. D. Butler, *Chemistry in Britain*, 11, 1975, p. 358)

Table 23.2   Deaths Caused by Lung Cancer in Britain
(number of deaths per 10^5 males per year)

Category	Ages 45–54			Ages 55–64			Ages 65–74		
	Rural	Mixed	Urban	Rural	Mixed	Urban	Rural	Mixed	Urban
Nonsmokers	0	0	31	0	0	147	70	0	336
Pipe smokers	0	0	104	34	59	143	145	26	232
Cigarette smokers:									
Light	69	57	112	70	224	378	154	259	592
Moderate	90	83	138	205	285	386	362	435	473
Heavy	117	214	205	626	362	543	506	412	588

NOTE: Data from J. D. Butler, *Chem. in Britain*, 11, 350 (1975).

been recognized. The main secondary effect in closed rooms is nausea in some people, and carbon monoxide levels are high enough to begin to affect efficiency in most. How many carcinogens are transferred probably will never be measured. There is so much general exposure that a zero-dose baseline cannot be established.

Table 23.2 makes it clear that both smoking (especially cigarettes) and city living increase the incidence of lung cancer. In fact, they tend to enhance one another. Positive interactions of this type are called *synergistic*—that is, mutually reinforcing. Some of the *carcinogens* in smoke have been identified, and there is no longer any question of the average effect of smoking American and Western European cigarettes; smoking considerably enhances the incidence of lung cancer. If all increased risks are lumped together, each minute of cigarette smoking, on the average, reduces the smoker's life by about 1 minute—according to an estimate made in 1976.

There are many synergistic effects among the carcinogens. There exist co-carcinogens that further enhance effects but are themselves harmless (much like coenzymes). All this makes the study of cancer difficult. Figure 23.3 shows some of the most powerful carcinogens in cigarette tar and/or city air.

There is evidence that the rapid oven-drying of cigarette tobacco practiced in the United States is a principal problem. Rapid drying kills the plant enzymes. The smoke from a cigarette has a high sugar content and is acid. These properties appear to enhance the thermal reactions *(pyrolysis)* producing carcinogens. Cigar tobacco is cured slowly, and the enzymes ferment the sugars and produce a safer, alkaline smoke. The Russians produce enzymatically cured cigarettes from tobacco grown in eastern Georgia that has in its tar no detectable carcinogenic activity toward laboratory animals.

Although cigarette smoking is undoubtedly the most serious form of air pollution in the United States, it mainly affects people who have chosen to smoke in spite of the warning from the Surgeon General on each pack. Many other people are affected by general atmospheric pollution of the NO_x–O_3 or SO_x smoke types I have discussed previously. And in many cases they have done little to produce the pollution—except for driving their cars to the store to buy the materials produced in the factories and carried in trucks.

You will recall that both common sense and the second law say that it is much easier to remove the pollutants at their source than to remove them after they have entered the atmosphere. Most of the pollutants come from combustion processes. Smoke consists partly of unburned fuels and partly of unburnable ash—as anyone who has burned a pile of leaves or wood knows. Smoke is easiest to control through more complete combustion of the fuel, and (industrially) by filtration or electro-

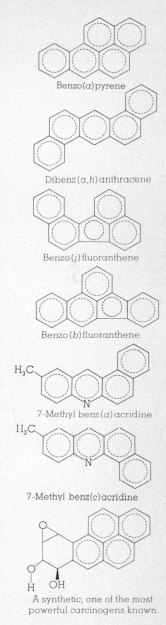

Benzo(*a*)pyrene

Dibenz(*a*,*h*)anthracene

Benzo(*j*)fluoranthene

Benzo(*b*)fluoranthene

7-Methyl benz(*a*)acridine

7-Methyl benz(*c*)acridine

A synthetic, one of the most powerful carcinogens known.

Figure 23.3
Some powerful carcinogens found in cigarette tar and/or in the urban atmosphere. Note the similarities. (Coal tar contains 1.5% benzo(*a*)pyrene, accounting for the endemic skin cancer of chimney sweeps.)

static precipitation of the unburnable ash. The filtering is usually done either with large fabric bags or (much less efficiently) with cyclone separators, which use centrifugal force to settle out the dust. Electrostatic precipitation is done by passing the smoke through an electric discharge that gives the particles a net electric charge. They are then passed over metal (say, chains to provide much surface) bearing the opposite electric charge. The charged dust is attracted to the metal, where it is electrically discharged and adheres to the surface, ready for later removal (say, by shaking the chains).

Sulfur is best removed before the fuel is burned. The volume that has to be treated before burning is much smaller than what must be treated after burning, because of the gases that form in the combustion process. Fortunately sulfur compounds are usually more reactive than the other molecules in the fuel. They can be extracted by preferential solution in a solvent if the fuel is a liquid or a very finely ground solid. Most of the sulfur in coal is in the form of iron pyrites (FeS_2, called *fool's gold*), and can be separated from finely ground coal by means of its higher density (4.9 g/cm^3 compared to 1.2 g/cm^3 for coal as mined). The ground coal is mixed with aqueous calcium chloride solution (1.35 g/cm^3) and suitable wetting agents. The pyrites settle to the bottom as the coal is washed away. Most U.S. communities had serious SO_x smoke problems from the 1920s to the 1940s. Pittsburgh and St. Louis were particularly bad examples. But methods similar to the one just described have cleaned up the effluents to an acceptable level.

The NO_x–O_3 problem is much more difficult to handle because the contaminants do not exist in the fuel. They form from the N_2 and O_2 present during combustion. They are hardest to control when the combustion is a rapid near-explosion and easier to control if a steady flame is used to generate the energy. This is the main reason for the interest in a steam car. Steam cars and turbines use steady flames. Internal combustion engines use explosions that give high local temperatures and fast temperature changes, both favorable to the formation of NO_x. Some atmospheric reactions that give smog are outlined in Table 14.5.

These combustion processes also produce unburned hydrocarbon fuel, identified as CH_x. The relation of $[CH_x]$ to $[NO_x]$ in untreated exhaust is unfortunate, as shown in Figure 23.4. The engine-design changes that lower one tend to raise the other. It is for this reason that most efforts are spent on treating the gases after they have left the engine. As usual, the problem is to find a catalyst. Some recent changes and the 1980 goals are shown in Table 23.3.

Evidence is accumulating that the drive to reduce CO concentrations may be foolish. At current atmospheric levels, even in congested traffic intersections, there are minimal physiological effects. One study has shown that London has eight times the CO concentration of the nearby island of Sark, where no cars are allowed. But London office workers

EXERCISE 23.1
The compound SO_2 can be removed from flue gas by passing the gas over $CaCO_3$. What is the net reaction? (Remember that SO_2 is more acidic than CO_2.)

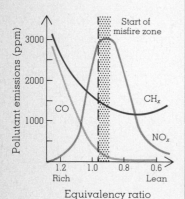

Figure 23.4
Relation of $[CO]$, $[CH_x]$, and $[NO_x]$ as a function of fuel-to-air ratio in a piston engine.

Table 23.3   Some Emission Levels for Motor Vehicles (grams per mile)

| Emission | United States | | | | Japan |
	Prior to control	1970	1977	Goal for 1981	1976
Hydrocarbons	11	2.2	1.5	0.41	1
CO	80	23	15	3.4	8
NO_x	4	—	2	1.0	2.4
Particulates	0.3	—	0.1	0.03	—

NOTE: Data from the Environmental Protection Agency.

have only twice as much CO in their blood (far below toxic levels), and outpatients in a clinic (where smoking is forbidden) have even less.

No one questions the desirability of lowering $[NO_x]$ and $[CH_x]$. Considerable progress has already been made, as you see in Table 23.3. Better tuning of the engine is the easiest way to make big gains. Redesigning engines to minimize hot spots and high temperatures is becoming effective. Two treatments of the exhaust (other than recycling through the engine) are being tried—the afterburner to lower $[CH_x]$ by further oxidation, and the catalytic converter to lower $[NO_x]$ by converting it to N_2. An especially promising three-birds-with-one-stone, single catalyst is

$$CO\ (g) + NO\ (g) = CO_2\ (g) + \tfrac{1}{2}N_2\ (g) \quad \Delta G^0 = -344\ kJ/mole \quad (23.1)$$

$$CH_2\ (g) + 3\ NO\ (g) = CO_2\ (g) + \tfrac{3}{2}N_2\ (g) + H_2O\ (g)$$
$$\Delta G^0 = -890\ kJ/mole \quad (23.2)$$

Sad to relate, most catalysts are badly poisoned by the lead from the $(C_2H_5)_4Pb$ (0.5 to 1 g/liter) used to increase the octane rating of the fuel. Thousands of substitutes have been tried. None is as effective, and several of the reasonably effective ones produce their own pollutants. Xylenes, for example, produce the carcinogen benzo(a)pyrene of Figure 23.3. Yet there is great pressure to "get the lead out" because of the possibility of lead poisoning in humans.

No one has yet suggested, or is likely to suggest, a better means of lowering air pollution from motor vehicles than minimizing the use of the vehicles. Pollution decreases as you move from cars to buses to trains to bicycles to walking. Such a move would do a great deal to convert the atmosphere from a drug to a pure food.

Water, Water Everywhere (But Not a Drop to Drink?)

More water is passed through your body than any other chemical except air. Most of it you drink, but a large portion is in your food and an almost equal portion is synthesized by you when you oxidize the hydrogen atoms in the food. The last two sources provide pure water, but most natural sources of water in the world provide water that is undrinkable,

EXERCISE 23.2
If 10^{12} liters of leaded gasoline is burned annually in the United States, what is the average daily output of lead from motor vehicles?

Fat
$(C_xH_yO_z)$
or
Carbohydrate
$[C_n(H_2O)_m]$
+
O_2
↓
$H_2O + CO_2$

EXERCISE 23.3
Write a net equation for the
reaction of $Ca(OH)_2$ (c) with
ionized aqueous $Ca(HCO_3)_2$.

usually because of the salt in it. You have read about reverse osmosis as one way to get pure water from salt water. The other principal method of purifying salt water is distillation, though selective crystallization and even the melting of icebergs have been suggested.

But even most of the fresh water in the world is dangerous to drink. Microorganisms, including those from animal and human excrement, and agricultural and industrial wastes befoul most of the world's rivers and lakes.

We learned a long time ago how to remove the merely annoying chemicals causing hard water and stains in laundry—Ca^{2+}, Mg^{2+}, Fe^{2+}, Fe^{3+}, HCO_3^-, and SO_4^{2-}. Calcium hydroxide—$Ca(OH)_2$, slaked lime— is added to convert the HCO_3^- to CO_3^{2-}, and $MgCO_3$, $CaCO_3$, $CaSO_4$, $Fe(OH)_2$, and $Fe(OH)_3$ precipitate. Some sodium carbonate is added at the same time to precipitate the insoluble carbonates of the remaining Ca^{2+} and Mg^{2+}, and to adjust the pH. The result is "soft" water with small amounts of innocuous Na^+ and Cl^- as the principal contaminants.

We have also learned to chlorinate the water to kill (by oxidation) most of the microorganisms that would otherwise "get us." Some cities use aeration, in which oxygen is blown through the water or the water is sprayed through the air for the same purpose. Aeration is what keeps mountain brooks so fresh-tasting. Bromine and ozone are now being actively explored as oxidizing additives. Addition of NH_3 with the chlorine produces chloramine ($ClNH_2$) which gives longer lasting residual protection than Cl_2 alone.

Today the use of oxidizing agents other than chlorine is being investigated, primarily because of the effect chlorine has on industrial wastes. It often chlorinates aromatic phenols (compounds similar to C_6H_5OH) to give the sharp taste that most people find offensive in drinking water. The chlorinated phenols are difficult to remove. There is also growing evidence that other chlorinated compounds that are carcinogenic are produced.

Recent data suggest that about 10% of the cancer cases in New Orleans may be attributable to drinking contaminated Mississippi River water. Another 10% appear to be due to factors found in the city but not in rural areas. It is somewhat calming to note that these differentials are not higher and that many more people would die if the waters were merely allowed to return to the conditions existing before chlorination was introduced. In 1900 the death rate in the United States from typhoid fever (largely carried by water) was about 100 per 100,000 people. Today the rate is less than 0.1 per 100,000. We find again that every solution to one set of problems poses its own threats. We must always decide what is the tolerable level. Zero pollution is unattainable.

Especially in waste water, there can be high concentrations of reducing agents. *Biological oxygen demand (BOD)* is defined as the quantity

of O_2 in milligrams required to react with the reducing agents in 1 liter of the waste water (1 mg/liter equals 1 ppm of O_2 that will react). "Good" water has a BOD of less than 1; "bad" water, a BOD of greater than 5. The usual concentration of O_2 in water varies from 7 to 15 ppm (fish require 5 to 6), so a BOD of more than 15 requires further dissolving of O_2. Waterfalls, ripples, and wind all help. One of the problems in a "meat-raising" economy is the high BODs resulting from animal wastes in the water. If humans equal 1, then pigs equal 2, sheep equal 3, and cattle equal 16 in terms of waste generation. Even chickens equal 0.1. The total animal waste in the United States equals that of 2×10^9 people!

Sometimes there are dramatic cases of water poisoning—as in Minamata in 1953 and Niigata in 1965, both in Japan. Both cases were apparently due to the presence in the water of dimethyl mercury, $(CH_3)_2Hg$. It formed by bacterial action on the metallic mercury in the effluent from a plant making polyvinyl chloride. In 1970 the waters of Lake St. Clair opposite Detroit were found to contain 7 ppm of mercury, presumably from the effluent of a plant making sodium hydroxide.

Intensive control methods are now in force at plants using large quantities of mercury, but it will take time for the bottom sludges containing the accumulated mercury to be flushed to the ocean. More encouraging is the pressure on the industries, and the response of most of them, to monitor their effluents to see what problems might arise. But it should be clear that even the best detection methods may not find the things that can produce new and unexpected results, especially those with long lag-times between exposure and symptomatic changes.

In a growing number of places, the biggest problem with water supply is contamination due to animal excrement and agricultural fertilizer. Nitrogen compounds, present in both, are oxidized to highly soluble nitrate, which enters the groundwater. If taken internally, this water may be reduced to nitrite (NO_2^-) by bacteria in the intestine. The nitrite can then oxidize the iron in hemoglobin to Fe(III), which will not carry oxygen. The disease, called *methemoglobinemia,* is one source of the blue-baby syndrome—the water and oxygen requirements of the infants are high, and their supply of hemoglobin is low because of the nitrite. Using bottled water is the only current solution.

The Chemical Feast

A recent feature story (under a *New York Times* News Service byline) was headlined "The Taste-Testing Goes on for a Chemical-Free Meat." The writer would do well to look at data such as those in Table 23.4. *Every food is 100% chemical.* You have already learned that foods (other than air and water) can be classified as carbohydrates, fats (now

EXERCISE 23.4
A river upstream from a city contains 10 ppm O_2. The city introduces a BOD of 8. What will be the concentration of O_2 downstream from the city? Any effect on the fish?

Animal wastes contaminate the water supply with nitrogen compounds.

Table 23.4 Your Breakfast—As Seen by a Chemist[a]

Chilled Melon		Coffee Cake	
Starches	Anisyl Propionate	Gluten	Methyl ethyl ketone
Sugars	Amyl acetate	Amylose	Niacin
Cellulose	Ascorbic acid	Amino acids	Pantothenic acid
Pectin	Vitamin A	Starches	Vitamin D
Malic acid	Riboflavin	Dextrins	Acetic acid
Citric acid	Thiamine	Sucrose	Propionic acid
Succinic acid		Pentosans	Butyric acid
		Hexosans	Valeric acid
Scrambled Eggs		Triglycerides	Caproic acid
		Sodium chloride	Acetone
Ovalbumin	Lecithin	Phosphates	Diacetyl
Conalbumin	Lipids	Calcium	Maltol
Ovomucoid	Fatty acids	Iron	Ethyl acetate
Mucin	Butyric acid	Thiamine	Ethyl lactate
Globulins	Acetic acid	Riboflavin	
Amino acids	Sodium chloride	Mono- and	
Lipovitelin	Lutein	diglycerides	
Livetin	Zeazanthine		
Cholesterol	Vitamin A		

Coffee		Tea	
Caffeine	Acetone	Caffeine	Phenyl ethyl
Essential oils	Methyl acetate	Tannin	alcohol
Methanol	Furan	Essential oils	Benzyl alcohol
Acetaldehyde	Diacetyl	Butyl alcohol	Geraniol
Methyl formate	Butanol	Isoamyl alcohol	Hexyl alcohol
Ethanol	Methylfuran		
Dimethyl sulfide	Isoprene		
Propionaldehyde	Methylbutanol		

NOTE: Data from the Manufacturing Chemists Association, Washington, D.C.

[a]All the chemicals listed are found naturally in the food. None is an "additive" or "synthetic." The lists certainly do not include all the natural substances present.

called lipids), proteins, vitamins, or minerals—in decreasing order of the amount you need of each in a well-balanced diet. (Review Table 1.2.)

Figure 23.5 traces the chemistry of the first three classes to two of their main products, CO_2 and H_2O. All the paths converge on the citric acid cycle, a principal means of generating both heat and stored energy to run the body metabolism.

Most of these steps have other steps that effectively reverse them. Your body chemistry can interconvert many carbohydrates, lipids, and proteins, though you cannot synthesize the essential lipids and amino acids. In time of great body drain, the fats are consumed first, followed by the carbohydrates. Finally the body begins to cannibalize the protein in the muscles—producing the emaciation usually associated with starvation.

Most of the emergency actions of the body are reversible. One time that these actions appear *not* to be reversible is during the formation of

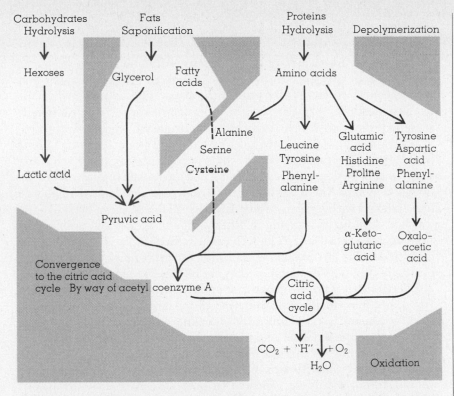

Carbohydrates
Hydrolysis

Fats
Saponification

Proteins
Hydrolysis

Depolymerization

Hexoses

Glycerol   Fatty
            acids

Amino acids

Alanine

Serine

Cysteine

Lactic acid

Leucine
Tyrosine

Phenyl-
alanine

Glutamic
acid
Histidine
Proline
Arginine

Tyrosine
Aspartic
acid
Phenyl-
alanine

Pyruvic acid

α-Keto-
glutaric
acid

Oxalo-
acetic
acid

Convergence
to the citric acid
cycle   By way of acetyl coenzyme A

Citric
acid
cycle

CO_2 + "H" $+ O_2$

H_2O Oxidation

Figure 23.5
Some main paths in converting food to energy. Don't worry about names you don't recognize. They indicate molecules involved in the sequences, but the formulas (though known and available in any text on biochemistry) are not essential to your following the paths from foods to CO_2 and H_2O. Note that all these foods converge to a single end mechanism—a highly efficient scheme.

the brain, which begins in the fetal stage and continues until the infant is about 18 months of age. Malnutrition of the mother or of the young infant appears to cause irreversible brain damage in the infant. This is the period in which all the brain cells the individual will ever have are laid down. Deprived children appear to have both fewer brain cells and less-capable ones.

It is quite likely that the best expenditure of funds in many places is not on school buildings but on feeding pregnant women and infants. If an educated population is the desired goal, it must first be educable, which means well-fed during the early months.

Until fairly recently, eating other than freshly cooked food was a gamble in most places in the world, including the United States. So preservatives were sought and introduced. The rate of food poisoning dropped dramatically, and the availability of a large variety of foods throughout the year rose just as dramatically. Talk with your grandparents or someone who remembers what the early years of the century were like—ask them what vegetables they ate in the winter, what fruit they had in the summer, and how much meat was available. Or talk to someone who is attempting to farm without fertilizers or to prepare foods without preservatives.

We now have some 2000 additives on the "generally regarded as safe"

EXERCISE 23.5
Why, at a molecular level, does cooking render food safe to eat?

BUT IT SURE STAYS FRESH!

list (GRAS list) in the United States. Most are flavors, some are colorants, some are preservatives, and others modify texture.

The ultimate, some say, is a loaf of bread that will keep fresh forever and taste good never. The nutritious germ is removed from grains to produce a flour easier to store and use. (Lipids in the germ oxidize and become rancid on storage, so they must be treated or removed.) This removes most of the nutrients other than carbohydrate. So we invent "enriched" flour to which some of the nutrients have been returned.

A varied diet balances out the lacks, but the narrow selection of food eaten by many young people and by many uneducated people can cause noticeable dietary problems. People may be on semistarvation diets even though they have the money to buy enough food and have available to them markets with a great variety of healthful foods. It is a little like the situation of diabetics starving to death—though their bodies are flooded with glucose—because they lack the insulin to catalyze the digestion of the glucose. The phenomena of limiting reagents and rate-determining steps apply to diets as much as they do to other things.

There is little if any evidence that organic farming (however the term is used) produces food better than that of the large commercial growers. What is needed is (1) informative labeling as to contents, (2) more government-provided staff and standards to ensure that the foods are safe to eat, and (3) as varied a diet as possible, supported by some education on what constitutes a balanced diet. Neither government nor manufacturers should be expected to protect you from yourself, but they should provide safe alternatives and information on which to base the choices.

POINT TO PONDER
The food situation in the United States mirrors in many ways a major problem of our times. How do you allow and even encourage people to make choices while minimizing (or preventing?) their making bad ones? One way, of course, is not to deceive them.

I Don't Feel Well, Doc

Diseases are often classified into the following groups:

1. *metabolic,* due to gene deficiencies;
2. *invasive,* due to bacterial or viral infections;
3. *poisoning,* due to harmful chemicals from outside;
4. *deficiency,* due to dietary lacks;
5. *eliminatory,* due to inability to get rid of some waste;
6. *degradative,* due to some body part showing excessive wear;
7. *mental,* due to most of the foregoing plus social pressures;
8. *mechanical,* due to blows from outside.

Let's look at the molecular nature of some of these problems.

Genes Can Be Predictable

More and more genetic diseases can be diagnosed and treated as our knowledge of enzyme chemistry improves. Over 2000 genetic diseases have been identified. About 200 can now be chemically diagnosed in the fetal stage by sampling the amniotic fluid in which the fetus is suspended. If the free cells there show enzyme deficiencies, so probably will the fetus. The technique is called *amniocentesis*. The parents can then decide, on the basis of known predictions for that disease, whether to consider abortion, the care of a permanently handicapped child (often of short life), or possible treatment at birth to allay development of the disease.

I strongly encourage you to get genetic counseling before marriage, and to repeat the procedure after marriage when planning your family. A great deal of very helpful information is available, and the amount is increasing at a rapid rate. Currently 1 in 30 infants is born with serious deficiencies. The percentage of healthy babies could be increased appreciably with good counseling. Many unfavorable predictions can be made with a probability ranging from 1 in 10 to near certainty—far better than the 1 in 30 of chance.

Analysis after birth is even more certain. Figure 23.6 shows the analysis of urine from an infant with phenylketonuria compared to that from a normal child. This disease, caused by the lack of a single enzyme, can be controlled by selecting food low in phenylalanine for the infant. Without such dietary control, permanent brain damage occurs. About 1 in 10,000 infants lacks this enzyme, and 1% of hospitalized mental patients owe their condition to untreated phenylketonuria.

Repelling the Invaders

Figure 23.7 shows what has happened to four diseases in the United States. The story could be repeated with other viruses. Even though new viruses (and mutations of old ones) keep appearing, the available techniques have enormously decreased the probability of the massive, lethal virus epidemics that used to be common. The general technique is to produce chemically modified viruses that cannot reproduce. They are not toxic, but they can initiate the *chemical immune response* in humans. The immune responses are complicated and far from understood, but we do know that they involve the generation within the body of large protein molecules, called *antibodies,* which attach tightly to the virus and render it unable to reproduce. The presence of the modified virus provides the pattern for the generation of antibodies against the virus. If the individual is later infected by the active virus, the preformed antibodies attach themselves to and disable the virus before it can reproduce.

There is a good chance that cases of some diseases will be reduced to zero. No cases of virulent smallpox have been detected anywhere since 1974 thanks to a worldwide campaign coordinated by the World Health Organization, WHO. Now WHO has laid plans to immunize, by 1990,

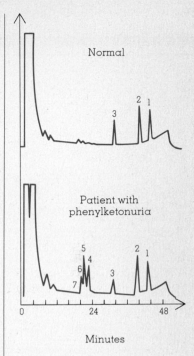

Figure 23.6
Urine analysis for phenyl-ketonuria. Urine was extracted using organic solvents. Peaks: (1) hexacosane (internal standard); (2) tetracosane (internal standard); (3) undecandioate (internal standard); (4) 3-phenylpyruvate; (5) 3-phenyllactate; (6) 2-hydroxy-phenylacetate; (7) mandelate. (Adapted from R. A. Chalmers and A. M. Lawson, *Chemistry in Britain,* 11, 1975, p. 290)

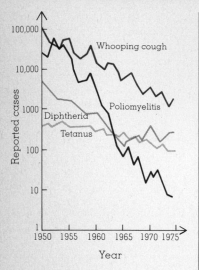

Figure 23.7
The recent history of four
diseases in the United States.
(Data from the annual reports of
the Center for Disease Control,
National Center for Health
Statistics)

every child in the world from all four of the diseases reported in Figure 23.7.

But no public or private health system can reduce contacts with all infectious bacteria and viruses to zero. So medical research tries to establish what level of attack individuals can probably resist. Table 23.5 is a summary of some allowable levels for contaminants in food, in air, and in water. It may or may not make you feel better to realize that normal individuals have much higher concentrations of highly infectious bacteria on their skin than they are likely to find in their food. But the normal body defenses handle the potential invaders quite well.

Sulfa drugs and penicillins were among the early "wonder drugs" effective against invasive diseases. The sulfa drugs mimic the essential molecule para-aminobenzoic acid, PABA (Figure 23.8), occupy its normal site on enzymes, and stop enzyme action. The invader dies.

The penicillins take advantage of the fact that bacteria have a network wall around each cell. Human cells have only a membrane. The

Table 23.5   Some Allowable Contaminant Levels

Contaminant	Limit
Microorganisms (number per gram of food)	
Salmonella	0.02
Staphylococcus	1
Enterobacteriaceae	10
Viable yeasts and molds	10^2
Clostridia	10^3
Lancefield Group D streptococci	10^4
Viable bacteria	10^5
Air Pollutants	
Asbestos	No visible emissions
Mercury	2300 g/day from a stationary source
Beryllium	10 g/day or level such that ambient concentration in vicinity of source does not exceed 0.01 μg/m³ averaged over a 30-day period

Water Pollutants (ppm discharged in effluent)

	Fresh water	Salt water
Aldrin–dieldrin	0.0005	0.0055
Benzidine	0.0018	0.0018
Cadmium	0.040	0.320
Cyanide	0.10	0.100
DDT, DDE, DDD	0.0002	0.0006
Endrin	0.0002	0.0006
Mercury	0.02	0.100
PCBs	0.28	0.010
Toxaphene	0.001	0.001

NOTE: Data from U.S. Public Health Service.

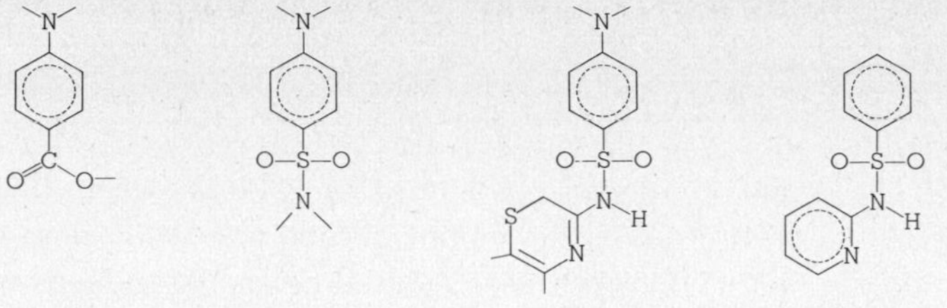

Figure 23.8
Some sulfa drugs and para-
aminobenzoic acid, PABA.

PABA   Sulfanilamide   Sulfathiazole   Sulfapyridine

penicillins prevent the synthesis of network cell walls, so the invaders die (Figure 1.3). But some bacteria have "learned" to synthesize penicillinase, an enzyme that deactivates penicillin. So now researchers look for a molecule to block penicillinase. The battle of synthesis between the scientists and the invaders continues.

One Person's Waste, Another Person's Poison

There are thousands of known chemical poisons. In fact, you may respond that everything is poisonous in a large enough dose. You are correct. A large number of substances that are poisonous in rather small doses are ions of metals. The most discussed now are mercury (as you saw under water purity), lead (as mentioned under air purity), and cadmium. These ions bind strongly to sulfide, which is present in all enzymes. Different enzymes are susceptible to each of these elements, and the symptoms of poisoning are quite different. There is evidence that the three metals have synergistic effects.

The worldwide annual commercial production of Cd, Hg, and Pb is 18×10^6, 10×10^6, and 8000×10^6 kg/year, respectively. Some of this (the second law again) is bound to become widely distributed in wastes. About half again as much mercury enters the atmosphere from the burning of coal. Tobacco smoke contains appreciable amounts of Cd, though in less than toxic doses (we think).

Water from lead plumbing (rare in the United States today), contains appreciable amounts of Pb. Concentrations are higher in areas having soft water. In fact, it has been suggested that the decline of Rome was at least partly due to the lead plumbing in the houses of the rulers. Lead is a poison of the central nervous system. It dulls the ability to arrive at clear conclusions.

Many other data are available to indicate that the situation with respect to all three metals is dangerous, that the current allowable levels in food, water, and the atmosphere are probably too high, that young people with their higher metabolism are more at risk than adults, but that toxic levels are still beyond likely doses. Heavy-metal dosage in water and food will probably be harder to control than air pollution,

POINT TO PONDER
The number of completely safe substances is still what it was in Chapter 1—zero.

EXERCISE 23.6
Assume that all the world's annual production of mercury enters the atmosphere and is uniformly distributed. If none leaves the atmosphere, how many years will elapse before the concentration becomes 0.001 ppm, one-fiftieth of the "allowed" level?

POINT TO PONDER
Man has learned to live, often
unconsciously, with many
poisons. All the following garden
plants may be fatal if eaten in
moderate amounts: azalea, black
locust, castor bean, daffodil
(bulbs), lantana (berries),
oleander, and rhubarb (leaves).
Those lovely Christmas poin-
settias have a sap that blinds if
it gets in the eyes. Many other
plants can cause stomach upset.
So we restrain children and eat
with care.

but modern detection techniques at least allow us to know what is
going on.

A recent British study suggested that tolerance levels for Cd, Hg,
and Pb in food should be 0.008, 0.014, and 10 ppm, respectively, and that
in water the levels of Cd and Hg should each be less than 0.001 ppm.
(The present FDA standard is 0.5 ppm in food and 0.005 ppm in water
for Cd and Hg.) The effect of the suggested British standards would be
the exclusion from markets of most ocean fish, especially those caught
in coastal waters off rivers bearing industrial discharges. Exclusion of
fish caught in the open ocean seems extreme. Humans have eaten such
fish for thousands of years, and old museum specimens show mercury
contents close to those found now. Most land-based food, water, and
air meet the suggested standards. Twenty years ago, it would have been
impossible to consider setting or monitoring even the present FDA
levels. If toxic levels were exceeded, the victims never knew what hit
them—just like the Roman senators with their "modern" plumbing.

Metal poisoning can be treated by administering chemicals (medicines) that combine with the threatening metal more tightly than do enzymes. The chemicals are chosen for easy elimination and, of course, low
toxicity. One such chemical (called BAL, formula $HOCH_2$—CHSH—
CH_2SH) was originally synthesized as an antidote to the war gas Lewisite ($ClHC=CHAsCl_2$), and it has been found to be a good antidote for
poisoning by As, Au, Te, Tl, or Bi—but not by Pb. So, another example
of an "antihuman" chemical (Lewisite) leading to something useful
for humans!

It is common to express the toxicity of substances in terms of the
dose that is lethal to 50% of individuals exposed to it. This dose, the
LD_{50}, is often given in terms of milligrams of poison per kilogram of
body weight.

When the Second Law Takes Over

"The statistics are grim. Each year in the U.S. more than 50% of all
deaths are due to cardiovascular diseases. Of these deaths, nearly 84%
are directly linked to arteriosclerosis, the hardening of the arteries that
produces degenerative heart disease, stroke, and other arterial diseases.
Cancer, on the other hand, accounts for about 18% of the deaths in the
United States each year." So read an article in *Chemical and Engineering News* (September 15, 1976, p. 20). About the same time, a Reuter's
News Agency report said that "death from cancer in the U.S. has doubled during the last four decades, and 60 to 90% of all cancer was related
to environmental factors." Both articles had a surprised tone, even a
hurt tone. Yet neither report should surprise you.

Most cardiovascular disease and probably most cancers are degener-

EXERCISE 23.7
Find the cause of and age
at death of as many of your
ancestors (including their
brothers and sisters) as you can.
Any trends in disease or age
at death?

ative—enhanced by the accumulation of environmental insults to which we subject our bodies. There are always more ways available for things to go wrong than to go right. As the death rates from genetic, invasive, poisoning, deficiency, eliminatory, mental, and mechanical diseases fall, the rate for degenerative must rise—unless we find the fountain of youth and perpetual life, a doubtful gain. Death rates from environmental effects with a long lag between exposure and effect will also rise. The time is rapidly approaching when individuals will ask themselves—with the possibility of influencing the result—"Of what do I wish to die?" Already cardiovascular failures (the principal present cause of death) seem the most likely choice.

It is not widely recognized that the great gains in life expectancy have occurred for those in their early years. The life expectancy at 60 has risen very slowly, that at 70 even more slowly. Body systems wear out, and the replacement problems become overwhelming. One of the principal reasons the cost of medical care has tripled in the last 30 years in the United States is that the cheap solutions are fewer. Shots and pills are cheap, but repair and maintenance jobs are difficult and expensive (Figure 23.9).

The Wonder Drugs

I have already described briefly the action of two of the earliest wonder drugs (medicines?), the sulfas and the penicillins. There are many others. If you are interested, get a good book on pharmacology and medicine.

But I would nominate for the highest category of "wonder" the medicines for controlling reproduction—"the pill" (actually a variety of pills). Here, for almost the first time, and on a very large scale, humans are deliberately using chemicals to control a normal human function. The women using the pill are not ill. They just wish to reproduce less often than they are biologically able. In effect, they are questioning and modifying the normal operation of their bodies. And well they might question, as we shall discuss in some detail in the next chapter. Whether —and, if so, how—they modify their behavior involves vital social and political questions.

It used to be said that a woman could not be partly pregnant. But no longer. Some 10 million women in the United States and millions more abroad are regularly "partly pregnant." The most common pills contain chemicals closely related to human hormones that are normally released during pregnancy to prevent further ovulation. No ovulation, no fertilizable egg, no reproduction.

There are many steps in the reproductive process, in both the female and the male, that seem susceptible to control. So far the best results

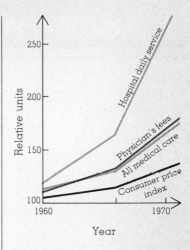

Figure 23.9
The rising cost of medical care.
(Data from U.S. Department of Labor, Bureau of Labor Statistics)

POINT TO PONDER
How many times have you heard a person say, "I must go take my drug." Seldom! We buy drugs at drugstores, but physicians prescribe (and we take) "medicines."

Figure 23.10
Some steps in developing the
birth-control pill. The first four
all prevent ovulation.

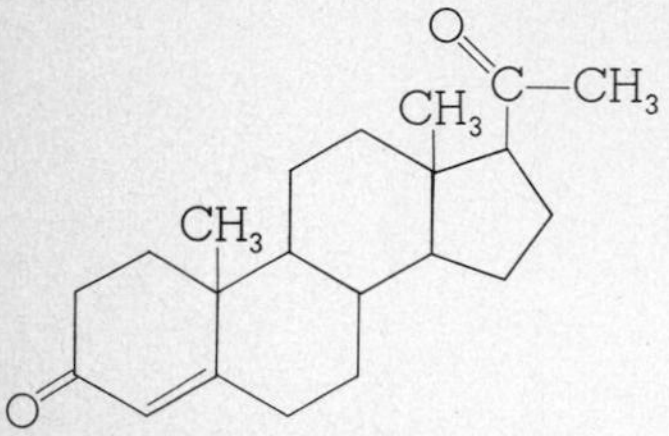

Natural progesterone; given by
injection; prevents ovulation

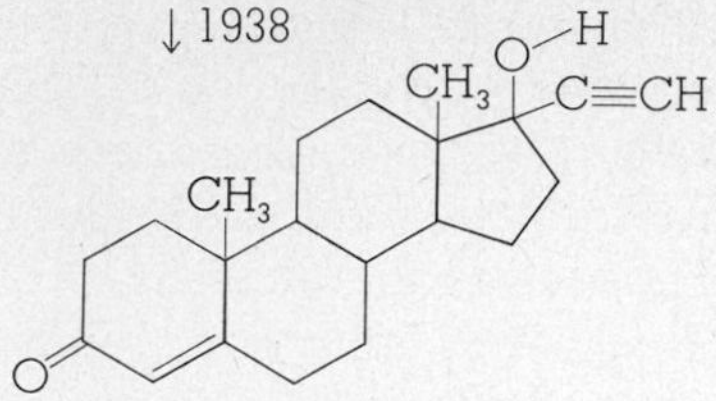

↓ 1938

Ethisterone; given orally, but
large doses needed

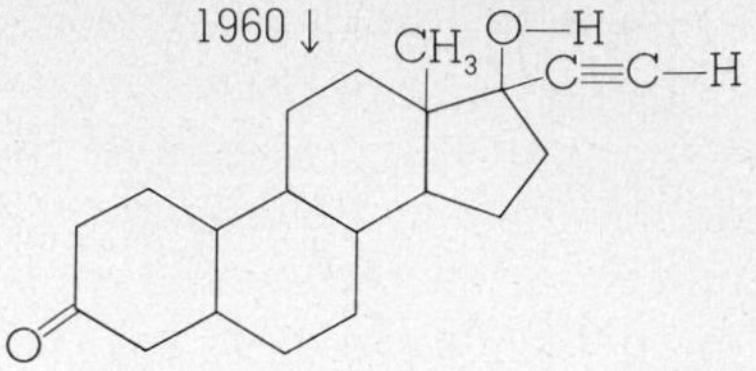

1951 ↓

no CH₃

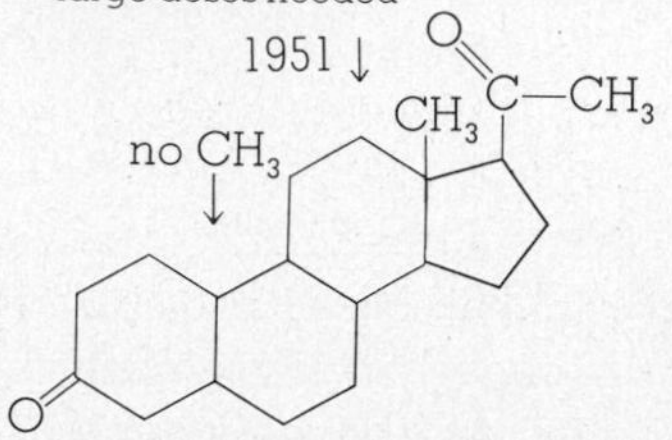

19-Norprogesterone; five times as
effective by injection as
progesterone

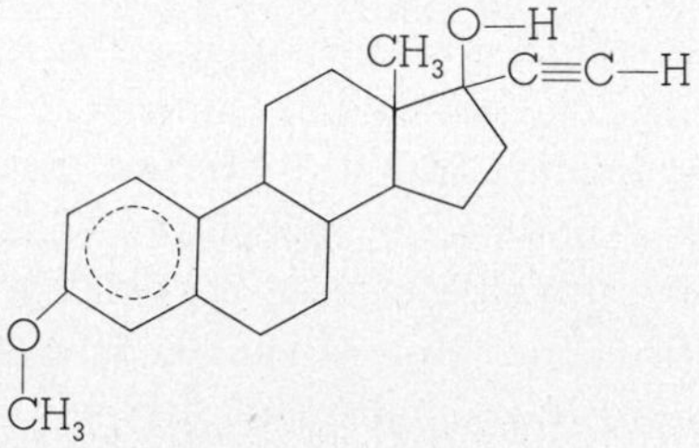

1960 ↓

Norlutin; small oral doses plus

mestranol (included to regulate
menstrual cycle)

have been produced with derivatives of progesterone (an ovarian hormone) taken orally by the female. It is currently easier to prevent the release of one of some 400 eggs than to affect its further change or to control successfully the production of millions of sperm. But all these steps are under active investigation to see whether they could be effective as rate-determining steps for controlling human reproduction.

Figure 23.10 summarizes the development of one of the birth-control drugs. Molecular modification of substances found to be active in nature is a major means of drug "discovery." Replacement of the —COCH₃ and —H combination with —C≡CH and —OH made oral administration possible, and removal of one —CH₃ group increased the effectiveness by a factor of about five (thus reducing the dose). Pills normally contain two substances. In one pill, Norlutin prevents ovulation, and mestranol controls the womb cycle. (Trade names, like Norlutin, are almost universal in medicine. The scientific name for Norlutin, for example, is 17-α-ethynyl-19-nortestosterone.)

The effectiveness of modern family planning in the United States is consistent with the observation that whereas 60% of the births in the United States in 1960 were unplanned, only 39% were unplanned in 1970. The percentage is still decreasing.

As with all chemicals, birth-control chemicals affect some people badly. Abnormal bleeding and hypertension are sometimes observed. Blood clots causing stroke or heart failure have occurred occasionally. But pregnancy itself is risky. Present birth-control dose levels cause fewer fatalities and complications than would occur if the women involved were exposed to the normal risks of pregnancy.

Additional birth-control chemicals have been and will continue to be synthesized. They are being tried in various parts of the world. Yet, in spite of the importance of knowledge concerning human reproductivity, money budgeted by the United States government for research in this area has been decreasing since 1973. Even in that peak year, less than 2% of government funds for medical research went into this area.

The benefits of successful control are great, but the threats of introduction of a dangerous drug or concoction are always present. In 1938 in the United States an inappropriate solvent, $HOCH_2CH_2OCH_2CH_2OH$, was used to dispense sulfanilamide to children. Eighty patients died, and the repercussions led to amendments to the 1906 Food and Drug Act.

Screening with animals also has its problems. Neither quinine, which causes blindness in dogs, nor aspirin, which harms rat embryos, would probably pass today's screens, though these effects are not found in humans.

The 1961 thalidomide debacle in Europe, in which a new drug for morning sickness in pregnant women resulted in the birth of more than 3000 deformed children, again struck home. Tragedy was prevented in the United States because the Food and Drug Administration tests uncovered *teratogenic* (birth-defect) effects in rats, so permission to sell the drug in the United States was refused. This experience provided further dramatic proof that extensive testing of both short- and long-term effects is necessary. Further revisions to the Act were introduced in 1962. It is now estimated that at least 6 years, and sometimes 20 years, plus a pre-use investment of $10 million or more may be required to develop a new drug from the laboratory to large-scale use in humans.

Drugs—The Mind Benders

For thousands of years humans have explored different means of facing (some say escaping from) "reality." We meditate, fast, do breathing exercises, and use yoga, tai chi chuan, chants, and drugs (chemicals that alter mood). Table 23.6 lists the common classes of drugs. The original member of each type of drug in a class is usually discovered by accident or by screening (testing on animals) thousands of the compounds newly sythesized each year. Screening produces about 10 interesting compounds per 1000 tried, but usually less than one of these is finally marketed. Usually it has by then undergone further molecular modification.

Alcohol (ethanol, C_2H_5OH) is not listed in Table 23.6, but alcohol is, by far, the most widely used (and abused) drug. As Table 23.7 shows, all simple alcohols are relatively toxic—ethanol is merely the least toxic. It is a relaxant in small quantities, then a stimulant, and a narcotic and hallucinogen at still higher doses. Two-thirds of all adults in the United States drink some alcohol, and an estimated 9 million are alcoholics unable to perform adequately in a full-time job. Over half of our fatal

POINT TO PONDER
Risks involved in new drugs may be put in perspective by a 1977 article in the *Journal of the American Medical Association*. According to J. Porter and H. Jick, one hospital patient in every 3600 dies because a doctor prescribed already existing drugs incorrectly. The *Washington Post* quotes Jick as saying that "very serious adverse reactions are about as infrequent as one could possibly expect, given the enormous amount of exposure to drugs."

Table 23.7
Toxicity of Some Simple Alcohols

Alcohol	Toxicity (g/kg)
Methanol, CH_3OH	a
Ethanol, C_2H_5OH	13.7
Propanol, C_3H_7OH	1.87
Butanol, C_4H_9OH	4.36
Hexanol, $C_6H_{11}OH$	4.9

[a] CH_3OH is not in itself very toxic but it is oxidized to formaldehyde, $H_2C{=}O$ (LD_{50}, 0.1 g/kg), which is toxic. Ingested, CH_3OH probably kills more people than any other alcohol except C_2H_5OH.

Table 23.6   Some Common Drugs Other Than Alcohol[a]

Class	Examples	Comments[b]
Analgesics (painkillers)	Aspirin Acetanilide 	
Tranquilizers	Meprobamate (Miltown, Equanil) $C_3H_8C(CH_3)(CH_2OCONH_2)_2$ Phenothiazines	
Sedatives	Barbiturates 	Possibly addictive; LD_{50}, 1.5 g/kg. Synergistic with C_2H_5OH; LD_{50} may become 0.01 g/kg

	R_1	R_2	
Barbital (Veronal)	$-C_2H_5$	$-C_2H_5$	Kills animals quickly (600)
Pentobarbital (Nembutal)	$-C_2H_5$	$-CH(CH_3)C_3H_7$	Short-acting hypnotic (130)
Phenobarbital (Luminal)	$-C_2H_5$		Long-acting anticonvulsant (660)
Amobarbital (Amytal)	$-C_2H_5$	$-C_2H_4CH(CH_3)_2$	Intermediate-acting (575)
Secobarbital (Seconal)	$-CH_2CH{=}CH_2$	$-CH(CH_3)C_3H_7$	Short-acting hypnotic (125)

Class	Examples	Comments[b]
Anesthetics	Diethyl ether $(C_2H_5)_2O$ Cocaine Halothane $F_3CCHBrCl$	Compare the structure of morphine with that of cocaine

Table 23.6 (continued)

Class	Examples	Comments[b]
Narcotics	Morphine Methadone (not a narcotic)	Essential features of all morphine-like substances: 1. Aromatic ring; 2. C bonded to 4 C's; 3. Tertiary N removed from feature 2 by 2 carbon atoms; 4. —O—R group

	R₁	R₂	
	R_1	R_2	
Morphine	—OH	—OH	(500)
Codeine	—OCH₃	—OH	(120)
Thebaine	—OCH₃	—OCH₃	
Heroin	—OCOCH₃	—OCOCH₃	(150)

Class	Examples	Comments[b]
Stimulants	Ephedrine	(5)
	Amphetamine	
	Caffeine (No-Doz)	(200)
Hallucinogens	LSD Tetrahydrocannibinol (active marijuana)	Colored part of structure also in mescaline, psilocybin, serotonin; see Figure 23.14.

[a]In the structures neither H atoms nor ring C atoms are usually shown
[b]Numbers in parentheses are rough LD_{50} values in milligrams drug per kilogram body weight, usually for oral administration in rats

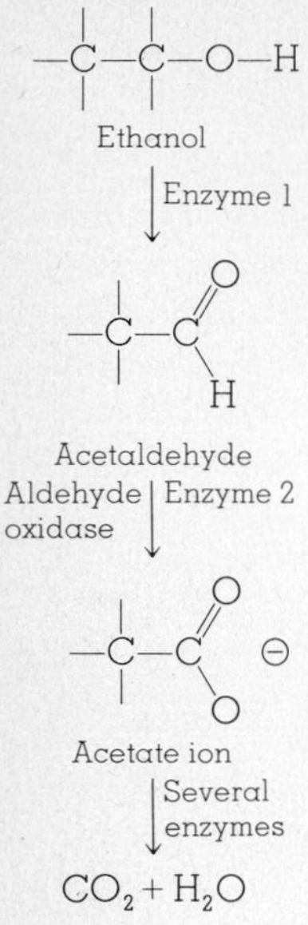

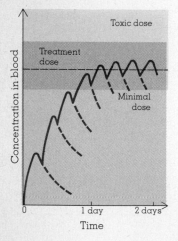

Figure 23.11
The metabolism of ethanol in humans.

EXERCISE 23.8
What is there about the formula of Antabuse that suggests it might be effective in blocking a metalloenzyme?

Figure 23.12
The effect of "Take one pill four times a day" on dosage level. Uptake is rapid. Loss is usually first order.

auto accidents have at least one of the drivers "under the influence."

As with most drugs and medicines, the detailed effects of alcohol on physiology are unknown. We do know that alcohol can cause irreversible damage to brain and liver, and we also know how alcohol is metabolized. Alcohol enters the blood rapidly and then disappears in a zero-order reaction. As Figure 23.11 shows, it is enzymatically converted to acetaldehyde, CH_3CHO (LD_{50}, 1.9 g/kg), which is further converted enzymatically to acetate, a common body metabolite with no further bad physiological effects at reasonable levels.

One treatment for those who wish to "kick the habit" has been to take $[(C_2H_5)_2N{-}CS]_2$, a drug known as Antabuse. It blocks the metalloenzyme that catalyzes the oxidation of acetaldehyde. Acetaldehyde builds up in the blood, causing the drinker to feel miserable and possibly conditioning him to give up drinking. However, the drug is seldom used now because of its own side effects.

The drugs listed in Table 23.6 are but a small sample of the available arsenal. All can be dangerous. Some of the most dangerous, like heroin, are also highly addictive (habit forming).

The safety of a drug can be expressed as its *therapeutic index,* the ratio of the toxic dose to the treatment dose. A high therapeutic index indicates a safe drug, and when it is coupled with a low dosage rate we have a "good" drug. One of the problems is to keep the dose level in this range with prescriptions such as "Take one pill four times each day" (Figure 23.12).

The United States produces about 2×10^7 kg of aspirin per year, 3×10^5 kg of barbiturates, 5×10^5 kg of sedatives, and (if the figures in the *Statistical Abstract of the United States* are to be believed) 2×10^7 kg of stimulants. The population of the United States is about 2×10^8 people, so we produce about a quarter of a kilogram (half a pound) per person per year of the types of drugs listed in Table 23.6.

We know detailed physiological action of none of these drugs. Until the prostaglandins were discovered, there was no hint as to how aspirin controlled pain as well as it does. However, with new analytical techniques and great interest in physiological effects, we can expect a rapid growth of detailed information.

Even a casual examination of Table 23.6 shows that many drugs related in structure are also related in action. In fact, it is generally assumed that drugs that are active in small doses attach to a limited number of *reactive sites* and block or stimulate a biological process at that site.

In some cases, as with the morphine-like drugs, good guesses can be made as to the shape of the site. It must fit the structure outlined in Table 23.6, probably with a negative charge at one end to hold the positive nitrogen, a groove in which the carbon chain fits, and a π-bonding site at the other end to interact with the aromatic ring. These suggestions of site geometry should call to mind the Amoore theory of odor.

The announcement in 1976 of natural "opiates" in humans offers further hope of finding an ideal pain deadener. One, called β-endorphin, contains 31 amino acids. A second, called methionine–enkaphelin, contains the first five amino acids in β-endorphin. Both are produced in the pituitary glands of humans. Both have also been synthesized in the laboratory. There is every reason to anticipate the discovery of the molecular basis of their properties and modes of action.

Study of much pharmacological data suggests that drugs must match reactive sites in geometry, in polarity, and (often) in chemical reactivity (acid–base, redox, and bond-breaking properties). As we learn more about sites, it will become possible to design medicines to fill the deficiencies of the natural system. As we learn where the sites are, it will be possible to administer the drug more effectively. What a waste to put it in the mouth if it is needed in a tiny identifiable site in the arm! It is in this feature, among others, that the oriental system of acupuncture has caused so much excitement. Great progress has already been made in this direction in the use of anesthetics. Local anesthetics coupled with relaxants or acupuncture have replaced general anesthetics in many cases.

One of the great problems in intelligently discussing the use of tobacco, alcohol, marijuana, street LSD, and other mass-produced drugs is their variation in chemical identity. Unlike most medicines and drugs, they are far from pure. Figure 23.13 shows analyses of three different samples of marijuana, for example. About all a scientist can say is that this is an easy way to trace a sample to its origin, because the analysis is characteristic of the source. It does not predict the physiological activity.

Brain, Nerves, and Mind

Although the modes of action of the drugs in Table 23.6 are still widely debated and largely unknown, it is clear that they interact strongly in most cases with the nervous system. I have already pointed out that the hormonal system is closely linked to the nervous system, and that enzymes are intimately involved as well. In fact, it is only for convenience in learning and discussing that we segregate the systems at all. Body systems all operate in a coordinated, thermodynamically downhill manner at rates controlled by numbers, energies, and geometries of collisions between molecules with rather fixed structures and polarities.

Figure 23.14 illustrates some of the known interrelationships in two common chemical systems—those of phenylalanine and tryptophan.

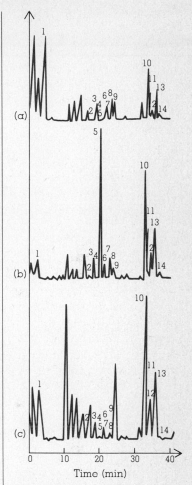

Figure 23.13
Analyses of samples of marijuana from (a) Turkey, (b) Mexico, and (c) Indiana-grown Mexican stock. Each of the numbered peaks comes from molecules of known structure. For example, peak 5 is $C_{12}H_{22}O_3$, and peak 13 is $C_{20}H_{38}$. The separation is achieved by a kind of distillation called vapor phase chromatography. (Reprinted with permission from M. Novotny, M. L. Lee, Chow-Eng Low, and A. Raymond, *Anal. Chem.*, **48**, 1976, p. 26; copyright by the American Chemical Society)

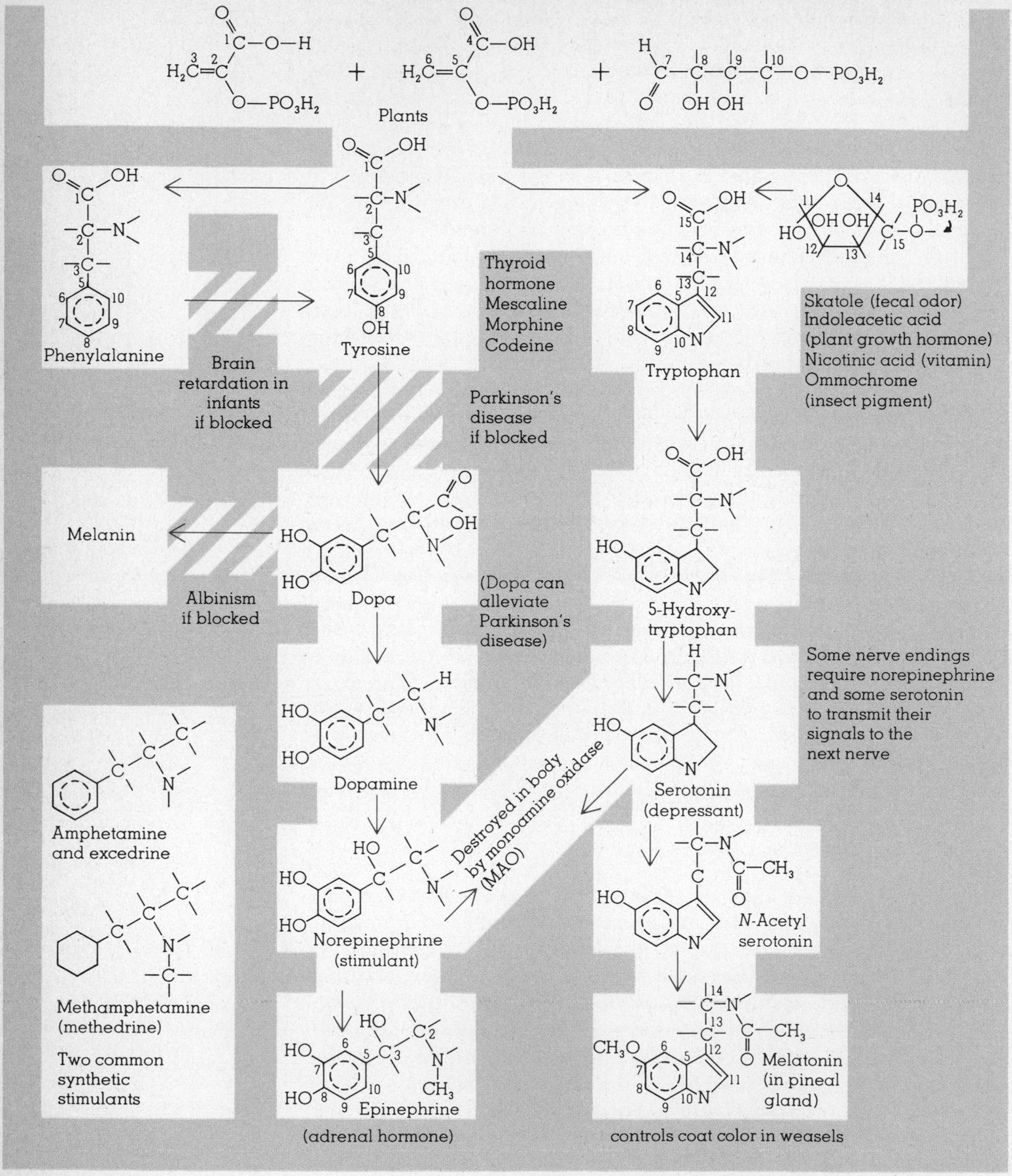

Plants

Phenylalanine

Tyrosine

Brain retardation in infants if blocked

Thyroid hormone
Mescaline
Morphine
Codeine

Parkinson's disease if blocked

Tryptophan

Skatole (fecal odor)
Indoleacetic acid (plant growth hormone)
Nicotinic acid (vitamin)
Ommochrome (insect pigment)

Melanin

Dopa

Albinism if blocked

(Dopa can alleviate Parkinson's disease)

5-Hydroxy-tryptophan

Dopamine

Some nerve endings require norepinephrine and some serotonin to transmit their signals to the next nerve

Amphetamine and excedrine

Destroyed in body by monoamine oxidase (MAO)

Serotonin (depressant)

Norepinephrine (stimulant)

N-Acetyl serotonin

Methamphetamine (methedrine)

Two common synthetic stimulants

Epinephrine (adrenal hormone)

Melatonin (in pineal gland)

controls coat color in weasels

Both these chemicals are essential amino acids for humans. Tyrosine would also be essential except that it can be synthesized from phenylalanine. Plants can synthesize all three from the chemicals listed. The numbered atoms allow you to trace the origin of the frameworks in the amino acids.

Your body has enzymes that then convert these two amino acids into a host of physiologically important molecules. Only a few are shown. You have probably heard the names of some of them. The chemical basis of three enzymatic diseases is shown. For example, albinism is caused by the lack of one of the enzymes necessary to convert dopa into the pigment melanin.

Phenylketonuria is caused by a deficiency of the enzyme that converts phenylalanine into tyrosine. When this enzyme, phenylalanine-4-monooxygenase, is deficient in newborns, they convert the phenylalanine in their diet to phenylpyruvic acid, which builds up and can cause permanent brain damage unless the phenylalanine level of the diet is lowered.

Parkinson's disease can be treated by oral administration of dopa. So something must be wrong with the tyrosine metabolism in victims of this disease, but what is wrong is not yet known. However, it is clear to the most casual observer that Parkinson's disease shows up as a nerve disorder—an uncontrolled oscillatory twitch. So we come to the brain and nerves.

Norepinephrine (from tyrosine), serotonin (from tryptophan), and acetylcholine are the chemicals that appear to carry the nerve impulses from one neuron (nerve cell) to the next across the synaptic gap (Figure 23.15). The details are not known, but the sodium ion transport signal from the axon causes the release of whichever of these chemicals is appropriate to the nerve involved. It moves across to the next nerve and there initiates a sodium ion transport signal in that nerve. These three substances are all monoamines (they have one exposed, ammonia-like nitrogen). All are deactivated by the enzyme monoamine oxidase (MAO), so they stimulate the nerve ending only once. They may then be transported back, regenerated, stored, and used to carry a later signal.

What is of immediate interest here is that many of the stimulants, narcotics, and tranquilizers (plus LSD and cocaine) contain similar nitrogens. Note that LSD actually has the aromatic ring and adjacent nitrogen-ring structure of serotonin. Similarly the amphetamine and ephedrine stimulants are much like norepinephrine. All are amplifications of acetylcholine. Whether the normally smaller effects of the barbiturates are due to less exposed nitrogens and the very heavy effects of the narcotics are due to more reactive nitrogens remains to be proved. But a coherent picture is beginning to form, and the reactive sites may

Figure 23.14 (opposite) Some metabolic relations in the tyrosine and tryptophan systems. The changed group in each step is colored.

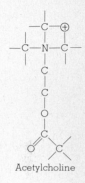

Acetylcholine

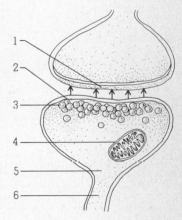

1. Postsynaptic membrane (site of absorption)

2. Presynaptic membrane (site of release)

3. Synaptic vesicles (site of storage)

4. Mitochondrion

5. Synaptosomal sap (site of synthesis)

6. Axon

Figure 23.15 A synaptic gap.

POINT TO PONDER
The real goal of pharmacology is to find what normally produced substance is missing (if a deficiency is involved), and then to introduce that substance at the appropriate site. Very little medicine has achieved this highly desirable state. We must now settle for substitutes.

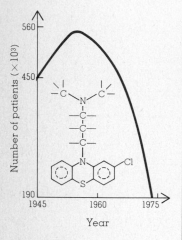

Figure 23.16
Number of resident patients in U.S. mental hospitals from 1945 to 1975. Chlorpromazine was introduced in therapy in 1952, and the widespread use of psychopharmaceuticals started in 1956. (Based on National Institute of Mental Health data)

POINT TO PONDER
A story is told in Hong Kong of the man who could not bring himself to kill the large African snails in his garden. So he spent an hour each day gathering them and taking them to a desert island where there was nothing for them to eat.

soon be identified as to geometry, polarity, and location. There is every reason to believe that mental diseases will be found to involve similar reactive sites.

In the meantime, it is wise to remember that minor changes in molecules can produce substances of considerable difference in physiological action. For example, all the molecules in Table 23.6 have similarities, yet each has a different function in the body. The problem in drug design is to produce molecules that will give the desired effect where wanted—without any undesired side effect elsewhere. Most drugs are prone to the latter. And none of the drugs mentioned, except dopa, appears to be a substance the body would normally synthesize.

In spite of the lack of detailed knowledge as to the nature of mental disease, considerable progress has been made in minimizing its effects through use of chemicals (Figure 23.16). A principal reason for the decline in mental-hospital inpatients starting in 1956 was the introduction of chlorpromazine as a tranquilizer. This has been followed by a widening selection of drugs, including lithium compounds highly effective in some manic–depressives. It now also seems that a major factor in the onset of schizophrenia is genetic. Perhaps it, as with some other genetic diseases, can be treated in terms of the enzymes involved. There is a growing group of scientists who feel that every mental disease will turn out to have a chemical basis, often (but not always) activated by social situations. The search for useful changes at the molecular level will join the search for useful changes at the social level in their treatment.

Poisons for All—Herbicides, Insecticides, and Other "Cides"

The world is a highly competitive place. All but the simplest forms of life digest other living systems in order to survive. Sometimes they must ingest certain specific molecules (what we have called *essential*), but sometimes they can settle for a wide range of foods that they can convert to meet their needs. Omnivores, to which class humans belong, can eat a wide variety of substances.

Sometimes two or more individuals try to digest the same thing. Or one individual tries to digest the other and finds resistance. Then the competition gets very direct. When a plant competes with a crop you wish to raise, you call the plant a weed. When an insect or other animal tries to eat the crop, you call it a pest. When an insect tries to eat you, you call it a bigger pest (with emphatic language, no doubt)—and when a larger animal tries to eat you, you may call for a gun and set out to rid the area of "vicious beasts." Humans have invented herbicides and pesticides to help them in their competitions with other living systems.

Table 23.8 lists some recently used herbicides and pesticides. (Note the suffix *-cide,* from the Latin *cido,* "to kill.") The early ones, such as DDT, were discovered by accident or by trial and error. Today more and more are being discovered by studying healthy plants and pests to see what they need to live. They are then fed the same material in excessive doses (the juvenile-insect and plant-growth hormones), or are fed substitute materials that will attach to the enzymes but will not be metabolized properly (the newer sulfa drugs). Or we flood a field with pheromones but place a trap rather than a female at the origin of the odors. Wide-area field surveys have been coupled with weather data and the factors that are known to influence pest life-cycles. A computer analyzes the data and prints out a map showing areas in which pesticides can be most effectively used. This procedure can reduce annual pesticide applications 85%, yet increase the control over pests. Many other techniques are also used.

These measures can be very effective. One estimate is that over half the food raised in the United States would not come to market without herbicides and insecticides. It is almost certain that, were all of them banned from use, a massive famine would engulf the world, and that no nation would eat as well as it does now.

The ideal herbicides and insecticides would be easy to apply, widely effective against all the species humans wished to kill, harmless to all other species, and cheap to manufacture. But even if the ideal formulations were discovered, the path to use is involved and long (Figure 23.17).

Sometimes only one species is involved, and then a specific poison (specific to that species) would be ideal. But as Table 1.2 shows, living systems have remarkably similar biochemistries. About the only chemicals that meet the specificity criterion are the insect hormones and the pheromones. So very active research is under way in these areas. Field tests are highly promising, but major successes have been limited.

But the genes are not quiescent, especially in insects. Mutations occur, and the breeding rate of insects is so high that many mutations occur in each generation. Some of these turn out to be resistant to poisons. Resistance to DDT was an early example. A large fraction of the houseflies in some areas are resistant to DDT. Penicillin is another example. Some bacteria, as I have already pointed out, now have an enzyme called penicillinase that can inactivate the drug. In 1976 the first insects resistant to malathion were discovered. This genetic mutability has further increased the pressure to find hormonal methods of controlling specific species. It is extremely doubtful that a species will develop a viable defense against its own hormones.

An equally serious problem is the resistance of many of the early pesticides to degradation. So they (or some of their early degradation products) accumulate in the environment. Although they may not be

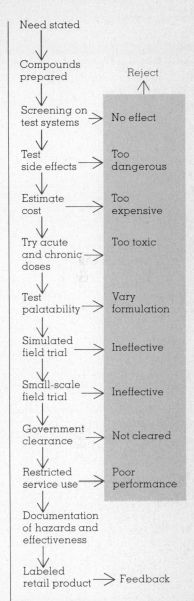

Figure 23.17
A typical development path for a pesticide. The minimum time from start to finish is probably 5 years, the average about 10 years.

Table 23.8   Some Herbicides and Pesticides[a]

HERBICIDES		
Common name	Formula[a]	Comments
2,4D		Pregnant women should avoid contact with 2,4D
2,4,5T		Liquid formulations illegal, granular form possibly dangerous
Calcium cyanamide	$Ca^{2+}(N{=}C{=}N)^{2-}$	Used to defoliate the cotton plant before the cotton is picked
PESTICIDES		
Pyrethrin		Safest; effective
Malathion		Used for aphids and scale insects; has low toxicity
Parathion		Half-life 1–10 weeks; toxic

[a] H atoms and ring C atoms are not usually shown.

Table 23.8   (continued)

PESTICIDES		
Common name	Formula[a]	Comments
Carbaryl, Sevin		Used for chiggers and Japanese beetles; half-life 1 week; little toxicity
Lindane		Possibly dangerous
DDT		Illegal
Chlordane		Illegal, except for underground termites
Heptachlor		Illegal
Dieldrin		Illegal; general insecticide; kills fish

EXERCISE 23.9
Malaria has recently increased
dramatically in frequency in
both Ceylon and India. Suggest
some reasons.

toxic at the original dose, they may become toxic at the concentrations that build up as one predator eats another in the food chain.

Egg-laying reptiles and birds seem susceptible to DDT poisoning at DDT concentrations reached near the top of the food chain. The DDT appears to interfere with the enzyme carbonic anhydrase, which catalyzes the formation of calcium carbonate eggshells. The thin shells that result account for poor hatching rates and a low number of offspring. In many areas this effect seems to have hit a maximum. With the now diminished use of DDT, many bird and reptile populations are recovering. For many species it was a close call, and it is still too early to know whether all will recover. But the warning is clear, and it has been heeded. Yet total elimination of DDT would result in the death of millions of people from insect-borne diseases like malaria.

Massive public-health campaigns have eliminated many diseases in limited target areas, as you saw in Figure 23.7. Other campaigns appear capable of eliminating some diseases on a worldwide basis. Virulent smallpox has been eliminated, and the less virulent form now exists only in parts of Ethiopia. Similar gains against bacterial diseases and specific insect and plant pests seem currently unlikely, due to the genetic mobility of these fast-breeding species. But selective elimination is the goal. Many feel that gains are more apt to come from a thorough study of the species and their detailed internal chemistry, rather than from the synthesis and trial of a large number of compounds. The possible number of compounds is just too large. In some cases the "perfect" synthesized medicine may have been missed by the "misplacement" of a single atom. The sensitivity of living systems to molecular modification is sometimes an advantage—and sometimes not.

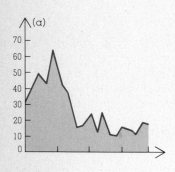

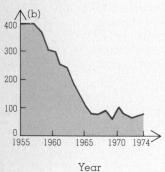

Year

Figure 23.18
(a) Number of newly synthesized medicines introduced
annually in the U.S. market.
(b) Total new medical products.
(Adapted from A. Burger,
Chem. Engr. News, September 22, 1975, p. 39; reprinted
with permission of the copyright
owner, the American Chemical
Society)

Summary

Figure 23.18 may well be the best possible summary to this chapter. The number of newly synthesized chemicals introduced in the drugstores each year has dropped remarkably from its 1959 peak. The reasons are manifold, most often appearing finally in the cost of introducing a new drug. This cost has mounted rapidly because of several factors. Many of the epidemic diseases can be handled with existing medicines, so the anticipated volume of sales of new medicines is more limited. There is a greater awareness of side reactions, and the public has a desire (which will never be fulfilled) for a no-risk treatment. There is growing concern at the environmental buildup of residues from past treatments. And there is the increased paperwork to be submitted in obtaining FDA approval—three copies of 167 volumes and 72,000 pages in the case of the drug ketamine, an anesthetic. In the case of ketamine, the elapsed time from the start of clinical studies to approval for use was 7 years.

On the other hand, clearance for cancer treatments is more rapid—perhaps because the life expectancies of the patients are low in any case and the tolerable risks can be higher. Antibiotics also tend to get more rapid clearance.

There is a widespread feeling among research scientists that a new surge in discoveries of important drugs is not far away. The many studies on basic physiology and detailed analyses of healthy and sick people are beginning to allow chemicals to be designed to treat physiological inadequacies. The almost random screening of chemicals against their effects on symptoms will continue. But molecules designed for known needs of particular species will become more common.

The probable rate-determining step will be the extent of risk taking the public is willing to accept. Currently the acceptable risks are low, and the rate of new-drug introduction is correspondingly low.

HINTS TO EXERCISES
23.1 Forms $CaSO_3$. 23.2 2×10^8 g Pb/day. 23.3 $CaCO_3$ and H_2O form. 23.4 Fish will die. 23.5 Cooling decomposes enzymes. 23.6 10^3 years. 23.7 Both genetics and living conditions. 23.8 Sulfur forms strong bonds to many metals. 23.9 Higher population density and lower DDT use.

Problems

Note: You will have to consult sources such as handbooks or encyclopedias to arrive at answers to many of these questions.

23.1   In the 1920s, most midwestern cities occasionally smelled of SO_2 from the burning of soft coal. The odor is apparent at 3 ppm. Compare this level to the toxic dose. Have you ever smelled SO_2 in city air? Comments?

23.2   The hospital practice of placing both non-smokers and smokers in the same double room has been questioned. How would you proceed to advise the hospital director on this policy?

23.3   Outline some of the reasons people are much less willing to accept polluted atmospheres now than they were even 50 years ago.

23.4   If you had a garden, would you use synthetic fertilizers? Pesticides? Be specific.

23.5   Nicotine is much more toxic taken intravenously than orally. Suggest a possible reason.

*23.6   Prior to 1900, most white paints had basic lead carbonate, $Pb_2(OH)_2CO_3$, as pigment. It turned black in coal-burning communities. Lithopone, formed by mixing BaS and $ZnSO_4$, was next as a white pigment. Now TiO_2 is the overwhelming favorite with 5 to 10 times the covering power of its predecessors, so giving "one-coat painting." How could BaS and $ZnSO_4$, both soluble in water, give a good pigment? Barium is as poisonous as lead. Why is there great concern over children chewing lead-painted articles, but not those painted with lithopone?

*23.7   The patient whose intestinal tract is to be x-rayed is commonly fed $BaSO_4$, even though Ba^{2+} is a poison. There are no bad effects, and the Ba causes heavy x-ray shadows, thanks to its high atomic number. Every so often a patient is fed $BaCO_3$ by mistake and becomes very ill, even dies. Why the difference?

23.8   Examine labels for the following products and list the ingredients. Compare costs. Use the *Merck Index* to identify some properties of the ingredients. Laxatives, monosodium glutamate, antacids, sodium benzoate, pain relievers, sodium nitrite, cough medicines, relaxants, sleeping pills.

23.9   Use the *Merck Index* to find the LD_{50} or toxicity of the following:
   a. ethanol (as in beer, wine, and liquor);
   b. acetylsalicylic acid (as in aspirin);
   c. ascorbic acid (as in vitamin C);
   d. sodium benzoate (as in catsup);
   e. sodium nitrite (as in artificially reddened hot dogs).
Comment on the possible toxicity of these compounds.

*23.10 Assume that an auto exhaust containing 80 ppm of CO and 4 ppm of NO comes to the catalyzed equilibrium described in equation 23.1. Calculate the concentrations of CO and NO. Remember that exhaust gases contain about 65% N_2 (g) and 20% CO_2. [$p(NO) \cong 10^{-56}$ atm.]

23.11 Figure 23.4 suggests that a fuel-to-air equivalency ratio of 0.6 would give minimum pollutants. Would you recommend such an engine?

23.12 The presumed tolerable body burden of Hg is 0.3 mg. Assume that the Hg is uniformly distributed into all 10^{14} cells of your body (a most unlikely situation). How many atoms are present in each cell? [$\sim 10^4$ Hg atoms/cell.]

23.13 A buildup of Cu(II) in the brain leads to irreversible damage unless early treatment is given. One treatment is with a penicillamine:

$$(H_3C)_2-\overset{\overset{\displaystyle SH}{|}}{\underset{\underset{\displaystyle H}{|}}{C}}-\overset{\overset{\displaystyle NH_2}{|}}{\underset{\underset{\displaystyle H}{|}}{C}}-\overset{\overset{\displaystyle O}{\|}}{C}\diagdown_{O-H}$$

Suggest a possible acid–base type interaction. (Hint: How many Lewis-base atoms in penicillamine?)

23.14 Smallpox vaccine was developed by Edward Jenner in the eighteenth century from innoculated cows, after he noticed that milkmaids had beautiful complexions. Outline a line of reasoning he might have followed.

23.15 Medical geography leads to some strange correlations. Esophageal cancer is found at an above-average frequency in men living near illicit liquor stores *(shebeens)* in Africa, and in Turkoman women of northern Iran who eat raisins ground with pomegranate seeds and black peppercorns (N. D. McGlashan, *Medical Geography,* Methuen, 1972). The cause in both cases is probably dimethyl-*N*-nitrosoamine. The empirical formula is $C_2H_6N_2O$. Suggest a structural formula and its electron structure. [30 valence electrons]

*23.16 Scientists normally analyze herbs, natural medicines and drugs, and poisons to isolate *the* active ingredient. This is done by bioassay, where each separated fraction of the material is tested for physiological activity. Discuss some of the problems inherent in this technique.

23.17 Alpha-amino acids, the building units of proteins, all have the structure

$$R-\overset{\overset{\displaystyle H}{|}}{\underset{\underset{\displaystyle N}{|}}{C}}-\overset{\overset{\displaystyle O}{\nearrow}}{\underset{\underset{\displaystyle }{}}{C}}\diagdown_{O-H}$$
$$\qquad\quad \overset{|}{\underset{}{N}}$$
$$\qquad H\quad\ H$$

The NH_2 is a weak base (K $\sim 10^{-9}$), the CO—H hydrogen a not-very-weak acid (pH ~ 2). What would be the structure of the molecule in strong acid, say in the stomach? What would the structure be in strong base? At pH ~ 6 both groups are electrically charged, but the molecule as a whole is electrically neutral. How can that be? It is said to be at its isoelectric point (that is, zero net charge).

*23.18 A very large number of drugs used on humans are eliminated from the body by a first-order mechanism. L. Dettle and P. Spring (*Physico-Chemical Basis of Drug Action,* Ariens, 1968) claim to be astonished by this. Suggest why first order is reasonable and zero order is not unlikely. For example, ethanol (C_2H_5OH) elimination is zero order in humans.

23.19 Should chemical agents be used to control riots or unruly prisoners always, sometimes, or never? Be specific as to when and as to alternatives.

23.20 a. A Compoz ad says, "This modern relaxant tablet contains no barbiturates so it helps you sleep more naturally." Does this sentence make sense?

b. An Anacin ad says, "Anacin gives you 23% more pain reliever/anti-inflammatory medication than Bayer." Is this necessarily better?

c. A Bufferin ad says, "Unlike Bayer or Anacin, Bufferin has special stomach protection ingredients." Cite and comment on some possible interpretations.

23.21 One of the proposed amendments to the Clean Air Act provides that where air quality is better than required by the standards there must be "no significant deterioration" of that quality as a result of any facility constructed in the future. Discuss the merits and demerits of this proposal and make a recommendation as to its passage.

23.22 On May 18, 1976 the U.S. Senate voted to give $1,250,000 compensation to the family of Frank Olson. Olson was an Army chemist who in 1953 was given LSD without his knowledge during a drug experiment sponsored by the Central Intelligence Agency. He developed serious effects and jumped to his death from a hotel room while under psychiatric treatment for the effects. Why would anyone give such a drug to a person without prior consent? Do you agree with the Senate action? What action would you recommend toward someone who spiked a party drink with LSD?

23.23 The World Health Organization (WHO) in 1976 called on governments to draw up programs to control and prevent smoking and to strengthen antismoking education, based on "indisputable scientific evidence that tobacco smoking is a major cause of chronic bronchitis, emphysema, and lung cancer." Outline possible reasons that only a few countries have taken any action in this area.

23.24 Drinking water in the Chicago area contains 0.018 ppm of Hg. Does this concentration seem dangerous if the toxic dose is assumed to be 300 μg/day? [$\sim$20 μg Hg/day]

23.25 The assumed toxic dose for mercury is 300 μg/day. Dental offices sometimes contain 180 μg/m^3 of air. If a human inhales 2×10^4 liters of air per day, is a job in a dentist's office hazardous? [$\sim 10^3$ μg Hg enter nose each day]

23.26 A dieting lady in New York City ate 0.35 kg of tuna per day. The tuna contained 1 ppm Hg. How does her daily dose, which led to obvious poisoning symptoms, compare with the presumed toxic dose of 300 μg/day? [$\sim$ 300 μg Hg/day]

23.27 Roger J. Williams and colleagues at the University of Texas in 1971 tested the effectiveness of each of the following as a sole diet for weanling rats: roasted peanuts, milk, puffed rice, hamburger, eggs. Arrange these foods in the order you think they would most closely completely fit the needs of growing rats, and defend your selections. The Texas group has produced an enrichment additive that materially improves the worth of bread to rats but does not improve puffed rice. Suggest an interpretation at the molecular level.

23.28 Kangaroo rats of North America never drink water, yet their body water content is about the same as that of other rats. Their diet is high in carbohydrates, low in fats, and as low as possible in proteins, although it supplies the required amino acids. How do these facts tie in with a desirable water balance?

*23.29 Bernard Rimland at the Institute for Child Behavior Research at San Diego writes (in S. C. Plog and R. B. Edgerton, eds., *Changing Perspectives in Mental Illness,* Holt, Rinehart and Winston, New York, 1969) as follows: "One needs only cite cretinism, phenylketonuria, galactosemia, epilepsy, and diabetes among the many organic diseases with clear mental or behavioral involvement which are readily amenable to medical control. Those who believe that psychological problems are necessarily more hopeful than physical ones seem oblivious to history, which shows that centuries of lawmaking, teaching, preaching, threatening, punishing, explaining, persuading, and cajoling have not resulted in a notably more exemplary man. Preventative and remedial medicine, on the other hand, have made memorable strides, even in many disorders that defied solution while they were called 'functional' " (p. 720). Comment on this situation in molecular terms.

23.30 Philip A. Butler points out in "Monitoring Pesticide Pollution" (*Bioscience,* **20,** 1970, pp. 889–891) that approximately 100 miles of beaches on an estuary in northwest Florida were lightly sprayed with DDT three or four times during early summer to control stable fly. If the DDT had been evenly dispersed in the estuary the concentration would have been 0.001 ppm. Butler found the following DDT concentrations (in parts per million) in some of the local living systems: plankton, 0.07; pinfish, 0.1–0.5; loon liver, 180; mullet gonads, 3–10; bottlenose dolphin blubber, 800. Can you suggest a reason for the differences in these numbers?

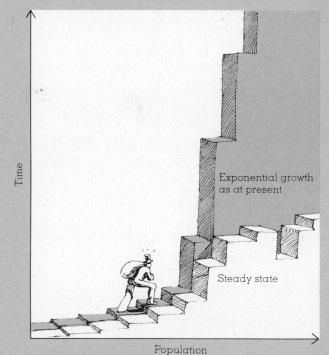

Figure 24.1
Which is preferable (possible)?

24 STEADY - STATE SOCIETY?

The world's population became 4 billion people on March 28, 1976, according to one count. It had reached 1 billion in 1850, 2 billion in 1930, 3 billion in 1960. The 5 billion mark will probably be reached by 1990. One projection suggests 8 billion by 2010—about the time your children will be having (and you will be enjoying) your grandchildren. These figures are based on a current estimate of 60 million deaths, 135 million births, and a net increase of 75 million people annually. The actual increases in 1970 and 1974 were 70 million and 63 million, respectively. The annual increase when I was born was about 20 million, slightly more than a fourth of the present value. This exponential growth in number of people over the last 200 years is the basis of much of the current discussion on the carrying power of the earth. Is there a reasonable total world population? (See Figure 24.1.)

The reaction of many population experts to such questions is similar to that in 1972 of Frank W. Notestein, longtime professor of population studies at Princeton University and the first director of the Population Division of the United Nations: "It is irrelevant to ask how much the globe can carry. This is not a meaningful question. If there were no intervening political, social, and economic frictions, so that man could do the best that man knows, then the carrying capacity becomes indefinite.... The problem is, here we are, where are we going? What are going to be the constraints? What advances of civilization, of rationality, of compassion, and of lowering of pain can be made, and how? What is the course, among the conceivable courses, that we want to take?"

In my opinion you will make few choices in your life more important than your decisions on these questions. A knowledge of chemistry will help you winnow out the false claims you will read. This chapter and the next one will set up a possible framework concerning chemical constraints and conceivable courses of action, based on the preceding chapters. Will some chemical be the limiting reagent, or will human planning set the limits?

Will We Run Out of Oxygen?

In the late 1960s it was suggested that consumption of fossil fuels would deplete the oxygen supply in the atmosphere. This suggestion—and many others like it—can be answered quite adequately with "back-of-

It is scarcely necessary to remark that a stationary condition of capital and population implies no stationary state of human improvement.
—*John Stuart Mill, 1867.*

POINT TO PONDER
Exponential growth is analogous to exponential decay. The latter has a half-life, the former a doubling time given by dividing 70 years by the rate of growth. (For example, 2% annual growth means a doubling time of 70/2 or 35 years.) No system has ever maintained exponential growth indefinitely. If it did it would approach infinity.

Column of
atmosphere

15 lb/in.²
at the surface
of the earth,
or 1.0
kg/cm²

Figure 24.2
The weight of the atmosphere.

the-envelope" calculations. It is surprising how often the calculations are not done. Let's try this one. We will calculate first the total atmospheric oxygen and then the amount of oxygen that would be consumed if all available fossil fuels were burned. For our initial purposes a calculation to one significant figure is quite sufficient.

You know from Chapter 12 that the atmospheric pressure is about 15 lb/in.² (Figure 24.2.) In other words, the weight of atmosphere above each square inch of surface is 15 lb. You also know that 20% of this air is oxygen, so the weight of oxygen is 0.20×15 lb/in.² $= 3$ lb/in.² How many square inches of surface are there?

The diameter of the earth is 8000 miles. There are 5000 feet per mile, and 10 inches per foot (remember, we are using one significant figure). So $8000 \times 5000 \times 10 = 4 \times 10^8$ in. The radius of the earth is half the diameter, or 2×10^8 in. To find the area of a sphere we use the formula $4\pi r^2$. So,

$$\text{Area of earth} = 4 \times 3 \times (2 \times 10^8)^2$$
$$= 5 \times 10^{17} \text{ in.}^2 \qquad (24.1)$$

The weight of gaseous oxygen is 3 lb/in.², so

$$\text{Weight of total atmospheric oxygen} = 3 \text{ lb/in.}^2 \times 5 \times 10^{17} \text{ in.}^2$$
$$= 15 \times 10^{17} \text{ lb} \qquad (24.2)$$

A pound is about 0.5 kg, so we get 8×10^{17} kg of O_2. We convert this to moles of dioxygen, to get

$$8 \times 10^{17} \text{ kg } O_2/0.03 \text{ (kg/mole } O_2)$$
$$= 3 \times 10^{19} \text{ moles } O_2 \text{ in the atmosphere} \qquad (24.3)$$

Now we need to know how much fossil fuel is available for burning. This figure is more uncertain than any of the foregoing figures, which is one reason we do the calculation to only one significant figure.

Various estimates of fossil fuel quantities (including fuel probably usable but not yet discovered) have been made. The largest such estimate for recoverable coal, oil, and gas is 9×10^{15} kg of carbon and 0.5×10^{15} kg of hydrogen. It is easier to do the calculations if we convert to moles:

$$9 \times 10^{15} \text{ kg C}/0.01 \text{ (kg/mole C)} = 9 \times 10^{17} \text{ moles C} \qquad (24.4)$$
$$0.5 \times 10^{15} \text{ kg } H_2/0.002 \text{ (kg/mole } H_2) = 3 \times 10^{17} \text{ moles } H_2 \qquad (24.5)$$

We will assume that the carbon burns to carbon dioxide, the hydrogen to water.

$$C \text{ (c)} + O_2 \text{ (g)} = CO_2 \text{ (g)} \qquad (24.6)$$

$$H_2 \text{ (g)} + \tfrac{1}{2}O_2 \text{ (g)} = H_2O \text{ (}\ell\text{)} \qquad (24.7)$$

Each mole of C requires 1 mole of O_2 (g); each mole of H_2 requires $\tfrac{1}{2}$ mole of O_2 (g). To find the total amount of O_2 required in the combustion of all the C and H_2, we calculate as follows:

Available O_2 . . . . 3×10^{19} moles O_2

9×10^{17} moles C require $\quad$ 9×10^{17} moles O_2

3×10^{17} moles H_2 require $\quad$ $\underline{2 \times 10^{17}\text{ moles }O_2}$

Total O_2 consumed $\quad\quad$ 11×10^{17} moles O_2 $= \underline{0.1 \times 10^{19}\text{ moles }O_2}$

% Change in atmospheric O_2 . . . . -3%

Thus, even if all the available oxygen-consuming fuels were burned (the figures include forests, plants, and grasses), only 3% of the present atmospheric oxygen, or $0.03 \times 20 = 0.6\%$ of the atmosphere, would be consumed. This loss of O_2 would have no detectable effect on life. Even burning five times the known reserves (this would cover all the available oil shale and tar sands) would leave plenty of oxygen. Oxygen would still be $20 - (5 \times 0.6)$ or 17% of the atmosphere.

A more recent worry has been that ocean contamination would kill the plankton, which are responsible for about half the oxygen regeneration from carbon dioxide. Suppose we go to the extreme and assume that all the plankton and all the plants lose their ability to regenerate oxygen. How would that affect the atmospheric oxygen?

To answer this question we need an estimate of the rate of cycling of oxygen through the atmosphere. The cycling time seems to be about 8000 years (Figure 2.11). So, if all means of oxygen regeneration were destroyed, the oxygen content would drop 1% in 80 years. If only the plankton were destroyed, the 1% drop in oxygen would take place in 160 years. We have seen that burning all the available carbon and hydrogen fuels would use up 3% of the oxygen. This would require almost 500 years (at present rates) and would give 0.6% CO_2 in the atmosphere. Our fuel would be gone (it will get used up anyway) but loss of plankton would not appreciably affect the oxygen balance in the near future. Of course, the loss of plankton would mean loss of almost all fish (which feed on them). It's hard to say why people don't concentrate on large effects (loss of fish) rather than small ones (loss of atmospheric oxygen).

The Hothouse Effect

However, recall our previous discussion about suffocation. You have seen that it is not usually the loss of oxygen that is toxic or fatal; it is the CO_2 buildup. How about the buildup of atmospheric carbon dioxide? Actually, if all oxygen regenerators were destroyed, we would die of starvation—but if only the plankton were destroyed, and if the plants

EXERCISE 24.1
The atmosphere of the early earth contained no free O_2. Most of the oxygen present now has been formed from CO_2 by photosynthesis. Some of the carbon compounds formed simultaneously are in available fossil fuels. Most are buried in the ocean sediments. Calculate the fraction of the reduced carbon that appears to be in these sediments, using the data from the available-O_2 calculation.

POINT TO PONDER
Weathering of rock also uses up O_2 (principally in oxidizing pyrites, FeS_2). Such weathering could use up half the atmospheric oxygen—in about 5 million years. Not an immediate problem.

Figure 24.3
Some present uses of sunlight,
its conversion to heat, and its
storage by means of photosynthesis. Essentially all the 7%
absorbed in photosynthesis is
eventually converted to heat
and radiated to the sky. Earth
stores much less than 1% of
the incoming solar energy.

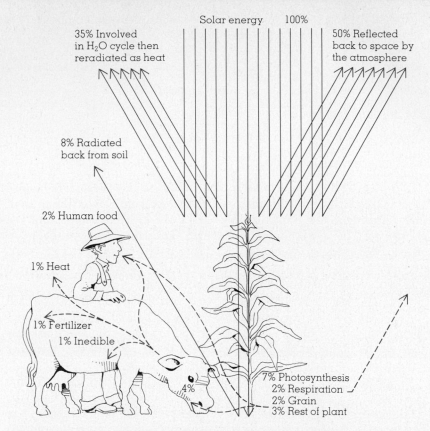

continued their cycling unchanged, the carbon dioxide content would
increase at the same rate at which the oxygen decreased, as you can see
from equation 24.6. The toxic level of CO_2 in the atmosphere due to
reaction of $\frac{3}{20}$ or 15% of the oxygen present would be reached in 15×160
or 2500 years. The "uncomfortable" level would be reached sooner. This
estimated time is a bit too short, because we know that the plants would
begin to grow faster on the higher CO_2 content and would offset the loss
of plankton to some extent. This should remind you of kinetics and the
effects on rates of varying concentrations as long as they are not zero
order.

But just a minute! You just saw that if all the available fossil fuels
were burned the oxygen content of the atmosphere would drop less than
1% (3% of its 20% total in the atmosphere). So the carbon dioxide content
could never exceed 1% of the atmosphere if it came only from oxidation
of available carbon–hydrogen compounds. This is below lethal levels.
The world will neither run out of oxygen nor suffocate if the plankton
are destroyed. Oxidizable CH_x is the limiting reagent.

A potentially more serious effect is that of increased carbon dioxide

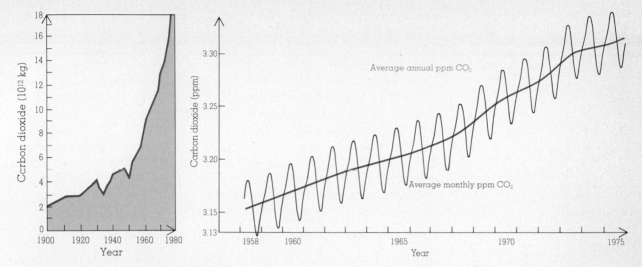

on air and surface temperatures. About 50% of solar energy is immediately reflected back to space. Most of the other 50% is absorbed and then reradiated as infrared energy (Figure 24.3). But carbon dioxide is a good absorber of infrared energy, so a higher concentration of CO_2 in the atmosphere would impede the loss of this radiation, and the air and surface temperatures would increase.

If the atmosphere contained as much as 1% CO_2 (30 times the present level), the changes in temperature would cause severe problems. One of the first effects would probably be the melting of the polar ice, a corresponding rise in ocean levels of 100 to 200 m, and the inundation of most of the major cities of the world because most are on coasts or on river plains of low altitude. Much of the United States and most of Europe and Southeast Asia would disappear beneath the flood.

Even our present practices in burning fossil fuels have caused concern. Figure 24.4 shows the increasing rate of CO_2 inputs to the atmosphere. Figure 24.5 shows the changing concentration of CO_2 in the atmosphere. (Any upset in the plankton population would increase the rate of CO_2 accumulation.) Figure 24.6 shows some data on average temperatures. Much of the temperature rise from 1910 to 1945 is attributed by many to the rise in atmospheric carbon dioxide.

But after 1945 there is a decrease in average temperature, as you can see. This decrease is attributed to the increasing particulate matter (dust and smoke) that man has introduced into the atmosphere, thus blocking some of the sunlight from reaching the earth. So, there is an argument as to which is more important in the long run—the "hothouse" effect of CO_2, or the reflective effect of dust. Perhaps we can titrate the atmosphere and control the heat effect in that way!

We started this discussion by considering whether there was a danger

Figure 24.4 (above left)
Rate of injection of CO_2 into the atmosphere from 1900 to 1970.

Figure 24.5 (above right)
Change in concentration of atmospheric CO_2. Note that it is still low, but growing.

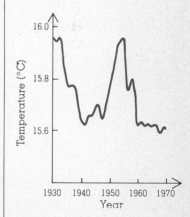

Figure 24.6
Average annual temperature in the United States. See also Figure 20.8. (Data from *Understanding Climatic Change*, National Academy of Science, Washington, D.C., 1975)

EXERCISE 24.2
One of the most common spray-
can propellants is Freon 12,
CF_2Cl_2. It absorbs ultraviolet
light and dissociates, yielding a
Cl atom. Why not a fluorine
atom, F? (Hint: See Table 9.1.)

EXERCISE 24.3
What is the rate-determining
step in the water cycle? How
can it be altered appreciably?

of running out of oxygen, and we end up by finding that temperature problems are the most likely symptom of meteorological difficulty if the ocean plankton are destroyed. There are few better illustrations of the ecological principle that everything is tied together. It is this tying together that makes accurate estimates of effects difficult, if not impossible. But do note that the very simple calculations above show that there is undoubtedly cause for concern about ocean pollution. Even with the plankton working, $[CO_2]$ is increasing in the atmosphere.

It also suggests where we should try to improve our pollution data, and what steps are likely to be rate-determining. This is a very great gain over blithe ignorance and over assumption of infinite space in which to dump wastes.

How About the Ozone Layer?

One big controversy during the 1970s has concerned the ozone layer in the stratosphere and its vulnerability to supersonic-transport (SST) exhausts, spray-can propellants, and agricultural fertilizers. It has even been suggested that the most serious effect of a nuclear war would be the damage inflicted on the ozone layer. Here is another mixed bag of effects. Who would have ever linked them?

The link is provided by the fact that SSTs, fertilizers, and nuclear explosions can introduce oxides of nitrogen into the upper atmosphere. Many spray cans introduce fluorocarbons, which are decomposed by stratospheric ultraviolet light to give chlorine atoms. Both chlorine atoms and the oxides of nitrogen have unpaired electrons (they are free radicals) and are excellent catalysts for the conversion of ozone (O_3) to dioxygen (O_2). Yet it is the ozone in the stratosphere that absorbs most of the ultraviolet light from the sun. This ultraviolet light is toxic and even lethal to life. If the ozone layer is destroyed, there would be little (if any) life on the surface of the earth. If the concentration of ozone is diminished, toxic effects (for example, increase in the incidence of skin cancer) can be predicted with confidence.

Figure 24.7 gives two 1974 estimates of the increase in skin cancers in the United States as a function of the number of Concorde-type SSTs flying. The estimated rate is 8000 additional cases of skin cancer per 1% reduction in ozone, or 50 cases annually per Concorde. Such an increase is not directly detectable. Predicted effects of agricultural nitrogen use by the end of the century range from 2 to 20% reduction in the ozone; spray cans may account for from 5 to 10% reduction. The effects on plants and plankton could be worse than this.

There is almost no disagreement that there are appreciable effects, even though it may be impossible to measure them directly. The question, as you have seen before, is how tolerable the effects are as balanced against gains. Another 10 years of collecting data should give ample evidence. This gives time to make the decisions before the effects become directly measurable. Thanks to combinations of laboratory research and engineering design, "preventive medicine" is replacing "disease treatment" in industrial planning as well as in the hospital.

Water Is More Than Something to Drink

So far we have discussed water for humans primarily as an essential food. But drinking water is much less than 1% of the total used. You drink about 1000 liters annually, but the total United States consumption is 10 million liters per person. Almost half is now used in steam-electric plants, over one-third in agriculture, and most of the rest in industrial processes. The current recycling rate in the United States is about 2.5 (every liter gets used 2.5 times between rain and return to the sea). This number will probably rise to more than 4.

Because the biggest use is in cooling steam-electric plants, there is great interest in conserving water there. If present practices are continued, the United States would "run out" of water by the end of the century. One suggestion is to circulate the hot water from the turbines through pipes buried in sand while blowing air through the sand. The net effect is to transfer the heat to the atmosphere (without requiring cooling water, or large atmospheric heat exchangers). The turbine water can then be recycled to the plant.

Agricultural water can be a much more serious limit. There are good years and bad years (as shown in Figure 24.8 for India). Usually local carrying capacity is taxed by the worst years, unless the existence of large-scale water-distribution systems can make up differences. But the world carrying capacity must consider a vital chemical, rain, as a possible limit. In much of the world, water is already the limiting factor to population—in deserts, for example. In fact, worldwide records indicate that the average climate has been exceptionally favorable for the last 50 years and no reason is known that it should remain so. The droughts and blizzards in the United States in 1976–1977 are potent evidence of the variations found in weather. The period from 1958 to 1976 was exceptionally stable (Figure 24.6). We can expect larger fluctuations now.

Water in all categories will get more expensive. Marked changes in water use will occur. But contamination is a more serious problem than running out. As you have already seen, the place to reduce contamination is at the source. The long-term problem is the ocean. The rivers and streams flush themselves. The rain is pure as a distillate from the sea. But the ocean has only biological degradation and precipitation

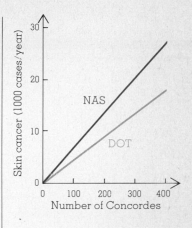

Figure 24.7
Effects of Concorde-type SSTs on the number of additional cases of skin cancer in the United States per year. Both National Academy of Sciences (NAS) and Department of Transportation (DOT) figures are shown. The Environmental Protection Agency estimates 375 Concordes in the air by 1990. (Reprinted with permission from H. S. Johnston, *Accounts of Chemical Research*, **8**, 1975, p. 289); copyright by the American Chemical Society)

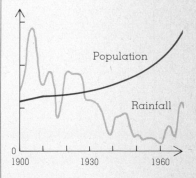

Figure 24.8
Rainfall and population in India, 1900–1972. The opposing trends make it difficult to supply adequate food.

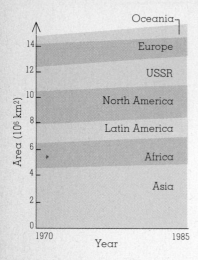

Figure 24.9
World arable land by region,
1970–1985. (Adapted from
*A Hungry World: The Challenge
to Agriculture*, University of
California Task Force, 1974)

EXERCISE 24.4
It has been suggested that we
relieve pressure on land area
by colonizing the moon. Comments? The radius of the moon
is 1000 miles.

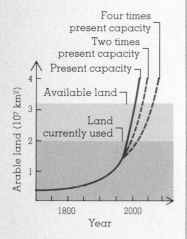

Figure 24.10
Available and needed arable
land if current population and
land-use trends continue.

with which to purify itself. Because 90% of the life in the ocean is near
the shores, contamination effects from human wastes are much greater
than if uniform dilution occurred.

Don't Fence Me In

One resource that is easy to measure is land area. The total surface of
the earth is, as you saw a few pages back, 5×10^{17} in.^{2}, or 5×10^8 km^2.
Of this total 1.5×10^8 km^2 is land, and 6×10^7 km^2 is usable land
(reasonably flat and livable). The principal large-scale use is agriculture.
The United Nations Food and Agriculture Organization (FAO) estimates that there are about 3.2×10^7 km^2 of agricultural land, half of
it currently in production. FAO suggests that the costs to settle, clear,
irrigate, and fertilize the other half are so high that it is far better to
concentrate on improving yields from the present land. So we have
perhaps 2×10^7 km^2 of land worth farming, distributed as shown in
Figure 24.9. The present population of 4×10^9, as you see, is being fed
from this land at the rate of 2000 people/km^2. If the best available agricultural technology were applied to each piece of land, the yield could
be doubled, possibly quadrupled, so it is not land that is currently the
limiting reagent. It is lack of seed, water, fertilizer, and distribution
facilities. But Figure 24.10 shows that land itself could become limiting
by about 2000 to 2070, regardless of increasing agricultural yields, if
population growth continues unabated.

Food

Food can also be estimated closely. The amount required varies with the
age, size, sex, and metabolic rate of each person, but the figures we will
use are about average for a large population. The basal metabolic rate
in kilojoules per day is 100 times the body weight in kilograms. (This is
the amount of energy required just for the body to function when you
rest in bed, not counting digestion.) Assuming an average weight of 70
kg, we get 7000 kJ/day. An additional 750 kJ is needed for food digestion. Our average working individual will need a total of about 9000–
10,000 kJ/day. Note that the majority of the energy goes into bodily
functions, not into external work. Human beings are less efficient than
machines at converting energy to external work (15–20% efficiency for
humans, 20–25% for machines). This is one reason that we should use
and improve machines. You will remember that fuel cells could be very
efficient (60–80%) converters of energy to work.

All our food is the direct or indirect product of photosynthesis, with
sunlight providing the energy at the rate of 8 J/min per cm^2 at the top of
the atmosphere. You read earlier that half of this energy never gets to

the earth's surface. Furthermore, half the earth is dark at any one time, so the average rate of receipt of solar energy is $\frac{1}{4} \times 8 = 2$ J/min per cm^2 on a surface perpendicular to the direction to the sun. Average food needs are 10,000 kJ/day $= 10^7$ kJ/day $= 7 \times 10^3$ J/min for each person. This requires the energy received on

$$7 \times 10^3 \text{ (J/min)}/2 \text{ (J/min} \times \text{cm}^2) = 4 \times 10^3 \text{ cm}^2$$

or 0.4 m^2. Thus, an area of less than half a square meter of land would provide your daily food energy requirements if conversion were complete. This may give you some perspective on why the direct capture of solar energy is such an important goal. However, photosynthesis of the most efficient crop in the world, corn, only produces a 1% conversion of energy to food, so you will require 40 m^2 as a minimum. Now compare this to "your" fraction of the arable land:

$$\text{Your arable land} = 2 \times 10^7 \text{ km}^2/4 \times 10^9 \text{ people} = 5 \times 10^{-3} \text{ km}^2$$
$$= 5 \times 10^{-3} \times (10^3)^2 (\text{m}^2) = 5000 \text{ m}^2$$

This figure should startle you. It indicates that mankind is already using almost 1% of the available photosynthetic solar energy to feed the world, even with the most efficient food. The actual percentage of solar energy used by humans is higher. (The actual figure is 5–10%.) There really isn't much margin to increase efficiency of food production if sunlight in present photosynthesis is the limiting reagent. Unless, of course, the efficiency of photosynthesis can be increased above 1%. Plenty of people are trying.

The Green Revolution

Most projections indicate that, if present population and agricultural trends continue, Australia, Canada, and the United States—the three countries that consistently have food surpluses—will be unable to satisfy the minimal food needs of the countries having food deficits. The options are these: (1) the population can level off; (2) local agriculture can become more efficient; and (3) transport and storage problems can be solved. The alternative is more widespread hunger than ever before. The most conservative projections suggest that, unless present trends change, just barely enough food will be available at the end of the century. Figure 24.11 gives one projection.

Norman Borlaug was given the Nobel Peace Prize in 1970 for his part in the development of high-yield hybrids. He was then one of the first to point out that the hybrids are not a long-term solution. They will give us a decade or two in which to take more fundamental steps. One decade has passed. Table 24.1 indicates part of the history of the Green Revolution in India. Just as Borlaug suggested (except faster) other factors

EXERCISE 24.5
We have left out several factors, each having values of approximately 2, in this calculation. They cancel partially so they do not affect the result much. Can you identify any of them? One has to do with area, another with the absorption of sunlight.

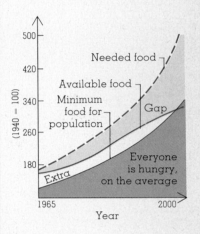

Figure 24.11
One projection of present population and food trends to 2000 (1940 index = 100). The food gap between supply and adequacy grows.

Table 24.1
Recent History of High-Yield Varieties (HYV) in India (as multiples of yields of normal crops)

Year	HYV crop	
	Wheat	Rice
1967	2.87	2.58
1968	3.70	2.18
1969	3.49	2.05
1970	3.68	2.26
1971	3.44	2.27
1972	2.50	2.03
1973	2.35	1.76
1974	2.59	1.71

Table 24.2   Status of Some Raw Materials

| Material | Reserves (kg) | Years until reserves used up if demand is | |
		Static	Dynamic
Aluminum	2×10^{12}	100	55
Coal	10^{16}	2300	150
Copper	6×10^{11}	36	98
Iron	6×10^{14}	240	173
Mercury	6×10^{8}	13	41
Oil	3×10^{14}	31	50
Tungsten	6×10^{9}	40	72

NOTE: Data from U.S. Bureau of Mines, 1970.

caught up. Some of the main factors in India have been an inability to maintain the seed stock, and the high degree of irrigation and the large fertilizer input these hybrids require. Costs, due partly to the fivefold increase in oil prices, make it impossible to run the irrigation pumps or to buy the oil-based fertilizer (remember the ammonia synthesis). But note that, even so, the high-yield crops still outproduce others.

Borlaug, probably as much as any scientist in history, both made a remarkable contribution to mankind and clearly and forcefully predicted its meaning in terms of required decisions and planning. But other, "more immediate" problems intervened. One of the decades he allotted still remains. It will be a tighter squeeze, but no one can blame a lack of time as the limiting reagent—yet.

Raw Materials

The U.S. Bureau of Mines annually publishes a list of known worldwide reserves of natural resources. Some of these are collected in Table 24.2. The static index is the time before exhaustion if present use rates remain constant and if no new reserves are developed. The last column (dynamic) gives the time before exhaustion if five times the present reserves turn up but use continues to rise at its present exponential rate. In Table 24.2, and on a longer list of 20 other essential materials, over half will be exhausted within your expected lifetime if the second projection is correct.

It is particularly appropriate for citizens of the United States to consider raw-material reserves. On the average, we use about one-third of the world's annual production of raw materials. We are about 6% of the world's population. Each U.S. citizen accounts for the use of 40,000 lb of minerals per year (Table 24.3). If the whole world were to use raw materials at the rate we do, the use rate would immediately increase by a factor of 5. Another way to look at this is to say that we would have to

Table 24.3
Annual Mineral Use
in the United States
(pounds per person)

Substance	Quantity used
Sand and gravel	9000
Stone	8500
Cement	800
Clay	600
Salt	450
Other nonmetals	1200
Iron	1200
Aluminum	50
Copper	25
Zinc	15
Lead	15
Other metals	35
Oil	7000
Coal	5000
Gas	5000
Uranium	0.05

reduce our use to one-fifth the present level to be at the world average— a comfortable 1920 U.S. level. Subsistence would be about one-fortieth our present level, or about where India now is.

All the reserves in Table 24.2 are nonrenewable. When they are gone, they are gone. Many are of materials that can be recycled. We have already looked at some of the problems of recycling, including the necessity of designing goods with recycling in mind.

Classical economics suggests two solutions to resource depletion: rolling substitution, and increased prices. By *rolling substitution* is meant moving from one raw material to another as availability and price become problems. But Table 24.2 suggests that a limit on substitutions can also occur. As to price rises, they really mean less widespread use—and use only by those who can afford the new price in time, labor, and investment. Rising prices do not increase the availability of a truly limiting reagent, though they may encourage the development of poorer and poorer sources. Figure 24.12 illustrates these effects, using the recent history of copper supplies.

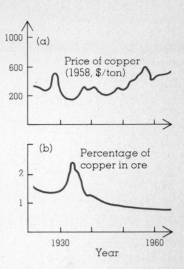

Figure 24.12
(a) Price of copper, and
(b) average percentage of copper in processed ore, 1925–1965.

Energy Is a One-Time Thing

Most of the raw materials in Table 24.2 are nonrenewable but recyclable. Coal and oil and many other energy sources are neither renewable (on a human time scale) nor recyclable. Table 24.4 presents a few estimates of

Table 24.4   Energy Sources in the United States and the World

Nonrenewable: Years that each fuel could provide the total present:	United States demand[a]		World demand[b]	
	Usable	Total	Usable	Total
Coal	125	1300	70	700
Oil	5	250	20	70
Gas	5	110	10	50
Shale	?	2500	50	300
Fission	2	15	4	10
Breeder reactors	113	750	150	400
Fusion	?	10^9	0	10^9
Geothermal	0.2	100	1	50

Renewable: Relative amount of present United States demand that each fuel could supply annually if used alone with 100% efficiency	
Solar	740
Wind	5
Ocean, thermal	6
Hydro (dams)	0.1
Photosynthesis	0.2
Wastes	0.1
Tides	0.1

[a]Data from Allen L. Hammond, "Energy Options: Challenge for the Future," *Science*, **177**, 1972, p. 875.
[b]Adapted from Imperial Chemical Industries, CHEMTECH, May 1977, p. 295.

Figure 24.13
A projection of total energy
demand and possible energy
sources for the world. 1975–2200.

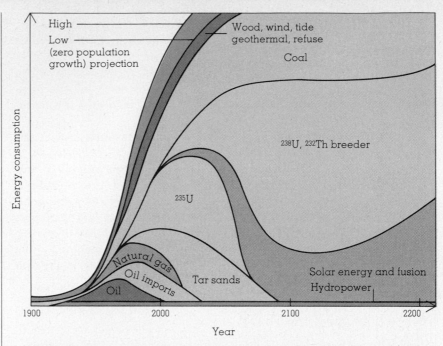

some renewable and nonrenewable energy sources in the United States.
Each number represents the number of years of present energy use in
the United States (6.6×10^{19} J/year) which that source could supply
if no other energy source were available. All the renewable resources
are supplied by the sun. It is clear why coal, solar, and nuclear (especially fusion) fuels are of so much interest.

One difficulty with many renewable resources is that they are periodic and require energy-storage devices. The sun is gone half the time,
the wind fluctuates, the tides occur on a 12-hour rhythm.

Nuclear fusion is one of the most attractive long-term sources. It
would provide essentially unlimited energy. Table 24.5 outlines recent
federal research and development funds. You can see some big shifts.
Here is one place that your *informed* opinion can lead to immediate as
well as long-term effects. Let your opinion be known!

A second thing that should strike you is the small fraction of each
nonrenewable energy supply that we are able to use. New extraction
techniques for oil, gas, coal, and oil shale would alter the resource picture remarkably. The same statement applies to solar energy, because
no one yet knows how to capture sunlight as efficiently as photosynthesis
does. Yet you have seen how relatively poor even photosynthesis is as
a large-scale energy source.

Figure 24.13 gives one projection of total energy demand and potential sources for the world. You should view it with several grains of salt,

Table 24.5
Federal Environmental Research
and Development Authority
(ERDA) Funds for Work on
Energy Sources
(millions of dollars)

Fuel	1973	1977	1978
Fossil	136	444	494
Fission	356	1013	1362
Fusion	65	322	431
Solar	4	183	235
Geothermal	3	49	68
Other	55	440	450
Total	619	2451	3040

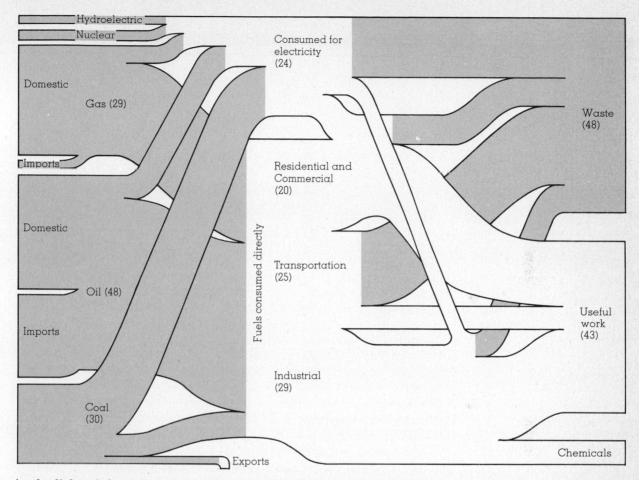

Figure 24.14
Current sources and uses of energy in the United States. Only half of the energy accomplishes useful work. The other half is converted to waste heat, eventually transmitted to the surroundings. (Adapted from Earl Cook, Texas A & M University)

in the light of the 2-year decline in energy use in the United States in 1974–1975. The 1974 drop was 4.9%, with 1975 showing an additional 2.5% drop. Similarly, the projected need for electrical energy in the United States in 1985 is now 25% less than the figure estimated in 1972.

The desirability of energy sources other than human labor is underlined by comparing the $1–2 cost of 1 million kJ of energy in the late 1970s with the $10,000 cost of the same amount of human energy if the wages are $3/hour.

Where Does All That Energy Go? Heat!

Figures 24.14 and 24.15 analyze present energy consumption in the United States in terms of sources and uses. The entries are self-explanatory, but most people are surprised by the fact that only half the energy generated accomplishes work; the other half must be dissipated

Industrial
41.2%

Process steam   16.7

Direct heat   11.4

Electric drive   7.9

Feed stocks   3.6
Other   1.6

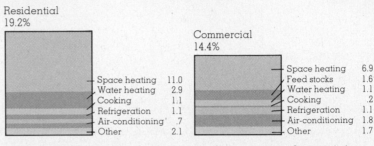

Transportation
25.2%

Fuel   24.9

Raw materials   .3

Residential
19.2%

┌ Space heating   11.0
├ Water heating   2.9
├ Cooking   1.1
├ Refrigeration   1.1
├ Air-conditioning   .7
└ Other   2.1

Commercial
14.4%

┌ Space heating   6.9
├ Feed stocks   1.6
├ Water heating   1.1
├ Cooking   .2
├ Refrigeration   1.1
├ Air-conditioning   1.8
└ Other   1.7

Figure 24.15
A more detailed analysis of
energy use in the United States.
Note that the "waste heat" of
Figure 24.14 is more than
enough to supply all the space
heating described here.

POINT TO PONDER
The more you air-condition, the
more you need to air-condition.

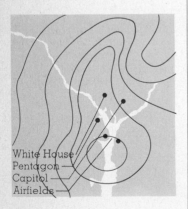

White House
Pentagon
Capitol
Airfields

Figure 24.16
Temperature contours (°C) for
Washington, D.C.

directly as heat. (Take another look at equations 17.14 and 17.15.) Actually, essentially all the energy generated ends up as heat eventually.

You may have noted that all methods of handling problems involve the expenditure of energy. As described by the first law of thermodynamics, the amount of energy remains constant. But, as described by the second law, the energy "degrades" to heat (the random motion of molecules) and is then less available for further use. Part of this degradation is inherent in the limitations of heat engines (as you learned in Chapters 17 and 20), part is due to frictional effects, and the rest is due to the general tendency of all forms of organized energy to end up as disorganized heat as described by the second law. So we have come to accept the "once-through" nature of energy flow.

But with our very large rates of energy flow, we are coming to face a second problem, heat buildup—that is, rises in temperature. Figure 24.14 shows the effect of human activity on temperature in Washington, D.C. (Note that the maximum effect is not caused by the hot air from the Capitol or the White House.) Most cities now show a similar effect; the temperature in the city averages several degrees hotter than the surrounding rural areas.

These local effects are strongest near power plants, because that is where most of the energy is generated. Heating of streams and ocean bays by power plants is common in occurrence and appreciable in size. It is already a cause of concern and of changes in design so that the heat will be spread more widely.

Of much greater significance in the long run is general heating of the surface of the earth. You have read of some of the problems (such as flooded cities) that would result from CO_2 buildup in the atmosphere.

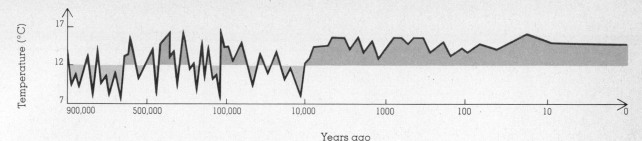

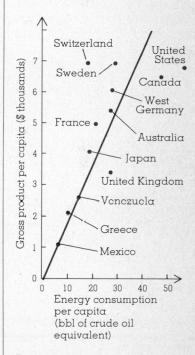

Look at Figure 24.17. You will see that the average temperature of the surface of the earth has fluctuated between 8 and 17°C (with a mean of 12°C) for hundreds of thousands of years. We now seem to be near 15°C. This is one reason for the predicted imminence of a new ice age. The principal effects we know about are the rise and fall of ocean levels and the decrease and increase of the ice fields at the poles. So we know what to expect from "normal" fluctuations. And they are not conducive to continual growth, or even to maintenance of our present life style.

But the historical changes have occurred gradually over thousands of years. They allowed time for living systems to adapt, and that is what happened. And there was time and space for populations to move to more favorable climates.

The point to note here is the effect of a rapid change in the temperature of the earth, such as would follow the use of energy without limit. If we assume that the rate of energy generation will continue to increase at its present 4%/year, mankind will achieve in 130 years an energy output 250 times the present one, and equal to 1% of the incoming solar radiation. The temperature increase has been estimated to be just under 1°C, if the heat is uniformly distributed (an unlikely event). The world would be hotter than it ever has been and presumably the ocean surface would be higher than ever before.

The real point of this discussion is that a limitless supply of energy cannot be used in a limitless fashion. There will be a limit set by the necessity of getting "rid" of the resulting heat. Further, whether we need additional energy per person is put in question by Figure 24.18. In 1974 the United States used energy at twice the rate of Sweden and three times that in Switzerland, though each of these countries has a standard of living at least equal to that in the United States.

Chemistry and Crime

One cause for concern on the part of many people is the crime rate in most countries of the world, especially for violent crime against persons. Much of this is due to political instability, but outbreaks occur in many

Figure 24.17
World temperature as a function of time. (Data from *Understanding Climatic Change*, National Academy of Science, Washington, D.C., 1975)

Figure 24.18
The United States exceeds all other nations in energy consumption per person; 1974 data. (Data from United Nations Statistical Yearbook, 1975)

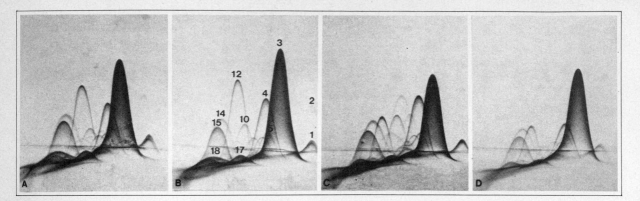

Figure 24.19
Analysis of bloodstains from
three individuals. Pattern A and
pattern B resulted from the
blood of a single individual.
The analyses were made 17
days apart. Patterns C and D
are from two other individuals.
All individuals tested, even
when genetically related, have
shown different patterns. (From
G. H. Sweet and J. W. Elyins,
Science, **192**, 1976, p. 1012)

places having quite stable governments. In these cases chemistry is becoming very useful in reconstructing the events and helping to determine guilt or innocence. Let me quote from a report in *Chemistry in Britain* in 1975 (pp. 434–437):

Some time ago the body of a young girl was found partly concealed behind a hawthorn hedge at the end of a rough track leading into some fields. Her clothing was in disarray and only one of her shoes was found.

. . . The remains of a girl's black shoe and some underwear was found in a field two days later, seven miles away. They had been burnt but the ashes remained, and these included metal components which matched those present on the shoe found near the girl's body.

Scientific examination of the scene, the girl's clothing and post-mortem samples yielded valuable scientific evidence involving biological, chemical, and tyre-mark examinations and led to the laboratory examination of over 10,000 saliva samples, 15 cars, 50 samples of car paint, 35 samples of car carpets, 10 samples of car seats and numerous tyres.

Much of the evidence had to do with careful examination of the clothing and tiny paint chips and carpet fragments from the car used to transport the body. By matching manufacturers' records, the car was identified as a repainted 1960 MG Magnette, with carpets from an Austin A60. The car was found in a scrap yard. Semen found on the girl allowed identification of the blood type of the attacker. (Recent discoveries suggest that he now could be identified from unique proteins in semen.) Four months after the crime the murderer was apprehended and he was subsequently convicted.

This unquestionably is a particularly dramatic case of forensic chemistry. It required an enormous amount of work. However, as more and more manufacturers' information gets filed in computers, the identification of manufactured products will become easier and easier. And there is no scientific reason that identification based on protein type should not be added to fingerprints as a means of tracing individuals (Figure 24.19).

But there are always "on the other hand" arguments. This information could greatly simplify the detection of criminals. It could also make easier the control of individuals in a much more general way. Another case of balancing gains against costs.

Summary: Projection Is Not Prediction

No intelligent and informed person believes that exponential growth can proceed forever. There is always a limiting reagent. For most of human history these limits were set by local events—disease, local crop failure, natural disasters, crippling or lethal accidents. The limits were usually well known and accepted. Even when the actual limits were not known (as with the Roman plumbing), the results caused little surprise. There were many ways to attribute responsibility for the unknown. "Acts of God" were invoked early in human history. They are not limited to current insurance policies.

The limits we have explored in this chapter are not controlled locally, and they are not attributable to unknown factors. They are subject to scientific (and often quite accurate) evaluation. They include oxygen, carbon dioxide, the stratosphere, water, chemical pollutants, land area, transportation, high-yield crops and their requirements, a long list of raw materials, energy sources, and heat disposal. Most of these limits were unknown to anyone until recently and were completely beyond the comprehension of most people even 20 years ago. But now many know, or at least they hear. And they confuse projections with predictions.

All the projections I have presented in figures, tables, and discussions are based on assumptions. Most of the assumptions include continuation of present rates and patterns of growth. The generally dismal resulting projections will certainly be wrong. The rates and patterns will change. The real questions are how and when? Up or down? In an acceptable manner or a catastrophic one? Very great social and political questions are involved. These I have not discussed.

The point I do want to make is that the conservation laws and the second law are among the best descriptions we have of the behavior of changing systems. It is extremely unlikely that violating them is a possible choice. Don't forget the power of back-of-the-envelope calculations. But couple the calculations with knowledge, comprehension, and wisdom.

One safe prediction is that exponential growth cannot continue indefinitely. An approximately steady state could. It should be set by human choice of limits, not ones imposed by "nature." A projection is a basis for decision, not grounds for acquiescence.

HINTS TO EXERCISES
24.1 97% not recoverable. 24.2 C—Cl bond is weaker. 24.3 Overall, it's the rate of evaporation of water; what goes up must come down—but how about locally? 24.4 If all the increase in the earth's population could get to the moon, the moon would have the same population density that the earth has in about 2 years. 24.5 The earth is a sphere, not a plane surface; over half is covered by water; about half of the sunlight must be used to evaporate water; changes in altitude have effects.

Problems

24.1 Count your breathing rate and estimate the volume of air you inhale with each breath. If the concentration of an air pollutant averages 0.1 ppm over 24 hours, how many moles of the pollutant will have entered your lungs? [$\sim 10^{-3}$ moles/day]

*24.2 About 10% by volume of the exhaust from a car is CO_2. The U.S. maximum emission level (1976) for CO is 15 g/mile. About what will the ratio of CO_2 to CO be in exhausts that just meet this standard? [~ 1 mole CO/20 moles CO_2]

24.3 About 5×10^{12} kg coal is burned per year. The coal has an average sulfur content of about 3% by weight. The atmosphere has a mass of 5.2×10^{18} kg. Assuming uniform mixing, what would be the increase in SO_2 content due to 1 year of burning? Sulfur dioxide is not known to have any toxic effects unless its concentration is greater than 0.5 ppm. Comment on SO_2 pollution. [~ 0.1 ppm SO_2]

24.4 There seems little chance that your grandchildren will have access to cheap oil and gas as fuels. Which substitute sources would you wish for them? Be specific as to reasons. Who should make the decision?

24.5 It used to be common to reclaim the lead from old storage batteries by making a bonfire of them. The plastic cases burned off. The lead melted and collected in a pool. This practice has been stopped because it pollutes the air in two ways. What are they?

24.6 The energy balance of the earth involves a yearly flow of 1.73×10^{17} watts (99.98% solar, 0.02% internal, 0.002% tidal). The energy effects are 30% reflection, 47% heating, 23% water cycle, 0.2% wind, and 0.02% photosynthetic. What happens to the energy absorbed in the last four processes listed?

*24.7 Until the mid-1800s paper was made almost entirely from used linen and cotton rags. Why not from raw cotton or flax?

24.8 The U.S. Army stored many tires at the end of World War II in 1945, only to find out with the Korean War in 1950 that the tires had cracked and now lasted only a few hundred miles. Suggest a possible cause and a solution. (Tires can now be stored successfully.)

24.9 Natural asphalt is usually heated and "blown" with air to raise its melting point. What probably happens during the blowing process (air bubbles through the asphalt to modify the asphalt—a hydrocarbon)?

24.10 The biochemical oxygen demand (BOD) in a river is 4.8 mg/liter because of waste from a paper mill. The river above the mill is saturated with oxygen at 8.9×10^{-3} g/liter. To what level will the downstream oxygen concentration fall, assuming no replenishment of oxygen? Any problem for the fish? How could replenishment be accomplished?

24.11 It has been claimed that if you hammer a long, thin nail of pure copper into an oak tree, the tree will die. Comment.

24.12 Professor Bert Bolin of the University of Stockholm stated in May 1976 before a House of Representatives subcommittee that "Man may not be able to use all available fossil fuel because of sensitivity of climate," according to an Associated Press news release. Write an interpretation of the probable background of his remark that would be understood by the average adult in the United States.

24.13 Current computers require about one-millionth the energy of the 1946 models. It has been suggested that similar reductions in other energy requirements can help solve our energy problem. Use Figure 24.13 as a basis of discussion of such possibilities.

24.14 There is about 3% sulfur in much coal that sells for $6/ton. The sulfur, which forms SO_2 when the coal burns, can be reclaimed as the element if H_2S is mixed with the hot stack gases. This process—called the Claus process: $2 H_2S (g) + SO_2 (g) = 3 S (c) + 2 H_2O (g)$—is similar to one that precipitates sulfur around volcanoes. What fraction of the world demand for sulfur (currently 10^{10} kg/year) could be met by reclaiming the sulfur from the 4×10^{12} kg of coal burned annually? [~ 10 times present needs]

24.15 The recoverable fossil fuels for the world are estimated (in terms of 10^{15} kg C) to be as follows: coal, 5.7; oil shale, 0.4; petroleum, 0.3; gas, 0.1; tar sands, 0.1; total, 6.5×10^{15}. The United States (with 6% of the world population) consumes about 3×10^{12} kg of carbonaceous fuel annually. If all the world's population consumed these fuels at the present U.S. rate, how long would the fossil fuel supply last? [~100 years]

24.16 About 10^{12} kg of gasoline (C_8H_{18}) is burned every year in motor vehicles in the United States. The average vehicle emits 5 g NO per mile, which is oxidized by moist air to HNO_3. Estimate the weight of nitric acid, HNO_3, formed annually from motor vehicles in the United States. [~4×10^{10} kg HNO_3/year]

24.17 The chemicals in each cubic mile of seawater are worth about $300,000. Why don't the developing nations with ocean coastlines claim this "bounty of the sea"?

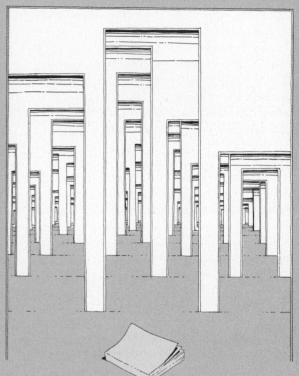

Figure 25.1
Many doors, one exit—and a
guidebook.

25 MANY CHOICES, ONE OUTCOME

Most textbooks just stop. The author has "covered the subject." But science is not just a collection of observations and correlations. It is also a process. I'd like to speculate a bit on the uses you may make of what you have learned.

My intention has been to provide a few bases and methods for judgment and decision making. Everything you will ever deal with is made of molecules. Chemists specialize in interpreting observations in terms of molecular behavior. My belief is that many decisions will be better based if you at least consider the limitations imposed and the possibilities provided at the molecular level.

Probability and Certainty

Scientists are convinced that the conservation laws are very good laws. For example, no exception to the conservation of electric charge has ever been observed in thousands and thousands of experiments, some especially designed to look for exceptions. Yet the lack of success in seeking exceptions in the past does not keep very good scientists from continuing to look for them. Intelligent scientists are never certain. There have been too many instances where beliefs held strongly over long periods of time have been shown to be inadequate.

Measurements are always inaccurate, and we can never fully describe an occurrence. Thus we can never predict with certainty what will happen next time. So the predicted *a priori* probability of a future occurrence must be somewhere between zero (no chance at all) and one (complete certainty). However, it is worth noting that, in science, uncertainties less than 1 in 1000 are routine, less than 1 in 10^5 are common, and less than 1 in 10^{12} are sometimes achievable. Careful work yields very small uncertainties in scientific statements.

Scientists learn to live with these uncertainties, even with some that are higher. They have invented methods that reduce the uncertainties. The resulting rise in the success rate of their projections and predictions has given humans unprecedented control over their surroundings and their future. At the same time there has been a rise in both the desire and the capability of humans to make decisions involving long-range problems.

My interest is in the future because that is where I am going to spend the rest of my life.
—*Charles F. Kettering.*

POINT TO PONDER
One scientist I know defines a miracle as something having an *a priori* probability of less than 0.001 (1 in 1000) and a sure-fire thing as one having an *a priori* probability greater than 0.999 (999 in 1000). The definitions are at least symmetrical. Better yet, they make a good point, and you can always provide your own numbers.

Problems, Problems Everywhere— Yet Never Time to Think

One of the major difficulties with decision making is that almost everyone wants to be free to make decisions, but very few actually enjoy the process. A principal reason, as I suggested at the very beginning of this book, is that many persons do not know the choices. Here science can help, by exploring choices (and showing that some are impossible to a high degree of probability).

For example, no one has found exceptions to the second law or to the conservation laws. So, thoughtful decisions do not try to "violate" these laws. Le Châtelier's principle seems to apply to systems at equilibrium, and states of change seem adequately described by collision, orientation, and energy effects. These and many other scientific generalizations include a good deal of common knowledge, but they do so in such a general form that the power of the knowledge increases. Less time is required for humans to think through presumed alternatives, evaluating the likely ones and rejecting those whose probability of success is so low they can be called impossible. Yet it must be at all times remembered that *some* things that have low probabilities, or even seem impossible, will turn out to be feasible.

Prove It!

Scientific proofs always start with a set of postulates or assumptions. The Euclidean postulates, such as "A straight line is the shortest distance between two points," are examples. They summarize a great deal of experience and are very useful. But even postulates must have a context. The one just quoted works very well on plane surfaces, but on a spherical surface it gives way to "The shortest distance between two points is along a great circle." A straight line is not useful on a spherical surface. Even the great-circle definition requires you to decide which way to travel on the great circle.

A study of science should help you establish a context for your decisions. What postulates apply? It then helps select the high-probability possibilities. But it goes further.

Scientists have biases and prejudices. They are human. But they try to design experiments whose results will not be influenced by their own wishes. It is called "appealing to the system." And, if they wish their results to withstand critical examination by others, scientists try to recognize their own biases and to take them into consideration when evaluating observations and results.

Scientists try to allow for their biases, not because they are more ethical or intelligent than other people, but because scientific results

HOW FAR FROM LONDON TO TOKYO?

can be checked. Once a result is announced, anyone is free to repeat the experiment and to interpret it from a different point of view. It is a great pleasure to receive agreement from a colleague on results and interpretations. And it is embarrassing to be found wrong.

Furthermore, scientific results lead to predictions. The most severe test of a scientific theory is the set of predictions it makes concerning the results of experiments that have not yet been performed. If the results of the suggested experiments agree with the predictions, the experience is highly rewarding. The reverse experience is humbling.

So a main reason for honesty and the search for truth among scientists is the ease with which they can be caught lying or being sloppy in their work.

Of course, natural scientists deal with relatively simple systems compared to many. But their successes with increasingly complicated systems (say, more and more complex living systems) suggest that the scientific approach has not yet reached its limits.

One result of the winnowing of useful from nonuseful ideas in science has been an increasing ability to predict outcomes successfully. In fact, many people agree that there is no method more powerful than the scientific "appeal to the system" (and its mandatory minimizing of human biases) in outlining likely outcomes of many decisions and courses of action.

Where science is not very useful is in defining which outcome is to be preferred. Science can often prove (predict successfully) what will happen if certain decisions and actions are taken. It cannot, in general, prove that one outcome is better than another. Scientists, as scientists, study the way things are, not the way they ought to be. They answer questions of *what* and *how*. To questions beginning with *why* they eventually must reply, "That's just the way things are." But they often greatly reduce the frequency of this answer by finding relationships. The number of "proved" scientific theories is much, much less than the number of observations they correlate and predictions they allow. Let me give you an example in terms of population growth and its control.

Projections and Preferences

Figure 25.2 shows graphs called *population trees* for Great Britain and the island of Mauritius, also administered by Great Britain. The data are comparable. The population tree for Britain is typical of many industrialized countries. That for Mauritius is typical of many agricultural countries. In fact, because many more people live in agricultural than in industrialized countries, the Mauritius tree is not far from that for the world.

Each tree shows the percentage of the population in each 5-year age

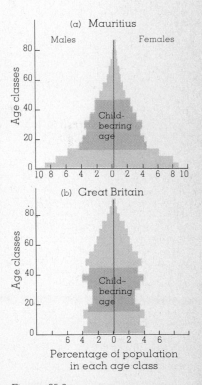

Figure 25.2
Population trees for (a) Mauritius, and (b) Great Britain.

class—females on the right, males on the left. These trees are based on 1970 data. The indentation in the British tree for age group 20–29 is a result of the low birth rate during and after World War II. The indentation for those about 20 years older is partly the result of casualties in the war and partly the low birthrate during the depression of 1930.

The most obvious difference between the trees is the rapid taper of the Mauritius one, whereas the British tree is nearly rectangular, with a triangle at the top. The trees differ in shape because the chance of death is almost independent of age in Mauritius but does not become appreciable in Britain until after age 50 or so, after which the rate is fairly constant.

So we can understand the data and the general shapes of the distributions in terms of death rates and birth rates. There is at least one other factor to consider—migration. Both these populations have been rather nonmobile during the 80 or so years represented, so migration does not have a big effect here—but it does in many countries.

Which tree represents the better set of conditions? You can deduce what the living conditions probably are, and both you and I have preferences for some over others. But at this point science, as science, does not speak. It presents the situation as clearly and correctly as possible in terms of carefully collected data; science describes the way things are. It provides the basis for decisions, not the decisions themselves.

Scientists have as strong opinions, and are as willing to express them, as anyone. But—if they have done their jobs well, so that you are fully informed—they are in no better position than you to reach a decision as to which type of population tree is to be preferred. Scientists have a strong ethical code in carrying out their work, but it cannot automatically extend to decisions on the uses of their results.

Suppose you as Chief Administrator prefer the British tree and wish to approximate it in Mauritius. Again the scientists can make a big contribution. There are three factors to consider, and many of the effects of each are predictable. You can encourage migration to countries like Britain that have rectangular population trees. You can decrease the death rate at all feasible levels. You can also modify the birth rate to produce a rectangular tree with as many people as desired in each age group. Many further details as to possible (and impossible) procedures and the probable results of each could be presented.

Figure 25.3 presents a United Nations projection of world population, based on trends since 1850 and estimates to 2120. Good or bad? My own conviction is that the prospect outlined is not only terrible, but not reasonably achievable. But the projection is a valid one within its assumptions. It can lead to excellent planning and action. The scientists, dealing with the way things are, will point out many possible methods

Figure 25.3 (opposite)
A United Nations projection of world population by area, to 2120. The United Nations calls it "Our cup runneth over." (Adapted from the UNESCO *Courier*, May 1974, p. 17)

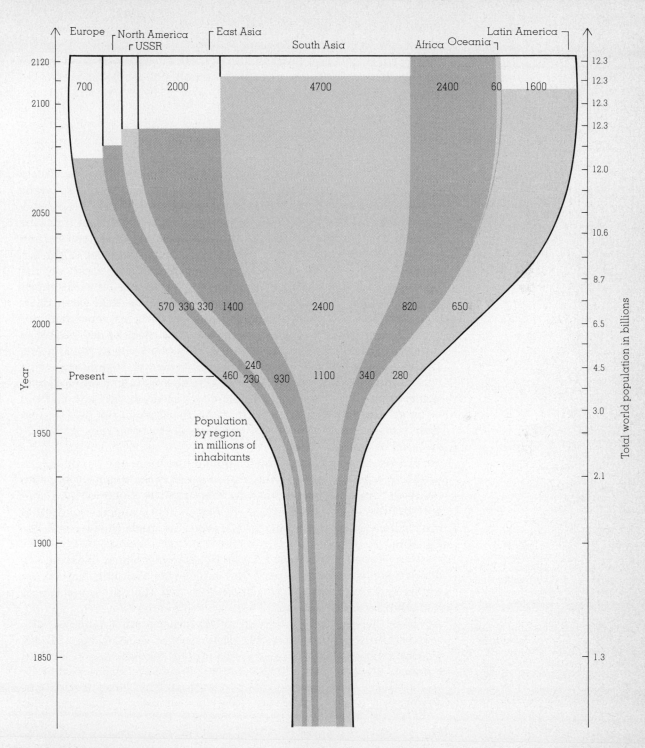

Europe   North America   East Asia   Latin America
         USSR   South Asia   Africa Oceania

2120   12.3
         700   2000   4700   2400   60   1600   12.3
2100   12.3
                                                                        12.3

                                                                        12.0
2050   10.6

                                                                        8.7
         570 330 330 1400   2400   820   650   6.5
2000

                                                                        4.5
                240
Present   460
                230 930   1100   340   280
                                                                        3.0
1950

Population
by region   2.1
in millions of
inhabitants

1900

1850   1.3

Year

Total world population in billions

EXERCISE 25.1
The present doubling time for the human population of the earth is about 35 years. Suppose all the additional population each year were sent to the moon. How long before the moon would have the same population density as the present earth? Try the same question for Mars.

of proceeding (together with outlining why some are impossible, that is, have a low likelihood of success).

Scientists will also point out that there are four possible types of actions that can be taken to limit the number of people on earth:

1. Decrease the birth rate.
2. Increase the death rate.
3. Foster migration (the moon and Mars have been suggested).
4. Let nature take its course; do "nothing."

One thing is a certainty. There *will* be a solution, an outcome. Most scientists would argue strongly for option 1. They would so argue because they know the means are available (some countries have already achieved dramatic results), and they are sure there is sufficient time (the rates at which people can change have been studied). Scientists are also repelled as individuals by options 2 and 4, though in terms of the unbiased approach of describing "the way things are" they must consider all four in equal detail. Finally they realize that option 3 is a very short-range "solution." The moon is one-sixteenth the area of the earth. Mars is one-fourth. Neither is as hospitable as the Sahara here on earth.

The scientists can also show unequivocally that the death rates in many countries, including the United States, are going to rise anyway. In these countries, which have had high birth rates, high immigration rates, and increasingly good public health, there is a high proportion of youth whose health is such that their death rates are very low, and will be for another 20 years or so. Then the death rates will rise. In a steady-state population, the death rate *must* equal the birth rate. At steady state, input equals output.

Because scientists are human, and because more and more humans use the methods of science, the distinction between scientific evidence (the way things are) and human decision (the way things ought to be) is often fuzzy. We would probably be better off (an "ought to be" statement) if we more clearly realize just when we move from evidence to decision.

Many people are offended at the rigidity with which scientific evidence is sometimes presented. They might view the four choices presented and say, "How do you know those are the only possibilities? Something else will turn up." There is, of course, no convincing answer. Accusing them of being like Dickens' Micawber is not productive. The scientists must admit the possible existence of other solutions, of which we have no knowledge. Scientists must also be open to suggestions from others.

EXERCISE 25.2
Go to the *Readers' Guide to Periodical Literature* and search for articles by scientists, astrologers, and clairvoyants on futurology. Compare the "batting averages." See John Von Neumann in *Fortune* magazine, June 1955, or Linus Pauling in *Scientific American*, September 1950, and again in the book *General Chemistry* (edited by J. Ifft and J. Hearst, W. H. Freeman, San Francisco, 1974). Also see *The Next Hundred Years* by Harrison Brown, James Bonner, and John Weir, Viking Press, New York, 1963.

Science and Magic

Many suggestions are of the type I shall call magic. Only a few people can accomplish these effects, and even they cannot do so at all times. I do not refer to stage magicians. Everyone knows they have secret devices—though many in the audience would prefer to regard the performance as true magic. I refer to so-called witch doctors, mediums, people with "superhuman" powers, astrologers, extrasensory perceptors, and clairvoyants. In other words, people who sincerely believe they have insights and powers but do not understand how they work.

I do not know whether there is "anything in these fields." My own minimal contacts with them have always uncovered either doubtful sincerity or a low *a priori* probability of correctness. All the thorough investigations that I have seen have come to similar conclusions. In every case the *a priori* probabilities of success are below those accepted in the natural sciences. So I prefer not to base decisions solely on such suggestions, but I'm interested in receiving and evaluating them for possibilities.

Most people enjoy magic. And many look askance at science and scientists for trying to investigate it, to remove the mystery, and to explain how magic "works." For many people this is a major seat of the distrust they have in science.

The Chemistry of Choice

You have already seen that chemical messengers are highly effective in determining the behavior of insects (with pheromones) and birds (migratory behavior). You have seen that humans use chemical messengers to control their physiology. Many other examples could be given at all levels of living systems.

The question of ability to choose is one of the most ancient questions asked by humans. One conclusion reached after several thousand years of discussion is that it is unlikely that we will ever know just how free we are to choose. There are too many interrelationships—many apparently unmeasurable and inseparable—to arrive at a firm answer.

For example, suppose someone eventually finds out how free choices are made and can describe the mechanism. It should then certainly be possible to influence the choice, and so to destroy the freedom of choice that had been so avidly studied. It is interesting to note that almost everyone acts as though his or her own choices were free in many cases. But most people feel that others have a more limited range of choice. They even feel they can control others' actions.

One of the causes of nervousness and lack of ease with science and scientists is the feeling that scientists are becoming more and more proficient at controlling behavior, including human behavior. And much of the concern about "mind-bending" drugs (as opposed, for example, to sleeping pills) is the loss of behavioral control and the onset of antisocial behavior they often foster. Let me give one or two examples of

POINT TO PONDER
According to José M. R. Delgado, one of the pioneer experimenters with electrical stimulation of the brain, "We can stimulate a point in the brain to . . . cause a patient to talk, but naturally he will use words he has learned in the past. Brain stimulation cannot create a new individual . . . it cannot change personality." (From *On Growth,* edited by W. L. Oltmans, Capricorn Press, Santa Barbara, CA., 1974, p. 261.)

Table 25.1
Some Genetically Controlled
Behaviors in Fruit Flies

Category	Variants
Locomotor	Sluggish
	Hyperkinetic
	Uncoordinated
Stress response	Easily shocked
	Paralyzed
	Shaker
Circadian rhythm	Arrhythmic
	Short period
	Long period
Sexual	Savoir faire
	Fruity
	Stuck

NOTE: Adapted from Seymour Benzer, *J. Am. Med. Assn.*, **218**, 1971, p. 1017.

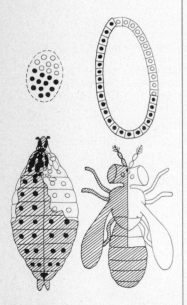

Figure 25.4
The development of a mosaic insect. Four parents give two sets of behaviors. Some cells in the developing egg come from one set of parents, some from another. (Adapted from Y. Hotta and S. Benzer, *Proc. Nat. Acad. Sci.*, **67**, 1970, p. 1157)

behavioral studies with possible interpretations at the molecular level.

Table 25.1 lists some behavioral patterns that can be selectively bred in fruit flies. In other words, the patterns are controlled by the genes, not by learning. Figure 25.4 shows how fruit flies can be bred from four, rather than two, parents. The fertilized and developing egg is essentially dissected at an early stage, and some of its cells are removed and replaced with cells from another set of parents. The resulting individual is called a *mosaic*. Suppose the left legs came from a hyperkinetic set of parents and the right ones from a sluggish set. The fly might only be able to walk in tight circles. By breeding many flies with this mosaic technique and noting the division line in each individual, it is sometimes possible to locate quite accurately the control center for some behavioral pattern. Sometimes it is in the brain, sometimes not. It is clear that very careful analysis of such results may reveal the molecular methods of behavioral control.

Mosaicism is known in humans as well. It is not due to four parents, but to genetic scrambling at some early stage in the development of the fetus. Albinism over only some section of the body is an example. All human females are thought to be mosaics, because of the activity of only one of their X chromosomes in each cell. The difficulties involved in using mosaics as a means of studying human behavior are obvious.

However, studies have been done on relationships between IQ and number of chromosomes in an individual. The number of X chromosomes may go as high as four per cell instead of one. According to a Belgian study, the greater the deviation from the usual number, the lower the average IQ.

Recent studies on lead concentrations in humans have shown, both in the United States and in Britain, that prisoners average about twice the concentration of lead compared to nonprisoners. The higher concentration is not due to exposure in prison. High lead levels are known to lead to hyperactivity and poor judgment, both often found in criminals. Correlations do not establish cause-and-effect relationships, but many such correlations between chemical concentrations and behavioral patterns are beginning to be noticed.

Figure 25.5 presents some data from Belgium, correlating behavior in seven areas with socioeconomic status. (You may read the original article if you want the terms defined more clearly.) The correlations are quite remarkable. But do remember there is great overlap. The averages show many trends, but some individuals from every socioeconomic group are found performing at any ability level. How much the correlations are attributable to genetics, diet, and/or physical or social environment is not known, but something other than chance seems to be involved. Many such investigations are under way.

One possibility is that genetics determines most of your physiological

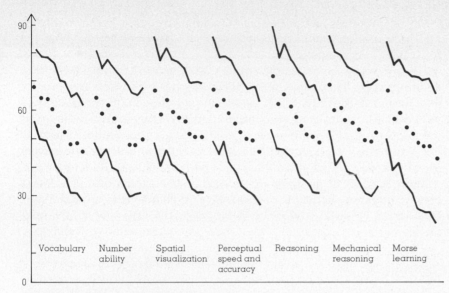

Figure 25.5
Distributions of abilities with
socioeconomic status. Flemish
military recruits from nine socioeconomic levels scored as
shown on seven measures of
ability. Dots give mean scores
in each group (note that the
high socioeconomic group is
highest in all ability categories).
Lines bracket ranges in scores
of the middle half of each
group. (From S. G. Vandenberg,
after Cliquet, in *Genetics,
Environment, and Behavior:
Implications for Educational
Policy*, edited by L. Ehrman,
et al., Academic Press, New
York, 1972)

characteristics and limitations, and some fraction of your behavioral patterns. The rest may be genetically limited, yet influenced by environmental, learned effects. But no one is happy with our present state of ignorance in this field.

One of the great current controversies, for example, is over the degree to which the ability to learn is genetically controlled as compared to environmentally controlled. This is an extremely difficult area to study (some say impossible), but you can be sure that both the studies and the controversies will continue for many years into the future.

One thing seems quite certain, even now. Limits are being found to the areas in which we have the freedom to choose. More and more physiological and behavioral patterns (assuming there is a valid distinction between the two) can be interpreted in terms of molecular effects and feedback systems.

Some people, like the boy in the frontispiece, feel that their freedoms are being compressed. They fear the scientists who find molecular limits to human potentialities. They suffer from a crisis of closure. But discovery of a limit does not create the limit. It shows the existence of a limit and makes it possible to move in another direction with a more distant limit. The range of real choices is not decreased at all, only made more clear. And, of course, the only persons free to choose are those who know the choices.

POINT TO PONDER
Correlations do not establish cause-and-effect relations, but cause-and-effect relations can lead to correlations.

POINT TO PONDER
Discovery of a limit does not create the limit. Discovery increases knowledge and lays a foundation for wisdom and progress.

Knowledge, Wisdom, and Progress

We are continually offered more and more choices. Consider, for example, the number of options (compared to 50 or more years ago) you have today in buying clothing, in traveling from your home to a place 1000 miles away, or in selecting fresh fruits and vegetables from the supermarket throughout the year.

Unquestionably, our knowledge has increased and now allows us to do things quite impossible only a short time ago. Yet mere availability does not make life easy. Many people do not enjoy making choices, as I have said. Many are more confused or irritated than they are pleased at the opportunity. To choose requires knowledge that choices exist. To use knowledge well requires wisdom. To utilize wisdom effectively requires a sense of what progress is. Where are the goals? Knowledge, wisdom, a sense of progress—hard as they are to attain and to use—greatly increase the *a priori* probability of satisfaction with the outcome. And, regardless of the number of choices, there can be only a single outcome.

Solutions Work

One of the reasons most people dislike making choices is the association of choice making with interruptions in the smooth flow of habitual (non-choice) life. Choices most commonly occur in connection with unforeseen problems that require a solution. Usually time is pressing and insufficient information is available. But persevere. The gains can be worth it.

For the first time in human history, some of our rivers are becoming purer. The Thames, Ohio, and Housatonic rivers—among others—are purer than they were 50 years ago. Tokyo Bay, Lake Erie, and San Francisco Bay are becoming cleaner. Others are on the schedule, and we know success is possible. Rivers have less of a foaming problem than they had 20 years ago, now that we control that property of detergents. Of the 20,000 major stationary sources of air pollution in 1975, 82% met emission limits or compliance schedules.

Atmospheric pollution due to cars and industry is on the downtrend in many cities, and is probably less serious now than the horse and smoke pollution found in large cities just before cars were introduced. The daily pollution from horses in New York City at the end of the nineteenth century was 2.5 million lb of manure and 60,000 gallons of urine. The air of many cities—Pittsburgh, St. Louis, even Los Angeles—has become cleaner and clearer. London hasn't had a "pea-soup" fog since December 1962, and 80% more sunshine than in 1955 now reaches its streets. Average year-round levels for SO_2 in the United States have declined more than 50% since 1964, and all air pollutants except oxides of nitrogen continue to drop, according to the Environmental Protective Agency (Figure 25.6). Water pollution is also decreasing.

The chlorination of water supplies has saved millions of lives. We

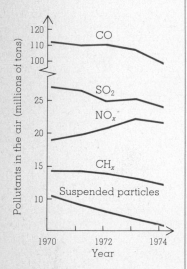

Figure 25.6
U.S. national trends in some average atmospheric pollutant levels. (Data from Environmental Protection Agency)

suspect it may be causing some cancer now. So that problem will be attacked. It does pay to note that no one has suggested that the cancer threat is as great as the bacterial threat used to be—and would be still without water treatment. Fluoridation has lowered the incidence of tooth decay. None of the predicted bad effects has as yet been detected.

The ospreys, the California pelicans, the Indian tigers, the Midwest robins (to name a few threatened species) have made good comebacks.

It is almost certain that none of these improvements would have resulted had there been no concern on the part of individuals. It is also quite likely that none of the problems would have arisen had humans been able to foresee the results of some early decisions. But in many cases the feedback systems seem to have worked. A gain in one place led to a threat in another. When the threat became obvious, the system was reexamined and adjustments were made. Both in 1974 and in 1975 some $6 billion was spent on pollution abatement in the United States. Improvement occurred.

As for human populations, all developed countries (except Ireland) experienced a fertility decline between 1950 and 1974. The average change was −29%. Furthermore, the seven largest of the developing countries (with a total current population of 2 billion) showed an average change of −17%. In less than one-fourth of the world population has there been negligible change in fertility.

Of course, not all goals have been achieved. Nor will they ever be. Nor will all our threats be as easy to diagnose and treat as those just listed. Some threats will lead to damage beyond the reversible level. However, we now have more sensitive analytical techniques and a greater awareness of long-term effects. The general acknowledgment that everything is tied together and that there are no "free lunches" has introduced a new element into planning. The headlong rush to progress is now seen to have been confused with a headlong rush to change. There is a good chance that the new awareness will spread widely enough to lead to the adoption of long-term points of view. Mistakes will continue. But, if feedback is built into the original design, the chance of mistake becoming catastrophe is diminished.

Epilogue

The last illustration, Figure 25.7, is a combination figure, exercise, and point to ponder. What do you observe? Now show it to others. What do they observe?

This is one of a large set of "ambiguous pictures." Some people observe nothing but black and white areas. Some observe one lady, or another lady, or two ladies. Is the figure just some cellulose partially covered with graphite? Hardly. But the message varies with the indi-

POINT TO PONDER
It has been estimated that the maximum size of a Western-style city would have been limited to a few hundred thousand if autos and trucks had not replaced the horse. A main cause of the limit would have been the manure output of the horses and the pollution problems resulting.

Welcome BACK!

Figure 25.7
What do you see?

HINTS TO EXERCISES
25.1 About 3 years using only
the moon, 8 years using Mars.
25.2 No one is always correct, but
the sources mentioned do pretty
well over the 15–25 year period.

vidual. So, in fact, do most messages. At least some of this variation is due to chemical differences between individuals.

Chemical differences in humans will always exist. Genes, varying inputs, and the second law all lead to that belief. Our observations of the past and present and our plans for the future should recognize the existence of these differences. But aren't the similarities greater? Studies of living systems indicate that they are. And the evidence is overwhelming that the same scientific laws describe *all* systems—living and not living, regardless of differences.

If you have studied this book successfully, you should have gained enough background in science (especially a molecular point of view) to increase your ability to explore and understand problems, including human problems, and so to reach reasonable decisions—decisions that have a high probability of working well in practice. Have a good life!

Appendix A   Numerical Calculations

I have used only six types of mathematical operations in this book. You have done addition, subtraction, multiplication, and division for a long time. The use of exponents and logarithms is probably less familiar to you, but both are variations of the same operation. The logarithm of a number to the base 10, abbreviated *log,* is just the exponent to which 10 must be "raised" to equal that number: $\log 100 = \log 10^2 = 2$. But let's approach the subject slowly and systematically, significant figures first, then exponential notation.

Significant Figures

You will save yourself a great deal of time (and avoid a great deal of self-deception) if you use significant figures in your calculations.

All measured numbers in science are uncertain, but usually the uncertainty is only in the last digit of the number as written. For example, the measured number 452 meters is known to be more than 451 and less than 453. In fact, it may be listed as 452 ± 1 meter to indicate that the uncertainty is 1 meter. In any case the 4 and 5 are "certain" and the 2 is in doubt. The number 452 is said to have three significant figures. Numbers are generally written with only one doubtful figure, so the last significant figure is doubtful. Later digits are not known at all. They are no more apt to be 0 than to be 1, 4, 7, 2, or any other digit.

Consider the problem of subtraction and addition: $12.4 + 0.2 = 12.6$; $768 - 567 = 201$. What about $12.4 - 0.02$? We do not know the fourth digit of 12.4. So we *round off* 0.02 to 0.0 and get $12.4 - 0.0 = 12.4$. Similarly, $8.36 - 0.0244 = 8.36 - 0.02 = 8.34$. Sometimes you round off upward: $6.87 + 1.428 = 6.87 + 1.43 = 8.30$. If the number being dropped is 5, 6, 7, 8, or 9, the preceding digit is increased by 1; if the number being dropped is 0, 1, 2, 3, or 4, the preceding digit is left unchanged. So $1.87 \rightarrow 1.9$, and $0.038454 \rightarrow 0.03845 \rightarrow 0.0385 \rightarrow 0.038 \rightarrow 0.04 \rightarrow 0.0$, depending on the number of figures to be retained. Note that $0.0385 \rightarrow 0.038$ because we know the original number to be 0.038454, otherwise $0.0385 \rightarrow 0.039$.

Rounding off for multiplication and division is done until each number has the *same* number of significant figures as the least well known number. If one of a set of measured numbers is known to only two significant figures, all the other numbers in the set are rounded to two *before* calculations are done:

$$\frac{2.794 \times 6.138}{2.84 \times 2.6} = \frac{2.8 \times 6.1}{2.8 \times 2.6} = 2.3461538 \qquad incorrect$$

I have written down all the digits my pocket calculator gives. In so doing I wasted a lot of time and have misled you. The answer cannot be more accurate than the data. The data were known to only two significant figures. The answer also will be known to only two significant figures. So the correct formulation is

$$\frac{2.794 \times 6.138}{2.84 \times 2.6} = \frac{2.8 \times 6.1}{2.8 \times 2.6} = 2.3$$

Don't let your hand calculator, or long division, waste your time and lead you into self-deception. Round off at the *beginning* of problems. If only multiplication and division are involved, round all numbers off to the *same* number of digits as in the least well known number. *Then* do the calculations.

If addition or subtraction is involved, round off to give a match in the decimal places being added and subtracted. *Do not assume that missing digits are zero!* Save time. Get truthful answers. And simplify the calculations.

Exponential (Scientific) Notation

The range in scientific measurements is so great that very large and very small numbers occur frequently. National budgets and even corporate finances are getting enormous. We talk of thousands, millions, billions, or even trillions of dollars. (British papers call $4,000,000,000 four thousand million dollars!) To grasp what quantities are being discussed, you must remember the relative size of each and how to convert one to the other. A scientist might express the same amounts as 10^3, 10^6, 10^9, 10^{12}, and 4×10^9 dollars. Note the saving in space, in symbols, in reading time, and in readiness of interconversion. Perhaps you don't see that the interconversion is easy. Let's look at the rules for the use of exponents.

Exponents indicate the number of digits the decimal point must be moved to convert a "regular" number to a number with a single digit to the left of the decimal point. Positive exponents indicate that the decimal has been moved to the left; negative exponents indicate that the decimal has been moved to the right:

$$1000 = 1 \times 10^3 = 10^3 \qquad 0.0002 = 2 \times 10^{-4}$$
$$5,000,000 = 5 \times 10^6 \qquad 167,000 = 1.67 \times 10^5$$
$$0.004 = 4 \times 10^{-3} \qquad 208 = 2.08 \times 10^2$$
$$0.048 = 4.8 \times 10^{-2}$$

Note that little space is saved unless the decimal point is moved four or more positions. But multiplication and division become easier if all the numbers have just one digit to the left of the decimal point.

Exponential numbers are *multiplied* by adding the exponents:

$$10^2 \times 10^5 = 10^7 \qquad 10^{-12} \times 10^{-4} = 10^{-16}$$
$$10^{-2} \times 10^3 = 10 \qquad 10^{-5} \times 10^5 = 10^0 = 1$$

Exponential numbers are *divided* by subtracting the exponents:

$$1/10^n = 1 \times 10^{-n}$$

So,

$$10^3/10^5 = 10^3 \times 10^{-5} = 10^{-2}$$
$$10^{-4}/10^5 = 10^{-4} \times 10^{-5} = 10^{-9}$$
$$10^{12}/10^{-8} = 10^{12} \times 10^8 = 10^{20}$$

Exponential numbers can be added and subtracted *if* the exponential terms are the same:

$$4.0 \times 10^{-3} - 2 \times 10^{-4}$$
$$= 4.0 \times 10^{-3} - 0.2 \times 10^{-3} = 3.8 \times 10^{-3}$$

$$7 \times 10^7 + 8.0 \times 10^8$$
$$= 7 \times 10^7 + 80. \times 10^7 = 87 \times 10^7 = 8.7 \times 10^8$$

Dividing an exponent by 2 gives the square root:

$$\sqrt{10^{-4}} = 10^{-2}$$

Dividing an exponent by 3 gives the cube root:

$$\sqrt[3]{10^{-9}} = 10^{-3}$$

You can check this by reversing the procedure:

$$(10^{-3})^3 = 10^{-3} \times 10^{-3} \times 10^{-3} = 10^{-9}$$

For complicated fractions (1) separate exponents from the rest of the number, (2) multiply the numbers in numerator and denominator, (3) multiply the exponential terms, and (4) do the separate divisions of both the numbers and the exponential terms:

$$\frac{500 \times 8 \times 10^{-5}}{4 \times 10^{-6} \times 50,000} = \frac{(5 \times 10^2) \times (8 \times 10^{-5})}{(4 \times 10^{-6}) \times (5 \times 10^4)}$$
$$= \frac{40 \times 10^{-3}}{20 \times 10^{-2}} = 2 \times 10^{-1} = 0.2$$

Base Units

Most numbers in science have units, such as meters or seconds, attached to them. Table A.1 lists the seven base units in which all quantities can be expressed. Combinations of the base units and exponential numbers allow simple descriptions of measured quantities.

Table A.1 Base Units for Scientific Measurements

Dimension	Name	Symbol
length	meter	m
mass	gram	g
time	second	s
temperature	Kelvin	K
amount of substance	mole	mole[a]
electric current	ampere	A
light intensity	candela	cd[b]

[a]Because it is short, *mole* is spelled out in this book.
[b]This book does not use light intensity, cd.

Exponential Prefixes

Table A.2 gives the internationally accepted set of prefixes for exponential numbers. They are used to specify large and small values of the scientific units. The base unit of length is the meter (m); I refer frequently to centimeters (10^{-2} m = 1 cm), nanometers (10^{-9} m = 1 nm), and kilometers (10^3 m = 1 km). Similarly, we talk of grams (g), nanograms (10^{-9} g = 1 ng), and milliseconds (10^{-3} s = 1 ms). Scientists sometimes refer to $1 million as a megabuck (10^6 bucks), or the population of the world as 4 gigapeeps (4×10^9 people). Peeps and bucks are not, of course, widely accepted units. But *mega-* and *giga-* are accepted and understood prefixes.

(Those used in this book are in the boxes.)

Notation	Prefix	Symbol	Amount
10^{12}	tera	T	trillion
10^9	giga	G	billion
10^6	mega	M	1,000,000
10^3	kilo	k	1,000
10^2	hecto	h	100
10^1	deka	da	10
10^0	—	—	1
10^{-1}	deci	d	0.1
10^{-2}	centi	c	0.01
10^{-3}	milli	m	0.001
10^{-6}	micro	μ	0.000 001
10^{-9}	nano	n	billionth
10^{-12}	pico	p	trillionth
10^{-15}	femto	f	—
10^{-18}	atto	a	—

Derived Units and Constants

Convenience dictates some specially named units derived from the base units. The dimensions of these derived units can be expressed in terms of the base units as shown in Table A.3.

Table A.3   Some Derived Units for Scientific Measurement

Quantity	Unit	Symbol	Dimensions
energy	joule	J	$kg \times m^2/s^2$
pressure	atmosphere	atm	$1.01 \times 10^5 \ kg/s^2 m$
electric potential	volt	V	$kg \times m^2/s^3 \ A$
power	watt	W	$kg \times m^2/s^3$

To express measurement in this book I have used six of the base units of Table A.1 plus other standard time units—year (yr), day (dy), hour (hr), minute (min)—and the four derived units of Table A.3. I have used one other derived unit (not in Table A.3 because it is not accepted by the International Committee on Weights and Measures). It is the liter which equals 1 cubic decimeter, $1 \ dm^3$. But liter volumes are so common that I have used this unofficial, but very widely used, unit. There are many other derived units that I have not used here.

Finally, there are some fundamental constants that occur so frequently that they are given special

Table A.4   Some Constants

Name	Symbol	Value
Avogadro's number	N_0	6.02×10^{23} particles/mole
Boltzmann constant	k	1.38×10^{-23} J/K
faraday	F	96,500 A × s
		96.9 kJ/(V × mole e⁻)
ideal gas constant	R	0.0821 (liter × atm)/(mole × K)
		8.31 J/(mole × K)
speed of light	c	3.00×10^8 m/s
pi	π	3.14

names, as shown in Table A.4. The dimensional relationships between the base units, derived units, and numerical constants are also shown.

All of the units mentioned are related to the International System of Units recommended by the International Committee on Weights and Measures. They are usually called SI units (from the French, Système International). You need a few conversion factors to get from the English units you may have been using to SI units. These are given in Table A.5.

Table A.5   Some Conversion Factors

Quantity	Conversion factor
length	1 in. = 2.54 cm
mass	1 lb = 454 g
temperature	1°F = 5/9°C or K
volume	1 quart = 0.946 liter

Logarithms

I use logarithms in determining pH (p. 294), entropy (p. 374), and the relationships between equilibrium constants and free energy changes (p. 350). If the hydrogen concentration in moles per liter, [H⁺], equals $10^{-5} \ M$, then the pH is 5. The pH is defined as the negative of the logarithm of [H⁺]:

$$pH = -\log [H^+]$$

For the example, $pH = -\log [H^+] = -\log 10^{-5} = -(-5) = 5$.

The logarithm of a number is the power to which 10 must be raised to give that number. Table A.6 lists the logarithms of the digits. Note that if you remember the log values for 1, 2, 3, and 7, you have the whole set. From Table A.6:

$$2 \log 2 = \log 4 = 0.602 \qquad 4 = 10^{0.602} \cong 10^{0.6}$$

$$\log 2 + \log 3 = \log 6 = 0.778 \qquad 6 = 10^{0.778} \cong 10^{0.8}$$

Suppose [H$^+$] is 4×10^{-8} M. Then

$$\mathrm{pH} = -\log (4 \times 10^{-8}) = -\log (10^{0.6} \times 10^{-8})$$

$$= -\log^{-7.4} = -(-7.4) = 7.4$$

As you may recall, ΔG^0 (kJ/mole) $= -5.708 \log K$. If $K = 3 \times 10^6$, we get

$$\Delta G^0 = -5.708 \log (3 \times 10^6)$$

$$= -5.708 \log (10^{0.477} \times 10^6)$$

$$= -5.708 \log (10^{0.5} \times 10^6)$$

$$= -5.708 \log 10^{6.5}$$

$$= -5.7 \times 6.5 = -37 \,\mathrm{kJ/mole}$$

Note that exponents that indicate the position of the decimal point, like 10^6, are exact numbers (not measured ones), so you can add and subtract other numbers with complete confidence that the exponent is not at all uncertain. It is a small counted number and is known to as many figures as you may need. This is also typically true of other small counted numbers, as opposed to measured numbers. Thus, 1 mole in a chemical equation means *exactly* 1 mole (to as many significant figures as you wish).

Table A.6   Some Useful Logarithms

Number	Logarithm
1	0 (exactly)
2	0.301
3	0.477
4	0.602; (2 $\times$ log 2)
5	0.699; (1 $-$ log 2)
6	0.778; (log 2 $+$ log 3)
7	0.845
8	0.903; (3 $\times$ log 2)
9	0.954; (2 $\times$ log 3)
10	1.000 (exactly)

Dimensional Analysis

Most sample calculations in this appendix have involved numbers only. But, as I have said, most numbers in science are associated with dimensional quantities: 1.2 atm, 3 m, 4.82 s, 298 K, and so forth. The unit is part of the quantity and should *always* be written with it. You will save yourself from many mistakes, and increase your comprehension of many problems, if you include the units or dimensions with every number. For example, the ideal gas equation is $PV = nRT$. Suppose you wish to calculate P in atmospheres (atm) when $V = 10$ liters, $n = 1$ mole, and $T = 300$ K.

$$PV = nRT$$

$$P = nRT/V = 1 \text{ mole} \times R \times 300 \text{ K}/10 \text{ liters}$$

What do we do about R? Table A.4 lists two values:

$$R = 0.0821 \text{ (liter} \times \text{atm)/(mole} \times \text{K)}$$

and

$$R = 8.31 \text{ J/(mole} \times \text{K)}$$

If you use the first value, you get

$$P = 1 \, \cancel{\text{mole}} \times 0.0821 \, [(\text{liter} \times \text{atm})/(\cancel{\text{mole}} \times \cancel{K})]$$
$$\times 300 \, \cancel{K}/10 \, \cancel{\text{liters}}$$

or

$$P = 1 \times 0.0821 \text{ atm} \times 300/10$$

$$= 1 \times 0.08 \text{ atm} \times 300/10 = 2.4 \text{ atm}$$

Or, to one significant figure, 2 atm. Try the other value of R. Does it give P in atmospheres? Is its answer incorrect? The answer is *not* incorrect, but it is in strange and essentially useless units.

Note that I "canceled units" just as you have learned to cancel numbers in fractions. The uncancelled units give the units of the answer—in this case, atmospheres. Dimensional analysis is an easy way to save time, to increase accuracy, and to enhance your security!

Summary

Your use of significant figures, exponential notation, scientific units, and dimensional analysis will greatly minimize the problems encountered in dealing with quantitative data. Because this book deals with the science of chemistry, not arithmetic or mathematics, many of the calculations are done with one significant figure. I do this to encourage you to think about molecular properties, not about arithmetical operations. You should be able to do the arithmetic for many of the exercises in your head.

Appendix B   Some Simple Chemical Nomenclature

I have kept the discussion of chemical names brief in this book and have frequently used structural formulas to carry the messages. (You will find such discussions on pages 24, 71, 97, 133, and 142.) However, as laws requiring full disclosure of contents of packages begin to go into force, everyone will be inundated with chemical nomenclature. I'm not sure much thought has been given to how citizens are supposed to interpret this new information. Is sodium acetylsalicylate better than aspirin? The present practice of giving a chemical three names—a fully systematic scientific name, a trivial name (simpler, but used by everyone to describe the compound), and a proprietary name (copyrighted by a firm to identify its package version of the compound)—will almost certainly continue.

But it is fun to decipher puzzles, so I write with that aim in view. You can use the many named structural formulas to test your skill as a decoder. If you are interested in a compound you cannot decode, look in the *Handbook of Chemistry and Physics* or the *Merck Index*.

Binary Compounds

In the name of a binary compound, the element that is more metallic is written and named first with its elemental name. The less metallic element follows, and the suffix *-ide* replaces the element's normal terminal syllable: NaI, sodium iodide; ScP, scandium phosphide; NO_2, nitrogen dioxide; PI_3 could be phosphorus iodide or phosphorus tri-iodide (but see the next paragraph).

Most pairs of elements form more than one binary compound. They are usually differentiated by giving the oxidation state of the element with variable oxidation state in Roman numerals: PI_3, phosphorus(III) iodide; PCl_5, phosphorus(V) chloride; PbO, lead(II) oxide; Pb_3O_4 (or Pb_2PbO_4), dilead(II) lead(IV) oxide.

Carbon–hydrogen compounds (hydrocarbons) are a special case. "Straight-chain" compounds are named in terms of the longest chain of carbons present. Side chains are then identified by length and by position on the main chain (H atoms are indicated by a bond; by convention the symbol H is often not included):

pentane   2,3-dimethylpentane

Compounds that have one saturated ring are named from the corresponding straight-chain compound:

cyclopropane   cyclohexane

Compounds with unsaturation, either straight chain or ring, must have the position of unsaturation noted:

but-1-ene   1,4-cyclohexadiene

Ring compounds that can be written as though they had alternating double and single bonds throughout, like benzene, are called *aromatic*. Each ring system has its own name, the positions being numbered sequentially around the rings to identify where substituents may be:

1,3-dimethylbenzene   2-ethyl-6-propylnaphthalene

Substituted Compounds With Catenated Carbons (Organic Compounds)

It is clearly impossible to learn systematic nomenclature for the millions of carbon compounds quickly.

The general technique is to consider the straight-chain and ring carbon structures just discussed as frameworks on which substitutions have been made in positions identified by number or element:

| 4-chloro-nitrobenzene | 2-bromobutane |

CCl_2F_2, dichlorodifluoromethane (Freon 12). For further examples, look in Chapters 23 and 24 especially. Note how many trivial (nonsystematic) names are used when the structures get complicated. It is easier to memorize trivial names than to learn and continually write out the full scientific names.

Isomerism gives rise to many structures with the same empirical formulas and interesting problems in nomenclature, as discussed briefly on pages 133 and 433.

Compounds Not Based on Carbon Atoms (Inorganic Compounds)

There are several hundred thousand known compounds that contain little or no carbon. Many contain groups of atoms that frequently are found together, such as NO_3^-, nitrate; SO_4^{2-}, sulfate; PO_4^{3-}, phosphate; NH_3, ammonia; and NH_4^+, ammonium. These are called *radicals*. Several hundred are known. All the ions, of course, are found in combination with other ions and are named accordingly: $NaNO_3$, sodium nitrate; $(NH_4)_2SO_4$, ammonium sulfate; $Cu(NO_3)_2 \cdot 5\,H_2O$, copper(II) nitrate pentahydrate; $Ni(NH_3)_4^{2+}$, tetrammine nickel(II); H_2SO_4, dihydrogen sulfate or sulfuric acid; and HPO_4^{2-}, mono-hydrogenphosphate.

Computer-Based Nomenclature

You might be interested to know that the advent of computers has led to the invention of systematic nomenclature that can be printed on one line (a limitation of computers) and can unambiguously describe most known structures, even three-dimensional structures. The symbolism bears little relationship to what you have seen. For example: CO_2, OCO; HCl, GH; $CHCl_3$, GYGG; CH_3COOH, QV1; $Cl_2C=CHCl$, GYG:1G; H_2SO_4, WSQQ.

Appendix C   Suggested Additional Sources

I list here some sources of further readings that will allow you quick access to information in almost every area of chemistry. These will give you a much better start for digging deeper or for term papers than any dated bibliography of actual articles I could include here. If you start with the most recent edition of the reference works, and the most recent issues of the journals (don't forget the indexes to each volume), you will quickly get access to current information in this rapidly changing human activity called chemistry. The card file in your local library is a much better source of information readily available to you than any list of books I could assemble.

General References

A good recent encyclopedia is usually the most available source of summaries of the recent status of a field and its historic background. Don't forget to use its index and bibliography. The *Kirk-Othmer Encyclopedia of Science and Technology* (Wiley, New York, 1963, with a supplementary volume in 1971) is an excellent summary up to its publication date. The *Handbook of Chemistry and Physics* (issued annually by CRC Press, Cleveland, Ohio) is a massive handbook, and the best and widest ranging collection of general data on chemicals, definitions, and conversion factors. The *Merck Index* (published occasionally by Merck and Company, Rahway, N.J.) is an excellent source of structural and health-related information, such as toxicity and lethal doses. Numerical data on the United States is best sought in the annual editions of the *Statistical Abstract of the United States* put out by the U.S. Bureau of the Census. Similar material is available from commercial publishers under titles such as *The U.S. Fact Book* and *The American Almanac.* For biochemical information it is hard to beat the 1000-page compendium *Biochemistry* by A. L. Lehninger (Worth, New York, 1975).

Current Periodicals

The *Readers' Guide to Periodical Literature* is the place to start a general search (unless you are looking for original research data). For original data, go to *Chemical Abstracts.* But for quick entries to gold mines of current treatments, jump directly into the journals listed next (in rough order of probable yield per issue):

Science (weekly; American Association for the Advancement of Science). First-rate articles (many over your head), book reviews, and research notes (almost certainly over your head). *Science*'s greatest virtue for you lies in the three or four articles each week on current tensions, controversies, and discussions in science at the individual, institutional, governmental, and international level. Excellent, fact-filled reporting.

Chemistry (monthly; American Chemical Society). Excellent articles, almost all easily readable and understandable by you, on various aspects of science.

Scientific American (monthly; Scientific American, Inc.) Good articles on a wide range of scientific issues—research, socioeconomic—political implications, some news, good book reviews.

Journal of Chemical Education (monthly; Division of Chemical Education, American Chemical Society). Excellent articles aimed at chemistry teachers. Most will be over your head, but a good source for skimming.

Chemistry in Britain (monthly; Chemical Society of Britain). Excellent discussions of the state of chemistry in Britain; reports similar to those in *Science,* including first-rate, highly readable articles on current research results of interest to the general public.

Chemical and Engineering News (weekly; American Chemical Society). Mostly news articles, with an occasional excellent review article in an area of general interest.

The two journals, *Accounts of Chemical Research* and *Quarterly Reviews,* have excellent articles. Most articles are far beyond beginners, but a quick scan will reveal two or three articles per year of the highest possible quality by experts in their fields. The first is put out by the American Chemical Society, the second by the British Chemical Society. If you get really serious about your research, there are some 5000 periodicals published with articles in chemistry!

Appendix D Supplementary Problems

Here are some further problems, numbered sequentially for the Chapters they accompany. They cover areas not stressed in earlier problems and can serve either for extra assignments or for review.

1.18 Paracelsus, in the sixteenth century, divided diseases into five categories based on their causes: (a) those caused by the influence of the stars; (b) those caused by poisons taken into the body; (c) those caused by the natural constitution of the patient's body; (d) those caused by mental or spiritual upsets; and (e) those sent by God and incurable by medical treatment. Was this a scientific theory? Was it useful in suggesting possible cures and hypotheses for future studies? If you object to some or all of these categories, what evidence can you cite to support your objections?

1.19 After a careful study of last season's race results, a horseplayer comes up with a system calling for a bet on any roan filly starting from position 5. He shows from the results that 80% of such horses won their races. Does this evidence prove that his system would be a good one to use the next season?

2.16 When you become hot, you sweat. Does this sometimes embarrassing process do any good? Many furry animals cannot sweat. They breathe rapidly (pant) when hot. What good does this do?

2.17 In 1838 J. B. Boussingault in France planted seeds in a pot containing no nitrogen fertilizer and obtained the following data. He got similar results in field experiments. Interpret the results in terms of the law of conservation of atoms.

	Wheat	Oats	Clover	Pea
N in seed, mg	57	59	114	47
N in 3-month-old plant, mg	60	53	156	100

4.13 President Harry Truman knew the United States had two nuclear bombs available when he decided to drop them on Japan with an interval of a few days between. The U.S. rate of production was about one bomb per month. Outline some of the alternatives available to him. Which would you pick? Each bomb killed about 100,000 people. The casualty rate for the war at that time was about 10,000 per week, and it was estimated that invading Japan would cause about 500,000 deaths.

*4.14 Many modern analytical methods were developed as part of the effort to produce nuclear bombs—mass spectrometry, electroanalysis for minute amounts of metals, ion exchange, solvent extraction, paper chromatography, emission spectroscopy, among other methods. Why was good analysis so important to the project?

*4.15 An estimated 25% of all patients admitted to U.S. hospitals are given treatments (diagnostic or therapeutic) involving radioactive tracers. Look up and list some possible tracers. Appendix C gives some guidance in looking things up.

4.16 Outline some of the effects of radioactivity and/or nuclear energy on (a) surgery, (b) military methods, (c) relative chances of peace and large-scale war, (d) relative chances of small- and large-scale war, and (e) improved food plants.

*4.17 Plutonium-238 (half-life 86 years), is used as a power source in heart pacemakers implanted under the skin. Does it probably emit α, β, or γ rays? Write a likely nuclear reaction. What property of the daughter nucleus makes it safe to the owner of the pacemaker?

4.18 The history of human development shows that, in most cases, "reasonable" experts have underestimated the bad effects of new developments (for example, automobiles, cigarettes, or DDT). Is the current expression of strong concern about nuclear power an indication that we have learned our lesson? Or that things will be even worse than claimed by the pessimists? Or perhaps that the critics are like those "crackpots" (many of them later proven wrong) who forecast evil results from the earlier developments?

*4.19 Outline a possible argument between a Stone Age discoverer of fire and an opponent who argues that the dangers of this new energy source outweigh its benefits.

4.20 There is about 0.2 ppm ^{238}U in most coal. How does the fission energy potentially available from this ^{238}U compare with the thermal energy available from burning the coal? (One kilogram ^{238}U is equal to about 10^9 kg coal.) [$\sim$200 times more than burning coal]

12.19 It is possible to separate ^{235}U and ^{238}U by converting uranium to UF_6 and allowing the gas to diffuse through many, many porous barriers. Calculate the relative diffusion rates to see why so many barriers are needed. [diffusion ratio $\cong$ 1.004]

*12.20 Estimate the amount of air over Los Angeles that passes through automobile engines each week, as a percentage of the air in a layer 100 ft thick over the city. The Los Angeles basin, an area of 10^4 square miles, contains half the cars in California. About 10^{11} gallons of gasoline (octane, C_8H_{18}) is consumed in the United States each year, 10% of it in California. [$\sim$70% air]

*12.21 Look up what an atmospheric inversion is, and account for its properties and effects in terms of the ideal gas law.

12.22 This chapter (and most general discussions) deals with atmospheric pollution. Yet many scientists are much more worried by solid and liquid pollutants. Suggest some specific reasons.

*12.23 Pollutants produced in the Northern Hemisphere spread through the atmosphere there long before they reach and spread through the Southern Hemisphere. The same effect is seen for those produced in the Southern Hemisphere. Explain this in terms of atmospheric effects at the molecular level.

12.24 The temperature of a satellite shaded from the sun by the moon, stars, or another planet would drop to about 4 K unless heated from its own power. Yet the gas molecules in the vicinity have velocities corresponding to a temperature of about 1000 K. How can the satellite get so cold? (Hint: By what methods is heat gained and lost by the satellite?)

12.25 The most common gas-welding outfit burns acetylene and oxygen from two separate cylinders of the pressurized gases. There are about 10 kg of acetylene, C_2H_2, in a standard cylinder. Oxygen cylinders usually contain about 40 liters of oxygen at 130 atm pressure. If the welder starts with the two cylinders full, which will be used up first? (Hint: First write the chemical equation.) [more than 4 O_2 cylinders for 1 C_2H_2 cylinder]

*12.26 Insects do not have lungs. Instead, their body surfaces are studded with many tiny tubes extending from the surface into the body. Each cell gets its oxygen from air that diffuses along nearby tubes. Would such a system work for a man-sized insect? (Explain at the molecular level.) Suppose that a human were shrunk to the size of a fly. Would the lungs still function appropriately?

12.27 A moth weighing 2 g uses up 100 cm^3 of oxygen per hour while flying. If sugar, $(CH_2O)_n$, is the source of energy, what is the minimum amount of sugar the moth must gather from flowers in an hour? [$\sim$100 mg]

12.28 The design rule for blimps and dirigibles is that 1000 liters of helium will lift 1 kg. Show that this is reasonable.

12.29 Death by suffocation in a sealed container is normally due to CO_2 poisoning (which occurs at about 7% CO_2 by volume), not to oxygen deficiency. For what length of time would it be safe to be in a sealed room 3 m $\times$ 3 m $\times$ 6 m? [about 10 days for one person]

*14.20 The maximum rate at which humans can run (as measured by world records in track) follows an S-shaped curve when rate is plotted versus distance. The rate is almost 11 m/s over the first 200 m, drops rapidly to 7.5 m/s at 1000 m, then drops slowly to about 5.5 or 6 m/s from 10,000 to 30,000 m. Suggest a possible interpretation at the molecular level.

17.19 Body metal on cars in the Midwest and Northeast corrodes much more rapidly than in the West and South, because of the $CaCl_2$ and NaCl put on icy roads. How do these chemicals enhance corrosion?

17.20 The tarnish (Ag_2S) on silver tableware can be converted back to silver by boiling the tarnished piece in an aluminum pan with some salt or baking soda in the water. Write a probable net equation. Outline the function of the aluminum, of the dissolved salt, and of the boiling. Any advantages over mechanical polishing? Pieces of metal (they look like aluminum and can be bought for a dollar or so) do this job when placed in the bottom of the pan. Would you recommend buying one?

17.21 Sodium vapor lamps have two deficiencies: (a) they emit a pure yellow light rather than a broad spectrum, and (b) the hot gaseous sodium attacks the SiO_2 in the glass envelope. In 1960 the General Electric Company sintered Al_2O_3 with 0.2% MgO to give a clear envelope they call Lucalox. It allows a high-pressure (0.5 atm) sodium light. The high pressure broadens the spectral emission, and the envelope is not attacked by the sodium as is SiO_2. Why is it not attacked?

17.22 "If a satisfactory technology of energy storage were available today, U.S. generating capacity could run at constant load and would not need to be expanded for the next seven years" (*Chemistry in the Economy,* American Chemical Society, 1973). Why not use lead storage batteries?

*17.23 We saw earlier in this chapter that electric power costs for production of aluminum are said to be 25% of the total cost. Alcoa in an ad in *Time* (February 9, 1976) said, "This remelting (of recycled aluminum cans) saves 95% of the energy it would take to make molten metal from virgin ore." Can you reconcile these two statements? If not, which seems more reasonable?

*19.13 Glucose exists in aqueous solutions in two ring-shaped forms: 36% α and 64% β. Calculate their differences in stability, ΔG^0, in kilojoules per mole. The β form has all —OH groups extended sideways roughly in the plane around the six-membered ring. The α form has one out-of-the-plane —OH group. Is this consistent with ΔG^0? [$\Delta G^0 = -1.42$ kJ]

19.14 Two basic human desires are for security and for freedom. Do you think there is any merit in

an analogy to enthalpy and entropy drives in molecules?

19.15 Suppose that the chemical reactions in a living system all achieved a state in which $\Delta G = 0$. What would be the effect on the system?

19.16 A very common reaction in living systems is the hydrolysis of a compound called ATP:

$$ATP^{2-} + H_2O = ADP^- + H_2PO_4^-$$
$$\Delta G^0 = -30 \text{ kJ/mole}$$

A common value for ΔG for this reaction in living systems is -20 kJ/mole. Why does this value differ from ΔG^0?

19.17 Ammonia, NH_3, has been predicted to be a "fuel of the future." Discuss some pros and cons in terms of Table 19.1.

*19.18 Where do you think butane, C_4H_{10} (g), would be found in Table 19.1? Guess, and justify, a value of ΔG_f^0 for C_4H_{10} (g). Also values for ΔH_f^0 and S^0. Check the consistency of your values with $\Delta G_f^0 = \Delta H_f^0 - T \Delta S^0$.

19.19 You have undoubtedly noted at the beach that wet sand is darker than dry sand. The wet sand is absorbing more of the incident radiant energy of the sun. Its ability to reflect light is about half that of dry sand. Why, then, is dry sand so much hotter than wet sand?

19.20 The heats of combustion for some molecules found in living systems are as follows: glucose (a carbohydrate; formula weight 180), 2800 kJ/mole; palmitic acid (a lipid or fatty acid; formula weight 256), $-10,000$ kJ/mole; and glycine (an amino acid; formula weight 75), -1000 kJ/mole. Adults in the United States generate about 10,000 kJ/day. What approximate weight would be lost per day if all the lost weight went into energy production? Is exercise the fastest way to lose weight?

*21.24 Wool is much slower to burn than cotton. Show why in terms of a possible equation. Urea polymers are much used to make cotton more flame-retardant. Does this fact fit your ideas?

21.25 Making, dyeing, and finishing fibers account for about 10% of the reatil cost of fabrics. Spinning, weaving, designing, and manufacturing account for about 50%. Fabrication costs five times as much as materials. Will the rise in petroleum prices have much effect on fabric costs?

*21.26 Butyl rubber is much used to line water reservoirs. A square meter of 0.75-mm film leaks about 0.05 cubic decimeters (dm^3) per year; 25-mm-thick concrete leads about 1000 times as much. Why the big difference?

*21.27 Polyethylene made at 80 to 300°C and 1000 to 3000 atm is branched and low in density. Polyethylene made at 0 to 300°C and 1 to 500 atm is linear and high in density. Why do the densities differ?

*21.28 Use bond energies to estimate ΔG^0 for the addition polymerization of ethylene or a vinyl monomer. Is enthalpy or entropy the larger driving force? Do the same thing for the condensation polymerization of nylon or glucose. Any differences? [$\Delta H_{reac} \simeq 0$ for condensation polymerization]

21.29 What occurs at the molecular level to limit the actual breaking strength of nylon to one-tenth of the theoretical value?

21.30 Automobile safety glass consists of a glass sandwich with a polyvinyl butyral polymer between. What properties must this polymer have for this use? The main monomer is $CH_2=CHC_4H_9CHO$, with some —OH side groups as well in the polymer.

21.31 Electron beams produce hard (cured) polyester films in a fraction of a second, much faster than the usual peroxide treatment. Suggest possible simple mechanisms that account for the similar result but different rates.

21.32 Polyethylene cannot be vulcanized, but it can be cross-linked by bombardment with high-energy radiation. Why?

*21.33 There is great interest in electrolytic polymerization directly onto an object to be coated with polymer. Compare this to a redox polymerizing catalyst.

21.34 No one has yet produced an accepted synthetic shoe leather, though Du Pont tried a large-scale marketing of Corfam. Suggest some of the problems, from what you know about leather.

22.14 An astronaut drinks about 2 liters of H_2O each day but excretes about 2.4 liters of H_2O each day. His weight does not change. Account for the difference.

22.15 According to Bernard Rimland (in S. C. Plog and R. B. Edgerton, eds., *Changing Perspectives in Mental Illness,* Holt, Rinehart and Winston, New York, 1969), "[with] IQ scores of identical twins, the greater the birth weight difference, the lower the IQ of the lightest twin" (p. 710). Suggest a molecular interpretation.

*22.16 Resting muscle consumes O_2 at about 2 ml/(kg × min), but a working muscle may use 100 times this and require a 50-fold increase in blood circulation. Maximum O_2 intake in a physically fit subject is 4–5 liters/min. What do these figures suggest must occur in a runner operating at maximum activity in a 440-yard footrace? [O_2 deficit $\simeq$ 10 liters]

22.17 It has been suggested that the oxygen consumption of the brain is remarkably uniform over the range of consciousness from sleep to severe intellectual activity (L. Sokoloff et al., *J. Clin. Invest.,* **34,** 1955, pp. 1092, 1101). How then can "thinking" be such hard work and require increased oxygen uptake in humans?

22.18 George Gaylord Simpson suggests that there are about 4×10^6 species in existence and about 4×10^9 extinct. He also suggests that the average lifetime of a species is about 10^6 years. Suggest an interpretation of these data at the molecular level.

*22.19 Hemoglobin, HHb, in the red blood cells is in equilibrium with both CO_2 and O_2, as shown schematically by the following equation:

$$O_2 + HHbCO_2 \rightleftharpoons HHbO_2 + CO_2$$

(Actual detailed equilibria are more complicated.) Discuss the evolutionary advantages to humans of having a single molecule, like hemoglobin, that reacts with both CO_2 (through its amino groups) and O_2 (through its iron ions).

Appendix E   Reference Table of Contents

Glossary of Fundamental Terms

Note: You can find discussions of each term on the pages numbered in parentheses. Use the index, also.

absolute scales (p. 206) Based on absolute zero; the quantity is really zero at the specified condition and there are no negative values: zero Kelvin, zero atm.

acid (p. 287) Donates protons to bases, accepts a pair of electrons to form a covalent bond: H^+.

activated complex (p. 231) An unstable molecule that forms when reactants collide, then dissociates into products: $Cl—Cl—Br^-$.

activation energy (p. 232) Difference in energy between reactants and activated complex: ΔE_a

amphoteric (p. 293) Acts as an acid toward stronger bases, as a base toward stronger acids: HCO_3^-.

aromatic (p. 155) Contains at least one benzene-type ring of delocalized electrons: C_6H_6, ⬡.

atom (p. 46) Atomic nucleus plus its surrounding electrons, if any: Cl, Na^+, H^+, H.

base (p. 287) Accepts protons from acids, donates a pair of electrons to form a covalent bond: OH^-.

bond, chemical (p. 135) Force due to the simultaneous attraction of one or more electrons by more than one nucleus: $H—H$, $N≡N$.

buffer (pp. 299–300) Mixture of acid and base that resists changes in acidity upon dilution or addition of other acids or bases: blood.

catalyst (p. 175) Reactant that is regenerated in a later mechanistic step: enzyme.

charge density (p. 289) Ratio of electric charge to molecular volume; greater for O^{2-} than for F^-.

compound (p. 22) Two or more kinds of atoms in a stable combination with a fixed melting point and boiling point: H_2O.

concentration (p. 294) Number of moles of substance per liter (moles/liter, M).

crystal (p. 112) Collection of molecules arranged in a regularly repeating pattern: diamond, graphite, quartz.

electron (p. 37) Particle with unit negative charge sometimes emitted by, but more often found distributed around, atomic nuclei: e^-, β ray.

element (pp. 22, 42) Collection of atoms, all of which have the same nuclear charge: O_2.

energy (p. 50) Measure of the amount one system could heat another, colder system, and/or the amount of work it could do: joules.

enthalpy (p. 358) ΔH, amount of heat gained by a closed system in a constant pressure change.

entropy (p. 360) Measure of the number of ways, W, the system can distribute its atoms and energy; increases for all total changes, $S = 2.303\ k \log W$.

enzyme (p. 123) Long-chain polymer of α-amino acids; catalyst for biochemical reactions.

equation, chemical (p. 169) Symbolic summary of conservation of atoms, charge, and mass: $H_2(g) + \frac{1}{2}O_2(g) = H_2O(l)$.

equilibrium, dynamic (p. 266) Final state of no further change for any chemical system; each reaction is exactly balanced by a reverse reaction of the same rate: $H_2O(c) = H_2O(l)$.

escaping tendency (p. 266) Tendency of a molecule to change phase or react

free energy, Gibbs (p. 350) Measure of the tendency of change to occur in a closed system at constant temperature and pressure; ΔG must be negative for such a change.

free radical (p. 401) Molecule containing one or more unpaired valence electrons.

gas, ideal (p. 210) A system obeying the equation $PV = nRT$.

gene (p. 419) A long molecule that guides the synthesis of an enzyme and so controls the inheritance of certain traits.

half-equation (p. 165) Equation that includes electrons as a reactant or product: $\frac{1}{2}H_2(g) = H^+(aq) + e^-$.

half-life (p. 59) Time required for half the initial molecules to react.

hydration (p. 286) Reaction with water: $Cl^-(g) + H_2O(l) \rightarrow Cl^-(aq)$.

hydrolysis (p. 252) Breaking apart (*lysis*) involving water: $CO_3^{2-}(aq) + H_2O(l) = OH^-(aq) + HCO_3^-(aq)$.

ion (p. 135) Electrically charged molecule: Na^+, Br^-, NO_3^-, H_3O^+.

joule (p. 50) Unit of energy, heat, or work. The heat that can be delivered by an electric current of 1 ampere flowing for 1 second at a voltage of 1 volt: $J = A \times V \times s$. One joule will heat 1 g (1 cm^3) of water approximately 0.25°C; so 1 kJ will heat 1 liter of water approximately 0.25°C.

mass (p. 43) Measure of resistance to motion and/or of gravitational force: gram.

mechanism, reaction (p. 226) Series of collisions (usually bimolecular) in which reactants change to products.

model (p. 73) Simplification of certain features of a real system that allows us to think about it, and remember and correlate properties.

mole (p. 84) 6.02×10^{23} particles of a listed formula: e^-, H_2, $C_6H_{12}O_6$.

molecule (p. 7) Several atoms bonded together long enough to make it convenient to think of them as a unit; also, an isolated atom: HCl, C_6H_6, Br_2, Ar.

nucleus (p. 40) Positively charged, high-density particle at the center of every atom, usually considered as made up of protons, $\frac{1}{1}p$, and neutrons, $\frac{1}{0}n$.

orbital (p. 72) Energy level in which

521

electrons exist in an atom or molecule: $1s$, $2p$.

oxidation (p. 95) Apparent loss of electrons: $I^- (aq) = \frac{1}{2}I_2 (c) + e^-$.

oxidation state (p. 290) Apparent charge on an atom or molecule: $1-$ for I^-, 0 for I_2.

periodic table (p. 70) Serial arrangement of chemical elements in order of increasing nuclear charge in families of similar valence electron structures.

pH (p. 294) The negative of the exponent of the hydrogen ion concentration: $[H^+] = 10^{-2}$, $pH = 2 = -\log[H^+]$.

phase (p. 16) Homogeneous region having the same properties throughout: gas, liquid, crystal.

photon: (p. 58) Particle of radiant energy: light energy.

probability (p. 492) Likelihood of occurrence, usually expressed as a number between zero (no likelihood) and 1 (complete certainty).

rate of reaction (p. 236) The change in concentration per unit time.

reaction, chemical (p. 365) Change in the bonding arrangement of the molecules: $H_2 (g) + Cl_2 (g) = 2 HCl (g)$.

reactive Able to undergo some *specified* change; sodium is reactive with oxygen.

redox (p. 279) Simultaneous occurrence of reduction and oxidation: $Na (c) + \frac{1}{2}Br_2 (l) = NaBr (c)$.

reduction (p. 95) Apparent gain of electrons: $Fe^{2+} (aq) + 2\,e^- = Fe (c)$.

reversible (p. 382) Capable of being returned completely to its initial condition. Closed systems can be reversible; changed isolated systems cannot be reversible.

second law (p. 372) The second law of thermodynamics: ΔS must be positive for any total change. $\Delta S = 0$ at equilibrium in an isolated system.

significant figures (Appendix A) The digits in a number that are based on measurement and do not merely indicate the position of the decimal point: 127, 0.0348, 2.03×10^2, and 387,000 all have three significant figures.

stable Unable to undergo some *specified* change. Argon is stable with respect to tungsten lamp filaments. Not merely: argon is stable.

steady state (p. 264) No net change because each reaction in the system is preceded and followed by another with the same rate; input equals output.

system, closed (p. 373) Constant mass, variable energy: hot-water bottle, ice bag.

system, isolated (p. 373) Constant mass, constant energy: perfect thermos bottle.

system, open (p. 373) Variable mass, variable energy: cup of hot coffee.

temperature (p. 207) The property that becomes the same in two systems separated by a nonreactive, heat-conducting wall: °C or K.

Index

References to figures are in italics; table references are indicated by (T)

Coenzymes, 8, 421
Coinage metal, 109(T), 110
Coke, 96, 102
Collision orientation, 227
Collisions, molecular, 235
Collisions per second, 235
Color, 74, 119
Combustion, 20, 442
Competition, chemical, 279
Composition
  of atmosphere, 92(T)
  of humans, 6(T), 7, 413
  isotopic, 45
  of living systems, 6(T)
  of ocean, 93(T)
Compounds
  Bethollide, 35
  binary, 129
  chemical, 22
  defect, 35
  essential, 8
  formation of, 141
  ubiquitous, 8
Concentration cell, *311*
Concentrations, 294
Conception, 415
Conchiolin, 121
Condensation, 18, 257
Condensation polymerization, 400
Condensed rings, *134*
Conductivity, electrical, 78, 136, 288
Conservation
  of atoms, 23
  of energy, 24, 373(T)
  of mass, 23, 373(T)
  of mass-energy, 373(T)
Constant composition law, 21, 23
Constant *m,P,T* systems, 356
Contaminants, allowable, 450(T)
Coordination number, 119(T), *138*
Copolymers, 399, 406
Copper
  cost of, *481*
  crystal, *187*
  mining of, 96
  production of, *481*
Corrosion, 253, 254(T), 323
Corticosterone, 434
Cotton, 198, 390, 394, 395
Cottonseed oil, 193
Coupled reactions, 321
Covalence, 136
Crewe, Albert, 10
Crick, Frances, 98, 418
Crime, and chemistry, 485
Cristobalite, 116
Critical experiments, 9
Cross-links, 401
Cryolite, 98

Crystal, 112, 187, 405
Curie, Marie and Pierre, 37
Cycles, 27, 381
  carbon, 27, *28*
  nitrogen, *29, 174*
  oxygen, *29,* 473
  water, *28*
Cylinders
  gas, 212
  high pressure, 211
Cytochrome c, 130, 421, *423*
  synthesis of, *422*
Cytosine, 418

D

Dacron, 399
Dalton (unit), 23, 43
Dalton, John, 21, 34, 211, 214
Davy, Humphrey, 36
DDT, 463, 466, 469
De Caelo, 305
De Chardonnet, Hilaire, 395
Decay, radioactive, 57, 59
Defect, mass, 44, 44(T), 45(T), 51
Delgado, José M. R., 497
Delocalized electrons, 153
Democritus, 15
Deoxyribonucleic acid (DNA), 11, 418, *420*
Desalination, 151
Detergents, 192
Deuterium, 42(T), 56, 269
Diamond, 110, *124, 131,* 153(T)
  synthetic, 114
Diamond doublets, 121
Diet, 414, 446
Diffusion
  gas, *218*
  molecular, 125
Diffusion-controlled rate, 237
Digestion, 341, 446, *447*
Dihydrogen, 163
  see also Hydrogen
Dinitrogen (N_2). *See* Nitrogen
Diogenes, 161
Dioxygen, 162
  see also Oxygen
Disorder, and entropy, 361, 364
Distillation, 19, 444
Disulfide bonds, 397
Diving, 217
DNA, 11, 418, *420*
Dolomite, 296
Dopa, *460, 461*
Doping, 321
Dosage level, *458*

Dose
  threshold, 60
  tolerable, 60
Double bonds, 146, 408
The Double Helix (Watson), 417
Doubling time, 471, 496
Drugs, 455
  common, 456–457(T)
  developing new, 462
  resistance to, 463
  risks in new, 455
Dry cells, *315,* 317
Ducks, 187
Dulong and Petit, law of, 379
Dust explosion, 238
Dyes, 395
  proton-sensitive, 291
Dynamic equilibrium, 264, *267*
Dynamite, 242

E

$\mathscr{E}, \mathscr{E}^0$ (electric cell voltage), 312
  calculation of, 314(T)
E. coli, 8, 367
Earth, as closed system, *381*
Earthquakes, 98, 197
Eddington, Arthur, 372, 380
Edison, Charles A., 397
Efficiency
  electric cell, 319
  heat engine, 319, 478
  human, 478
Eggs, 416
Eggshell, 25(T)
Einstein, Albert, 46, *48, 50,* 109
Electric cells, construction of, *308*
Electrical conductivity, 78, 136, *288*
Electricity, 36
Electrode potential, 310, 312(T)
Electrodes, 306
Electrolysis, 36, 97, 98, *99*
Electrolyte, 308
Electrolytic cells, 307
Electromotive force, 311
Electron bombardment, 91
Electron energy levels, *71,* 141
Electron microscope, 11
Electron structure, 106–107
Electronegativity, relative, 136(T)
Electrons, 22, 36, 37, 41
  delocalized, 153
  displaced, 113
  localized, 81
  nonbonding, 139
  pairs, 136
  reactive, 76

Smoking, and cancer, *440,* 440(T)
Soap, 191, 192, *193,* 296, 385
Soddy, Frederick, 38
Sodium chloride, *138*
Sodium lamps, 74
Solar cells, 319
Solar energy, *474,* 475, 479
Soldering, 189
Solubility, 149, 327
   limited, 332
   of gases, 278
   unlimited, 332
Solubility-product, 329
Solutes, 333
   boiling temperature of, 335
   melting point of, 335–336
   osmosis in, 336–339
   vapor pressure of, 333–334
Solutions, 16, 17, *19,* 327–343
   enthalpy of, 403
   formation of, *329*
   ideal, 329
Solvents, 152, 332
Sound, velocity of, 206, 218
*sp*³ set, 141
Sparrows, 434
Specific heat, 378
Spectator molecules, 285
Spectrometer, mass, *41,* 91
Spectrum, mass, *44*
Sperm, 416
Spontaneous reactions, 354
SST, 477
Stability, 351
   relative, 348
Stalactites, 273, *297*
Standard state, 315
   for electric cells, 312
   for elements, 348
Starch, 238
Static, 384
Static state, 263
Steady state, 242, 263, 264
Steel mills, *103*
Steel
   price of, 101(T)
   production of, 103(T)
Stickiness, 155, 185
Stimulants, 458, 461
Stomach, 299
Stoney, George, 36
Storage batteries, 308–310, 377
Stress, on a system, 271
Strip-mining, 94
Strong acids and bases, 289(T)
Strong bonds, 357, 364
Strong inference, 9
Styrene, 408

Sublimation, 277
Subscripts, 25
Sucaryl, 429
Sucrose, 238, 429(T)
Sugar, 8
Sulfa drugs, 450, *451,* 455
Sulfate process, 392
Sulfite process, 391
Sulfur, *81,* 179, 408, 412, 442
   brimstone, 179
   as by-product, 181
sulfur dioxide, 179, 442, 467
Sulfuric acid, 179, *180*
Sunburn, 261
Suntan, 124
Supercooled, 341
Supersaturated, 343
Surfaces, 185
   molecular area of, 156(T)
   nonstick, 405
Swan, Joseph, 395
Sweet, G. H., 486
Sweet, polymer, 409
Sweetness, *429*
Symbols, chemical, 3, 24, 90
   evolution of, 24
   sources of, 24(T)
Synaptic gap, *461*
Syndiotactic isomers, *405*
Synergism, 441, 451
Synthesis, 9, 10, 23, *175,* 389, 403
Systems, 15, 16
   closed, 26, 372, 379, 382
   conservation in, 373(T)
   heterogeneous, 16
   homogeneous, 16
   human, 415
   isolated, 372, 382
   open, 357
Szent-György, Albert, 413
Szilard, L., *50*

T

T (Kelvin temperature), 205
Taste, 428
Taylor, F. M., *175, 270*
Teeth, 340
Teflon, 192, 405, *406*
Television, 46
Temperature, 17, *207*
   absolute, 207
   air, *382*
   Celsius, 17
   contours, *484*
   Fahrenheit, 17

   heat, 378
   Kelvin, 207
   United States, 475
   world, *485*
Tendency to randomness, 357
Tendency to spread out, 360, 371
Tendency to strong bonds, 357
Tensides, 193
Teratogen, 455
Tetrahedral holes, 119(T), 138
Tetrahedral structure, *112*
Thalidomide, 455
Theory
   atomic, 34
   neutron–proton, 42(T), 57
Therapeutic index, 458
Thermal chains, 241
Thermodynamics
   first law, 373
   second law, 372, 395, 403, 452
   third law, 361
Thermoplastic, 401, 402
Thermoset, 401, 402
Third law of thermodynamics, 361
Thomson, G. P., 33
Thompson, J. J., 38
Thorium, 37
"Three-log effect," 241
Threonine, 433
Thymine, 418
Thyroid releasing factor (TRF),
   433, *434*
Timber, 393
Tin plating, 323
Tires, *407*
Titanium dioxide, 169, 393
Titrations, *170, 300*
TLV (threshold limit value), 60
TNT, 242
Tobacco mosaic virus, *11,* 367
Tolerance levels, 452
Tooth decay, 13
Toothpastes, 25
Topaz, 118
Toxic levels, 452
Toxicity, 171
Tracers, radioactive, 62
Trans isomers, 408
Transmutation, 37
Transport, fat, 340
Trash, 394, 435
Tridymite, 116
Triple bonds, 146
Triple point, 209, 270
Tris, 396
Tritium, 45, 63, 269
Trouton's rule, 366
Tryptophan, 459, *460*

RELATIVE SIZES OF ATOMS AND IONS

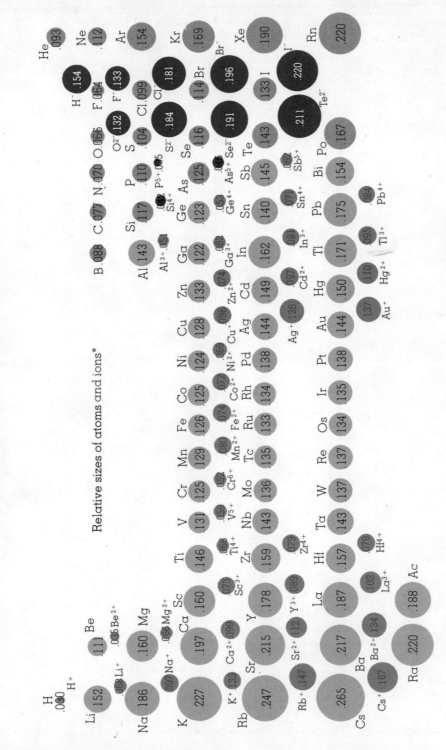

Relative sizes of atoms and ions*

*In nanometers.